DNA Tumor Viruses
Oncogenic Mechanisms

INFECTIOUS AGENTS AND PATHOGENESIS

Series Editors: Mauro Bendinelli, *University of Pisa*
Herman Friedman, *University of South Florida*

COXSACKIEVIRUSES
A General Update
Edited by Mauro Bendinelli and Herman Friedman

DNA TUMOR VIRUSES
Oncogenic Mechanisms
Edited by Giuseppe Barbanti-Brodano, Mauro Bendinelli, and Herman Friedman

FUNGAL INFECTIONS AND IMMUNE RESPONSES
Edited by Juneann W. Murphy, Herman Friedman, and Mauro Bendinelli

MYCOBACTERIUM TUBERCULOSIS
Interactions with the Immune System
Edited by Mauro Bendinelli and Herman Friedman

NEUROPATHOGENIC VIRUSES AND IMMUNITY
Edited by Steven Specter, Mauro Bendinelli, and Herman Friedman

***PSEUDOMONAS AERUGINOSA* AS AN OPPORTUNISTIC PATHOGEN**
Edited by Mario Campa, Mauro Bendinelli, and Herman Friedman

PULMONARY INFECTIONS AND IMMUNITY
Edited by Herman Chmel, Mauro Bendinelli, and Herman Friedman

VIRUS-INDUCED IMMUNOSUPPRESSION
Edited by Steven Specter, Mauro Bendinelli, and Herman Friedman

A Continuation Order Plan is available for this series. A continuation order will bring delivery of each new volume immediately upon publication. Volumes are billed only upon actual shipment. For further information please contact the publisher.

SBHMC BOOK

54009000111381

$114
1995

QW166
DNA

WITHDRAWN
FROM STOCK
QMUL LIBRARY

DNA Tumor Viruses
Oncogenic Mechanisms

Edited by
Giuseppe Barbanti-Brodano
University of Ferrara
Ferrara, Italy

Mauro Bendinelli
University of Pisa
Pisa, Italy

and

Herman Friedman
University of South Florida
Tampa, Florida

Plenum Press • New York and London

Library of Congress Cataloging-in-Publication Data

```
DNA tumor viruses : oncogenic mechanisms / edited by Giuseppe Barbanti
 -Brodano, Mauro Bendinelli, Herman Friedman.
       p.   cm. -- (Infectious agents and pathogenesis)
    Includes bibliographical references and index.
    ISBN 0-306-45151-4
    1. Oncogenic viruses.    I. Barbanti-Brodano, Giuseppe.
  II. Bendinelli, Mauro.  III. Friedman, Herman, 1931-     .
  IV. Series.
     [DNLM: 1. DNA Tumor Viruses--pathogenicity.  2. TumorVirus
  Infections--complications.  3. Neoplasms--etiology.  4. Cell
  Transformation, Neoplastic.  5. Viral Vaccines.   QW 166D6289 1995]
  QR372.06D595  1995
  616.99'4071--dc20
  DNLM/DLC
  for Library of Congress                                        95-42518
                                                                     CIP
```

ISBN 0-306-45151-4

©1995 Plenum Press, New York
A Division of Plenum Publishing Corporation
233 Spring Street, New York, N.Y. 10013

10 9 8 7 6 5 4 3 2 1

All rights reserved

No part of this book may be reproduced, stored in a retrieval system, or transmitted in any form or by any means, electronic, mechanical, photocopying, microfilming, recording, or otherwise, without written permission from the Publisher

Printed in the United States of America

Contributors

LAURE AURELIAN • Virology/Immunology Laboratories, Department of Pharmacology and Experimental Therapeutics, University of Maryland School of Medicine, Baltimore, Maryland 21201-1192; and Departments of Biochemistry and Comparative Medicine, The Johns Hopkins Medical Institutions, Baltimore, Maryland 21205

GIUSEPPE BARBANTI-BRODANO • Institute of Microbiology, School of Medicine, University of Ferrara, I-44100 Ferrara, Italy

MAURO BOIOCCHI • Division of Experimental Oncology 1, Centro di Riferimento Oncologico, 33081 Aviano (PN), Italy

MARIE ANNICK BUENDIA • Unité de Recombinaison et Expression Génétique, INSERM U163, Département des Rétrovirus, Institut Pasteur, 75724 Paris Cedex 15, France

M. SAVERIA CAMPO • The Beatson Institute for Cancer Research, CRC Beatson Laboratories, Bearsden, Glasgow G61 1BD, Scotland

ANTONINO CARBONE • Division of Pathology, Centro di Riferimento Oncologico, 33081 Aviano (PN), Italy

MICHELE CARBONE • Department of Pathology, University of Chicago, Chicago, Illinois 60637

E. CASELLI • Institute of Microbiology, University of Ferrara, I-44100 Ferrara, Italy

E. CASSAI • Institute of Microbiology, University of Ferrara, I-44100 Ferrara, Italy

CHRISTA CERNI • Institute of Tumor Biology—Cancer Research, University of Vienna, 1090 Vienna, Austria

DOROTHY H. CRAWFORD • Department of Clinical Sciences, London School of Hygiene and Tropical Medicine, London WC1E 7HT, England

LIONEL CRAWFORD • ICRF Tumour Virus Group, Department of Pathology, University of Cambridge, Cambridge CB2 1QP, England

VALLI DE RE • Division of Experimental Oncology 1, Centro di Riferimento Oncologico, 33081 Aviano (PN), Italy

DARIO DI LUCA • Institute of Microbiology, University of Ferrara, I-44100 Ferrara, Italy

JOSEPH A. DiPAOLO • National Cancer Institute, Laboratory of Biology, Bethesda, Maryland 20892

RICCARDO DOLCETTI • Division of Experimental Oncology 1, Centro di Riferimento Oncologico, 33081 Aviano (PN), Italy

M. EICKMANN • Institut für Virologie, Philipps-Universität, D-35037 Marburg, Germany

JEAN FEUNTEUN • Institut Gustave Roussy, Laboratoire d'Oncologie Moleculaire, 94805 Villejuif, France

ANNUNZIATA GLOGHINI • Division of Pathology, Centro di Riferimento Oncologico, 33081 Aviano (PN), Italy

MARIA E. JACKSON • The Beatson Institute for Cancer Research, CRC Beatson Laboratories, Bearsden, Glasgow G61 1BD, Scotland

LAYLA KARIMI • Department of Clinical Sciences, London School of Hygiene and Tropical Medicine, London WC1E 7HT, England

H. KERN • Institut für Zellbiologie, Philipps-Universität, D-35037 Marburg, Germany

LAURA de LELLIS • Institute of Microbiology, School of Medicine, University of Ferrara, I-44100 Ferrara, Italy

PETER G. MEDVECZKY • Department of Medical Microbiology and Immunology, University of South Florida, Tampa, Florida 33612-4799

T. MOCKENHAUPT • Institut für Virologie, Philipps-Universität, D-35037 Marburg, Germany

PAOLO MONINI • Institute of Microbiology, School of Medicine, University of Ferrara, I-44100 Ferrara, Italy

ANDREW J. MORGAN • Department of Pathology and Microbiology, School of Medical Sciences, University of Bristol, Bristol BS8 1TD, England

MEIHAN NONOYAMA[†] • Tampa Bay Research Institute, St. Petersburg, Florida 33716

JOSEPH S. PAGANO • Departments of Microbiology and Immunology and Medicine, UNC Lineberger Comprehensive Cancer Center, University of North Carolina, School of Medicine, Chapel Hill, North Carolina 27599-7295

HARVEY I. PASS • Thoracic Oncology Section, National Cancer Institute, National Institutes of Health, Bethesda, Maryland 20892

PASCAL PINEAU • Unité de Recombinaison et Expression Génétique, INSERM U163, Département des Rétrovirus, Institut Pasteur, 75724 Paris Cedex 15, France

K. RADSAK • Institut für Virologie, Philipps-Universität, D-35037 Marburg, Germany

F. ANDREW RAY • Life Sciences Division, Los Alamos National Laboratory, Los Alamos, New Mexico 87545; *present address*: Department of Microbiology, Immunology and Molecular Genetics, The Albany Medical College, Albany, New York 12208-3479

[†]Deceased.

B. REIS • Institut für Virologie, Philipps-Universität, D-35037 Marburg, Germany

M. RESCHKE • Institut für Virologie, Philipps-Universität, D-35037 Marburg, Germany

ROBERT P. RICCIARDI • Department of Microbiology, School of Dental Medicine, and Graduate Program in Microbiology and Virology, University of Pennsylvania, Philadelphia, Pennsylvania 19104

PAOLA RIZZO • Department of Pathology, University of Chicago, Chicago, Illinois 60637

SIEGFRIED SCHERNECK • Max-Delbrück-Centrum for Molecular Medicine, Tumorgenetics, 13122 Berlin, Germany

CHRISTIAN SEELOS • Institute of Tumor Biology—Cancer Research, University of Vienna, 1090 Vienna, Austria

DANIEL T. SIMMONS • Department of Biology, University of Delaware, Newark, Delaware, 19716

NANCY S. SUNG • Departments of Microbiology and Immunology and Medicine, UNC Lineberger Comprehensive Cancer Center, University of North Carolina, School of Medicine, Chapel Hill, North Carolina 27599-7295

AKIKO TANAKA • Tampa Bay Research Institute, St. Petersburg, Florida 33716

CRAIG D. WOODWORTH • National Cancer Institute, Laboratory of Biology, Bethesda, Maryland 20892

Preface to the Series

The mechanisms of disease production by infectious agents are presently the focus of an unprecedented flowering of studies. The field has undoubtedly received impetus from the considerable advances recently made in the understanding of the structure, biochemistry, and biology of viruses, bacteria, fungi, and other parasites. Another contributing factor is our improved knowledge of immune responses and other adaptive or constitutive mechanisms by which hosts react to infection. Furthermore, recombinant DNA technology, monoclonal antibodies, and other newer methodologies have provided the technical tools for examining questions previously considered too complex to be successfully tackled. The most important incentive of all is probably the regenerated idea that infection might be the initiating event in many clinical entities presently classified as idiopathic or of uncertain origin.

Infectious pathogenesis research holds great promise. As more information is uncovered, it is becoming increasingly apparent that our present knowledge of the pathogenic potential of infectious agents is often limited to the most noticeable effects, which sometimes represent only the tip of the iceberg. For example, it is now well appreciated that pathologic processes caused by infectious agents may emerge clinically after an incubation of decades and may result from genetic, immunologic, and other indirect routes more than from the infecting agent in itself. Thus, there is a general expectation that continued investigation will lead to the isolation of new agents of infection, the identification of hitherto unsuspected etiologic correlations, and, eventually, more effective approaches to prevention and therapy.

Studies on the mechanisms of disease caused by infectious agents demand a breadth of understanding across many specialized areas, as well as much cooperation between clinicians and experimentalists. The series *Infectious Agents and Pathogenesis* is intended not only to document the state of the art in this fascinating and challenging field but also to help lay bridges among diverse areas and people.

Mauro Bendinelli
Herman Friedman

Preface

DNA tumor viruses have long been useful experimental models of carcinogenesis and have elucidated several important mechanisms of cell transformation. Research in recent years has shown that human tumors have a multifactorial nature and that some DNA tumor viruses may play a key role in their etiology. The aim of this book is to assess our knowledge of DNA tumor viruses by reviewing animal models, mechanisms of transformation, association with human tumors, and possibilities of prevention and control by vaccination.

Animal models of tumor virology have contributed significantly to our understanding of the epidemiology and pathogenesis of virus-induced tumors. Bovine papillomaviruses induce papillomas in the intestine of cattle. The papillomas undergo a transition to carcinomas in cows feeding on bracken fern, which produces a toxin with radiomimetic and immunosuppressive functions. This example of cooperation between a virus and chemical carcinogens parallels the cooperative role of human papillomaviruses (HPVs) and herpes simplex virus type 2 (HSV-2) with environmental carcinogens in the pathogenesis of cervical cancer. Likewise, hepatocarcinomas appearing in woodchucks chronically infected by woodchuck hepatitis virus (WHV) provide strong support for the relationship between hepatitis B virus (HBV) infection and human hepatocellular carcinoma. Also, the fact that WHV DNA integrates closely to cellular oncogenes suggests a possible molecular mechanism for the tumorigenesis induced by HBV. Two animal herpesviruses inducing lymphoproliferative diseases, Marek's disease virus and *Herpesvirus saimiri*, are presented as models of lymphomas associated with the human herpesviruses, Epstein–Barr virus (EBV), and human herpesvirus 6. Of particular interest is that the SCOL gene of *Herpesvirus saimiri* behaves as a viral oncogene like the LMP-1 gene of EBV.

Several chapters discuss the molecular mechanisms of transformation by DNA tumor viruses. Some important results have been contributed recently in this field. It was discovered that simian virus 40 (SV40) T antigen, a DNA-binding protein with pleiotropic functions, induces extensive and severe chromosomal lesions before the appearance of the transformed phenotype, suggesting that genomic damage is a relevant factor in SV40-induced transformation and tumorigenesis. Binding of polyomavirus, papillomavirus, and adenovirus transforming proteins to the products of the tumor suppressor genes Rb-1 and p53 has emerged as a common mechanism of transformation by small DNA tumor viruses. Blockade of Rb-1 and p53 functions removes the checkpoints that control the cell cycle by arresting cells in G_0–G_1 and allows cells to proliferate indefinitely; p53 maintains genome stability and may induce apoptosis. Therefore, removal of its function would generate genomic instability leading to a higher probability of mutations and chromosomal rearrangements. Moreover, cells normally programmed to apoptotic death may survive and become prone to transformation. The molecular mechanisms of transformation by large DNA tumor viruses are less well understood. HSV-1 and HSV-2 have mutagenic activity and induce gene amplification in the host genome. Several functions of HSV are potentially involved in transformation, and HSV-2 carries its own oncogene, encoding the protein ICP10 PK, which has the characteristics of the classical retrovirus oncogenes. In fact, ICP10 PK is appropriated from the host cell and functions as a growth factor receptor kinase that signals through c-*ras* to stimulate cell proliferation.

The complex mechanisms of transformation by EBV highlight two viral genes, EBNA-2 and LMP-1, that are able to confer immortalization on B lymphocytes. EBNA-2 behaves as a transcription factor that activates transcription of several cellular genes, including oncogenes, whereas the most intriguing effect of LMP-1 is its ability to up-regulate expression of the *bcl*2 oncogene. Because *bcl*2 inhibits apoptosis, EBV latently infected B cells are protected from programmed cell death and survive for a long time, a step that may be a prelude to immortalization.

The epidemiologic correlation between HBV infection and hepatocellular carcinoma has long been in contrast with the lack of transforming and oncogenic ability by HBV, so that mostly indirect mechanisms of carcinogenesis have been postulated during the chronic evolution of liver inflammation leading to cirrhosis and hepatocarcinoma. However, it has recently been shown that the product of the HBV X gene is a transcription factor with the ability to bind p53 and to activate transcription of a wide variety of genes, including the c-*myc*, c-*fos*, and c-*jun* oncogenes. These observations suggest a direct oncogenic role of HBV functions in the pathogenesis of hepatocellular carcinoma.

The etiological role of DNA tumor viruses in human tumors is still a matter of discussion. However, epidemiologic, experimental, and clinical evidence indicates that some DNA tumor viruses cooperate with genetic and environmental factors to induce human tumors of specific histotypes. Surprisingly, SV40, which is not a human virus, has been detected in human ependymomas, choroid plexus papillomas, and mesotheliomas. SV40 was massively introduced in the human population 40 years ago with contaminated polio vaccine. The SV40 sequences detected in human tumors may belong to an SV40 variant adapted to human cells or to a recombinant virus bearing the regulatory sequences of human polyomaviruses BK or JC and the T-antigen coding sequences of SV40. BK and JC are ubiquitous in humans, and their regulatory sequences would confer on the recombinant viruses the tropism for human cells. BK virus sequences are detected in human brain tumors, tumors of pancreatic islets, and osteosarcomas, three tumor histotypes that are induced by BK virus when it is inoculated into experimental animals. However, the ubiquitous distribution of BK virus in normal human tissues renders it difficult to draw any conclusion about a possible etiological role of BK virus in human tumors.

The HPVs immortalize human keratinocytes *in vitro* and are associated with cervical and anogenital cancer at high frequency. Although HPVs are not tumorigenic by themselves, they probably act as promoting agents in synergy with carcinogenic initiators. The HSV can be considered one of the candidates for the latter function because of its mutagenic activity on host cells. The possibility that chromosomal alterations or insertional activation of cellular oncogenes arise as a consequence of integration of HSV sequences is in contrast with the absence of HSV footprints in cervical tumors and HSV-transformed cells. To explain this discrepancy, the "hit-and-run" theory was proposed, whereby integrated HSV sequences would be excised from the host genome after induction of irreversible DNA damage. This hypothesis is supported by the observation that HSV-2 transforming regions can assume the conformation of transposon-like insertion elements. The EBV is associated with Burkitt's lymphoma and non-Hodgkin's B-cell lymphoma in the course of AIDS. Recent results indicate that EBV is also frequently associated with Hodgkin's lymphoma and that expression of viral antigens may be responsible for the reactive T-cell component present in this neoplasia.

The information already available on the mechanisms of transformation by DNA tumor viruses and on the viral products responsible for the transformed phenotype supports hopes about future prospects of immunologic intervention. Two chapters are therefore devoted to discussions of humoral and cell-mediated immunity to HPVs and EBV in relation to vaccine design and other possibilities of immunologic control of the associated tumors.

Giuseppe Barbanti-Brodano
Mauro Bendinelli
Herman Friedman

Contents

1. The Hamster Polyomavirus
   SIEGFRIED SCHERNECK and JEAN FEUNTEUN

   1. Introduction 1
   2. The HaPV Is a Polyomavirus 1
   3. Hamster Polyomavirus Infections in Syrian Hamsters 3
      3.1. Natural Infections 3
      3.2. Infection of Newborn Syrian Hamsters 4
   4. Transgenic Mice as Models for HaPV Pathogenesis 6
   5. Replication of the Viral Genome and Productive Cycle 7
   6. The HaPV Transforming Properties *in Vitro* 9
   7. Concluding Remarks 11
      References 12

2. Simian Virus 40 Large T Antigen Induces Chromosome Damage That Precedes and Coincides with Complete Neoplastic Transformation
   F. ANDREW RAY

   1. Introduction 15
   2. Chromosome Damage Precedes Immortalization and Transformation . 17
   3. Chromosome Damage Continues in Newly Immortal and High-Passage Cells 19
   4. T-Antigen-Expressing Cells Eventually Become Tumorigenic 22
   5. Summary and Future Directions 22
      References 23

3. Transformation by Polyomaviruses: Role of Tumor Suppressor Proteins
   DANIEL T. SIMMONS

   1. Introduction 27
   2. Structure and Function of SV40 Large T Antigen 28

2.1. Replication Activities		28
2.2. Transformation Activities		28
3. Association of T Antigen with Rb		29
3.1. Functions of Rb		29
3.2. Role of T Antigen Binding to Rb		30
3.3. Rb-Related Proteins		31
4. Association of T Antigen with p53		31
4.1. Functions of p53		31
4.2. Role of T Antigen Binding to p53		35
5. Association with p300		35
6. Transformation and Immortalization Domains of SV40 Large T Antigen and Their Relationship to the Binding of Cellular Proteins		36
7. Human and Mouse Polyomaviruses		37
7.1. Transformation by Human Polyomaviruses		37
7.2. Transformation by Mouse Polyomaviruses and Functions of Middle T Antigen		38
8. Role of Small t Antigen in Transformation by SV40		38
9. Summary and Evaluation of the Roles of Small t Antigen and of the Binding of p53, Rb, and p300 to Large T Antigen in Transformation by SV40		39
10. Future Directions		39
References		40

4. Association of BK and JC Human Polyomaviruses and SV40 with Human Tumors

PAOLO MONINI, LAURA de LELLIS,
and GIUSEPPE BARBANTI-BRODANO

1. Introduction		51
2. General Characteristics of BKV and JCV		51
3. Natural History of BKV and JCV Infection		52
3.1. Epidemiology		52
3.2. Primary Infection, Latency, and Reactivation		52
4. Oncogenicity of BKV and JCV		53
4.1. Experimental Tumorigenesis with BKV		53
4.2. Experimental Tumorigenesis with JCV		54
5. *In Vitro* Transformation by BKV and JCV		54
5.1. Transformation of Rodent Cells by BKV		54
5.2. Transformation of Human Cells by BKV		55
5.3. State of BKV DNA in Transformed Cells		55
5.4. Role of BKV Large T Antigen in Transformation		56
5.5. *In Vitro* Transformation by JCV		56
6. Presence of BKV and JCV in Human Tissues		57
6.1. Presence of BKV and JCV in Nonneoplastic Tissues		57
6.2. Presence of BKV and JCV in Neoplastic Tissues		58

7. Presence of SV40 in Human Tumors	60
8. Conclusions	61
References	63

5. Association of Simian Virus 40 with Rodent and Human Mesotheliomas
MICHELE CARBONE, PAOLA RIZZO, and HARVEY I. PASS

1. Simian Virus 40	75
1.1. General Characteristics	75
1.2. The SV40 Large T Antigen	76
1.3. The SV40 Small t Antigen	76
1.4. Wild-Type SV40 Oncogenicity	76
1.5. Oncogenicity of SV40 Small t Antigen Deletion Mutants	78
2. The SV40 Virus in Humans	79
2.1. Human Infection by SV40	79
2.2. Follow-up Studies of Populations Injected with SV40-Contaminated Polio Vaccines	80
2.3. SV40-like Sequences in Human Tumors	81
3. Mesotheliomas	81
3.1. General Characteristics	81
3.2. SV40-like Sequences in Human Mesotheliomas	83
4. Should SV40 Be Considered a Potential Health Hazard to Humans?	83
5. Conclusions	86
References	86

6. Immortalization of Keratinocytes by Human Papillomaviruses
CRAIG D. WOODWORTH and JOSEPH A. DiPAOLO

1. Introduction	91
2. Genome Organization and Biology	92
3. Immortalization	95
4. Differentiation of HPV-Containing Cells	98
5. Cytogenetic Alterations in Cervical Neoplasia	99
6. Risk Factors for Cervical Neoplasia	101
References	103

7. Cooperation between Bovine Papillomaviruses and Dietary Carcinogens in Cancers of Cattle
MARIA E. JACKSON and M. SAVERIA CAMPO

1. Introduction	111
2. The Identification of Papillomaviruses and Bracken Fern as Cofactors in Tumorigenesis	112
3. Carcinogenic and Mutagenic Components of Bracken	113

4. Genomic Organization of BPV4 114
5. The Viral Contribution to Carcinogenesis 115
6. Cooperation between BPV4 and Chemicals in Experimental Systems 116
7. Discussion 118
   References 118

8. Papillomaviruses as Promoting Agents in Human Epithelial Tumors
   CHRISTA CERNI and CHRISTIAN SEELOS

1. Introduction 123
2. Human Papillomaviruses 124
   2.1. Classification of HPVs 124
   2.2. Organization of the HPV Genome 125
   2.3. Upstream Regulatory Region of HPVs 126
   2.4. The Viral Proteins 129
3. Interaction of Viral Proteins with Cellular Factors 138
   3.1. Binding of E7 to pRb and p107 138
   3.2. Binding of E6 to p53 140
4. Epithelial Cells 141
   4.1. Differentiation of Epithelial Cells 141
   4.2. Effects of HPV Infection on Cellular Response to Growth
        Factors 142
   4.3. Aberration of Cellular Genes 143
   4.4. HPV-Negative Cervical Carcinomas 143
5. Immune Response 144
   5.1. To Oncogenic HPV Proteins 144
   5.2. To Capsid Proteins L1 and L2 144
6. Conclusion 144
   References 145

9. Vaccines against Human Papillomaviruses and Associated Tumors
   LIONEL CRAWFORD

1. Introduction 157
2. Infection by HPV 158
3. Transformation and Tumorigenesis 160
4. Immune Response to HPV Infection 161
5. Requirements for Generation of Cell-Mediated Immunity 162
6. Prophylactic Vaccines 162
7. Therapeutic Vaccines 163
8. Delivery 164
9. Validation of Vaccine Efficacy 166
10. Current Vaccine Trials 166
11. Closing Remarks 167
    References 168

10. The Complex Role of Hepatitis B Virus in Human Hepatocarcinogenesis
    MARIE ANNICK BUENDIA and PASCAL PINEAU

1. Introduction 171
2. Indirect Mechanisms: Importance of the Immune Response in
   Necroinflammatory Liver Disease 172
   2.1. Physiopathological Features 172
   2.2. Immunopathogenesis 173
   2.3. Molecular Aspects 174
3. Does the HBV Genome Contain Any Conclusively Transforming
   Gene? 175
   3.1. The Genetic Organization of HBV 175
   3.2. Potential Oncogenicity of the Viral X Trans-Activator 177
   3.3. Endoplasmic Reticulum Localization of HBV Surface Proteins .. 180
4. Mutagenic Action of HBV DNA Integration in the Host Genome 180
   4.1. Direct Mechanisms 181
   4.2. Indirect Mechanisms 181
5. Importance of myc Family Genes in Carcinogenesis Induced in
   Rodents by Hepatitis-B-like Viruses 182
6. Genetic Alterations and Tumor Suppressor Genes in Liver Cancer .. 184
7. Conclusions 186
   References 186

11. Transformation and Tumorigenesis Mediated by the Adenovirus E1A
    and E1B Oncogenes
    ROBERT P. RICCIARDI

1. Introduction 195
2. Adenovirus Transformation 196
3. The Role of E1A Proteins in Transformation 196
   3.1. E1A General Overview 196
   3.2. The Zinc Finger Trans-Activating Domain Binds to TBP and
        Other Transcription Factors 199
   3.3. The CR1 and CR2 Domains Bind to the pRb Family of Proteins
        Causing Deregulation of the Cell Cycle 199
   3.4. The Amino-Terminal and CR1 Domains Bind to p300, which
        May Block Differentiation 203
4. The Role of E1B Proteins in Transformation 204
   4.1. E1B General Overview 204
   4.2. E1B 55-kDa Protein Binds to p53 and Blocks Growth Arrest 206
   4.3. E1B 19K Protein Blocks Apoptosis in Response to E1A-Mediated
        Destruction of DNA 206
5. Basis for the Collaboration between E1A and E1B in Transformation 207
6. Tumorigenesis: E1A Repression of MHC Class I Transcription as an
   Immune Escape Mechanism 207

6.1. Diminished Class I Expression in Ad12-Transformed Cells Provides a Means of Immune Escape from CTLs 207
6.2. Ad12 E1A Mediates Down-Regulation of Class I Expression 209
6.3. In Ad12-Transformed Cells, Class I Transcription Is Down-Regulated by Global Repression of the Class I Enhancer 210
6.4. Viral Persistence as the Biological Basis for Ad12 E1A-Mediated Down-Regulation of Class I Transcription 212
7. Summary 214
References 214

12. Current Developments in the Molecular Biology of Marek's Disease Virus

MEIHAN NONOYAMA and AKIKO TANAKA

1. Introduction 227
2. Genomic Structure of MDV DNA Serotype I 228
3. Interaction with Retrovirus 228
4. Marek's Disease Virus Gene Expression in Lytically Infected Cells ... 229
5. Viral Gene Expression in Latently Infected Cells 233
6. Conclusion 234
References 235

13. Oncogenic Transformation of T Cells by *Herpesvirus saimiri*

PETER G. MEDVECZKY

1. Infection of T Cells *in Vivo* and *in Vitro* by *Herpesvirus saimiri* 239
2. The Structure of the Viral DNA: Arrangement and Origin of Viral Genes   240
3. Circularization, Deletions, and Methylation of the Viral Genome in Transformed T Cells 242
4. DNA Variability and Mapping of a Region of the Viral Genome Involved in Oncogenic Transformation 243
5. Expression of an mRNA and Its Collagen-like Oncoprotein Product SCOL and an IL-11-like Protein in Transformed T Cells 244
6. Viral Small RNAs in Transformed Cells and Identification of Cellular Proteins That Bind Both Viral AUUUA Repeats and the 3' End of Unstable mRNAs of Lymphokines 245
7. Secretion of Lymphokines and Expression of Their Receptors 247
8. Conclusions: Possible Mechanisms of Transformation 248
References 249

14. Transformation and Mutagenic Effects Induced by Herpes Simplex Virus Types 1 and 2

LAURE AURELIAN

1. Introduction 253
2. Transformation by Inactivated HSV 253

3. Transforming HSV Genes 255
4. RR2 Does Not Cause Neoplastic Transformation 257
5. Neoplastic Transformation Is a Multistep Process 258
6. HSV-2 Genes That Cause Cellular Immortalization 259
7. The ICP10 PK Oncogene 260
8. ICP10 PK Is a Novel Kinase 261
9. ICP10 PK Is a Growth Factor Receptor 264
10. Signaling Pathways in ICP10 PK Minigene Transformed Cells .. 265
11. The ICP10 PK Oncogene Is an HSV-Appropriated Cellular Gene .. 266
12. ICP10 PK Expression in HSV-2-Infected Cells 267
13. Mutagenesis and Gene Amplification 267
14. Activation of DNA Synthesis and Induction of Cellular Genes .. 268
15. Activation of Endogenous Viruses 269
16. Homology of HSV DNA with Cell DNA Sequences 269
17. Animal Models of HSV Carcinogenesis 270
18. Conclusions 270
    References 271

15. Herpes Simplex Virus as a Cooperating Agent in Human Genital Carcinogenesis

DARIO DI LUCA, E. CASELLI, and E. CASSAI

1. Introduction 281
2. Cell Transformation 282
   2.1. Transforming Potential 282
   2.2. Interactions with Cell DNA and Other Viruses 283
3. Herpes Simplex Virus and Genital Neoplasms 284
   3.1. Seroepidemiologic Evidence 284
   3.2. Herpes Simplex Virus-Specific Macromolecules in Genital
        Neoplasms 285
4. Conclusions 287
   References 289

16. Human Cytomegalovirus: Aspects of Viral Morphogenesis and of Processing and Transport of Viral Glycoproteins

K. RADSAK, H. KERN, B. REIS, M. RESCHKE, T. MOCKENHAUPT, and M. EICKMANN

1. Introduction 295
2. Egress of HCMV Nucleocapsids through the Nuclear Envelope:
   Transport Budding 297
   2.1. HCMV-Induced Alteration of the Nuclear Envelope 297
   2.2. Human CMV Maturation in the Presence of Inhibitors of
        Glycoprotein Processing and Transport 298
3. Identification of Cytoplasmic Cisternae Engaged in HCMV
   Maturational Budding 299

3.1.	Visualization of Early Endosomal Cisternae in Human Fibroblasts	299
3.2.	Herpesvirus Envelopment at Cisternae of the TE	299
4.	Cellular Transport and Functional Domains of HCMV Glycoprotein B	300
4.1.	Transport and Processing of HCMV Glycoprotein B	300
4.2.	Function of C-Terminal Hydrophobic Domains of HCMV gB	301
4.3.	Identification of HCMV gB Cysteine Residues Responsible for Oligomerization	304
4.4.	Proteolytic Cleavage of HCMV gB by a Cellular Furin-like Protease	305
4.5.	Cellular Pathway of HCMV gB into the TE	306
5.	Aspects of HCMV Biology Related to Its Putative Oncogenic Potential in Man	306
5.1.	*Trans*-Activation of Cellular Genes by HCMV	307
5.2.	Induction of Papovavirus DNA Replication by HCMV	307
5.3.	Induction of Chromosome Aberrations by HCMV	307
6.	Outlook	307
	References	308

17. Association of Human Herpesvirus 6 with Human Tumors
DARIO DI LUCA and RICCARDO DOLCETTI

1.	Introduction	313
2.	*In Vitro* Transformation Studies	314
3.	Human Herpesvirus 6 in Lymphoproliferative Diseases	315
3.1.	Non-Hodgkin's Lymphomas	315
3.2.	Hodgkin's Disease	317
3.3.	HIV-Associated Lymphoproliferations	319
3.4.	Leukemia	320
4.	Role of HHV-6 in Nonlymphoid Neoplasms	321
5.	Conclusions	322
	References	322

18. Molecular Mechanisms of Transformation by Epstein–Barr Virus
NANCY S. SUNG and JOSEPH S. PAGANO

1.	Introduction	327
2.	Latent Replication	329
3.	Cell Transformation by EBV	331
3.1.	Epstein–Barr Nuclear Antigen 2	331
3.2.	Latent Membrane Protein 1	333
3.3.	The EBERs	334
4.	Cytolytic Cycle	334
5.	Epstein–Barr Virus Inhibition of Apoptosis	335

19. Epstein–Barr Virus: Mechanisms of Oncogenesis
    LAYLA KARIMI and DOROTHY H. CRAWFORD

    1. Introduction 347
    2. The Virus 347
    3. Epstein–Barr Virus Infection *in Vitro* 348
    4. Epstein–Barr Virus Infection *in Vivo* 353
    5. Epstein–Barr Virus-Associated Malignancies 354
       5.1. Epstein–Barr Virus in the Immunocompromised Host 355
       5.2. Burkitt's Lymphoma 356
       5.3. Hodgkin's Lymphoma 360
       5.4. Nasopharyngeal Carcinoma 361
    6. Conclusions 364
       References 364

20. Association of Epstein–Barr Virus with Hodgkin's Disease
    MAURO BOIOCCHI, RICCARDO DOLCETTI, VALLI DE RE,
    ANTONINO CARBONE, and ANNUNZIATA GLOGHINI

    1. Introduction 375
    2. Epidemiologic Data 376
    3. Detection of EBV in HD Samples 377
    4. Epstein–Barr Virus Association with Specific HD Histological
       Subtypes 380
    5. Other Viruses 381
    6. Persistence of EBV in the Course of HD 381
    7. Pattern of EBV Gene Expression in HD 382
    8. Pattern of EBV Gene Expression in HD-Related Lymphoproliferative
       Disorders 384
    9. Concluding Remarks 385
       References 386

21. The Development of Epstein–Barr Virus Vaccines
    ANDREW J. MORGAN

    1. The Demand for an Epstein–Barr Virus Vaccine 395
    2. Selection of an EBV Vaccine Molecule 398
    3. An Animal Model of EBV-Induced Lymphoma 400
    4. Natural Product gp350 Subunit Vaccines 401
    5. Recombinant gp350 Subunit Vaccines 403
    6. Choice of Adjuvant 404
    7. Live Virus Vector Recombinants 405

6. Conclusion 336
   References 338

8. Cell-Mediated Immune Responses to gp350 407
9. T- and B-Cell Epitopes on the gp350 Molecule 409
10. Epstein–Barr Virus Latent Antigen Vaccines 410
11. Conclusions 411
    References 412

Index 421

1

The Hamster Polyomavirus

SIEGFRIED SCHERNECK and JEAN FEUNTEUN

1. INTRODUCTION

The hamster polyomavirus (HaPV) was originally described in 1967 by Graffi *et al.* as a virus associated with skin epithelioma of the Syrian hamster.[1–4] The tumors appear spontaneously in animals at about 3 months to more than 1 year of age in a laboratory colony bred in Berlin Buch, Germany (HaB). Virus-particles identified in cell extracts prepared from skin epitheliomas cause lymphoma and leukemia when injected into newborn hamsters from a distinct and practically tumor-free colony bred in Potsdam, Germany (HaP). In contrast to the skin epithelioma, the hematopoietic tumors are virus-free but accumulate large numbers of nonrandomly deleted extrachromosomal viral DNA. Although HaPV interaction with keratinized cells may be reminiscent of the papillomavirus life cycle, the recent characterization of the viral genome classifies it as a polyomavirus. However, the HaPV tumor spectrum, which reflects the capacity of the virus to infect both undifferentiated keratinocytes and lymphocytes, is unique within the papovavirus family and raises interesting questions concerning the expression and interaction of viral oncogenes in different cellular contexts (Fig. 1). It should be emphasized that the HaPV described in the review may not be a singular isolate: a closely related virus has been described as the etiological agent of Syrian hamster skin epithelioma in Alabama.[5]

2. THE HaPV IS A POLYOMAVIRUS

Virus particles accumulate in large amounts in skin epithelioma.[1,2] The morphology of the negatively stained virus particles, about 40 nm in diameter, is

---

SIEGFRIED SCHERNECK • Max-Delbrück-Centrum for Molecular Medicine, Tumorgenetics, 13122 Berlin, Germany.   JEAN FEUNTEUN • Institut Gustave Roussy, Laboratoire d'Oncologie Moleculaire, 94805 Villejuif, France.

DNA Tumor Viruses: Oncogenic Mechanisms, edited by Giuseppe Barbanti-Brodano *et al.* Plenum Press, New York, 1995.

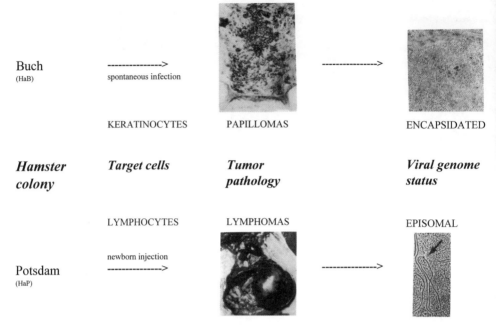

FIGURE 1. Pathologies induced by HaPV in the HaB and HaP Syrian hamster colonies.

indistinguishable from that of murine polyoma (Py) and simian virus 40 (SV40). Other characteristic properties such as molecular weight (27.5×10^6), sedimentation coefficient (223 S), and buoyant density (1.340 g/ml) are also within the range of SV40 and Py.[6,7] The viral genome has been molecularly cloned from a DNA preparation purified from a pool of epithelioma. The identity of the cloned genome as the major species present within the original pool of DNAs has been established by restriction enzyme analysis. This cloned genome, hereafter referred to as "wild-type HaPV," has been totally sequenced.[8] It is a double-stranded circular DNA molecule of 5366 base pairs, slightly larger than those of Py (5292 bp) and SV40 (5243 bp). This overall genetic organization immediately establishes HaPV as a member of the polyomavirus family but, in fact, very much resembles that of Py itself (Fig. 2). The open-reading-frame organization predicts the existence of an early and a late region transcribed on opposite strands and separated by a noncoding region displaying sequence motifs presumably involved in *cis*-regulation of transcription and replication of the viral genome. The putative early gene region has coding capacity for small T (ST), middle T (MT), and large T (LT) antigens, and the late genome region encodes the major capsid protein VP1 and the capsid proteins VP2 and VP3. In the noncoding sequence two consecutive near-perfect palindromic structures are located between bases 5320 and 5339 and between bases 5356 and 11. The second palindrome is highly homologous to the Py structure considered to be the origin of replication (nucleotides 5281 to 20) and plays the same role in HaPV. No extensive repeats are apparent in this region, and no TATA

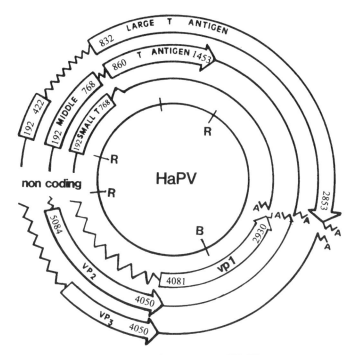

FIGURE 2. Genetic map of HaPV.

box is recognizable upstream of the early or late gene initiation codons. However, as in Py, enhancer core elements are located on the late side of the putative origin of replication. The sequence 5304 to 5312 (A core) fits with the consensus sequence proposed for the SV40 core enhancer. The B core sequences (5183 to 5195 and 5232 to 5244) are also in good agreement with the consensus adenovirus E1A enhancer.

HaPV is a second example of an MT-antigen-coding polyomavirus. The homology of organization with the Py physical map confirms the early morphological and biochemical observations and provides a definite basis for a classification as a polyomavirus despite the initial characterization as an epithelioma-inducing agent. A computerized analysis of the base sequence homologies designates Py as the closest relative. Among the primate polyomaviruses the lymphotropic papovavirus (LPV) shows the highest degree of homology. The homologies between HaPV and LPV are especially interesting to consider in view of the lymphotropism of these two viruses.[9,10]

3. HAMSTER POLYOMAVIRUS INFECTIONS IN SYRIAN HAMSTERS

3.1. Natural Infections

Five to ten percent of 6- to 9-month-old animals of the random-bred HaB hamster colony are affected by multiple skin epitheliomas. Beginning at the chin, the growth arises on the scalp, neck, back, and flanks and frequently also about the eyes and external ears. Numerous nodules arising multicentrically coalesce in the cutis and subcutis to form massive layers. Histologically the tumors result from the proliferation of hair root epithelium, which forms cyst-like masses filled with cornified material sometimes containing melanin. Virus particles especially occur in the cornified layer but are absent in the proliferating cells of the stratum basale and stratum spinosum.[1-4] This tight linkage between completion of the virus productive cycle and the terminal differentiation of the skin epithelium is strikingly reminiscent of the papillomavirus infection.[11] However, these two pathologies differ in regard to the nature of the respective virus target cells, i.e., the hair follicle keratinocytes for HaPV and the interfollicular epidermis keratinocytes for the papillomaviruses. A high incidence of hair follicle tumors has also been reported for the Py strain PTA, the tumor spectrum of which includes epithelioma of the hair follicles.[12]

Hamster polyomavirus particles were first detected on electron micrographs of primary skin epithelioma sections. Since the spontaneous appearance of epithelioma in the HaB hamster colony 20 years ago, an enzootic infection has been established, very probably via massive horizontal transmission. Injection of HaPV particles into newborn HaB animals yielded only 5–10% and rarely 30% incidence of epitheliomas, not substantially more than the incidence ascertained for spontaneous tumors.[3] A search for the virus reservoirs in HaB weanling animals well before the appearance of skin tumors by *in situ* hybridization of whole-body animal sections demonstrates that thymus and spleen are the most active virus reservoirs. The absence of detectable viral genomes in total embryo tissues supports the model of horizontal transmission. Autonomous growth potential has been shown by the transplantability of single tumor nodules. Different transplantation lines exhibited different latent periods ranging from 3 weeks to 8 months. In contrast to primary skin epitheliomas, transplanted skin tumors have not exhibited virus particles in electron micrographs but contain extrachromosomal viral DNA[13] (Figs. 1 and 3a,c).

3.2. Infection of Newborn Syrian Hamsters

Virus particles isolated from the skin epithelioma or DNA extracted from such virus preparations induces lymphoma and leukemia when injected into newborn hamsters from the separate, uninfected and quasi-tumor-free colony HaP. The tumor incidence is in the range of 30% to 80% with short latent periods (4–8 weeks). In most cases these tumors affect the liver, less frequently the thymus and the kidneys, and never the spleen. In HaB hamsters affected by skin epithelioma, the spontaneous incidence of lymphoma is low (1–3%). Both lymphoma and epithelioma are rarely found in the same animal, and the frequency of double tumors does not increase on inoculation of virus particles. This suggests that most HaB animals could be "protected" against the leukemogenic activity of the virus via an immune response elicited by virus structural antigens or virus-induced tumor antigens or both.

This lymphotropism is shared by two other polyomaviruses: SV40 can induce a

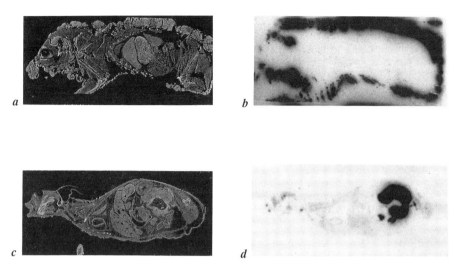

FIGURE 3. Whole-body sections of an HaB hamster with skin epitheliomas (*a*) and of an HaP hamster bearing a lymphoma (*c*). The respective autoradiographs of these whole-body sections photographed after hybridization with ^{32}P-labeled HaPV DNA are demonstrated in *b* and *d*, respectively.

broad spectrum of hematopoietic tumors when injected intravenously into Syrian hamsters,[14] and LPV, initially isolated from a monkey B lymphoid cell line, replicates only in primate hematopoietic cells.[9–15] By contrast, the murine hematopoietic organs seem to be strictly refractory to the tumorigenicity of Py. A large number of lymphomas have been carefully examined for the presence of virus particles by electron microscopy; none of them showed either papovaviruses or C-type virus; HaPV is present in these tumors as abundant (1000 to 5000 copies per cell) extrachromosomal circular molecules resulting in each tumor from the clonal amplification of a single DNA species.[16] These free genomes differ from the wild type by a single deletion that is unique to each of them but that always removes on the late side of the origin of replication the B enhancer sequences, the late transcription signals, and part of the late coding sequences (Fig. 4). These deletions, which are certainly deleterious for the productive cycle, may create conditions for an efficient autonomous replication. Injection of cloned wild-type genome does not induce lymphoma, whereas deleted species cloned from the lymphoma cells are very active, indicating that the capacity to induce tumors is elicited by the deletion of specific sequences. Such deletions may occur naturally during the replication of the viral genome within the epithelioma and may generate lymphomatogenous variants in virus stocks. The lymphoma can be serially transplanted in hamsters. At late passages we have observed that the viral genomes have lost their capacity to replicate autonomously and have integrated into the cell genome. This

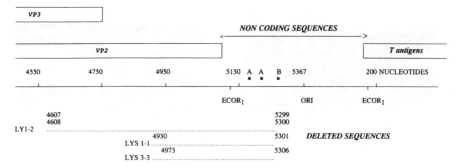

FIGURE 4. Physical maps of deletions characterized in the HaPV genomes cloned from some lymphoma DNAs. The top part represents the region of the viral genome affected by the deletions; it includes the noncoding region and the 5' ends of both early and late coding sequences. The two adenovirus core (A) and the SV40 core (B) enhancer sequences are indicated. The putative origin of the DNA replication (ORI) and the ECORI site systematically deleted in the mutants are also marked.

tendency to integrate on serial passages in vivo has also been reported for cottontail rabbit papillomavirus-induced tumors.[17]

The differential susceptibility of the hamsters from the HaB and HaP colonies, illustrated by the radically different diseases caused by HaPV infection, raises intriguing questions concerning the genetic control of the virus–host interaction. In the animals of the HaB colony, natural horizontal transmission early in life generates reservoirs of infectious virus particles, essentially in the thymus and the spleen, without pathological consequences. Not only does the incidence of spontaneous hematopoietic tumors remain low, but the animals also seem to be immunologically protected against induction of lymphoma by high-dose inoculation of purified virus. The viremia remains asyptomatic until the age of 3 months, when the hair root keratinocytes become infected and start to proliferate and produce large amounts of virus. By contrast, in the virus-free HaP hamsters, the de novo inoculation of HaPV causes a lymphoproliferative disease with tumor cells containing massive amounts of free defective viral genomes (Figs. 1 and 3b,d). The HaPV is also a potent lymphomatogenous agent in rats, but no tumors have been obtained after injection of virus and viral DNA into different mouse strains.[3]

4. TRANSGENIC MICE AS MODELS FOR HaPV PATHOGENESIS

Transgenic mice have been obtained by microinjection of HaPV viral supercoiled DNA into pronuclei of fertilized eggs of Gat:NMRI mice.[18] Two of seven founder mice have been bred over three generations. Analysis of different tissues in all three generations have established that the HaPV transgene is present as extrachromosomal DNA and expressed preferentially in the thymus and the spleen. At the age of 18 months, four of seven founders developed skin papillomas histologically similar to the epithelioma in the HaB hamster strain. No virus particles but extrachromosomal HaPV DNA can be detected in these tumors. Three of five mice of the first and second generation (5–9 months old) that died had developed lymphoma, also free of virus but containing extrachromosomal viral genomes (Fig. 5). F_2 litters are affected by severe developmental and often lethal damage. In some litters about 70% of the pups died between 2 days and 3 weeks after birth. All these animals showed almost complete thymus degeneration.

Transgenic mice in which the expression of the reporter gene *lac Z* is driven by the HaPV early promoter have shown that this promoter displays a rather specific pattern of expression restricted to hematopoietic organs (thymus and spleen).

5. REPLICATION OF THE VIRAL GENOME AND PRODUCTIVE CYCLE

Since the isolation of HaPV in 1967, no permissive host capable of supporting the full HaPV productive cycle *in vitro* has been described. Graffi *et al.*[3] reported that the viral infection of hamster kidney cells or newborn hamster thymus cells resulted in virus proliferation in a few cells, detectable by electron microscopy. However, no cytopathic effect of the cell layers was observed concomitantly. More recently, Barthold *et al.*[19] also described a similar observation of an acytopathic infection of primary total hamster embryo cells. The lack of a fully permissive host cell has hampered the detailed biological characterization of the virus for a long time. In a search for a HaPV permissive cell culture system, a panel of murine and hamster cell types has been assayed for capacity to replicate the viral genome and yield virus progeny. These experiments led to the conclusion that hamster cells represent the most permissive host for HaPV DNA replication.[20] However, hamster leukemia cell lines GD36 and GD2251, established by Dr. G. Diamandopoulos from a lymphoblastic leukemia induced in the Syrian hamster by SV40, are the only ones in which the virus achieves serial productive cycles. In other hamster cells, including BHK cells, despite a successful initial virus burst following DNA transfection, a block in the productive cycle caused by uncoating deficiencies prevents the spread of infection. The virions produced by GD36 cells are indistinguishable from the particles isolated from the epithelioma both structurally (sizes of the capsid, restriction of the DNAs) and functionally (they induce lymphoma in the HaP hamsters). Additional experiments have shown that serial passages of transfected GD36 cells establish a persistent infection yielding virus progeny without detectable cytopathic effect. Although these cells do not represent a permissive host comparable to murine fibroblasts for Py, they have been valuable in studying the replication of the viral genome *in vitro*.[20,21]

Sequence motifs required in *cis* for early transcription have been located in the HaPV noncoding region. Functional analysis of the regulatory region of the polyomavirus genome has defined three distinct elements involved in viral DNA replication and transcription: an upstream transcriptional enhancer, an origin of replication, and a downstream region containing LT antigen-binding sites.[22] The sequence of the HaPV regulatory region suggests a similar organization. Conserved

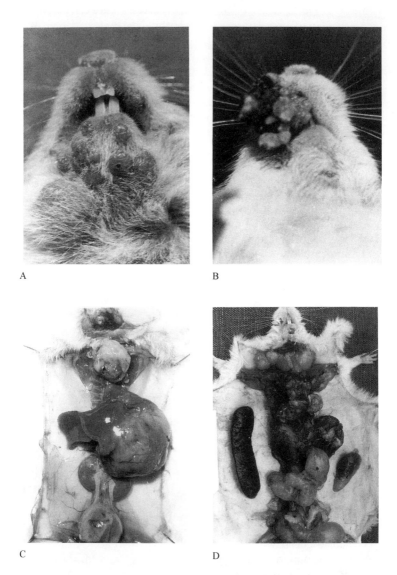

FIGURE 5. Comparison of pathologies induced by HaPV in the Syrian hamster and in the HaPV transgenic mice. (A) Spontaneously occurring multiple tumors in the skin of the lower jaw area in a HaB hamster. (B) HaPV-transgenic mouse with a squamous cell carcinoma in the jaw area. (C) HaPV-induced lymphoma, 6 weeks after injection of an HaP hamster with HaPV. (D) Interior of a lymphoma-bearing HaPV-transgenic mouse. Note the enormously enlarged tumorous spleen and lymph node as well as tumorous enlarged lung, liver, and kidney.

motifs are recognized in the region in which viral DNA replication putatively initiates. They include a GC-rich palindrome that contains a putative binding site for the LT antigen surrounded on its late side by an AT-rich sequence and on its early side by an inverted repeat sequence. Together, these three conserved motifs are referred to as the core component of the replication origin that is required for replication under all conditions. Thus, small deletions within the GC-rich palindrome or the adjacent AT-rich sequence inactivate the HaPV origin of replication in permissive cells *in vitro*.[20] The remainder of the noncoding region, containing promoter or enhancer elements, constitutes the auxiliary component of the replication origin, possibly dispensable under some conditions.

Unlike Py but like SV40, the initiation of HaPV DNA replication does not strictly require the upstream transcriptional elements, although they have a strong stimulatory effect. In contrast, part of the downstream transcriptional elements containing putative large T-antigen-binding sites is absolutely required for the initiation of HaPV DNA replication. These results allow the definition of the origin core component of HaPV, with an upstream boundary lying just beyond the AT-rich domain and a downstream boundary localized farther from the inverted repeat than that described for both Py and SV40. By contrast with the wild-type genome, which replicates in both lymphoblastic and fibroblastic cell lines, the lymphoma-associated HaPV genomes characterized by deletions affecting the late coding region as well as a specific part of the noncoding regulatory region replicate in lymphoblastic but not in fibroblastic cell lines.[23] The deletion acts in a *cis*-dominant manner and is the primary determinant of this host-range effect on replication The boundaries of the regulatory region necessary for viral DNA replication in the two cell contexts have been defined. The regulatory region can be functionally divided into two domains: one domain (distal from the origin of replication) is necessary for viral genome replication in fibroblasts, and the other domain (proximal to the origin of replication) is functional only in the lymphoblastoid cell context and contains the sequence specifically conserved in the lymphoma-associated genomes. This sequence harbors a motif recognized by a lymphoblastoid cell-specific *trans*-acting factor.[24]

6. THE HaPV TRANSFORMING PROPERTIES *IN VITRO*

The tumor specificity displayed by HaPV viral infections or the transgene *in vivo* can be bypassed *in vitro* because HaPV carries the full transforming properties of a polyomavirus. This can be demonstrated by the induction of unlimited proliferative capacity *in vitro* in transfected primary rat fibroblasts (immortalization) and of a spectrum of phenotypic alterations in transfected immortal rat fibroblasts such as focus formation or growth in semisolid medium (transformation)[8-25] (Fig. 6). The fragment of genomic viral DNA carrying the coding capacities for the three early antigens (nt 182 to nt 2998) contains all genetic information necessary for immortalization of primary rat embryo fibroblasts and transformation of F111 rat cells.[26] Similar observations have been reported for Py.[27-28] Attempts to assign

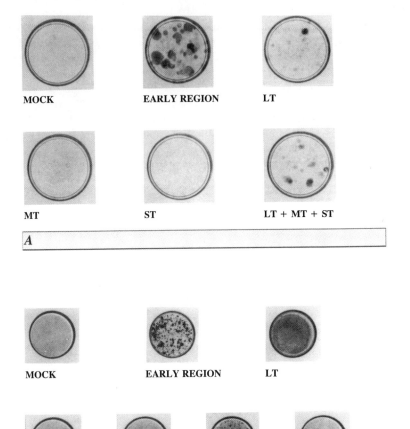

FIGURE 6. Transforming properties of HaPV *in vitro*. (A) Immortalization of primary rat embryo fibroblasts by HaPV early genes. (B) Transformation of F111 rat fibroblasts by HaPV early genes. The nature of each transfecting viral gene is indicated below each plate.

biological activities to each individual cDNA have demonstrated first that immortalization is essentially carried out by the LT antigen.

The LT antigen of most of the polyomaviruses has been shown to bind pRb.[29,30] In the case of Py, mutagenesis has demonstrated that the capacity of LT to bind pRb is essential for immortalization, in contrast to SV40.[31] In this case, pRb binding appears irrelevant for immortalization but absolutely required for transformation. The HaPV LT antigen can indeed complex the pRb polypeptide. Mutation in the HaPV LT antigen (^{134}Glu–Lys) that obliterates this binding is strongly deleterious to the immortalization capacity of the viral genome. This mutation has no effect on transformation of F111 cells. These results indicate that the interaction between HaPV LT antigen and pRb is required in the immortalization process but irrelevant to transformation.[32]

Although the MT and ST antigens are not strongly required to promote unlimited proliferation of primary cells, they stimulate the growth and modify the phenotype of immortal cell lines. Similar activities are classically associated with Py MT and ST antigen expression.[27–33]

A stringent cooperative effect is observed in the transformation of F111 cells, which requires the simultaneous presence of the MT and ST antigens. None of the T antigens alone is capable of inducing foci. This observation represents a clear discrepancy with the pathway of transformation by Py, which can be carried out by the MT antigen alone.[28] As predicted on the basis of the open reading frames, the respective early gene products share large regions of homologous sequences. However, the two MT antigens are highly divergent over a stretch of 190 amino acids, which may determine the specificity of binding to cellular tyrosine kinases. The Py MT antigen disturbs the cell cycle by altering the signal transduction pathway through its interaction with several cellular proteins within the cell membrane. These include tyrosine kinases of the *Src* family (c-*src*, c-*yes*, c-*fyn*,[34–37] phosphatidylinositol 3-kinase (PI kinase),[38] and the serine/threonine phosphatase 2A (PP2A).[39] The tyrosine kinases and PI kinase activities are strongly stimulated through these interactions. A parallel situation has been described for the HaPV MT antigen, which, in addition to PI kinase and PP2A, complexes almost exclusively c-*fyn* among the members of the *Src* family.[40] The fact that T lymphocytes express an appreciable level of p59c-*fyn* might account for the ability of HaPV to induce lymphomas.

7. CONCLUDING REMARKS

The different susceptibility of the hamsters from the HaB and HaP colonies, illustrated by the strikingly different tumor pattern caused by HaPV infection, raises numerous questions concerning the genetic control of the virus–host interaction and the genetic determinants responsible for the unique tumor specificity of HaPV. The first level of determination is illustrated by the response of the two hamster strains to HaPV infection, which leads to two completely divergent pathologies: skin epithelioma or lymphoma. At this level, the genetic constitution of the host, particularly the immune response, is likely to control the viremia. Once the virus successfully reaches a cell target, the functional identity of this cell provides a molecular context that controls the expression of the viral genes and in return responds specifically. In the HaB animals, despite a widely spread viremia, tumorigenicity remains restricted to the hair root epithelium. By contrast, in the infected HaP hamsters, hematopoietic cells seem to be the only tissue capable of replicating the viral genome and responding by tumor proliferation. It is noteworthy that the two pathologies observed in transgenic mice in the absence of virus particle production are closely related to those of the hamster epithelioma and lymphoma. The viral cis-regulatory sequences and the coordinated coding sequences must carry the adequate specificities for a given cellular context. Interestingly, the HaPV genome must undergo profound rearrangements of regulatory sequences to switch its cell specificity from skin epithelium to hematopoietic tissue. The identification and the functional characterization of these genetic determinants should be a goal for future work on HaPV.

REFERENCES

1. Graffi, A., Schramm, T., Bender, E., Bierwolf, D., and Graffi, I., 1967, Uber einen neuen virushaltigen Hauttumor beim Goldhamster, *Arch. Geschwulstforsch* **30:**227–283.
2. Graffi, A., Schramm, T., Graffi, I., Bierwolf, D., and Bender, E., 1968, Virus-associated skin tumors of the Syrian hamster: Preliminary note. *J. Natl. Cancer Inst.* **40:**867–873.
3. Graffi, A., Bender, E., Schramm, T., Graffi, I., and Bierwolf, D., 1970, Studies on the hamster papilloma and the hamster virus lymphoma, *Comp. Leukemia Res. Bibl. Haematol.* **36:**293–303.
4. Graffi, I., Bierwolf, D., Schramm, T., Bender, E., and Graffi, A., 1972, Elektronenmikroskopische Untersuchungen über das Papova (Papillom-) Virus des Goldhamsters, *Arch. Geschwulstforsch* **40:**191–236.
5. Coggin, J. M., Jr., Hyde, B. M., Heath, L. S., Leinbach, S. S., Fowler, E., and Stadtmore, L. S., 1985, Papovavirus in epitheliomas appearing on lymphoma-bearing hamsters: Lack of association with horizontally transmitted lymphomas of Syrian hamsters. *J. Natl. Cancer Inst.* **75:**91–97.
6. Böttger, M., and Scherneck, S., 1985, Heterogeneity, molecular weight and stability of an oncogenic papovavirus of the Syrian hamster, *Arch. Geschwulstforsch* **55:**225–233.
7. Böttger, M., Bierwolf, D., Wunderlich, V., and Graffi, A., 1971, New calibration correlations for molecular weights of circular DNA: The molecular weight of the DNA of an oncogenic papovavirus of the Syrian hamster, *Biochim. Biophys. Acta* **232:**21–31.
8. Delmas, V., Bastien, C., Scherneck, S., and Feunteun, J., 1985, A new member of the polyomavirus family: The hamster papovavirus. Complete nucleotide sequences and transformation properties, *EMBO J.* **4:**1279–1286.
9. Zur Hausen, H., and Gissmann, L., 1979, Lymphotropic papovavirus isolated from African green monkey and human cells, *Med. Microbiol. Immunol.* **167:**137–153.
10. Vogel, F., Rhode, K., Scherneck, S., Bastien, C., Delmas, V., and Feunteun, S., 1986, The hamster papovavirus: Evolutionary relationships with other polyomaviruses, *Virology* **154:**335–343.
11. Giri, I., and Danos, O., 1986, Papillomavirus genomes: From sequence data to biological properties, *Trends Genet.* **2:**227–232.
12. Dawe, C. J., Freund, R., Mandel, G., Balmer-Hofer, K., Talmage, D., and Benjamin, T. L., 1987, Variations in polyomavirus genotype in relation to tumor induction in mice: Characterization of wild type strains with widely differing tumor profiles, *Am. J. Pathol.* **127:**243–261.
13. Bender, E., Schramm, T., Graffi, A., and Schneiders, F., 1969, Transplantationseigenschaften der durch das Hamster-Papova-Virus induzierten Hauttumorendes Goldhamsters, *Arch. Geschwulstforsch* **34:**144–151.
14. Diamandopoulos, G. T., 1972, Leukemia, lymphoma and osteosarcoma induced in the Syrian golden hamster by simian virus 40, *Science* **176:**173–175.
15. Schöler, H. R., and Gruss, P., 1985, Cell type-specific transcriptional enhancement *in vitro* requires the presence of transacting factors, *EMBO J.* **4:**3005–3013.
16. Scherneck, S., Delmas, V., Vogel, F., and Feunteun, J., 1987, Induction of lymphomas by the hamster papovavirus correlates with massive replication of nonrandomly deleted extrachromosomal viral genomes, *J. Virol.* **61:**3992–3998.

17. Georges, E., Croissant, O., Bonneaud, N., and Orth, G., 1984, Physical state and transcription of the cottontail rabbit papillomavirus genome in warts and transplantable VX2 and VX7 carcinomas of domestic rabbits, *J. Virol.* **51**:530–538.
18. Hoffmann, S., Arnold, W., Becker, K., Rüdiger, K.-D., and Scherneck, S., 1989, The hamster papovavirus produces papillomas and lymphomas in transgenic mice, *Biol. Zentralbl.* **108**:13–18.
19. Barthold, S. W., Bhatt, P. N., and Johnson, E. A., 1987, Further evidence for papovavirus as the probable etiology of transmissible lymphoma of Syrian hamsters, *Anim. Sci.* **37**:283–287.
20. De la Roche Saint Andre, C., Harper, F., and Feunteun, J., 1990, Analysis of the hamster polyomavirus infection in vitro: Host-restricted productive cycle, *Virology* **177**:532–540.
21. De la Roche Saint Andre, C., Delmas, V., Bastien, C., Goutebroze, L., Scherneck, S., and Feunteun, J., 1989, Molecular aspects of pathogenesis in hamster polyomavirus infection, in: *Common Mechanism of Transformation by Small DNA Tumor Viruses* (L. P. Villareal and D. C. Washington, eds.). American Society for Microbiology, Washington, pp. 225–238.
22. DePamphilis, M. L., 1989, Transcriptional elements as components of eukaryotic origins of replication, *Cell* **52**:635–638.
23. De la Roche Saint Andre, C., Mazur, S., and Feunteun, J., 1993, Viral genomes maintained extrachromosomally in hamster polyomavirus-induced lymphomas display a cell-specific replication in vitro, *J. Virol.* **67**:7172–7180.
24. De la Roche Saint Andre, C., and Feunteun, J., 1993, Distinct segments of the hamster polyomavirus regulatory region have differential effects on DNA replication, *J. Gen. Virol.* **74**:125–128.
25. Bastien, C., and Feunteun, J., 1988, The hamster polyomavirus transforming properties, *Oncogene* **2**:129–135.
26. Goutebroze, L., and Feunteun, J., 1992, Transformation by hamster polyomavirus: Identification and functional analysis of the early genes, *J. Virol.* **66**:2495–2504.
27. Rassoulzadegan, M., Cowie, A., Carr, A., Glaichenhaus, N., Kamen, R., and Cuzin, F., 1982, The role of individual polyoma virus early proteins in oncogenic transformation, *Nature* **300**:713–718.
28. Treisman, R., Novak, U., Favaloro, J., and Kamen, R., 1981, Transformation of rat cells by an altered polyoma virus genome expressing only the middle T protein, *Nature* **292**:595–600.
29. DeCaprio, J. A., Ludlow, J. W., Figge, J., Shew, J.-Y., Huang, C.-M., Lee, W.-H., Marsilio, E., Paucha, E., and Livingston, D. M., 1988, SV40 large tumor antigen forms a specific complex with the product of the retinoblastoma susceptibility gene, *Cell* **54**:275–283.
30. Dyson, N., Bernards, R., Friend, S. H., Gooding, L. R., Hassell, J. A., Major, E. O., Pipas, J. M., Vandyke, T., and Harlow, E., 1990, Large T antigens of many polyomaviruses are able to form complexes with the retinoblastoma protein, *J. Virol.* **64**:1353–1356.
31. Pipas, J. M., 1992, Common and unique features of T-antigens encoded by the polyomavirus group, *J. Virol.* **66**:3979–3985.
32. Cherington, V., Morgan, B., Spiegelman, B. M., and Roberts, T. M., 1986, Recombinant retroviruses that transduce individual polyoma tumor antigens: Effects on growth and differentiation, *Proc. Natl. Acad. Sci. U.S.A.* **83**:4307–4311.
33. Goutebroze, L., De la Roche Saint Andre, C., Scherneck, S., and Feunteun, S., 1993, Mutations within the hamster polyomavirus large T-antigen domain involved in pRb binding impair virus productive cycle and immortalization capacity, *Oncogene* **8**:685–693.
34. Courtneidge, S. A., and Smith, A. E., 1983, Polyoma virus transforming protein associates with the product of the c-*src* gene, *Nature* **303**:435–439.
35. Bolen, J. B., Thiele, C. J., Israel, M. A., Yonemoto, W., Lipsich, L. A., and Brugge, J. S., 1984, Enhancement of cellular *src* gene product associated tyrosyl kinase activity following polyoma virus infection and transformation, *Cell* **38**:767–777.
36. Cheng, S. H., Piwnica-Worms, H., Harvey, R. W., Roberts, T. M., and Smith, A. E., 1988. The carboxy terminus of pp60$^{c\text{-}src}$ is a regulatory domain and is involved in complex formation with the middle-T antigen of polyomavirus, *Mol. Cell. Biol.* **8**:1736–1747.

37. Horak, I. D., Kawakami, T., Gregory, F., Robbins, K. C., and Bolen, J. B., 1989, Association of p60^{fyn} with middle tumor antigen in murine polyomavirus-transformed rat cells. *J. Virol.* **63:**2343–2347.
38. Whitman, M., Kaplan, D. R., Schaffhausen, B., Cantley, L., and Roberts, T. M., 1985, Association of phosphatidylinositol kinase activity with polyoma middle T component for transformation, *Nature* **315:**239–242.
39. Pallas, D. C., Cherington, V., Morgan, W., DeAnda, J., Kaplan, D., Schaffhausen, B., and Roberts, T. M., 1988, Cellular proteins that associate with the middle and small T antigens of polyomavirus, *J. Virol.* **62:**3934–3940.
40. Courtneidge, S. A., Goutebrouze, L., Cartwright, A., Heber, A., Scherneck, S., and Feunteun, J., 1991, Identification and characterization of the hamster polyomavirus middle T antigen, *J. Virol.* **65:**3301–3308.

2

Simian Virus 40 Large T Antigen Induces Chromosome Damage That Precedes and Coincides with Complete Neoplastic Transformation

F. ANDREW RAY

1. INTRODUCTION

In the early 1900s, Theodor Boveri made some rather remarkable predictions.[1] On the basis of his observations of mitoses in sea urchin eggs, he postulated the involvement of abnormal chromosomes in oncogenesis and even postulated the existence of "chromosomes which inhibit division." Such early work laid the foundation for the theory that tumorigenesis is a series of genetic mutations. In the early 1980s, actual mutations in growth regulatory genes, termed oncogenes, were described.[2,3] Pairs of oncogenes deregulated by mutation were found sufficient to induce tumorigenesis in rodent cells.[4,5] Then negative regulators of cell growth, the antithesis of the oncogenes, called tumor suppressor genes received renewed and intense attention on identification of the retinoblastoma gene.[6,7] It is now clear that the loss of function of tumor suppressor genes is as important, if not more so, in the tumor-forming process than the activation of oncogenes.[8–12] Thus, recently, heritable changes have been found at the level of single-base-pair changes

---

F. ANDREW RAY • Life Sciences Division, Los Alamos National Laboratory, Los Alamos, New Mexico 87545. *Present address*: Department of Microbiology, Immunology and Molecular Genetics, The Albany Medical College, Albany, New York, 12208-3479.

DNA Tumor Viruses: Oncogenic Mechanisms, edited by Giuseppe Barbanti-Brodano *et al.* Plenum Press, New York, 1995.

that, when combined to form a sufficient cadre of changes, enable a cell to become deregulated to the point of autonomy.

One tumor suppressor gene that is involved in the genesis of at least half of all human tumors is p53.[11,12] The p53 gene is located on the short arm of human chromosome 17.[13] The p53 protein causes cells to arrest in the G_1 phase of the cell cycle when genetic damage is detected and may even participate in the repair process.[14-16] Now it seems we have come full circle back to Dr. Boveri's proposal of inhibitory chromosomes, with the discovery that chromosome 17 is a chromosome that inhibits division, and this inhibition may prevent further abnormal chromosome formation and subsequent oncogenic evolution.

Why is it necessary to have an unstable karyotype in order to form a tumor? We now know that tumorigenesis is a multistep transition from normality to uncontrolled proliferation, and, as Boveri knew, it is a cell lineage evolution.[1,17] A cell must acquire anchorage independence, the ability to grow in reduced concentrations of serum growth factors, angiogenesis, the ability to metastasize, etc.[18-21] Many of these steps require a heritable change. A mechanism for mutation is therefore necessary. Chromosome mutation, defined as the structural loss, gain, or relocation of chromosome segments, is a legitimate mechanism of mutation.[22] Classic examples of this include the "Philadelphia" marker chromosome, where the *abl* oncogene is translocated from chromosome 9 to chromosome 22 and an aberrant protein is produced, found in chronic myelogenous leukemia, and the 8:14 translocation, where the *myc* oncogene is deregulated by juxtaposition into an active immunoglobulin locus, found in Burkitt's lymphoma.[23,24] A lineage with a selective advantage can survive and soon outgrow its contemporaries.

Soon after the discovery of the SV40 virus and the realization that it would cause tumors if injected into newborn baby hamsters, it was also discovered to cause chromosome damage in human cells.[25-27] The damage was dramatic and began within 48 hr of infection; it manifested both numerically in the form of aneuploidy and structurally in the form of aberrations. The most frequent aberration was the dicentric chromosome, but all types of chromosome and chromatid-type aberrations were observed. Virus-infected cells could be immortalized at low frequency, and the cells would acquire phenotypic traits associated with cancer cells.[28] Fortunately for man, the virus has only rarely been associated with cancer in man, presumably because of its strong immunogenicity.[29-31]

It was soon discovered that the "transforming" region of the virus lay in the early genes encoding the large T and small t proteins[32] (see below for discussion pertaining to a recently described third T antigen). Large T antigen is the only protein of viral origin that is required to replicate the viral genome.[33,34] The remaining replication proteins are provided by the host cells. Large and small t genes inhabit the same region of DNA, and through alternative splicing either one or the other protein is synthesized.[35,36] Therefore, in many studies, with plasmid constructs using the early genes, both T/t antigens were expressed.[37-41] Another confounding fact in these early plasmid studies was that they frequently contained the SV40 viral DNA replication origin.[42-46] In human cells the large T protein stimulates autonomous replication at integrated origin sequences.[46,47] Such exogenous DNA might cause chromosome instability. Also, integration of these sequences near cellular genes could theoretically contribute to the transformation process. When investigators had used constructs lacking small t, the *ori* were present, and vice versa. No one had conclusively shown that the T antigen protein was responsible for chromosome damage, nor was the temporal relationship between chromosome damage, a possible driver of mutation, and the acquisition of neoplastic traits known. The question remained: does T antigen cause cells to acquire neoplastic traits directly, or is the mechanism indirect? Work of Gotoh *et al.* hinted, more than 10 years ago, that the mechanism might indeed be indirect.[48] Which comes first, chromosome instability or transformation?

2. CHROMOSOME DAMAGE PRECEDES IMMORTALIZATION AND TRANSFORMATION

In order to address such questions, we constructed a plasmid designed to express large T only.[49] The origin of replication/promoter region was replaced by the strong Rous sarcoma virus promoter (RSV).[50] We chose a deletion mutant dl884 reported to have a 247-bp deletion encompassing a large percentage of the small t unique sequences and deleting the translational termination signal and splice donor site for small t.[51] We coelectroporated the plasmid with pRSVneo, another plasmid that contained the RSV promoter rather than the ubiquitous SV40 origin/promoter region. Using ten plasmids of T to one of neo we were able to assure that most G418^r colonies expressed a T-antigen protein.

The rationale for these experiments was simple. We wished to integrate the T antigen gene stably into an otherwise stable diploid human fibroblast strain and study the direct and indirect effects of T antigen as soon as a suitably large cloned population of cells could be derived. Human fibroblasts were used in order to avoid the spontaneous transformation and immortalization background observed with rodent fibroblasts.[52] During the expansion process from single cells to millions of cells, T-antigen expression was assayed using an indirect immunofluorescence assay. Shortly after expansion and at approximately the same population-doubling level, metaphase spreads were prepared for karyotype analysis, and plating experiments were done in reduced concentrations of serum and to determine anchorage independence. Cells resistant to G418 but not expressing T antigen served as controls. The results were clear. The expression of T antigen was sufficient to cause chromosome alterations in virtually every cell. Only rare cells had acquired the ability to grow in reduced concentrations of serum growth factors or to form anchorage-independent colonies. And yet, T antigen of the predicted molecular weight was being expressed in every cell of these clonally derived populations. This suggests that chromosome damage comes first, prior to transformation, and may provide a mechanism for acquisition of new traits.

The distribution of these chromosome aberrations is shown in Fig. 1. As our colleagues had found almost 30 years earlier, dicentric chromosomes were the most frequent aberration, but every type of aberration imaginable was present at some frequency.[26,27] Chromosome number was aberrant in most cells, with numerous cells with near-tetraploid chromosome contents. It was rare to find a cell with 46

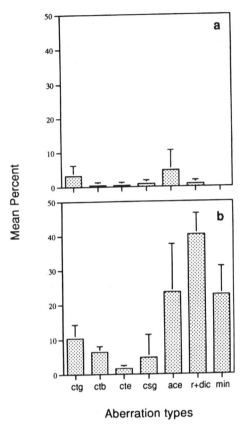

FIGURE 1. Mean percentage of cells with specific aberrations in T-antigen-expressing human fibroblasts. Chromatid aberrations: ctg, chromatid gap; ctb, chromatid break; cte, chromatid exchange. Chromosome aberrations: csg, chromosome gap; ace, acentric fragment; r+dic, ring and dicentric chromosomes; min, minute chromosomes, single and double. (a) Mean of five G418-resistant T-antigen-negative control strains. (b) Mean of five T-antigen-positive clones at approximately the same population-doubling level. The percentage of cells with at least one of a given aberration was determined ($n = 50$); thus, cells with more than one of the same aberration are not represented.

chromosomes and no structural aberrations. Chromosome and chromatid-type aberrations were both represented. Stewart and Bacchetti, using a similar T-only construct, confirmed this result shortly thereafter.[53]

Recently a third T antigen, called 17kT, has been described.[54] This protein is produced by a second splice 3' to the large T splice. The 17kT is identical to large T for 131 amino acid residues, and then four different residues are added before a translational termination signal is reached in a different reading frame. The 17kT could be produced from dl884 used in our study and from dl2005 used by Stewart and Bacchetti.[49,53] In immunoprecipitation experiments using the antibody that reacts with 17kT (PAb419), we have seen no evidence of the protein, even when large amounts of large T were evident. It is yet unclear whether this protein is produced in human fibroblasts at all. We have evidence that certain mutations in large T gene that are 3' to 17kT sequences can decrease the amount of chromosome damage observed (unpublished data). We also have evidence that chromosome damage is correlated with the amount of expressed large T antigen. These results imply that the 17kT, if present, is not important in producing chromosome damage.

As mentioned above, chromosome damage preceded acquisition of colony formation in low serum and anchorage independence. Two additional direct phenotypic traits were imparted by T-antigen expression. First, we noted a large number of dead and dying cells in T-antigen-positive cells.[49] An average of 20% of the cells in a T-antigen-positive culture were incapable of excluding propidium iodide and were therefore scored as dead. In fact, cell death has allowed us readily to discriminate a T-antigen-positive population of cells because T antigen alone had very little effect on the morphology of the cells.[55] As shown in Fig. 2, we continued to culture the cells to determine if they were immortal. At each passage we determined the number of population doublings accomplished and could therefore record cumulative population doublings (CPD). As described by numerous investigators, T-antigen-positive cells had an extended life-span.[57,58] The cells grew an extra approximately 20 CPD and reached a crisis stage of cell culture where cell death exceeded cell growth,[59] and then the number of cells decreased rapidly. Therefore, the direct effects of T antigen that we observed were chromosome damage, cell death, and extended life-span. T-antigen-positive cells were not anchorage independent, serum growth factor independent, or immortal. If anchorage independence or low-serum colony-formation assays were performed on higher-passage but still mortal cell populations, increased numbers of colonies were observed. The results suggested that the cells were acquiring these phenotypic traits through mutation and selection.

The cell number generally dropped from 10^7 to less than 10^5 within a period of 3 months (Fig. 2). In a few instances, rare colonies appeared that were expandable. The expanded colonies eventually were determined to be immortal. After expansion of these colonies to millions of cells, they were frozen and stored as newly immortal cells (Fig. 2). This was a tentative assessment, as putatively immortal cells have been known to cease division.[28] We therefore scored immortalization as those cells that had divided through more than twice their normal CPD. This scoring was therefore retrospective.

3. CHROMOSOME DAMAGE CONTINUES IN NEWLY IMMORTAL AND HIGH-PASSAGE CELLS

As mentioned above, immortalization was judged retrospectively. By thawing newly immortal cells and again assaying chromosome damage, we determined that chromosome damage was continuing, and the level of damage had increased.[60] The newly immortal cells frequently grew quite slowly (Fig. 2), and large numbers of dead cells continued to be generated.

Immortalization likely results from the loss of finite-life-span genes, not from the activation of an immortalization gene *per se*. The finite-life-span genes are believed to be autosomal dominant as judged by the low frequency of immortalization and the results of cell fusion experiments between immortal and mortal cells.[61–63] The finite-life-span gene loci are of great interest because of the importance of the immortalization trait to a neoplastic cell (Fig. 3). We therefore studied G-banded karyotypes of nine newly immortal cell lines and found several recurrent

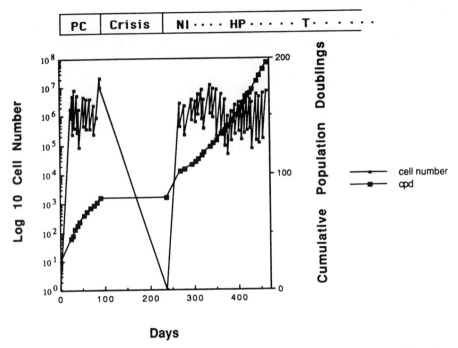

FIGURE 2. Cell growth data representative of the SV40 T-antigen human fibroblast model used to produce stage-specific cell populations of neoplastically transformed cells. The cell number and cumulative population-doubling data were from CT10-12 but were similar for all T-antigen-positive cell strains that became immortal cell lines and are plotted from the initial cloning (electroporation) through the *de facto* cloning (immortalization) until many doublings after crisis. The stages are PC (precrisis), crisis, NI (newly immortal), HP (high passage), and T (tumor derived). Cells were assayed for transformation and frozen at these stages of lineage evolution. The variable rate of evolution between immortal lineages is represented by the lack of clear boundaries between postimmortalization stages. Note that the X intercept for cell number at crisis is not precisely known. Reproduced from Ray and Kraemer (56) with permission from Oxford University Press.

FIGURE 3. Importance of immortalization to the neoplastic process. Multiple mutations are required for a cell lineage to evolve to tumorigenicity.[64–66] For example, to activate one oncogene requires at least one mutation, and to eliminate both copies of a tumor suppressor gene requires at least two mutations. If a normal finite-proliferative human fibroblast has a proliferative potential of 50 population doublings (PD), and if mutations are occurring in a given gene at $\sim 1 \times 10^{-6}$ per cell per generation (a high estimate for mutation rates in human cells), then at most three mutations could occur before proliferative potential expired. This model is portrayed in the figure. If mutation 1 occurred at time 0, then 20 PD later this cell would be expanded to a million cells. A second mutation could then be expected to occur. If the second mutation occurred, then the remaining 999,999 cells would be eliminated from the lineage evolution. At that point in the evolution of the lineage, there would be 30 PD remaining. After a second iteration of this process, there would be only ten PD remaining, and the cells would reach the end of their life-span prior to acquiring enough mutations to form a tumor.

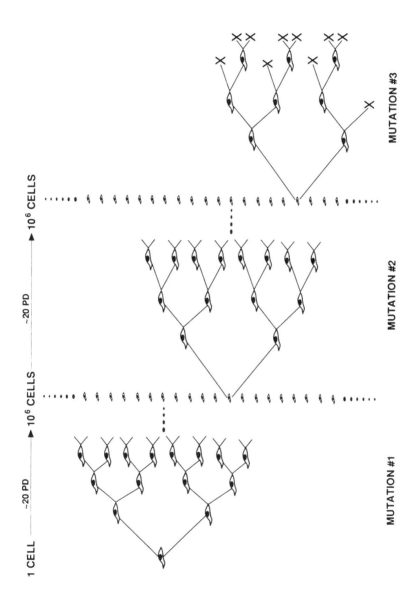

chromosome deletions.[67] The most frequent were deletions distal to 6q21, 3p24, 1p34, 4q25, and 5p14 found in the majority of the nine immortal lines. Hubbard-Smith *et al.* confirmed that the 6q21 locus is frequently associated with immortalization.[68]

As the immortal T-antigen-positive populations of cells continue to grow in culture and accumulate population doublings, their growth rate invariably increases (Fig. 2), and the percentage of dead cells decreases. Chromosome damage continues, albeit abated, and the number of cells that can form anchorage-independent colonies or colonies in low serum increases. These cells do not usually form tumors when injected subcutaneously into nude mice.

4. T-ANTIGEN-EXPRESSING CELLS EVENTUALLY BECOME TUMORIGENIC

After several hundred CPD, the cells became conditionally tumorigenic in nude mice.[56] The cells formed tumors if injected into preimplanted collagen sponges that were already vascularized.[69] If more than 10 million high-passage cells expressing T antigen were injected into these collagen sponges, then a tumor *might* form. Tumors formed infrequently and with long latencies, as if an additional step or steps were required for tumor growth. The additional steps were heritable. When G418^r human cells expressing T antigen were retrieved from these tumors and reinjected into new mice, tumors would form quickly at high incidence without the requirement for sponges and with fewer cells.

Other investigators have reported that tumors would form if T-antigen-positive immortal cells were treated with carcinogens or were transfected with activated oncogenes.[70–71] In our system, T antigen alone was sufficient to cause human cells to become tumorigenic, but extended time *in vitro* and then *in vivo* was required, presumably time to acquire additional mutations.

5. SUMMARY AND FUTURE DIRECTIONS

In summary, the SV40 T antigen causes severe chromosome damage in virtually every human fibroblast in which it is expressed. This damage precedes and accompanies transformation. There is evidence that this phenomenology is not unique to human cells.[72,73] The data therefore support the hypothesis that T antigen causes, at least partially, transformation of human cells by a mechanism of iterative chromosome mutation and selection of favored phenotypes.

We are attempting to determine how the T antigen causes chromosome damage. To accomplish viral replication, the small SV40 virus utilizes cellular enzymes. Therefore, it is necessary to stimulate the cells to enter the appropriate phase of the cell cycle when these enzymes are at their highest levels. Toward this end, T antigen has evolved the ability to bind cellular proteins involved in growth control.[74,75] One potential mechanism of damage is to bypass cellular checkpoints where DNA damage is monitored and repaired, such as by binding and inactivating the p53 protein.[14,15] We have evidence using the p53-binding T-antigen mutation (D402>H) described by Lin and Simmons[58] that p53 binding is indeed important in the induction of chromosome damage (unpublished data). T antigen is a DNA helicase that binds to viral origin of replication sequences, melts this region, and precedes the replication fork.[76,77] Recognizing similar sequences in the cellular DNA[78] and inappropriately unwinding them may cause chromosome damage.

ACKNOWLEDGMENTS. I wish to thank my frequent collaborator in these studies, Paul M. Kraemer, for critically reading this manuscript. I also wish to thank Miriam Smyth for critical comments.

REFERENCES

1. Boveri, T., 1914, *Zur Frage der Entstehung Maligner Tumoren*, Gustav Fischer, Jena.
2. Cooper, G. M., 1982, Cellular transforming genes, *Science* **218**:801–806.
3. Bishop, J. M., 1983, Cellular oncogenes and retroviruses, *Annu. Rev. Biochem.* **52**:301–354.
4. Land, H., Parada, L. F., and Weinberg, R. A., 1983, Tumorigenic conversion of primary embryo fibroblasts requires at least two cooperating oncogenes, *Nature* **304**:596–602.
5. Ruley, E. H., 1983, Adenovirus early region 1A enables viral and cellular transforming genes to transform primary cells in culture, *Nature* **304**:602–606.
6. Knudson, A. G., 1985, Hereditary cancer, oncogenes, and antioncogenes, *Cancer Res.* **45**:1437–1443.
7. Klein, G., 1987, The approaching era of the tumor suppressor genes, *Science* **238**:1539–1545.
8. Baker, S. J., Fearon, E. R., Nigro, J. M., Hamilton, S. R., Preisinger, A. C., Jessup, J. M., van Tuinen, P., Ledbetter, D. H., Barker, D. F., Nakamura, Y., White, R., and Vogelstein, B., 1989, Chromosome 17 deletions and p53 gene mutations in colorectal carcinomas, *Science* **244**:217–221.
9. Vogelstein, B., Fearon, E. R., Kern, S. E., Hamilton, S. R., Preisinger, A. C., Nakamura, Y., and White, R., 1989, Allelotype of colorectal carcinomas, *Science* **244**:207–211.
10. Fearon, E. R., Cho, K. R., Nigro, J. M., Kern, S. E., Simons, J. W., Ruppert, J. M., Hamilton, S. R., Preisinger, A. C., Thomas, G., Kinzler, K. W., and Vogelstein, B., 1990, Identification of a chromosome 18q gene that is altered in colorectal cancers, *Science* **247**:49–56.
11. Vogelstein, B., 1990, A deadly inheritance, *Nature* **348**:681–682.
12. Hollstein, M., Sidransky, D., Vogelstein, B., and Harris, C. C., 1991, p53 mutations in human cancers, *Science* **253**:49–53.
13. McBride, O. W., Merry, D., and Givol, D., 1986, The gene for human p53 cellular tumor antigen is located on chromosome 17 short arm (17p13), *Proc. Natl. Acad. Sci. USA* **83**:130–134.
14. Kastan, M. B., Onyekwere, O., Sidransky, D., Vogelstein, B., and Craig, R. W., 1991, Participation of p53 protein in the cellular response to DNA damage, *Cancer Res.* **51**:6304–6311.
15. Kuerbitz, S. J., Plunkett, B. S., Walsh, W. V., and Kastan, M. B., 1992, Wild-type p53 is a cell cycle checkpoint determinant following irradiation, *Proc. Natl. Acad. Sci. USA* **89**:7491–7495.
16. Bakalkin, G., Yakovleva, T., Selivanova, G., Magnusson, K. P., Szekely, L., Kiseleva, E., Klein, G., Terenius, L., and Wiman, K. G., 1994, p53 binds single-stranded DNA ends and catalyzes DNA renaturation and strand transfer, *Proc. Natl. Acad. Sci. USA* **91**:413–417.
17. Nowell, P. C., 1976, The clonal evolution of tumor cell populations, *Science* **194**:23–28.
18. Freedman, V. H., and Shin, S.-I., 1974, Cellular tumorigenicity in nude mice: Correlation with cell growth in semi-solid medium, *Cell* **3**:355–359.
19. Dulbecco, R., 1970, Topoinhibition and serum requirement of transformed and untransformed cells, *Nature* **227**:802–806.

20. Folkman, J., Watson, K., Ingber, D., and Hanahan, D., 1989, Induction of angiogenesis during the transition from hyperplasia to neoplasia, *Nature* **339:**58–61.
21. Fidler, I. J., 1990, Critical factors in the biology of human cancer metastasis: Twenty-eighth G. H. A. Clowes Memorial Award Lecture, *Cancer Res.* **50:**6130–6138.
22. German, J. (ed.), 1983, *Chromosome Mutation and Neoplasia*, Alan R. Liss, New York.
23. de Klein, A., van Kessel, A. G., Grosveld, G., Bartram, C. R., Hagemeijer, A., Bootsma, D., Spurr, N. K., Heisterkamp, N., Groffen, J., and Stephenson, J. R., 1982, A cellular oncogene is translocated to the Philadelphia chromosome in chronic myelocytic leukaemia, *Nature* **300:**765–767.
24. Croce, C. M., 1986, Chromosome translocations and human cancer, *Cancer Res.* **46:**6019–6023.
25. Eddy, B. E., Borman, G. S., Grubbs, G. E., and Young, R. D., 1962, Identification of the oncogenic substance in rhesus monkey kidney cell cultures as simian virus 40, *Virology* **17:**65–75.
26. Koprowski, H., Ponten, J. A., Jensen, F., Raudin, R. G., Moorhead, P., and Saksela, E., 1962, Transformation of human tissue infected with simian virus 40, *J. Cell. Comp. Physiol.* **59:**281–292.
27. Wolman, S. R., Hirschhorn, K., and Todaro, G. J., 1964, Early chromosomal changes in SV40-infected human fibroblast cultures, *Cytogenetics* **3:**45–61.
28. Huschtscha, L. I., and Holliday, R., 1983, Limited and unlimited growth of SV40-transformed cells from human diploid MRC-5 fibroblasts, *J. Cell Sci.* **63:**77–99.
29. Soriano, F., Shelburne, C. E., and Gokcen, M., 1974, Simian virus 40 in a human cancer, *Nature* **249:**421–424.
30. Bergsagel, D. J., Finegold, M. J., Butel, J. S., Kupsky, W. J., and Garcea, R. L., 1992, DNA sequences similar to those of simian virus 40 in ependymomas and choroid plexus tumors of childhood, *N. Engl. J. Med.* **326:**988–993.
31. Carbone, M., Pass, H. I., Rizzo, P., Marinetti, M., Di Muzio, M., Mew, D. J., Levine, A. S., and Procopio, A., 1994, Simian virus 40-like DNA sequences in human pleural mesothelioma, *Oncogene* **9:**1781–1790.
32. Topp, W. C., Lane, D., and Pollack, R., 1980, Transformation by SV40 and polyoma virus, in: *DNA Tumor Viruses*, 2nd ed., part 2 (J. Tooze, ed.), Cold Spring Harbor Laboratory, New York, pp. 205–296.
33. Li, J. J., and Kelly, T. J., 1984, Simian virus 40 DNA replication *in vitro. Proc. Natl. Acad. Sci. USA* **81:**6973–6977.
34. Kelly, T. J., 1988, SV40 DNA replication, *J. Biol. Chem.* **263:**17889–17892.
35. Noble, J. C., Pan, Z. Q., Prives, C., and Manley, J. L., 1987, Splicing of SV40 early pre-mRNA to large T and small t mRNAs utilizes different patterns of lariat branch sites, *Cell* **50:**227–236.
36. Fu, X.-Y., and Manley, J. L., 1987, Factors influencing alternative splice site utilization in vivo, *Mol. Cell. Biol.* **7:**738–748.
37. Major, E. O., and Matsumura, P., 1984, Human embryonic kidney cells: Stable transformation with an origin-defective simian virus 40 DNA and use as hosts for human papovavirus replication, *Mol. Cell. Biol.* **4:**379–382.
38. Murnane, J. P., Fuller, L. F., and Painter, R. B., 1985, Establishment and characterization of a permanent pSV ori- transformed ataxia–telangiectasia cell line, *Exp. Cell Res.* **158:**119–126.
39. Canaani, D., Naiman, T., Teitz, T., and Berg, P., 1986, Immortalization of xeroderma pigmentosum cells by simian virus 40 DNA having a defective origin of DNA replication, *Somat. Cell Mol. Genet.* **12:**13–20.
40. Daya-Grosjean, L., James, M. R., Drougard, C., and Sarasin, A., 1987, An immortalized xeroderma pigmentosum, group C, cell line which replicates SV40 shuttle vectors, *Mut. Res.* **183:**185–196.
41. Neufeld, D. S., Ripley, S., Henderson, A., and Ozer, H. L., 1987, Immortalization of human fibroblasts transformed by origin-defective simian virus 40, *Mol. Cell. Biol.* **7:**2794–2802.
42. Chang, L.-S., Pater, M. M., Hutchinson, N. I., and Di Mayorca, G., 1984, Transformation by purified early genes of simian virus 40, *Virology* **133:**341–353.

43. Chang, L.-S., Pan, S., Pater, M. M., and Di Mayorca, G., 1985, Differential requirement for SV40 early genes in immortalization and transformation of primary rat and human embryonic cells, *Virology* **146**:246–261.
44. Chang, P. L., Gunby, J. L., Tomkins, D. J., Mak, I., Rosa, N. E., and Mak, S., 1986, Transformation of human cultured fibroblasts with plasmids carrying dominant selection markers and immortalizing potential, *Exp. Cell Res.* **167**:407–416.
45. Mayne, L. V., Priestley, A., James, M. R., and Burke, J. F., 1986, Efficient immortalization and morphological transformation of human fibroblasts by transfection with SV40 DNA linked to a dominant marker, *Exp. Cell Res.* **162**:530–538.
46. Wood, C. M., Timme, T. L., Hurt, M. M., Brinkley, B. R., Ledbetter, D. H., and Moses, R. E., 1987, Transformation of DNA repair-deficient human diploid fibroblasts with a simian virus 40 plasmid, *Exp. Cell Res.* **169**:543–553.
47. Zouzias, D, Jha, K. K., Mulder, C., Basilico, C., and Ozer, H. L., 1980, Human fibroblasts transformed by the early region of SV40 DNA: Analysis of "free" viral DNA sequences, *Virology* **104**:439–453.
48. Gotoh, S., Gelb, L., and Schlessinger, D., 1979, SV40-transformed human diploid cells that remain transformed throughout their limited lifespan. *J. Gen. Virol.* **42**:409–414.
49. Ray, F. A., Peabody, D. S., Cooper, J. L., Cram, L. S., and Kraemer, P. M., 1990, SV40 T antigen *alone* drives karyotype instability that precedes neoplastic transformation of human diploid fibroblasts, *J. Cell. Biochem.* **42**:13–31.
50. Gorman, C. M., Merlino, G. T., Willingham, M. C., Pastan, I., and Howard, B. H., 1982, The Rous sarcoma virus long terminal repeat is a strong promoter when introduced into a variety of eukaryotic cells by DNA-mediated transfection, *Proc. Natl. Acad. Sci. USA* **79**:6777–6781.
51. Shenk, T. E., Carbon, J., and Berg, P., 1976, Construction and analysis of viable deletion mutants of simian virus 40, *J. Virol.* **18**:664–671.
52. Kraemer, P. M., Travis, G. L., Ray, F. A., and Cram, L. S., 1983, Spontaneous neoplastic evolution of Chinese hamster cells in culture: Multistep progression of phenotypes, *Cancer Res.* **43**:4822–4827.
53. Stewart, N., and Bacchetti, S., 1991, Expression of SV40 large T antigen, but not small t antigen, is required for the induction of chromosomal aberrations in transformed human cells, *Virology* **180**:49–57.
54. Zerrahn, J., Knippschild, U., Winkler, T., and Deppert, W., 1993, Independent expression of the transforming amino-terminal domain of SV40 large T antigen from an alternatively spliced third SV40 early mRNA, *EMBO J.* **12**:4739–4746.
55. de Ronde, A., Sol, C. J., van Strien, A., ter Schegget, J., and van der Noordaa, J., 1989, The small t antigen is essential for the morphological transformation of human fibroblasts, *Virol.* **171**:260–263.
56. Ray, F. A., and Kraemer, P. M., 1993, Iterative chromosome mutation and selection as a mechanism of complete transformation of human diploid fibroblasts by SV40 T antigen, *Carcinogenesis* **14**:1511–1516.
57. Shay, J. W., Pereira-Smith, O. M., and Wright, W. E., 1991, A role for both RB and p53 in the regulation of human cellular senescence, *Exp. Cell Res.* **196**:33–39.
58. Lin, J.-Y., and Simmons, D. T., 1991, The ability of large T antigen to complex with p53 is necessary for the increased lifespan and partial transformation of human cells by simian virus 40, *J. Virol.* **65**:6447–6453
59. Stein, G. H., 1985, SV40-transformed human fibroblasts: Evidence for cellular aging in precrisis cells, *J. Cell. Physiol.* **125**:36–44.
60. Ray, F. A., Meyne, J., and Kraemer, P. M., 1992, SV40 T antigen induced chromosomal changes reflect a process that is both clastogenic and aneuploidogenic and is ongoing throughout neoplastic progression of human fibroblasts, *Mut. Res.* **284**:265–273.
61. Shay, J. W., and Wright, W. E., 1989, Quantitation of the frequency of immortalization of normal human diploid fibroblasts by SV40 large T-antigen, *Exp. Cell Res.* **184**:109–118.

62. Norwood, T. H., Pendergrass, W. R., Sprague, C. A., and Martin, G. M., 1974, Dominance of the senescent phenotype in heterokaryons between replicative and post-replicative human fibroblast-like cells, *Proc. Natl. Acad. Sci. USA* **71**:2231–2235.
63. Goldstein, S., 1990, Replicative senescence: The human fibroblast comes of age, *Science* **249**:1129–1133.
64. McCormick, J. J., and Maher, V. M., 1989, Malignant transformation of mammalian cells in culture, including human cells, *Environ. Mol. Mutagen.* **16**:105–113.
65. Levine, A. J., and Momand, J., 1990, Tumor suppressor genes: The p53 and retinoblastoma sensitivity genes and gene products, *Biochim. Biophys. Acta* **1032**:119–136.
66. Loeb, L. A., 1991, Mutator phenotype may be required for multistage carcinogenesis, *Cancer Res.* **51**:3075–3079.
67. Ray, F. A., and Kraemer, P. M., 1992, Frequent deletions in nine newly immortal human cell lines, *Cancer Genet. Cytogenet.* **59**:39–44.
68. Hubbard-Smith, K., Patsalis, P., Pardinas, J. R., Jha, K. K., Henderson, A. S., and Ozer, H. L., 1992, Altered chromosome 6 in immortal human fibroblasts, *Mol. Cell. Biol.* **12**:2273–2281.
69. Kraemer, P. M., Travis, G. L., Saunders, G. C., Ray, F. A., Stevenson, A. P., Bame, K., and Cram, L. S., 1984, Tumorigenicity assays in nude mice: Analysis of the implanted gelatin sponge method, in: *Immune-Deficient Animals* (B. Sordat, ed.), S. Karger, Basel, pp. 214–219.
70. Rhim, J. S., Jay, G., Arnstein, P., Price, F. M., Sanford, K. K., and Aaronson, S. A., 1985, Neoplastic transformation of human epidermal keratinocytes by Ad12-SV40 and Kirsten sarcoma viruses, *Science* **227**:1250–1252.
71. Rhim, J. S., Fujita, J., Arnstein, P., and Aaronson, S. A., 1986, Neoplastic conversion of human keratinocytes by adenovirus 12-SV40 virus and chemical carcinogens, *Science* **232**:385–388.
72. Gorbunova, L. V., Varshaver, N. B., Marshak, M. I., and Shapiro, N. I., 1982, The role of the transforming A gene of SV40 in the mutagenic activity of the virus, *Mol. Gen. Genet.* **187**:473–476.
73. Liu, J., Li, H., Nomura, K., Dofuku, R., and Kitagawa, T., 1991, Cytogenetic analysis of hepatic cell lines derived form SV40-T antigen gene-harboring transgenic mice, *Cancer Genet. Cytogenet.* **55**:207–216.
74. Levine, A. J., 1990, The p53 protein and its interactions with the oncogene products of the small DNA tumor viruses, *Virology* **177**:419–426.
75. Levine, A. J., 1993, The tumor suppressor genes, *Annu. Rev. Biochem.* **62**:623–651.
76. Dodson, M., Dean, F. B., Bullock, P., Echols, H., and Hurwitz, J., 1987, Unwinding of duplex DNA from the SV40 origin of replication by T antigen, *Science* **238**:964–967.
77. Borowiec, J. A., Dean, F. B., Bullock, P. A., and Hurwitz, J., 1990, Binding and unwinding—How T antigen engages the SV40 origin of replication, *Cell* **60**:181–184.
78. Gruss, C., Wetzel, E., Baack, M., Mock, U., and Knippers, R., 1988, High-affinity SV40 T-antigen binding sites in the human genome, *Virology* **167**:349–360.

3

Transformation by Polyomaviruses
Role of Tumor Suppressor Proteins

DANIEL T. SIMMONS

1. INTRODUCTION

Since its identification as a tumor virus, SV40 has been a favorite model for understanding the mechanism by which a DNA virus can control the state and growth properties of susceptible cells. From its humble beginnings, the polyomavirus tumor biology field has mushroomed into several new directions, most notably into tumor suppressors. These new branches are now buzzing with activity and touching or merging with other established disciplines including cell-cycle controls, oncogenes, and differentiation.

The major transforming protein specified by the monkey and human polyomaviruses is large T antigen.[1–4] Small t antigen is needed in some cases for efficient transformation of nonpermissive cells (see Section 8) but cannot transform by itself. The situation is different for the mouse polyomaviruses (see Section 7), where a different protein (middle T antigen) is the primary transforming protein. Because only one or two proteins are needed for transformation of cells, there was initially reason to believe that the molecular mechanism of transformation was going to be relatively simple. Nothing has been further from the truth. These proteins have evolved to target and usurp several cellular regulatory pathways to the viruses' own purposes. Investigations into these areas paint a complex picture of molecular cycles and controls leading us to believe that these viruses have

---

DANIEL T. SIMMONS • Department of Biology, University of Delaware, Newark, Delaware 19716.

DNA Tumor Viruses: Oncogenic Mechanisms, edited by Giuseppe Barbanti-Brodano *et al.* Plenum Press, New York, 1995.

descended into the heart of the cell itself and that they manipulate several crucial regulatory activities. They have, in fact, led us directly to the essence of the cell.

In this chapter, I first describe the major features of the structure and function of the SV40 large T antigen (Section 2). Next, I discuss the interactions between T antigen and Rb, the product of the retinoblastoma susceptibility gene (Section 3), T antigen and the tumor suppressor protein p53 (Section 4), and T antigen and p300 (Section 5). In Section 6, I summarize the state of our knowledge about the regions of T antigen that are needed for cell transformation or immortalization. Transformation by human and mouse polyomaviruses are then discussed (Section 7), especially as they relate to SV40. Small t antigen and its role in transformation are covered in Section 8. Finally, I attempt to summarize the field in Sections 9 and 10.

2. STRUCTURE AND FUNCTION OF SV40 LARGE T ANTIGEN

2.1. Replication Activities

Large T antigen is a multifunctional protein possessing an impressive array of functions involved during virus replication and cell transformation. Most of these activities have been mapped to discrete regions of the polypeptide chain (Fig. 1). Of primary importance to T antigen's replication functions, the DNA-binding domain maps from residues 146 to 246,[5-10] its helicase domain extends from residues 131 to 616,[11] and the ATP-binding region maps from residues 418 to 528.[12,13] Because this protein not only initiates virus DNA replication but also turns on late gene expression,[14-16] it has the ability to activate transcription from the late viral promoter as well as from several cellular promoters.[17-20] This function has been mapped to residues 1–82.[21]

The protein is phosphorylated at several serine and threonine residues located between residues 106 and 124 near the NH_2-terminal end and between 639 and 701 near the COOH end.[22-24] Some of these phosphorylations regulate the replication activities of T antigen. In particular, phosphorylations at serines 121 and 123 inhibit DNA binding and/or DNA replication,[25-28] whereas phosphorylation at threonine 124 appears to be essential to the unwinding activity of full-length T antigen[29,30] and therefore to its DNA replication activity.[31,32]

2.2. Transformation Activities

The binding to at least three cellular proteins appears to be necessary for full transformation activity by SV40 T antigen.[33,34] The binding to the p53 tumor suppressor protein has been localized to residues 347–517 (Fig. 1), although additional COOH-terminal sequences may be necessary for efficient binding.[35-38] Because this is a large stretch of T antigen, p53 probably associates with some tertiary structure of T antigen. On the other hand, the tumor suppressor protein Rb appears to bind to a small primary or secondary region of T antigen located between residues 106 and 111.[39] The third protein p300, probably binds to the amino-terminal end of T antigen.[40,34]

Replication functions

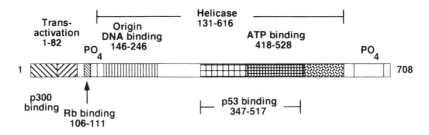

Transformation functions

FIGURE 1. Functional domains of SV40 large T antigen. Replication and transformation functions are mapped on the linear polypeptide chain. Replication activities localized on the protein chain are transactivation, origin binding, ATP binding, and helicase. Also shown are the locations of two clusters of phosphorylated residues. The mapped transformation functions are the binding sites for p300, Rb, and p53. See text for references.

3. ASSOCIATION OF T ANTIGEN WITH Rb

The adenovirus E1A proteins interact with a number of cellular proteins.[41] One of these cellular proteins, it turns out, is the product of the retinoblastoma susceptibility gene.[42] A related cellular protein, called p107, also binds to E1A.[41] Interestingly, the SV40 and mouse polyomavirus large T antigen has some sequence homology with E1A, and this led to the discovery that Rb (as well as p107) also binds to T antigen.[43,44] The region of homology between E1A and T antigen corresponds to amino acids 105 and 114 in T antigen.[45] By mutational analysis[43] and by the use of competing peptides,[39] it was demonstrated that a slightly smaller region (amino acids 106–111) corresponds to the Rb binding site. Rb-binding-negative mutants of SV40 T antigen with a deletion or amino acid substitutions in this region fail to transform susceptible cells,[46,47] although mutations generally have no effect on virus viability.[47] This demonstrates that Rb binding is essential for full transforming activity by large T antigen.

3.1. Functions of Rb

Rb, like T antigen, is a phosphoprotein. The phosphorylation state of Rb is tightly regulated during the cell cycle (Fig. 2) (see review by Riley *et al.*).[48] During G_0 and G_1, Rb is underphosphorylated; however, during S, G_2, and M, the protein becomes highly phosphorylated.[39,49–51] There are at least three stages of phosphorylation and dephosphorylation during the cell cycle.[39,50–52] The kinases that phosphorylate Rb are cell-cycle-regulated enzymes called cdc2 kinases.[53–55] The underphosphorylated form of Rb functions to inhibit the progression of cells out of the G_1 phase of the cell cycle into S.[42,44,56,57] When Rb becomes phosphorylated, its growth-suppression activity is inhibited, and cells are free to progress into S. To

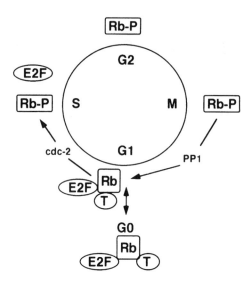

FIGURE 2. The relationship between the phosphorylation and dephosphorylation of Rb and the cell cycle. Rb protein is phosphorylated at several points during the cell cycle starting in early S by cyclin-dependent kinases. Rb is dephosphorylated during M by phosphatase 1 (PP1) to activate Rb into its growth-inhibitory state. Rb remains underphosphorylated in G_1 and G_0. The underphosphorylated form is the one that binds to SV40 T antigen and to E2F transcription factor. When Rb becomes phosphorylated in S, E2F is unbound and free to function in transcription.

complete the cycle, Rb is dephosphorylated by protein phosphatase 1 (PP1) during mitosis (M),[58–60] and this results in the reactivation of Rb in G_1 (Fig. 2).

3.2. Role of T Antigen Binding to Rb

Intriguingly, SV40 large T antigen binds only to the underphosphorylated form of Rb[61] (Fig. 2) and, in so doing, most likely inhibits its growth-suppression activity. This is inferred from an experiment demonstrating that the microinjection of underphosphorylated Rb inhibits the progression of cells from early G_1 to S in the absence of T antigen but has no effect in its presence.[62] It is therefore likely that T antigen maintains Rb in its underphosphorylated form. Rb appears to inhibit progression into the cell cycle, in part, by binding and presumably inactivating transcription factor E2F (Fig. 2).[63] Several growth-promoting genes (c-*myc*, N-*myc*, DNA polymerase α, and thymidilate kinase) have E2F recognition sites in their promoters,[63–66] so it is reasonable to assume that E2F has a role in promoting cells from G_1 to S (Fig. 2).

By binding to the underphosphorylated form of Rb, T antigen most likely selectively interferes with the growth-inhibitory properties of Rb while having no effect on growth-stimulating activities of the phosphorylated forms. Rb appears to have several targets in the cells, and T antigen may block all of these events simultaneously by binding to this single protein. It is intriguing that other viral transforming proteins, in particular E1A[42] and human papillomavirus (HPV) E7 protein,[67] also bind to the underphosphorylated form of Rb, strongly suggesting that three widely different viral proteins have found it beneficial (to the viruses) to target and knock out the same cellular protein.

It has been demonstrated that Rb positively regulates the expression of cyclin

D1, and this activation is prevented by T antigen and other viral oncogenes.[68–70] Cyclin D1 and the ensuing activation of cyclin-dependent kinases, followed by Rb phosphorylation, are, however normally required for cell cycle progression from G_1 into S, suggesting the existence of a regulatory loop between Rb and cyclin D1.[68] Antibody-mediated cyclin D1 knockout experiments demonstrate that the cyclin D1 protein is dispensable for passage through the cell cycle in cell lines whose Rb is inactivated through complex formation with T antigen,[69,70] suggesting that this regulatory loop is abrogated in cells expressing T antigen. Rb also normally binds to the nuclear matrix[71] and to proteins whose major function is in mitosis,[48] so that it appears to act in a number of ways to regulate the cell cycle. It is not known if these Rb functions are also inactivated by T antigen.

3.3. Rb-Related Proteins

At least two other cellular proteins (known as p107 and p130) are closely related to Rb.[72] These other two proteins have been found in association with E1A[41] and SV40 large T antigen[72,74] just like Rb. The three cellular proteins have extensive amino acid sequence similarities, and the region that binds to the viral proteins is especially conserved.[72,74] This region, although not identical in Rb-related proteins,[75] may serve as well as a general binding pocket for specific cellular proteins such as the E2F transcription factor and certain cyclins.[75]

4. ASSOCIATION OF T ANTIGEN WITH p53

The binding of SV40 T antigen to p53 was detected many years ago by immunoprecipitation experiments using SV40-transformed cells.[76–79] The importance of this complex in the transformation process by SV40 was not appreciated until much more recently.

The interaction between these two proteins is very stable, and the complex is easily detected in SV40-transformed cells of many species. The p53 levels are much higher in SV40-transformed cells than in untransformed cells, mostly because the p53 is greatly stabilized. Although early experiments suggested that this stability resulted from the protein's association with T antigen, more recent experiments[80] indicate that complex formation is not needed to induce stabilization of p53. Rather, T antigen indirectly alters p53 stability.[81]

In the context of full-sized large T antigen, p53 binding appears to be a strict requirement for the transformation of a variety of different cells.[38,82,83] Recent evidence has been obtained to support the hypothesis that metabolic stabilization of p53, and the ensuing higher levels of the protein, are important for the initiation and/or maintenance of transformation by SV40.[81,84]

4.1. Functions of p53

Wild-type p53 is an inhibitor of the transforming activity of many oncogenes both *in vitro*[85,86] and *in vivo*[87] In fact, transformed cells cannot generally tolerate high levels of functional wild-type p53.[88–91] Embryonic cells, nontransformed cells, and cells expressing some oncogenes such as *mdm2* can accumulate significantly more wild-type p53.[85,90,92,93] These properties and the identification of mutant forms of p53 in a variety of human cancers indicate that p53 is a tumor suppressor protein. In order to circumvent the normal growth-regulatory function of p53, oncogenes must either rely on the absence of functional p53 in the cell or inactivate the p53 directly.

How does p53 function in tumor suppression? There is overwhelming evidence that one of wild-type p53's functions is to negatively regulate the entry of cells into the S phase of the cell cycle. Various experiments have demonstrated that when p53 is introduced into growing cells, the cells stop dividing. This conclusion comes from a number of failed attempts to generate cells that continually express wild-type 53 after p53-expressing plasmids are introduced by transfection. Secondly, when the p53 gene is introduced under the regulation of an inducible promoter, and p53 expression is suddenly induced, cells quickly become blocked in G_1.[94] Similarly, cells transformed with a temperature-sensitive mutant form of p53 (135AV) grow quickly when the cells are incubated at 37°C, a temperature at which the mutant protein is nonfunctional, but become growth-inhibited at 32°C when the protein adopts a wild-type conformation.[95–97] Furthermore, in immortalization experiments, p53-transfected normal diploid fibroblasts usually undergo crisis and senesce at about 35–50 passages in culture, but the few cells that overcome the crisis period and become immortal for the most part do not express wild-type p53.[98] Finally, when high levels of p53 are induced in cells receiving ionizing radiation or ultraviolet light (i.e., by DNA damage), growing cells finish their round of division, but these cells do not reenter the S phase for at least 72 hr, at which point the levels of p53 have dropped to near normal.[99–101] Thus, high levels of wild-type p53 are usually incompatible with cell growth.

Although it is inferred from the above results that p53 is a critical cellular protein, experiments with null-p53 mice demonstrate that this protein is not essential to development (and differentiation) or cell division: p53-null mice appear to be normal at birth, and growth development is normal.[102] The major difference between these and normal mice is that the p53-null mice develop a number of tumors at an early age.[102] Even p53-heterozygous mice are much more likely to develop malignancies than p53-wild-type mice. The situation may be very similar in humans, where as many as 60% of cancers contain deleted or mutated p53.[103] Therefore, the inactivation of the resident p53 greatly increases the propensity for developing malignancies.

There appear to be a number of different ways that tumor cells inactivate their p53. First, the two alleles could be missing altogether, although in some cancers such as carcinomas, only one allele is usually deleted, and the other is mutated.[103] Second, wild-type p53 could be localized to the cytoplasm, where it is nonfunctional, rather than to the nucleus,[104] and third, it could be inactivated by a cellular oncogene product such as *mdm2*[93,105] or viral oncoproteins such as the SV40 large tumor antigen[76,79] or HPV E6[106] (Fig. 3).

How does p53 control cell division? We now know that p53 is a potent activator of transcription. This protein is able to activate transcription from a promoter containing a p53 binding site.[107,108] The p53 recognizes a number of different

TUMOR SUPPRESSOR PROTEINS

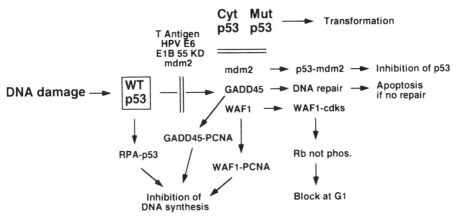

FIGURE 3. Activities of wild-type and mutant forms of p53. when the cell DNA is damaged such as after UV irradiation, higher levels of p53 accumulate. The p53 activates at least three cellular genes, *mdm2*, GADD45, and WAF1, using its ability to activate transcription. This activity is blocked in cells containing T antigen, HPV E6 protein, adenovirus E1B 55 KD protein, and cellular oncogene *mdm2*. Various mutant and cytoplasmic variants of p53 are unable to activate transcription of these genes and, instead, have the ability to induce transformation. Newly made *mdm2* binds to p53 and inhibits it, forming a regulatory loop. GADD45 induces excision repair of damaged DNA. If the DNA is not repaired, cells undergo apoptosis (programmed death). WAF1 protein is a potent inhibitor of cyclin-dependent kinases. These cdks are unable to phosphorylate substrate proteins, including Rb. As a consequence, cells become blocked at G_1. After the DNA has been repaired, levels of p53 lower, and the cdks are released from their inhibitory complex with WAF1, allowing Rb phosphorylation and the cells to reenter S. In addition to these events, elevated amounts of p53 inhibit DNA synthesis more directly by virtue of the binding to RPA. Induced GADD45 and WAF1 proteins also inhibit DNA synthesis by binding to the pol δ factor PCNA.

binding sites; however, a major consensus sequence is two or more copies of the 10-bp sequence PuPuPuC(A/T)(T/A)GPyPyPy.[109] The NH_2-terminal region of p53 contains an acidic domain[110,111] that can activate transcription through binding with the coactivators $TAF_{\|}40$ and $TAF_{\|}60$, two subunits of the basal transcription factor TFIID.[112] From its crystal structure, we can infer that the central region of p53 contains its DNA-binding domain,[113] and the portions that make contact with the DNA are those that are frequently mutated in tumors.[113] This same region also corresponds to the T-antigen-binding site of p53.

Which cellular genes are activated by p53? One such gene is *mdm2* (murine double minute), which contains a p53 binding site in the first intron[114,115] (Fig. 3). When the *mdm2* gene is activated by p53, the resulting protein binds to the NH_2-terminal region of p53 and inhibits its ability to activate transcription.[116] The *mdm2*, therefore, acts as a negative regulator of p53 and may provide a mechanism to release the p53-mediated block at G_1 following DNA damage by UV light.[117] However, the interaction of p53 with *mdm2* does not mediate the transformation suppression function of p53,[118] so it is unlikely to mediate p53's regulation of cell division. Another gene activated by p53 is WAF1,[119] also known as Cip-1[120] or p21[121] (Fig. 3). Unlike *mdm2*, this protein is closely linked to the cell cycle. WAF1 is a potent inhibitor of cyclin-dependent kinases (cdks).[120] Certain cdks are activated by cyclins to allow cells to progress into S from G_1 (for a review, see Norburg and Nurse).[122] Therefore, when p53 levels are stimulated, such as after UV or ionizing radiation, WAF1 is activated, which in turn inhibits G_1-specific cdks, and the cell is blocked at G_1. It is likely that inhibition of phosphorylation of Rb by cdc2, one of the members of the cdk family, plays a role in this block.[48] WAF1 also controls DNA replication by binding to and inactivating PCNA, which is needed for the stimulation of polymerase δ, the major polymerase in the cell[123] (Fig. 3). When p53 is inactivated by mutations or by oncogenes, WAF1 is absent, and the cdks are free to respond to cyclin activation, resulting in progression into S. This makes for an attractive scenario that links p53's ability to activate transcription and its suppression of cell growth (Fig. 3); p53 may also have a direct role in the inhibition of DNA synthesis (Fig. 3) by binding to the single-stranded DNA-binding protein RPA.[124–126]

The link between transcriptional activation and growth suppression is clearly seen with mutant forms of p53. In all cases, when a mutation knocks out transcriptional activity, it also knocks out growth suppression.[127] Interestingly, however, the correlation does not hold for suppression of transformation by oncogenes.[127] At least two p53 mutations that prevent the ability of p53 to suppress transformation by the oncogenes E1A and *ras* or HPV 16 E7 and *ras* retain the ability to activate transcription. This indicates that another activity distinct from transcriptional activation must be involved in tumor suppression. It is possible that the oligomerization domain, which is located near the COOH-terminal end, is involved.

A third gene activated by p53 is one involved in DNA repair called GADD45.[101] When the cell's DNA is damaged, p53 levels rise by a posttranslational mechanism. This high level of p53 activates a number of genes, probably including WAF1, *mdm2*, and GADD45 (Fig. 3), which contain activatable p53 binding sites, and as a consequence, cells are prevented from entering S. GADD45 stimulates excision repair and inhibits cell DNA replication, presumably by binding to the proliferating cell nuclear antigen (PCNA) (Fig. 3).[128] After the DNA has been repaired, p53 levels fall, and p53-dependent suppression of growth is relieved, perhaps by a mechanism involving *mdm2*.[129]

Another role of p53 is its involvement in programmed cell death (apoptosis) (Fig. 3). When cells are unable to repair their damaged DNA, or when they receive conflicting signals such as the signal to proliferate from an oncogene and the signal to suppress growth from wild-type p53, the apoptotic pathway may be activated.[130–132] In many cell types, this pathway is dependent on a functional wild-type p53. The genetic switches involved in apoptosis are yet to be elucidated, but it appears that one important function of p53 is to prevent cells from being overtaken by oncogenes, and this is accomplished by inducing cellular suicide. Hence, cells transformed by oncogenes must abrogate this activity of p53. Interestingly, tumor cells containing wild-type p53 respond much more effectively to chemotherapeutic agents than those lacking p53 or containing mutant forms of p53, demonstrating that wild-type p53 is usually required for apoptosis induced by these agents and that mutant p53s, have lost this activity.[133,134]

4.2. Role of T Antigen Binding to p53

The binding of p53 to T antigen results in its inactivation. The bound p53 is unable to bind to its cognate DNA sequence[135] and unable to activate transcription from a p53-responsive promoter.[136,137] Likewise, T antigen inactivates p53's ability to suppress growth[138] and overcomes p53's role in apoptosis.[139] Cells transformed by SV40 grow rapidly and display significant transformation properties, including an ability to form tumors in susceptible animals, indicating that T antigen has also inactivated p53's tumor-suppression activity. However, this is a give-and-take situation, since if sufficient p53 is introduced into cells simultaneously exposed to T antigen, the p53 can become dominant and inhibit the formation of transformed cell foci.[92] Thus, p53's multiple activities are altered by SV40 T antigen to induce a general change in the growth behavior of the infected or transformed cell.

In addition to its role as a dominant negative inhibitor of p53 function, T antigen may activate p53's transforming activity in transformed cells. Many mutant forms of p53 isolated from natural tumors have the acquired ability to transform cells in cooperation with activated *ras* oncogene.[85,86] In fact, it was this property of p53 that erroneously allowed it to be classified as an oncogene. In part, the transforming activity of transfected mutant p53s appears to be related to its ability to form heterooligomers with the resident wild-type p53.[140–142] The binding of mutant p53 inactivates wild-type p53's function, perhaps by inducing a conformational change.[141] In addition to the dominant negative behavior of certain mutant forms of p53, there is ample evidence that these mutants have also acquired a new activity, independent of the loss of wild-type 53, to transform cells to a malignant (tumorigenic) state.[87,143] An intriguing possibility is that this new gain of function property is induced by T antigen in SV40-transformed cells; however, there is no direct evidence for this at the moment.

Not only does T antigen have a dramatic effect on p53 function, but conversely, p53 impacts on T antigen's function, and these changes may be invoked during transformation of cells by SV40. Wild-type, but not mutant, human p53 proteins inhibit the replication activities of large T antigen.[144] The inhibition takes place when p53 blocks T antigen's ability to initiate DNA replication.[145,146] Interestingly, p107, an Rb-related protein, also inhibits T antigen's ability to initiate DNA replication.[147]

5. ASSOCIATION WITH p300

The p300 protein, like Rb and p107, complexes with adenovirus E1A proteins.[41] Deletion mutant analysis of E1A indicated that the p300 binding site mapped to the NH_2-terminal 25 residues.[148,149] This region of E1A has limited amino acid sequence homology with the NH_2-terminal end of SV40 T antigen. In fact, a mutant of SV40 T antigen containing two point mutations in this region is defective in transformation,[21] suggesting that p300 binding is important to the transforming activity of T antigen. Furthermore, an NH_2-terminal deletion mutant of E1A, which itself cannot transform cells and cannot bind to p300, can be complemented in *trans* by a transformation-defective mutant of SV40 large T antigen with a deletion in the Rb-binding region,[34] suggesting that T antigen contains p300 binding activity.

Little is known about the function of p300 and whether it, like p53 and Rb, is a tumor suppressor protein. Recently,[150] the p300 gene was cloned. Its presumed structure, consisting of a central bromo domain, and its ability to abrogate E1A-mediated repression of the SV40 enhancer suggest that p300 functions as transcriptional coactivator. This indicates that p300, like p53 and Rb, may function as a regulator of transcription, perhaps directed toward specific genes like Myo D,[151] a muscle-specific differentiation gene that is expressed in myoblasts as they differentiate into myotubes.

6. TRANSFORMATION AND IMMORTALIZATION DOMAINS OF SV40 LARGE T ANTIGEN AND THEIR RELATIONSHIP TO THE BINDING OF CELLULAR PROTEINS

As determined by mutational analysis of full-length T antigen, at least three regions of the protein are required for transformation of established and primary cells.[21,82,152,153] These regions correspond to the domains involved in binding to the three cellular proteins described above. These activities may not be, however, the only ones needed, because SV40-polyomavirus T-antigen chimeras, able to bind at least to Rb and p53, were still defective at transformation.[37] Furthermore, Cavender *et al.*[154] recently implicated an additional activity, presumably requiring sequences between 251 and 301, that may be needed for the full transformation of rat embryo fibroblasts. Additional evidence comes from the apparent requirement of four independent functions of T antigen to induce cells from G_1 into S.[155]

Transformation or immortalization assays performed with truncated T antigen polypeptides appear to paint a different picture, however. Sompayrac and Danna[156,157] have shown that an NH_2-terminal fragment of large T antigen consisting of residues 1 to 147, which presumably binds both Rb and p300 but not p53, transforms secondary rat embryo fibroblasts like wild-type protein. A slightly smaller protein containing only the first 137 amino acids was capable of immortalizing primary rat embryo fibroblasts.[158] In apparent conflict with these results, NH_2-terminal truncated proteins, such as those missing the first 127 amino acids or amino acids 127 to 250, transform primary mouse cells normally.[159] In a *ras* cooperation assay, Cavender *et al.*[154] extended the NH_2-terminal deletion to 175 amino acids, showing that this fragment (residues 176–708) immortalizes rat embryo fibroblasts about 20% to 50% as well as full-sized T antigen. A fragment consisting of the first 147 amino acids immortalized these cells about half as well as full-length T antigen in the presence of *ras*. These results were taken to suggest that T antigen has two independent immortalization functions, one located in the NH_2-terminal region and a second one located downstream and dependent on the binding to p53.

Although these two sets of data are difficult to reconcile with one another, there is at least one explanation, *viz.*, that the genetic "fragmentation" of T antigen unmasks activities sometimes absent in the full-length protein. It appears clear, for example, that two immortalization activities are uncovered when the protein is genetically "cleaved" into two parts at or around residue 147.[154,157] A similar situation exists when one measures the binding to DNA polymerase α: both an NH_2-terminal fragment of T antigen[160,161] and a more central region of the protein[162] bind independently. Unpublished data from my lab indicate that the same conclusion can be made for the binding of T antigen to topoisomerase I. Moreover, although the transactivation domain of T antigen has been mapped to the NH_2-terminal end,[21,163] deletion mutants lacking that region still exhibit the ability to transactivate.[164] It seems unlikely that these redundant activities would all be operational in the full-sized protein. If that were the case, small deletions or point mutations in any one region would not affect the function of the protein as a whole, and this is clearly not the case. Perhaps, during its evolution, large T antigen was formed by the fusion of NH_2-terminal and COOH-terminal regions possessing several similar or identical activities. When the polypeptides fused, one of the two duplicated activities was inactivated (i.e., by the new three-dimensional conformation); however, this activity can be unmasked by splitting one section away from the other. Although this is highly speculative, it does offer a possible way out of the apparent conflicting data dealing with transforming regions of SV40 T antigen. Whatever the explanation, it is clear from an analysis of small mutations made in full-length T antigen that interactions with at least three cellular proteins are required for full transformation activity. The binding to other yet to be identified cellular proteins may also be involved in this process.

7. HUMAN AND MOUSE POLYOMAVIRUSES

7.1. Transformation by Human Polyomaviruses

SV40 and the closely related human polyomaviruses BKV and JCV probably transform cells by very similar mechanisms. They share considerable homology in the p53, Rb, and p300 binding sites. The large T antigens of the human polyomaviruses have been shown to bind directly to p53 and Rb,[165–167] indicating that transformation by these viruses is unlikely to be different from transformation by SV40.

The mouse polyomaviruses, on the other hand, transform by a very different mechanism. Genetically, the mouse polyomavirus early region is significantly different from those of the primate polyomaviruses in that it codes for three early proteins. In addition to large and small T antigens, the mouse viruses code for a middle T antigen that contains sequences in common with small and large T antigens as well as unique sequences.[168] The DNA region corresponding to the middle T antigen gene is all that is necessary for mouse polyomavirus-dependent transformation and tumor induction.[169]

The large T antigen of mouse polyomavirus is capable of binding to and inactivating Rb[37,167,170] and presumably p300 as well. These NH_2-terminal activities of polyomavirus large T antigen appear to be required for the immortalization of primary cells.[171] The large T protein does not bind to p53. Hence, like SV40 large T antigen, the mouse polyomavirus large T antigen inactivates tumor suppressors. The Rb-binding activity, although required for immortalization, is not needed for transformation of mouse cells or tumor formation in the mouse.[171]

7.2. Transformation by Mouse Polyomaviruses and Functions of Middle T Antigen

Middle T antigen is the major transforming protein of the mouse polyomavirus. It is a membrane-associated protein that functions primarily as an activator of cellular protooncogene products. An extensive review of this field is beyond the scope of this chapter, however; the reader is directed to an excellent review by Benjamin and Vogt.[168] In brief, middle T antigen associates with several cellular protooncoproteins including c-*src*,[172,173] c-*yes*,[174] and c-*fyn*.[175–177] Middle T antigen binds to a negative regulatory region in the COOH-terminal region of c-*src*.[178] This activates the tyrosine protein kinase activity of the protooncogene by 10 to 50-fold,[178,179] thereby inducing its oncogenic activity. Phosphorylation of middle T antigen on serine residues appear to be important in the binding of the protein to c-*src*.[180] In complex with middle T antigen, c-*src* phosphorylates Tyr-315 and possibly Tyr-250 of middle T antigen.[181–183] These phosphorylations appear to act as a switch to promote the binding of this complex to phosphatidylinositol 3 (PIP3)-kinase[184–187] and inducing its activity. This generates phosphatidylinositol phosphorylated at the number-3 carbon,[188] which is believed to stimulate cell proliferation, presumably by a Ca^{2+}-efflux pathway. Thus, oncogenesis by middle T antigen is dependent on its ability to form complexes with both c-*src* and PIP3-kinase. In addition, middle T antigen binds to two subunits (60 and 35 kDa) of protein phosphatase 2A (PP2A).[189,190] Two downstream events are the activation of the c-*myc* oncogene[191] and p21c-*ras*,[192] which would also stimulate cell growth.

8. ROLE OF SMALL t ANTIGEN IN TRANSFORMATION BY SV40

Although large T antigen of SV40 is sufficient for the transformation of growing cells, efficient transformation of growth-arrested cells depends on the expression of the small T antigen protein.[193–196] This is especially true when large T antigen is in limiting amounts.[197] The mechanism by which transformation of these cells is enhanced has not been exactly determined, but it is likely to be related to the ability of small t antigen to stimulate transcription of certain genes[198] and to disrupt the network of actin cables.[199,200] These activities may be mediated through the binding of small t antigen with the cellular serine- and threonine-specific protein phosphatase 2A (PP2A).[201–204] Recent evidence suggests that small t antigen alters the ability of PP2A to dephosphorylate certain proteins such as the myosin light chain and myelin basic protein,[204] the mitogen-activated protein kinase 2, the extracellular-signal-regulated protein kinase 2, and the mitogen-activated protein/extracellular-signal-regulated protein kinase.[203] Small t antigen also interferes with the PP2A-mediated dephosphorylation of the cyclic-AMP-regulatory element binding protein (CREB), resulting in a constitutive activation of its ability to stimulate transcription.[205] Although there is no direct evidence that these interactions play a role in small t antigen-induced transformation, that possibility is intriguing. It is also noteworthy that mouse polyomavirus middle T antigen has the same PP2A-binding activity.[189–190]

9. SUMMARY AND EVALUATION OF THE ROLES OF SMALL t ANTIGEN AND OF THE BINDING OF p53, Rb, AND p300 TO LARGE T ANTIGEN IN TRANSFORMATION BY SV40

It is perhaps somewhat premature to sort out the relative roles of small t and of the three large T antigen-binding functions in transformation by SV40. However, several points can be made. First, the requirements for small t antigen in the transformation of primary cells may result from the two proteins' cooperation to stimulate DNA synthesis and to induce the hyperphosphorylation of Rb,[206] indicating that full transformation activity depends on the presence of both proteins. In sorting out the relative roles of p53 and Rb binding, it appears that Rb binding is not essential for growth stimulation by T antigen but is involved in the full growth induction potential of T antigen.[207] Although increased proliferative potential is primarily related to p53 binding, SV40-induced focus formation requires Rb binding.[208] Hence, both are required for maximal transforming activity. The role of p300 binding in transformation is the least understood, but this protein may very well be a major factor. Finally, the Rb- and/or p300-binding regions of T antigen are required to overcome wild-type p53-mediated suppression of growth, suggesting that these regions may communicate with wild-type p53 in the cell.[153] The apparent take-home message is that, under certain conditions (i.e., in primary cells), all transforming activities are needed.

10. FUTURE DIRECTIONS

The dominant negative functions of T antigen to inactivate the resident wild-type p53 and Rb proteins in SV40-transformed cells are essential for the maintenance of the full transformed state. These loss-of-function activities have been extensively studied in many laboratories and are probably the best understood of T antigen's requirements for transformation. There are several other areas where our level of understanding is quite limited and that merit additional investigations. First, is there any gain of function activities of T antigen–p53 or T antigen–Rb complexes in transformed cells? If T–p53 complexes do have some new activity, what is its relationship with those of mutant, transforming forms of p53? What is the function of p300 binding in transformation, and how does this relate with p53 and Rb binding? What other protein(s) does T antigen bind to during cellular transformation? Finally, how does small t participate in the transformation process, and why is it apparently not needed under certain conditions? In summary, although there has been significant progress in understanding how SV40 and other polyomaviruses transform cells, there are still many unsolved questions and likely to be many surprising answers.

REFERENCES

1. Brugge, J. S., and Butel, J. S., 1975, Role of simian virus 40 gene A function in maintenance of transformation, *J. Virol.* **15**:619–635.
2. Kimura, G., and Itagaki, A., 1975, Initiation and maintenance of cell transformation by simian virus 40: A viral genetic property, *Proc. Natl. Acad. Sci. USA* **72**:673–677.
3. Martin, R. G., and Chou, J. Y., 1975, Simian virus 40 functions required for the establishment and maintenance of malignant transformation, *J. Virol.* **15**:599–612.
4. Osborn, M., and Weber, K., 1975, Simian virus 40 gene A function and maintenance of transformation, *J. Virol.* **15**:636–644.
5. Arthur, A. K., Höss, A., and Fanning, E., 1988, Expression of simian virus 40 T antigen if *Escherichia coli*: Localization of T-antigen origin DNA-binding domain to within 129 amino acids, *J. Virol.* **62**:1999–2006.
6. Simmons, D. T., 1986, DNA-binding region of the simian virus 40 tumor antigen, *J. Virol.* **57**:776–785.
7. Simmons, D. T., 1988, Geometry of the simian virus 40 large tumor antigen–DNA complex as probed by protease digestion, *Proc. Natl. Acad. Sci. USA* **85**:2086–2090.
8. Simmons, D. T., Loeber, G., and Tegtmeyer, P., 1990, Four major sequence elements of simian virus 40 large T antigen coordinate its specific and nonspecific DNA binding, *J. Virol.* **64**:1973–1983.
9. Simmons, D. T., Wun-Kim, K., and Young, W., 1990, Identification of simian virus 40 T antigen residues important for specific and nonspecific binding to DNA and for helicase activity, *J. Virol.* **64**:4858–4865.
10. Strauss, M., Argani, P., Mohr, I. J., and Gluzman, Y., 1987, Studies on the origin-specific DNA-binding domain of simian virus 40 large T antigen, *J. Virol.* **61**:3326–3330.
11. Wun-Kim, K., and Simmons, D. T., 1990, Mapping of helicase and helicase substrate binding domains on simian virus 40 large T antigen, *J. Virol.* **64**:2014–2020.
12. Bradley, M. K., Smith, T. F., Lathrop, R. H., Livingston, D. M., and Webster, T. A., 1987, Consensus topography in the ATP binding site of the simian virus 40 and polyomavirus large tumor antigens, *Proc. Natl. Acad. Sci. USA* **84**:4026–4030.
13. Bradley, M. K., 1990, Activation of ATPase activity of simian virus 40 large T antigen by the covalent affinity analog of ATP, fluorosulfonylbenzoyl 5'-adenosine, *J. Virol.* **64**:4939–4947.
14. Brady, J. N., Bolen, J. B., Radonovich, M., Salzman, N., and Khoury, G., 1984, Stimulation of simian virus 40 late gene expression by simian virus 40 tumor antigen, *Proc. Natl. Acad. Sci. USA* **81**:2040–2044.
15. Keller, J. M., and Alwine, J. C., 1984, Activation of the SV40 late promoter: Direct effects of T antigen in the absence of viral DNA replication, *Cell* **36**:381–389.
16. Alwine, J. C., 1985, Transient gene expression control: Effects of transfected DNA stability and *trans*-activation by viral early proteins, *Mol. Cell. Biol.* **5**:1034–1042.
17. Schutzbank, T., Robinson, R., Oren, M., and Levine, A. J., 1982, SV40 large tumor antigen can regulate some cellular transcripts in a positive fashion, *Cell* **30**:481–490.
18. Scott, M. R. D., Westphal, K.-H., and Rigby, P. W. J., 1983, Activation of mouse genes in transformed cells, *Cell* **34**:557–567.
19. Segawa, K., and Yamaguchi, N., 1987, Induction of c-Ha-*ras* transcription in rat cell by simian large T antigen, *Mol. Cell. Biol.* **7**:556–559.
20. Hiscott, J., Wong, A., Alper, D., and Xanthoudakis, S., 1988, *Trans* activation of type I interferon promoters by simian virus 40 T antigen, *Mol. Cell. Biol.* **8**:3397–3405.

21. Srinivasan, A., Peden, K. W. C., and Pipas, J. M., 1989, The large tumor antigen of simian virus 40 encodes at least two distinct transforming functions, *J. Virol.* **63:**5459–5463.
22. Scheidtmann, K. H., Echle, B., and Walter, G., 1982, Simian virus 40 large T antigen is phosphorylated at multiple sites clustered in two separate regions, *J. Virol.* **44:**116–133.
23. Scheidtmann, K.-H., 1986, Phosphorylation of simian virus 40 large T antigen: Cytoplasmic and nuclear phosphorylation sites differ in their metabolic stability, *Virology* **150:**85–95.
24. Chen, Y.-R., Lees-Miller, S. P., Tegtmeyer, P., and Anderson, C. W., 1991, The human DNA-activated protein kinase phosphorylates simian virus 40 T antigen at amino- and carboxy-terminal sites, *J. Virol.* **65:**5131–5140.
25. Simmons, D. T., Chou, W., and Rodgers, K., 1986, Phosphorylation downregulates the DNA-binding activity of simian virus 40 T antigen, *J. Virol.* **60:**888–894.
26. Grasser, F. A., Mann, K., and Walter, G., 1987, Removal of serine phosphates from simian virus 40 large T antigen increases its ability to stimulate DNA replication in vitro but has no effect on ATPase and DNA binding, *J. Virol.* **61:**3373–3380.
27. Mohr, I. J., Stillman, B., and Gluzman, Y., 1987, Regulation of SV40 DNA replication of phosphorylation of T antigen, *EMBO J.* **6:**153–160.
28. Schneider, J., and Fanning, E., 1988, Mutations in the phosphorylation sites of simian virus 40 (SV40) T antigen alter its origin DNA-binding specificity for sites I or II and affect SV40 DNA replication activity, *J. Virol.* **62:**1598–1605.
29. Moarefi, I. F., Small, D., Gilbert, I., Hopfner, M., Randall, S. K., Schneider, C., Russo, A. A. R., Ramsperger, U., Arthur, A., Stahl, H., Kelly, T. J., and Fanning, E., 1993, Mutation of the cyclin-dependent kinase phosphorylation site in simian virus 40 (SV40) large T antigen specifically blocks SV40 origin DNA unwinding, *J. Virol.* **67:**4992–5002.
30. McVey, D., Ray, S., Gluzman, Y., Berger, L., Wildeman, A. G., Marshak, D. R., and Tegtmeyer, P., 1993, *cdc2* phosphorylation of threonine 124 activates the origin-unwinding functions of simian virus 40 T antigen, *J. Virol.* **67:**5206–5215.
31. McVey, D., Brizuela, L., Mohr, I., Marshak, D. R., Gluzman, Y., and Beach, D., 1989, Phosphorylation of large tumour antigen by *cdc2* stimulates SV40 DNA replication, *Nature* **341:**503–507.
32. Prives, C., 1990, The replication functions of SV40 T antigen are regulated by phosphorylation, *Cell* **61:**735–738.
33. Fanning, E., and Knippers, R., 1992, Structure and function of simian virus 40 large tumor antigen, *Annu. Rev. Biochem.* **61:**55–85.
34. Yaciuk, P., Carter, M. C., Pipas, J. M., and Moran, E., 1991, Simian virus 40 large-T antigen expresses a biological activity complementary to the p300-associated transforming function of the adenovirus E1A gene products, *Mol. Cell. Biol.* **11:**2116–2124.
35. Schmieg, F. I., and Simmons, D. T., 1988, Characterization of the *in vitro* interaction between SV40 T antigen and p53: Mapping the p53 binding site, *Virology* **164:**132–140.
36. Mole, S. E., Gannon, J. V., Ford, M. J., and Lane, D. P., 1987, Structure and function of large T antigen, *Phil. Trans. R. Soc. Lond. [B]* **317:**455–469.
37. Manfredi, J. J., and Prives, C., 1990, Binding of p53 and p105-Rb is not sufficient for oncogenic transformation by a hybrid polyomavirus–simian virus 40 large T antigen, *J. Virol.* **64:**5250–5259.
38. Kierstead, T. D., and Tevethia, M. J., 1993, Association of p53 binding and immortalization of primary C57BL/6 mouse embryo fibroblasts by using simian virus 40 T-antigen mutants bearing internal overlapping deletion mutations, *J. Virol.* **67:**1817–1929.
39. DeCaprio, J. A., Ludlow, J. W., Lynch, D., Furukawa, Y., Griffin, J., Piwnica-Worms, H., Huang, C.-M., and Livingston, D. M., 1989, The product of the retinoblastoma susceptibility gene has properties of a cell cycle regulatory element, *Cell* **58:**1085–1095.
40. Montano, X., Millikan, R., Milhaven, J. M., Newsome, D. A., Ludlow, J. W., Arthur, A. K., Fanning, E., Bikel, I., and Livingston, D. M., 1990, Simian virus 40 small tumor antigen and an

amino-terminal domain of large tumor antigen share a common transforming function, *Proc. Natl. Acad. Sci. USA* **87:**7448–7452.

41. Harlow, E., Whyte, P., Franza, B. R. J., and Schley, C., 1986, Association of adenovirus early region 1A with cellular polypeptides, *Mol. Cell. Biol.* **6:**1579–1589.
42. Whyte, P., Buchkovick, K. J., Horowitz, J. M., Friend, S. H., Raybuck, M., Weinberg, R. A., and Harlow, E., 1988, Association between an oncogene and an anti-oncogene; the adenovirus E1a proteins bind to the retinoblastoma gene product, *Nature* **334:**124–129.
43. DeCaprio, J. A., Ludlow, J. W., Figge, J., Shew, J.-Y., Huang, C.-M., Lee, W.-H., Marsilio, E., Paucha, E., and Livingston, D. M., 1988, SV40 large tumor antigen forms a specific complex with the product of the retinoblastoma susceptibility gene, *Cell* **54:**275–283.
44. Dyson, N., Buchkovich, K., Whyte, P., and Harlow, E., 1989, The cellular 107 kD protein that binds to adenovirus E1A also associates with the large T antigens of SV40 and JC virus, *Cell* **58:**249–255.
45. Kalderon, D., and Smith, A. E., 1984, *In vitro* mutagenesis of a putative DNA binding domain of SV40 large-T, *Virology* **139:**109–137.
46. Ewen, M. E., Ludlow, J. W., Marsilio, E., DeCaprio, J. A., Millikan, R. C., Cheng, S. H., Paucha, E., and Livingston, D. M., 1989, An N-terminal transformation-governing sequence of SV40 large T antigen contributes to the binding of both p110Rb and a second cellular protein, p120, *Cell* **58:**257–267.
47. Chen, S., and Paucha, E., 1990, Identification of a region of simian virus 40 large T antigen required for cell transformation, *J. Virol.* **64:**3350–3357.
48. Riley, D. J., Lee, E. Y.-H. P., and Lee, W.-H., 1994, The retinoblastoma protein: More than a tumor suppressor, *Annu. Rev. Cell. Biol.* **10:**1–29.
49. Buchkovich, K., Duffy, L. A., and Harlow, E., 1989, The retinoblastoma protein is phosphorylated during specific phases of the cell cycle, *Cell* **58:**1097–1105.
50. Chen, P. L., Scully, P., Shew, J.-Y., Wang, J. Y. J., and Lee, W.-H., 1989, Phosphorylation of the retinoblastoma gene product is modulated during the cell cycle and cellular differentiation, *Cell* **58:**1193–1198.
51. DeCaprio, J. A., Furukawa, Y., Ajchenbaum, F., Griffin, J. D., and Livingston, D. M., 1992, The retinoblastoma-susceptibility gene product becomes phosphorylated in multiple stages during cell cycle entry and progression, *Proc. Natl. Acad. Sci. USA* **89:**1795–1798.
52. Furukawa, Y., DeCaprio, J. A., Freedman, A., Kanakura, Y., Nakamura, M., Ernst, T. J., Livingston, D. M., and Griffin, J. D., 1990, Expression and state of phosphorylation of the retinoblastoma susceptibility gene product in cycling and non-cycling human hematopoietic cells, *Proc. Natl. Acad. Sci. USA* **87:**2770–2774.
53. Lee, W.-H., Hollingworth, R. E. J., Qian, Y.-W., Chen, P.-L., Hong, F., and Lee, E. Y.-H. P., 1991, Rb protein as a cellular "corral" for growth promoting proteins, *Cold Spring Harbor Symp. Quant. Biol.* **56:**211–217.
54. Lees, J. A., Buchkovich, K. J., Marshak, D. R., Anderson, C. W., and Harlow, E., 1991, The retinoblastoma protein is phosphorylated on multiple sites by human *cdc2*, *EMBO J.* **10:**4279–4290.
55. Lin, B. T.-Y., Gruenwald, S., Morla, A. O., Lee, W.-H., and Wang, J. Y. J., 1991, Retinoblastoma cancer suppressor gene product is a substrate of the cell cycle regulator *cdc2* kinase, *EMBO J.* **10:**857–864.
56. Ludlow, J. W., Shen, J., Pipas, J. M., Livingston, D. M., and DeCaprio, J. A., 1990, The retinoblastoma susceptibility gene product undergoes cell-cycle-dependent dephosphorylation and binding to and release from SV40 large T antigen, *Cell* **60:**387–396.
57. Chellappan, S., Kraus, V. B., Kroger, B., Munger, K., Phelps, W. C., Nevins, J. R., and Howley, P. M., 1992, Adenovirus E1A, simian virus 40 tumor antigen, and human papillomavirus E7 protein share the capacity to disrupt the interaction between transcription factor E2F and the retinoblastoma gene product, *Proc. Natl. Acad. Sci. USA* **89:**4549–4553.
58. Durfee, T., Becherer, K., Chen, P.-L., Yeh, S.-H., and Yang, Y., 1993, The retinoblas-

toma protein associates with the protein phosphatase type 1 catalytic subunit, *Genes Dev.* **7:** 555–569.
59. Alberts, A. S., Thorburn, A. M., Shenolikar, S., Mumby, M. C., and Framisco, J. R., 1993, Regulation of cell cycle progression and nuclear affinity of the retinoblastoma protein by protein phosphatases, *Proc. Natl. Acad. Sci. USA* **90:**388–392.
60. Ludlow, Y. W., Glendening, C. L., Livingston, D. M., and DeCaprio, J. A., 1993, Specific enzymatic dephosphorylation of the retinoblastoma protein, *Mol. Cell. Biol.* **13:**367–372.
61. Ludlow, J. W., DeCaprio, J. A., Huang, C.-M., Lee, W.-H., Paucha, E., and Livingston, D. M., 1989, SV40 large T-antigen binds preferentially to an underphosphorylated member of the retinoblastoma susceptibility gene product family, *Cell* **56:**57–65.
62. Goodrich, D. W., Wang, N. P., Qian, Y.-W., Lee, E. Y.-H. P., and Lee, W.-H., 1991, The retinoblastoma gene product regulates progression through the G_1 phase of the cell cycle, *Cell* **67:** 293–302.
63. Nevins, J. R., 1992, A link between the Rb tumor suppressor protein and viral oncoproteins, *Science* **258:**424–429.
64. Rustgi, A. K., Dyson, N., and Bernards, R., 1991, Amino-terminal domains of c-*myc* and N-*myc* proteins mediate binding to the retinoblastoma gene product, *Nature* **352:**541–544.
65. Dou, Q.-P., Markell, P. J., and Pardee, A. B., 1992, Thymidine kinase transcription is regulated at the G_1/S phase by a complex that contains retinoblastoma-like protein and a *cdc2* kinase, *Proc. Natl. Acad. Sci. USA* **89:**3256–3260.
66. Pearson, A. B., Nasheuer, H.-P., and Wang, T. S.-F., 1991, Human DNA polymerase a gene: Sequences controlling expression in cycling and serum-stimulated cells. *Mol. Cell. Biol.* **11:**2081–2095.
67. Dyson, N., Howley, P. M., Munger, K., and Harlow, E., 1989, The human papillomavirus-16 E7 oncoprotein is able to bind to the retinoblastoma gene product, *Science* **243:**934–936.
68. Mueller, H., Lukas, J., Schneider, A., Warthoe, P., Bartek, J., Eilers, M., and Strauss, M., 1994, Cyclin D1 expression is regulated by the retinoblastoma protein, *Proc. Natl. Acad. Sci. USA* **91:**2945–2949.
69. Lukas, J., Muller, H., Bartkova, J., Spitkovsky, D., Kjerulff, A. A., Jansen-Durr, P., Strauss, M., and Bartek, J., 1994, DNA tumor virus oncoproteins and retinoblastoma gene mutations share the ability to relieve the cell's requirement for cyclin D1 function in G_1, *J. Cell Biol.* **125:**625–638.
70. Bartkova, J., Lukas, J., Muller, H., Lutzhoft, D., Strauss, M., and Bartek, J., 1994, Cyclin D1 protein expression and function in human breast cancer, *Int. J. Cancer* **57:**353–361.
71. Mancini, M., Shan, B., Nickerson, J., Penman, S., and Lee, W.-H., 1994, The retinoblastoma gene product is a cell-cycle dependent, nuclear-matrix associated protein, *Proc. Natl. Acad. Sci. USA* **91:**418–422.
72. Ewen, M. E., Xing, Y. G., Lawrence, J. B., and Livingston, D. M., 1991, Molecular cloning, chromosomal mapping, and expression of the cDNA for p107, a retinoblastoma gene product-related protein, *Cell* **66:**1155–1164.
73. Hannon, G. J., Demetrick, D., and Beach, D., 1993, Isolation of the Rb-related p130 through its interaction with CDK2 and cyclins, *Genes Dev.* **7:**2378–2391.
74. Mayol, X., Grana, X., Baldi, A. M. S., Hu, Q., and Giordano, A., 1993, Cloning of a new member of the retinoblastoma gene family (pRB2) which binds to the E1A transforming domain, *Oncogene* **8:**2561–2566.
75. Ewen, M. E., Faha, B., Harlow, E., and Livingston, D. M., 1992, Interaction of p107 with cyclin A independent of complex formation with viral oncoproteins. *Science* **255:**85–87.
76. Linzer, D. I. H., Maltzman, W., and Levine, A. J., 1979, Characterization of a 54K dalton cellular SV40 tumor antigen present in SV40-transformed cells and uninfected embryonal carcinoma cells, *Cell* **17:**43–52.
77. Chang, C., Simmons, D. T., Martin, M. A., and Mora, P. T., 1979, Identification and partial characterization of new antigens from simian virus 40-transformed mouse cells. *J. Virol.* **31:** 463–471.

78. Kress, M., May, E., Cassingena, R., and May, P., 1979, Simian virus 40-transformed cells express new species of proteins precipitable by anti-simian virus 40 tumor serum, *J. Virol.* **31**:472–483.
79. Lane, D. P., and Crawford, L. V., 1979, T-antigen is bound to a host protein in SV40 transformed cells, *Nature* **278**:261–263.
80. Deppert, W., and Steinmayer, T., 1989, Metabolic stabilization of p53 in SV40-transformed cells correlates with expression of the transformed phenotype but is independent from complex formation with SV40 large T antigen, *Curr. Top. Microbiol. Immunol.* **144**:77–84.
81. Tiemann, F., and Deppert, W., 1994, Stabilization of the tumor suppressor p53 during cellular transformation by simian virus 40: Influence of viral and cellular factors and biological consequences, *J. Virol.* **68**:2869–2878.
82. Lin, J.-Y., and Simmons, D. T., 1991, The ability of large T antigen to complex with p53 is necessary for the increased life span and partial transformation of human cells by simian virus 40, *J. Virol.* **65**:6447–6453.
83. Zhu, J., Abate, M., Rice, P. W., and Cole, C. N., 1991, The ability of simian virus 40 large T antigen to immortalize primary mouse embryo fibroblasts cosegregates with its ability to bind to p53, *J. Virol.* **65**:6872–6880.
84. Tiemann, F., and Deppert, W., 1994, Immortalization of BALB/c mouse embryo fibroblasts alters SV40 large T-antigen interactions with the tumor suppressor p53 and results in a reduced SV40 transformation-efficiency, *Oncogene* **9**:1907–1915.
85. Finlay, C. A., Hinds, W., and Levine, A. J., 1989, The p53 protooncogene can act as a suppressor of transformation, *Cell* **57**:1083–1093.
86. Eliyahu, D., Michalovitz, D., Eliyahu, S., Pinhasi-Kimhi, O., and Oren, M., 1989, Wild-type p53 can inhibit oncogene-mediated focus formation, *Proc. Natl. Acad. Sci. USA* **86**:8763–8767.
87. Chen, P.-L., Chen, Y., Bookstein, R., and Lee, W.-H., 1990, Genetic mechanisms of tumor suppression by the human p53 gene, *Science* **250**:1576–1580.
88. Diller, L., Kassel, J., Nelson, C. E., Gryka, M. A., Litwak, G., Gebhardt, M., Bressac, B., Ozturk, M., Baker, S. J., and Vogelstein, B., 1990, p53 functions as a cell cycle control protein in osteosarcomas, *Mol. Cell. Biol.* **10**:5772–5781.
89. Baker, S. J., Markowitz, S., Fearon, E. R., Willson, J. K. V., and Vogelstein, B., 1990, Suppression of human colorectal carcinoma cell growth by wild-type p53, *Science* **249**:912–915.
90. Johnson, P., Gray, D., Mowat, M., and Benchimol, S., 1991, Expression of wild-type p53 is not compatible with continued growth of p53-negative tumor cells, *Mol. Cell. Biol.* **11**:1–11.
91. Yonish-Rouach, E., Resnitzky, D., Lotem, J., Sachs, L., Kimchi, A., and Oren, M., 1991, Wild-type p53 induces apoptosis of myeloid leukaemic cells that is inhibited by interleukin-6, *Nature* **352**:345–347.
92. Fukasawa, K., Sakoulas, G., Pollack, R. E., and Chen, S., 1991, Excess wild-type p53 blocks initiation and maintenance of simian virus 40 transformation, *Mol. Cell. Biol.* **11**:3472–3483.
93. Finlay, C. A., 1993, The *mdm-2* oncogene can overcome wild-type p53 suppression of transformed cell growth, *Mol. Cell. Biol.* **13**:301–306.
94. Mercer, W. E., Shields, M. T., Lin, D., Appella, E., and Ullrich, S. J., 1991, Growth suppression induced by wild-type p53 protein is accompanied by selective down-regulation of proliferating-cell nuclear antigen expression, *Proc. Natl. Acad. Sci. USA* **88**:1958–1962.
95. Michalovitz, D., Halevy, O., and Oren, M., 1990, Conditional inhibition of transformation and of cell proliferation by a temperature-sensitive mutant of p53, *Cell* **52**:671–680.
96. Martinez, J., Georgoff, I., Martinez, J., and Levine, A. J., 1991, Cellular localization and cell cycle regulation by a temperature sensitive p53 protein, *Genes Dev.* **5**:151–159.
97. Ginsberg, D., Michalovitz, D. M., Ginsberg, D., and Oren, M., 1991, Induction of growth arrest by a temperature-sensitive p53 mutant is correlated with increased nuclear localization and decreased stability of the protein, *Mol. Cell. Biol.* **11**:582–585.
98. Harvey, D., and Levine, A. J., 1991, p53 alteration is a common event in the spontaneous immortalization of primary BALB/C murine embryo fibroblasts, *Genes Dev.* **5**:2375–2385.

99. Kastan, M. B., Onyekwere, O., Sidransky, D., Vogelstein, B., and Craig, R. W., 1991, Participation of p53 protein in the cellular response to DNA damage, *Cancer Res.* **51**:6304–6311.
100. Kuerbitz, S. J., Plunkett, B. S., Walsh, W. V., and Kastan, M. B., 1992, Wild-type p53 is a cell cycle checkpoint determinant following irradiation, *Proc. Natl. Acad. Sci. USA* **89**:7491–7495.
101. Kastan, M. B., Zhan, Q., El-Deiry, W. S., Carrier, F., Jacks, T., Walsh, W. V., Plunkett, B. S., Vogelstein, B., and Fornace, A. J., Jr., 1992, A mammalian cell cycle checkpoint pathway utilizing p53 and GADD45 is defective in ataxia–telangiectasia, *Cell* **71**:587–597.
102. Donehower, L. A., Harvey, M., Slagle, B. L., McArthur, M. J., Montgomery, C. A., Jr., Butel, J. S., and Bradley, A., 1992, Mice deficient for p53 are developmentally normal but susceptible to spontaneous tumours, *Nature* **356**:215–221.
103. Levine, A. J., 1993, The tumor suppressor genes, *Annu. Rev. Biochem.* **62**:623–651.
104. Moll, U. M., Riou, G., and Levine, A. J., 1992, Two distinct mechanisms alter p53 in breast cancer: Mutation and nuclear exclusion, *Proc. Natl. Acad. Sci. USA* **89**:7262–7266.
105. Momand, J., Zambetti, G., Olson, D. C., George, D., and Levine, A. J., 1992, The *mdm2* oncogene product forms a complex with the p53 protein and inhibits p53-mediated transactivation, *Cell* **69**:1237–1245.
106. Werness, B. A., Levine, A. J., and Howley, P. M., 1990, Association of human papillomavirus types 16 and 18 E6 proteins with p53, *Science* **248**:76–79.
107. Weintraub, H., Hauschka, S., and Tapscott, S. J., 1991, The MCK enhancer contains a p53 responsive element, *Proc. Natl. Acad. Sci. USA* **88**:4570–4571.
108. Farmer, G., Bargonetti, J., Zhu, H., Friedman, P., Prywes, R., and Prives, C., 1992, Wild-type p53 activates transcription *in vitro*, *Nature* **358**:83–86.
109. El-Deiry, W. S., Kern, S. E., Pietenpol, J. A., Kinzler, K. W., and Vogelstein, B., 1992, Human genomic DNA sequences define a consensus binding site for p53, *Nature Genet.* **1**:44–49.
110. Fields, S., and Jang, S. K., 1990, Presence of a potent transcription activating sequence in the p53 protein, *Science* **249**:1046–1049.
111. Raycroft, L., Wu, H., and Lozano, G., 1990, Transcriptional activation by wild-type but not transforming mutants of the p53 anti-oncogene, *Science* **249**:1049–1051.
112. Thut, C. J., Chen, J.-L., Klemm, R., and Tjian, R., 1995, p53 transcriptional activation mediated by coactivators $TAF_{II}40$ and $TAF_{II}60$, *Science* **267**:100–104.
113. Cho, Y., Gorina, S., Jeffrey, P. D., and Pavletich, N. P., 1994, Crystal structure of a p53 tumor suppressor–DNA complex: Understanding tumorigenic mutations, *Science* **265**:346–355.
114. Barak, Y., Juven, T., Haffner, R., and Oren, M., 1993, *mdm2* expression is induced by wild-type p53 activity, *EMBO J.* **12**:461–468.
115. Wu, X., Bayle, H., Olson, D., and Levine, A. J., 1993, The p53–*mdm-2* autoregulatory feedback loops, *Genes Dev.* **7**:1126–1132.
116. Chen, C.-Y., Oliner, J. D., Zhan, Q., Fornace, A. J., Vogelstein, B., and Kastan, M. B., 1994, Interactions between p53 and MDM2 in a mammalian cell cycle checkpoint pathway, *Proc. Natl. Acad. Sci. USA* **91**:2684–2688.
117. Perry, M. E., Piette, J., Zawadzki, J. A., Harvey, D., and Levine, A. J., 1993, The *mdm-2* gene is induced in response to UV light in a p53-dependent manner, *Proc. Natl. Acad. Sci. USA* **90**:11623–11627.
118. Marston, N. J., Crook, T., and Vousden, K. H., 1994, Interaction of p53 with MDM2 is independent of E6 and does not mediate wild type transformation suppressor function, *Oncogene* **9**:2707–2716.
119. El-Deiry, W. S., Tokino, T., Velculescu, V. E., Levy, D. B., Parsons, R., Trent, J. M., Lin, D., Mercer, E., Kinzler, K. W., and Vogelstein, B., 1993, WAF1, a potential mediator of p53 tumor suppression, *Cell* **75**:817–825.
120. Harper, J. W., Adami, G. R., Wei, N., Keyomarsi, K., and Elledge, S. J., 1993, The p21 Cdk-interacting protein Cip1 is a potent inhibitor of G_1 cyclin-dependent kinases, *Cell* **75**:805–816.

121. Xiong, Y., Hannon, G. J., Zhang, H., Casso, D., Kobayashi, R., and Beach, D., 1993, p21 is a universal inhibitor of cyclin kinases, *Nature* **366:**701–704.
122. Norbury, C., and Nurse, P., 1992, Animal cell cycles and their control, *Annu. Rev. Biochem.* **61:**441–470.
123. Waga, S., Hannon, G. J., Beach, D., and Stillman, B., 1994, The p21 inhibitor of cyclin-dependent kinases controls DNA replication by interaction with PCNA, *Nature* **369:**574–578.
124. Dutta, A., Ruppert, J. M., Aster, J. C., and Winchester, E., 1993, Inhibition of DNA replication factor RPA by p53, *Nature* **365:**79–82.
125. Zhigang, H., Brinton, B. T., Greenblatt, J., Hassell, J. A., and Ingles, C. J., 1993, The transactivator proteins VP16 and GAL4 bind replication factor A, *Cell* **73:**1223–1232.
126. Rong, L., and Botchan, M. R., 1993, The acidic transcriptional activation domains of VP16 and p53 bind the cellular replication protein A and stimulate *in vitro* BPV-1 DNA replication, *Cell* **73:**1207–1221.
127. Crook, T., Marston, N. J., Sara, E. A., and Vousden, K. H., 1994, Transcriptional activation by p53 correlates with suppression of growth but not transformation, *Cell* **79:**817–827.
128. Smith, M. L., Chen, I.-T., Zhan, Q., Bae, I., Chen, C.-Y., Gilmer, T. M., Kastan, M. B., O'Connor, P. M., and Fornace, A. J. J., 1994, Interaction of the p53-regulated protein Gadd45 with proliferating cell nuclear antigen, *Science* **266:**1376–1380.
129. Zambetti, G. P., and Levine, A. J., 1993, A comparison of the biological activities of wild-type and mutant p53, *FASEB J.* **7:**855–865.
130. Debbas, M., and White, E., 1993, Wild-type p53 mediates apoptosis by E1A which is inhibited by E1B, *Genes Dev.* **7:**546–554.
131. Chiou, S.-K., Tseng, C.-C., Rao, L., and White, E., 1994, Functional complementation of the adenovirus E1B 19-kilodalton protein with Bcl-2 in the inhibition of apoptosis in infected cells, *J. Virol.* **68:**6553–6566.
132. Lowe, S. W., Jacks, T., Housman, D. E., and Ruley, H. E., 1994, Abrogation of oncogene-associated apoptosis allows transformation of p53-deficient cells, *Proc. Natl. Acad. Sci. USA* **91:**2026–2030.
133. Lowe, S. W., Ruley, H. E., Jacks, T., and Housman, D. E., 1993, p53-dependent apoptosis modulates the cytoxicity of anticancer agents, *Cell* **74:**957–967.
134. Lowe, S. W., Bodis, S., McClatchey, A., Remington, L., Ruley, H. E., Fisher, D. E., Housman, D. E., and Jacks, T., 1994, p53 status and the efficacy of cancer therapy *in vivo*, *Science* **266:**807–810.
135. Bargonetti, J., Reynisdottir, I., Friedman, P. N., and Prives, C., 1992, Site-specific binding of wild-type p53 to cellular DNA is inhibited by SV40 T antigen and mutant p53, *Genes Dev.* **6:**1886–1898.
136. Mietz, J. A., Unger, T., Huibregtse, J. M., and Howley, P. M., 1992, The transcriptional transactivation function of wild-type p53 is inhibited by SV40 large T-antigen and by HPV-16 E6 oncoprotein, *EMBO J.* **11:**5013–5020.
137. Jiang, D., Srinivasan, A., Lozano, G., and Robbins, P. D., 1993, SV40 T antigen abrogates p53-mediated transcriptional activity, *Oncogene* **8:**2805–2812.
138. Michalovitz, D. M., Yehiely, F., Gottlieb, E., and Oren, M., 1991, Simian virus 40 can overcome the antiproliferative effect of wild-type p53 in the absence of stable large T antigen-p53 binding, *J. Virol.* **65:**4160–4168.
139. McCarthy, S. A., Symonds, H. S., and Van-Dyke, T., 1994, Regulation of apoptosis in transgenic mice by simian virus 40 T antigen-mediated inactivation of p53, *Proc. Natl. Acad. Sci. USA* **91:**3979–3983.
140. Hinds, P., Finlay, C., and Levine, A. J., 1989, Mutation is required to activate the p53 gene for cooperation with the *ras* oncogene and transformation, *J. Virol.* **63:**739–746.
141. Milner, J., and Medcalf, E. A., 1991, Cotranslation of activated mutant p53 with wild-type drives the wild-type p53 protein into the mutant conformation, *Cell* **65:**765–774.

142. Milner, J., Medcalf, E. A., and Cook, A. C., 1991, Tumor suppressor p53: Analysis of wild-type and mutant p53 complexes, *Mol. Cell. Biol.* **11**:12–19.
143. Dittmer, D., Pati, S., Zambetti, G., Ghu, S., Teresky, A. K., Moore, M., Finlay, C., and Levine, A. J., 1993, p53 gain of function mutations, *Nature Genet.* **4**:42–45.
144. Friedman, P. N., Kern, S. E., Vogelstein, B., and Prives, C., 1990, Wild-type, but not mutant, human p53 proteins inhibit the replication activities of simian virus 40 large tumor antigen, *Proc. Natl. Acad. Sci. USA* **87**:9275–9279.
145. Kienzle, H., Baack, M., and Knippers, R., 1989, Effects of the cellular p53 protein on simian-virus-40-T-antigen-catalyzed DNA unwinding *in vitro*, *Eur. J. Biochem.* **184**:181–186.
146. Wang, E. H., Friedman, P. N., and Prives, C., 1989, The murine p53 protein blocks replication of SV40 DNA *in vitro* by inhibiting the initiation functions of SV40 large T antigen, *Cell* **57**:379–392.
147. Amin, A. A., Murakami, Y., and Hurwitz, J., 1994, Initiation of DNA replication by simian virus 40 T antigen is inhibited by the p107 protein, *J. Biol. Chem.* **269**:7735–7743.
148. Barbeau, D., Marcellus, R. C., Bacchetti, S., Bayley, S. T., and Branton, P. E., 1992, Quantitative analysis of regions of adenovirus E1A products involved in interactions with cellular proteins, *Biochem. Cell Biol.* **70**:1123–1124.
149. Wang, H. G., Rikitake, Y., Carter, M. C., Yaciuk, P., Abraham, S. E., Zerler, B., and Moran, E., 1993, Identification of specific adenovirus E1A N-terminal residues critical to the binding of cellular proteins and to the control of cell growth, *J. Virol.* **67**:476–488.
150. Eckner, R., Ewen, M. E., Newsome, D., Gedes, M., DeCaprio, J. A., Lawrence, J. B., and Livingston, D. M., 1994, Molecular cloning and functional analysis of the adenovirus E1A-associated 300-kD protein (p300) reveals a protein with properties of a transcriptional adaptor, *Genes Dev.* **8**:869–884.
151. Caruso, M., Martelli, F., Giordano, A., and Felsani, A., 1993, Regulation of MyoD gene transcription and protein function by the transforming domains of the adenovirus E1A oncoprotein, *Oncogene* **8**:267–278.
152. Zhu, J., Rice, P. W., Gorsch, L., Abate, M., and Cole, C. N., 1992, Transformation of a continuous rat embryo fibroblast cell line requires three separate domains of simian virus 40 large T antigen, *J. Virol.* **66**:2780–2791.
153. Quartin, R. S., Cole, C. N., Pipas, J. M., and Levine, A. J., 1994, The amino-terminal functions of the simian virus 40 large T antigen are required to overcome wild-type p53-mediated growth arrest of cells, *J. Virol.* **68**:1334–1341.
154. Cavender, J. F., Conn, A., Epler, M., Lacko, H., and Tevethia, M. J., 1995, Simian virus 40 large T antigen contains two independent activities that cooperate with a ras oncogene to transform rat embryo fibroblasts, *J. Virol.* **69**:923–934.
155. Dickmanns, A, Zietvogel, A., Simmersbach, F., Weber, R., Arthur, A. K., Dehde, S., Wildeman, A. G., and Fanning, E., 1994, The kinetics of simian virus 40-induced progression of quiescent cells into S phase depend on four independent functions of large T antigen, *J. Virol.* **68**:5496–5508.
156. Sompayrac, L., and Danna, K. J., 1988, A new SV40 mutant that encodes a small fragment of T antigen transforms established rat and mouse cells, *Virology* **163**:391–396.
157. Sompayrac, L., and Danna, K. J., 1991, The amino-terminal 147 amino acids of SV40 large T antigen transform secondary rat embryo fibroblasts, *Virology* **181**:412–415.
158. Asselin, C., and Bastin, M., 1985, Sequences from polyomavirus and simian virus 40 large T genes capable of immortalizing primary rat embryo fibroblasts, *J. Virol.* **56**:958–968.
159. Thompson, D. L., Kalderon, D., Smith, A. E., and Tevethia, M. J., 1990, Dissociation of Rb-binding and anchorage-independent growth from immortalization and tumorigenicity using SV40 mutants producing N-terminally truncated large T antigens, *Virology* **178**:15–34.
160. Dornreiter, I., Hoss, A., Arthur, A. K., and Fanning, E., 1990, SV40 T antigen binds directly to the catalytic subunit of DNA polymerase a, *EMBO J.* **9**:3329–3336.

161. Dornreiter, I., Erdile, L. F., Gilbert, I. U., von Winkler, D., Kelly, T. J., and Fanning, E. 1992, Interaction of DNA polymerase alpha-primase with cellular replication protein A and SV40 T antigen, *EMBO J.* **11:**769–776.
162. Gannon, J. V., and Lane, D. P., 1987, p53 and DNA polymerase a compete for binding to SV40 T antigen, *Nature* **329:**456–458.
163. Zhu, J., Rice, P. W., Chamberlain, M., and Cole, C. N., 1991, Mapping the transcriptional transactivation function of simian virus 40 large T antigen, *J. Virol.* **65:**2778–2790.
164. Gruda, M. C., Zabolotny, J. M., Xiao, J. H., Davidson, I., and Alwine, J. C., 1993, Transcriptional activation by simian virus 40 large T antigen: Interactions with multiple components of the transcription complex, *Mol. Cell. Biol.* **13:**961–969.
165. Bollag, B., Chuke, W. F., and Frisque, R. J., 1989, Hybrid genomes of the polyomaviruses JC virus, BK virus, and simian virus 40: Identification of sequences important for efficient transformation, *J. Virol.* **63:**863–872.
166. Haggerty, S., Walker, D. L., and Frisque, R. J., 1989, JC virus-simian virus 40 genomes containing heterologous regulatory signals and chimeric early regions: Identification of regions restricting transformation by JC virus, *J. Virol.* **63:**2180–2190.
167. Dyson, N., Bernards, R., Friend, S. H., Gooding, L. R., Hassell, J. A., Major, E. O., Pipas, J. M., Vandyke, T., and Harlow, E., 1990, Large T antigens of many polyomaviruses are able to form complexes with the retinoblastoma protein, *J. Virol.* **64:**1353–1356.
168. Benjamin, T., and Vogt, P. K., 1991, Cell transformation by viruses, in: *Fundamental Virology* (B. N. Fields, D. M. Knipe, R. M. Chanock, M. S. Hirsch, J. L. Melnick, T. P. Monath, and B. Roizman, eds.), Raven Press, New York, pp. 321–325.
169. Freund, R., Sotnikov, A., Bronson, R. T., and Benjamin, T. L., 1992, Polyoma virus middle T is essential for virus replication and persistence as well as for tumor induction in mice, *Virology* **191:**716–723.
170. Resnick-Silverman, L., Pang, Z., Li, G., Jha, K. K., and Ozer, H. L., 1991, Retinoblastoma protein and simian virus 40-dependent immortalization of human fibroblasts, *J. Virol.* **65:**2845–2852.
171. Freund, R., Bronson, R. T., and Benjamin, T. L., 1992, Separation of immortalization from tumor induction with polyoma large T mutants that fail to bind the retinoblastoma gene product, *Oncogene* **7:**1979–1987.
172. Courtneidge, S. A., and Smith, A. E., 1983, Polyoma virus transforming protein associates with the product of the c-*src* cellular gene, *Nature* **303:**435–439.
173. Courtneidge, S. A., and Smith, A. E., 1984, The complex of polyoma virus middle-T antigen and pp60c-*src*, *EMBO J.* **3:**585–591.
174. Kornbluth, S., Sudol, M., and Hanafusa, H., 1987, Association of the polyomavirus middle-T antigen with c-*yes* protein, *Nature* **325:**171–173.
175. Cheng, S. H., Harvey, R., Espino, P. C., Semba, K., Yamamoto, T., Toyoshima, K., and Smith, A. E., 1988, Peptide antibodies to the human c-*fyn* gene product demonstrate pp59c-*fyn* is capable of complex formation with middle-T antigen of polyomavirus, *EMBO J.* **7:**3845–3855.
176. Horak, I. D., Kawakami, T., Gregory, F., Robbins, K. C., and Bolen, J. B., 1989, Association of p60*fyn* with middle tumor antigen in murine polyomavirus-transformed rat cells, *J. Virol.* **63:**2343–2347.
177. Kypta, R. M., Hemming, A., and Courtneidge, S. A., 1988, Identification and characterization of p59*fyn* (a *src*-like protein tyrosine kinase) in normal and polyoma virus transformed cells, *EMBO J.* **7:**3837–3844.
178. Courtneidge, S. A., 1985, Activation of the pp60c-*src* kinase by middle T antigen binding or by dephosphorylation, *EMBO J.* **4:**1471–1477.
179. Bolen, J. B., Thiele, C. J., Israel, M. A., Yonemoto, W., Lipsich, L. A., and Brugge, J. S., 1984, Enhancement of cellular *src* gene product associated tyrosyl kinase activity following polyoma virus infection and transformation, *Cell* **38:**767–777.
180. Matthews, J. T., and Benjamin, T. L., 1986, 12-*o*-Tetradecanoylphorbol-13-acetate stimulates

phosphorylation of the 58,000-M_r form of polyoma virus middle T antigen *in vivo*: Implications of a possible role of protein kinase C in middle T function, *J. Virol.* **58**:239–246.
181. Harvey, R., Oostra, B. A., Belsham, G. J., Gillett, P., and Smith, A. E., 1984, An antibody to a synthetic peptide recognizes polyomavirus middle-T antigen and reveals multiple *in vitro* tyrosine phosphorylation sites, *Mol. Cell. Biol.* **4**:1334–1342.
182. Hunter, T., Hutchinson, M. A., and Eckhart, W., 1984, Polyoma middle-sized T antigen can be phosphorylated on tyrosine at multiple sites *in vitro*, *EMBO J.* **3**:73–79.
183. Schaffhausen, B., and Benjamin, T. L., 1981, Comparison of phosphorylation of two polyoma virus middle T antigens *in vivo* and *in vitro*, *J. Virol.* **40**:184–196.
184. Talmage, D. A., Freund, R., Young, A. T., Dahl, J., Dawe, C. J., and Benjamin, T. L., 1989, Phosphorylation of middle T by pp60c-*src*: A switch for binding of phosphatidylinositol 3-kinase and optimal tumorigenesis, *Cell* **59**:55–65.
185. Courtneidge, S. A., and Heber, A., 1987, An 81 kd protein complexed with middle T antigen and pp60c-*src*: A possible phosphatidylinositol kinase, *Cell* **50**:1031–1037.
186. Kaplan, D. R., Whitman, M., Schaffhausen, B., Pallas, D. C., White, M., and Cantley, L., 1987, Common elements in growth factor stimulation and oncogenic transformation: 85kd phosphoprotein and phosphatidylinositol kinase activity, *Cell* **50**:1021–1029.
187. Otsu, M., Hiles, I., Gout, I., Fry, M. J., Ruiz-Larrea, F., Panayotou, G., Thompson, A., Dhand, R., Hsuan, J., Totty, N., Smith, A. D., Morgan, S. J., Courtneidge, S. A., Parker, P. J., and Waterfield, M. D., 1991, Characterization of two 85 kd proteins that associate with receptor tyrosine kinases, middle-T/pp60c-*src* complexes, and PI3-kinase, *Cell* **65**:91–104.
188. Whitman, M., Downes, C. P., Keeler, M., Keller, T., and Cantley, L., 1988, Type I phosphatidylinositol kinase makes a novel inositol phospholipid, phosphatidylinositol-3-phosphate, *Nature* **332**:644–646.
189. Pallas, D. C., Shahrik, L. K., Martin, B. L., Jaspers, S., Miller, T. B., Brautigan, D. L., and Roberts, T. M., 1990, Polyoma small and middle T antigens and SV40 small t antigen form stable complexes with protein phosphatase 2A, *Cell* **60**:167–176.
190. Walter, G., Ruediger, R., Slaughter, C., and Mumby, M., 1990, Association of protein phosphatase 2A with polyoma virus medium tumor antigen, *Proc. Natl. Acad. Sci. USA* **87**:2521–2525.
191. Rameh, L. E., and Armelin, M. C. S., 1991, T antigens' role in polyomavirus transformation: c-*myc* but not c-*fos* or c-*jun* expression is a target for middle T, *Oncogene* **6**:1049–1056.
192. Jalinek, M. A., and Hassell, J. A., 1992, Reversion of middle T antigen-transformed Rat-2 cells by Krev-1: Implications for the role of p21c-*ras* in polyomavirus-mediated transformation, *Oncogene* **7**:1687–1698.
193. Feunteun, J., Kress, M., Gardes, M., and Monier, R., 1978, Viable deletion mutants in the simian virus 40 early region, *Proc. Natl. Acad. Sci. USA* **75**:4455–4459.
194. Martin, R. G., Setlow, V. P., Edwards, C. A. F., and Vembu, D., 1979, The roles of the simian virus 40 tumor antigens in transformation of Chinese hamster lung cells, *Cell* **17**:635–643.
195. Seif, R., and Martin, R. G., 1979, Simian virus 40 small t antigen is not required for the maintenance of transformation but may act as a promoter (cocarcinogen) during establishment of transformation in resting rat cells, *J. Virol.* **32**:979–988.
196. Sleigh, M. J., Topp, W. C., Hanich, R., and Sambrook, J., 1978, Mutants of SV40 with an altered small t protein are reduced in their ability to transform cells, *Cell* **14**:79–88.
197. Bikel, I., Montano, X., Agha, M. E., Brown, M., McCormack, M., Boltax, J., and Livingston, D. M., 1987, SV40 small t antigen enhances the transformation activity of limiting concentrations of SV40 large T antigen, *Cell* **48**:321–330.
198. Loeken, M., Bikel, I., Livingston, D. M., and Brady, J., 1988, Trans-activation of RNA polymerase II and III promoters by SV40 small t antigen, *Cell* **55**:1171–1177.
199. Graessmann, A., Graessmann, M., Tjian, R., and Topp, W. C., 1980, Simian virus 40 small-t protein is required for loss of actin cable networks in rat cells, *J. Virol.* **33**:1182–1191.
200. Hiscott, J. B., and Defendi, V., 1981, Simian virus 40 gene A regulation of cellular DNA synthesis. II. In nonpermissive cells, *J. Virol.* **37**:802–812.

201. Carbone, M., Hauser, J., Carty, M. P., Rundell, K., Dixon, K., and Levine, A. S., 1992, Simian virus 40 (SV40) small t antigen inhibits SV40 DNA replication *in vitro J. Virol.* **66:**1804–1808.
202. Scheidtmann, K. H., Mumby, M. C., Rundell, K., and Walter, G., 1991, Dephosphorylation of simian virus 40 large-T antigen and p53 protein by protein phophatase 2A: Inhibition by small-t antigen, *Mol. Cell. Biol.* **11:**1996–2003.
203. Sontag, E., Federov, S., Kamibayashi, C., Robbins, D., Cobb, M., and Mumby, M., 1993, The interaction of SV40 small tumor antigen with protein phosphatase 2A stimulates the MAP kinase pathway and induces cell proliferation, *Cell* **75:**887–897.
204. Yang, S., Lickteig, R. L., Estes, R., Rundell, K., Walter, G., and Mumby, M., 1991, Control of protein phosphatase 2A by simian virus 40 small-t antigen, *Mol. Cell. Biol.* **11:**1988–1995.
205. Wheat, W. H., Roesler, W. J., and Klemm, D. J., 1994, Simian virus 40 small tumor antigen inhibits dephosphorylation of protein kinase A-phosphorylated CREB and regulates CREB transcriptional stimulation, *Mol. Cell. Biol.* **14:**5881–5890.
206. Ogryzko, V. V., Hirai, T. H., Shih, C. E., and Howard, B. H., 1994, Dissociation of retinoblastoma gene protein hyperphosphorylation and commitment to enter S phase, *J. Virol.* **68:**3724–3732.
207. Dobbelstein, M., Arthur, A. K., Dehde, S., van-Zee, K., Dickmanns, A., and Fanning, E., 1992, Intracistronic complementation reveals a new function of SV40 T antigen that co-operates with Rb and p53 binding to stimulate DNA synthesis in quiescent cells, *Oncogene* **7:**837–847.
208. Maclean, K., Rogan, E. M., Whitaker, N. J., Chang, A. C., Rowe, P. B., Dalla-Pozza, L., Symonds, G., and Reddel, R. R., 1994, *In vitro* transformation of Li–Fraumeni syndrome fibroblasts by SV40 large T antigen mutants, *Oncogene* **9:**719–725.

4

Association of BK and JC Human Polyomaviruses and SV40 with Human Tumors

PAOLO MONINI, LAURA de LELLIS, and GIUSEPPE BARBANTI-BRODANO

1. INTRODUCTION

The human polyomaviruses BK (BKV) and JC (JCV) are ubiquitous in human populations and have a worldwide distribution.[1] They are oncogenic in rodents and monkeys and transform cells *in vitro* to a neoplastic phenotype. For all these reasons, BKV and JCV have been considered possible candidates in the etiology of human tumors. Association of BKV, but not of JCV, with human tumors has been described, although a formal proof for an etiological role of BKV in human oncogenesis is still lacking. SV40, which is not a ubiquitous human virus, has also been sporadically detected in human tumors. In this chapter, we consider their general characteristics, the natural history of infection, experimental transformation and oncogenicity by BKV and JCV, as well as the evidence linking BKV, JCV, and SV40 to human neoplasia.

2. GENERAL CHARACTERISTICS OF BKV AND JCV

Excellent reviews on BKV and JCV were published previously.[2,3] Both BKV and JCV, like SV40, belong to the family *Papovaviridae*. The virion is a 40 to 45-nm icosahedral particle with a density of 1.34–1.35 g/cm^3, and the genome is a circular,

---

PAOLO MONINI, LAURA de LELLIS, and GIUSEPPE BARBANTI-BRODANO • Institute of Microbiology, School of Medicine, University of Ferrara, I-44100 Ferrara, Italy.

DNA Tumor Viruses: Oncogenic Mechanisms, edited by Giuseppe Barbanti-Brodano *et al.* Plenum Press, New York, 1995.

double-stranded DNA molecule. Like SV40, BKV and JCV code for six viral proteins: two early nonstructural polypeptides, large T and small t antigens, an agnoprotein, probably involved in assembly of viral particles and processing of late mRNA,[4-6] and three capsid proteins, VP1, VP2, and VP3. In all three viruses the early and late genes are transcribed on different DNA strands, and, as a consequence, transcription proceeds divergently from the regulatory region and terminates within DNA sequences containing the polyadenylation signals. The large T antigens from BKV, JCV, and SV40 strongly cross-react with the same antisera,[7,8] and, although only a little cross-reactivity is observed in most structural antigenic determinants, a genus-specific capsid antigen located on viral peptide VP1 has been identified.[9] The DNA sequences of BKV and JCV share 75% homology,[10] and the homology with SV40 is 70% for BKV[11] and 69% for JCV.[10] The greatest homology is found in the early region, whereas the weakest homology is in the regulatory region. This probably reflects adaptation to *in vitro* cell culture,[12-16] and most laboratory strains may have evolved from a common, natural archetype.[17-21] However, the analysis of independent isolates by either direct cloning or sequencing of products obtained by polymerase chain reaction (PCR) amplification shows that different arrangements of the regulatory region are often detected *in vivo*.[22-26] Selection of variants with a particular cell specificity or transformation potential has been proposed as a possible outcome of such variability.[22,24,27]

3. NATURAL HISTORY OF BKV AND JCV INFECTION

3.1. Epidemiology

Both BKV and JCV were first isolated in 1971, BKV from the urine of a renal transplant recipient[28] and JCV from the brain of a patient with progressive multifocal leukoencephalopathy (PML), a rare demyelinating disease associated with impaired immunity.[29,30] The BKV and JCV are ubiquitous and infect a large proportion of humans all over the world, except for some segregated populations living in remote regions of Brazil, Paraguay, and Malaysia.[31] Primary infection occurs in childhood. At 3 years of age BKV antibodies are detected in 50% of children, and almost all individuals have been infected by the age of 10 years.[32-35] Infection with JCV develops later. Seroconversion is observed at highest rates during adolescence and continues afterward at lower frequency; by the age of 60 years, about 75% of adults show antibody evidence of JCV infection.[36,37]

3.2. Primary Infection, Latency, and Reactivation

Primary infections with BKV and JCV are usually inapparent and only occasionally associated with clinical conditions; BKV can cause upper respiratory or urinary tract disease,[38-42] and acute JCV infection has recently been associated with chronic meningoencephalitis.[43] Primary infection is followed by a persistent, latent infection that is reactivated under conditions of impaired immunosurveillance. Both viruses have been detected in the urine of renal and bone marrow transplant recipients undergoing immunosuppressive therapy,[44–53] in the urine of pregnant women,[54–57] and in patients with both hereditary and acquired immunodeficiency syndromes.[58–67] Reactivation of BKV and JCV has also been demonstrated in patients affected by a number of diseases, some of them related to immunosuppression: neoplastic disease (lymphoma and carcinoma),[68] systemic lupus erythematosus,[69] various forms of anemia,[70–72] nephrotic syndrome,[73] and Guillain–Barré syndrome.[74]

Little is known about the modality of virus transmission, though induction of upper respiratory disease by BKV and presence of latent BKV DNA in tonsils[40] may indicate a possible oral or respiratory route of transmission. The identification of viruses in the urine of pregnant women suggested a congenital transmission. However, early reports showing the presence of virus-specific IgM in umbilical cord sera[75,76] were not confirmed by other studies.[55,56,77,78]

4. ONCOGENICITY OF BKV AND JCV

4.1. Experimental Tumorigenesis with BKV

The oncogenic potential of BKV has been well established by inoculating, via different routes, young or newborn mice, rats, and hamsters. The BKV-induced tumors contained integrated and free BKV DNA sequences and large T antigen.[79–83] Fusion of tumor cells with permissive monkey or human cells yielded infectious virus.[79,80] The frequency of tumor induction in hamsters is highly dependent on the route of injection: BKV was weakly oncogenic when inoculated subcutaneously [79,84–86] but induced tumors in 73% to 88% of animals when inoculated intracerebrally or intravenously.[79–81,83,87,88] Tumors appearing in BKV-injected hamsters belong to a variety of histotypes, such as fibrosarcoma, ependymoma, neuroblastoma, pineal gland tumors, tumors of pancreatic islets, and osteosarcoma.[79–81,83–92] However, ependymoma, tumors of pancreatic islets, and osteosarcomas are the most frequent histotypes, suggesting that the virus may have a marked tropism for specific organs. Tumors induced by BKV in mice and rats were fibrosarcoma, liposarcoma, osteosarcoma, glioma, nephroblastoma, and choroid plexus papilloma, the latter arising only in mice.[79,93,94] Gardner's BKV strain seems more oncogenic than other isolates, such as RF or MM BKV.[87,89] It has been shown that the induction of different tumors may reflect the presence of several viral variants in the same inoculum.[81] In particular, an insulinoma-inducing variant has been associated with a viable deletion mutant originated in a Gardner BKV stock after several passages in culture.[91,92] A key role for viral genetic variability is further suggested by the tumorigenic properties of BKV-IR, a variant rescued from a human insulinoma.[95] This strain is associated with other human tumors and harbors an insertion sequence (IS)-like structure in the regulatory region (see below).[27,96] Tumors induced by BKV-IR develop at a lower frequency but display a more malignant phenotype than tumors induced by wild-type BKV.[27]

Purified BKV DNA is not oncogenic when inoculated intravenously or subcutaneously and induces tumors at a very low frequency when inoculated intracerebrally in rodents.[83] It displays, however, a strong synergism with activated oncogenes.[97] Newborn hamsters inoculated subcutaneously with a recombinant DNA molecule (pBK/c-*ras*A) expressing BKV early region gene and the c-Ha-*ras* oncogene yielded tumors within few weeks. Tumors developed at the site of injection and consisted of undifferentiated sarcomas expressing both BKV large T antigen and c-Ha-*ras* p21. Neither BKV DNA nor c-Ha-*ras* inoculated independently was tumorigenic. The same recombinant pBK/c-*ras*A induced brain tumors on intracerebral inoculation in newborn hamsters.[8] These data suggest an interaction of BKV transforming functions with human oncogenes.

4.2. Experimental Tumorigenesis with JCV

Intracerebrally and subcutaneously inoculated JCV produced in newborn hamsters brain tumors in 83% of animals.[99] Most tumors consisted of cerebellar medulloblastoma, but glioblastoma, astrocytoma, pineocytoma, and tumors of other histotypes were also observed. Tumors expressed JCV large T antigen and yielded viable JCV in culture on fusion with permissive human fetal glial cells. As with BKV, cell lines derived from hamster tumors contained integrated viral DNA in a tandem head-to-tail array.[100] Hamster brain tumors induced by JCV have been obtained by other authors. In these experiments, in addition to medulloblastoma, the frequent appearance of thalamic gliomas was reported.[101–103] Primitive neuroectodermal tumors were observed at a very low frequency,[104] and neuroblastoma and retinoblastoma have been consistently induced by intraocular inoculation.[105–107] Induction of pineal gland tumors was rarely observed with JCV strains Mad-1 or Tokyo-1,[108,109] but pineocytomas were described after inoculation of JCV Mad-4, suggesting that, as with BKV, different strains may display a different oncogenic potential or tropism.[102]

Owl monkeys inoculated with JCV by either the intracerebral, intravenous, or subcutaneous route developed cerebral tumors, mostly astrocytomas, within 14–36 months.[110–112] Derived tumors and cell lines contained integrated viral DNA and expressed JCV large T antigen.[110,112–114] No tumor induction has been described in primates inoculated with either BKV or SV40.[110]

5. *IN VITRO* TRANSFORMATION BY BKV AND JCV

5.1. Transformation of Rodent Cells by BKV

Complete BKV DNA or its early region or BKV itself can transform embryonic fibroblasts and cells cultured from the kidney and brain of hamster, mouse, rat, rabbit, and monkey.[86,115–128] The efficiency of *in vitro* transformation depends on the genetic characteristics of the viral strain and does not necessarily parallel the oncogenic potential of the viral isolate.[128] Cooperation in transformation of primary rodent embryo fibroblasts with the human c-*ras* oncogene has been observed with pBK/c-*ras*A,[129] the same recombinant DNA shown to induce malignant sarcomas in hamsters.[97] Transformation of rat pancreatic islet cells, a natural target of BKV tumorigenesis in rodents, has recently been described.[130] However, human pancreatic cells persistently infected with BKV did not display a transformed phenotype,[131] though BKV is frequently present in human pancreatic tumors.[132]

5.2. Transformation of Human Cells by BKV

Transformation of human cells by BKV is inefficient and often abortive.[123,133] Cells never display all the markers of malignant transformation (immortalization, anchorage independence, and tumorigenicity in nude mice), although they show morphological alterations and an increased life-span.[126–134] A fully transformed phenotype was induced in human embryo kidney (HEK) cells transfected with a recombinant plasmid expressing BKV early region and the adenovirus 12 E1A gene.[135] Cooperation of BKV with human *ras* and *myc* oncogenes has been demonstrated in human embryo fibroblasts and HEK cells, although it did not result in the induction of complete transformation.[136,137] Tumorigenic cell lines have only been established from human fetal brain cells persistently infected by BKV or after transfection of pBK/c-*ras*A in HEK-T cells (human embryo kidney cells from a fetus with Turner's syndrome).[137,138] Fetal brain cells had all the characteristics of transformed cells and retained viral DNA in an episomal state but, unlike HEK-T cells, were negative for large T antigen expression.[138] The reason for the higher susceptibility of HEK-T to transformation is unclear, but it may reflect the absence of genes on the Y sex chromosome, missing in Turner's syndrome cells. Indeed, cytogenetic analysis indicates that the Y chromosome is often deleted during progression of human solid tumors and leukemias and probably harbors tumor suppressor genes.

Further evidence that BKV-dependent transformation can be abolished by human chromosomes has recently been presented: BKV-transformed mouse and hamster cells were reduced or suppressed in both anchorage independence and tumorigenicity after transfer of human chromosomes 6 or 11.[139–141] Interestingly, one clone that lost the tumorigenic phenotype but maintained the ability to grow in soft agar had deleted the short arm of chromosome 11, suggesting that different human genes may control separate functions in BKV transformation.[139]

5.3. State of BKV DNA in Transformed Cells

The presence and physical state of BKV DNA in transformed cells have been studied by several authors.[82,142–145] BKV DNA is generally present in an integrated state in rodent cells, although variable amounts of free episomes have been reported. In one hamster osteosarcoma, viral DNA was found exclusively in the form of monomeric and polymeric extrachromosomal defective genomes.[146] Unlike rodent cells, transformed human cells harbor viral DNA mostly as unintegrated episomal molecules.[126,134,138]

5.4. Role of BKV Large T Antigen in Transformation

It has been demonstrated that BKV large T antigen, like SV40 and polyoma virus large T antigens, papillomavirus E6 and E7, and adenovirus E1A and E1B oncoproteins, interacts with p53 and p105Rb tumor suppressor products.[147–149] Furthermore, an interaction with the Rb-related p107 protein has been described for several transforming papovaviruses, including JCV.[148] Consistent with a continuous need for such interactions, persistent expression of a functional large T protein has been shown to be required for BKV transformation.[147] Expression of antisense large T antigen RNA in BKV-transformed rodent cells resulted in the abrogation of anchorage-independent growth.[147] However, BKV can occasionally transform cells via a "hit-and-run" mechanism, as proposed for herpesviruses.[150] Yogo et al. described a hamster choroid plexus papilloma containing one copy of BKV genome integrated within the early region, which implied that expression of a functional T antigen was no longer possible.[151] Recently, hamster cells were transfected with purified DNA obtained from a human tumor containing BKV DNA sequences. Although transfection resulted in the appearance of transformed cells, BKV DNA was absent in most clones.[152] Therefore, either BKV was irrelevant to the pathogenesis of this human tumor or genetic changes fixed in human cells after initiation of the oncogenic process by BKV were sufficient for expression and maintenance of the transformed phenotype. These changes may be caused by a mutagenic activity of the virus. Indeed, it was shown that BKV is mutagenic in human cells,[153] and BKV T antigen, like SV40 T antigen,[154,155] induces chromosomal aberrations (dichromatid gaps, breaks, dicentric chromosomes, triradial and quadriradial figures, sister chromatid exchanges) in human embryonic fibroblasts (G. Barbanti-Brodano, unpublished results).

5.5. *In Vitro* Transformation by JCV

Despite its high tumorigenic activity in rodents, JCV transformation of cells in culture is inefficient. Early studies failed to show induction of a fully transformed phenotype in human endothelial and fetal glial cells with JCV. These cells expressed JCV large T antigen and showed several markers of cell transformation, but they were not immortalized.[156,157] Few transformed cell lines were established in two studies by infection or transfection of viral DNA in primary hamster brain cells[158] and human amnion cells.[159] The molecular basis of the restricted transforming capacity of JCV has recently been investigated by Frisque and co-workers.[160,161] These authors constructed chimeric DNAs by exchanging the regulatory regions between JCV and SV40 or JCV and BKV; other recombinant DNAs contained JCV–BKV and JCV–SV40 large T antigen hybrids under the control of either BKV, JCV, or SV40 regulatory regions. In addition, the authors studied the transformation efficiency of wild-type versus variant or mutated JCV genomes.[162] The results of these experiments demonstrate that the JCV regulatory region and large T antigen sequences are both related to the low transforming efficiency of the virus. In addition, these studies showed that complexes between p53 and wild-type, mutant, or chimeric JCV large T antigen display a different stability, quaternary structure, and intranuclear concentration, suggesting that a specific modality of JCV T antigen interaction with p53 might be critical for its restricted transforming ability.

6. PRESENCE OF BKV AND JCV IN HUMAN TISSUES

6.1. Presence of BKV and JCV in Nonneoplastic Tissues

Virus isolation and Southern hybridization analysis established that the main site of BKV and JCV latency in healthy people is the kidney.[163–166] BKV sequences were also detected in other organs such as liver, stomach, lungs, parathyroid glands, and lymph nodes,[167,168] and JCV DNA was found in bone marrow and spleen mononuclear cells.[169] Polyomavirus virions were detected in peripheral blood lymphocytes,[170,171] and BKV was isolated from tonsils of children after mild upper respiratory disease.[40] In addition, restricted BKV replication was shown in human lymphocytes,[172] suggesting that the lymphoid tissue is another site of latency for BKV and that BKV infection of lymphocytes may favor virus spread to other tissues.

The application of PCR to the study of BKV and JCV latency revealed the presence of viral sequences in a variety of normal human tissues: BKV and JCV were detected by PCR in brain and peripheral blood lymphocytes.[173–175] The percentage of positive samples ranged from 30% for BKV and JCV in the brains of patients with neurological diseases other than PML[173] to 100% for BKV in normal brains.[174] In the study by Elsner and Dörries,[173] all the samples positive for JCV but one were also coinfected by BKV, suggesting a specific competence of certain brain cells for infection by human polyomaviruses or a need for cooperation or interference between the two viruses to establish latent infection in the brain. The amount of latent JCV DNA in the brain (1 to 500 genome equivalents in 100 cells) was greater than the amount of latent BKV DNA (1 to 10 genome equivalents in 100 cells), suggesting a reduced viral activity of BKV compared to JCV in the central nervous system. Cloning of BKV and JCV DNA sequences from latently infected brains led to isolation of full-length viral genomes, indicating that polyomavirus DNA is in an episomal state in latently infected brain. The percentage of samples positive for viral sequences in lymphocytes was 92.4% for BKV and 83.3% for JCV.[175] Moreover, BKV DNA sequences were detected by PCR in normal kidney, bladder, prostate, uterine cervix, vulva, lips, tongue (P. Monini, unpublished results), bone, and peripheral blood cells.[174] The frequency of positive samples ranged from 40% to 83% in different tissues. The JCV DNA was also detected by PCR in normal kidney, bladder, and prostate, but with a significantly lower frequency (17–25%) than BKV (P. Monini, unpublished results). The results of PCR analysis indicate that BKV and JCV can establish latent infection in many more organs than previously thought. This evidence may have important consequences for the routes and mechanisms of virus transmission as well as the epidemiology and the reactivation of BKV and JCV latent infection.

6.2. Presence of BKV and JCV in Neoplastic Tissues

Early studies by Southern hybridization with BKV DNA specific probes had shown the presence of BKV sequences in a variety of tumors of different histotypes, such as rhabdomyosarcoma, lung, kidney, liver carcinomas, and brain tumors.[168,176] These tumors contained full-length BKV genomes but also rearranged and defective BKV DNA molecules. Because BKV shows a specific oncogenic tropism for the ependymal tissue, endocrine pancreas, and bones in rodents,[79,80,83] BKV DNA sequences were searched by Southern blotting in primary human ependymomas and other brain tumors, insulinomas, and osteosarcoma cell lines. Episomal BKV DNA was detected in all these tumor types with a mean frequency of 28%. Both BKV early region RNA and T antigen were expressed in these tumors.[132] Furthermore, a BKV variant, BKV-IR, was rescued from a human insulinoma.[95] The BKV-IR variant was found to be associated with most human tumors positive for BKV sequences,[27] suggesting its specific involvement in certain human malignancies. The genome of BKV-IR contains an IS-like structure, a type of transposable element able to integrate and excise from the host genome.[177] The IS-like sequence of BKV-IR incorporates two of the early region transcriptional enhancers in its loop[96] and may promote cell transformation by insertional mutagenesis and activation of cellular oncogenes or more generally as a mutagen by random integration into cellular genes. In another study, BKV DNA was detected by Southern hybridization in 46% of brain tumors of the most common histotypes.[178] In this report BKV DNA sequences were found integrated into chromosomal DNA. Tumors typically associated with immunosuppression were also investigated by Southern hybridization, and BKV DNA was detected in Kaposi's sarcoma (KS) at a frequency of 60%.[179,180] Transfection of BKV-positive tumor DNA into human embryonic fibroblasts yielded a defective BKV variant with a deletion in a sequence of the early region coding for small T antigen.[27] Search for JCV sequences by Southern blotting in human brain tumors, insulinomas, and osteosarcomas was consistently negative.[132,178]

Neoplastic human tissue was investigated by PCR using specific primers for BKV DNA sequences covering the early region.[174] Twenty-eight primary brain tumors (eight glioblastomas, two spongioblastomas, three oligodendrogliomas, three meningiomas, six ependymomas, three choroid plexus papillomas, three astrocytomas) and six primary osteogenic sarcomas as well as five glioblastoma and eight osteosarcoma cell lines were analyzed. All primary brain tumors, three osteogenic sarcomas, and four glioblastoma and seven osteosarcoma cell lines were positive for BKV DNA sequences. All of 13 normal brains, two of five samples of normal bone, and 10 of 20 samples of peripheral blood cells from healthy donors were positive by PCR amplification with the same primers used for tumor DNA. Nucleotide sequence analysis of four brain tumors and one normal brain confirmed that the amplified sequences corresponded to the expected fragment of BKV early region. Expression of BKV early region was detected by reverse transcriptase PCR in one primary meningioma, three osteogenic sarcomas, four glioblastomas, and six osteosarcoma cell lines.

Amplification of DNA sequences from BKV early and regulatory regions was carried out by PCR in 15 kidney carcinomas, four ureter, 26 bladder, and seven prostate carcinomas. Positive samples ranged from 50% to 67% with an average positivity of 31 out of 52 samples (60%) (P. Monini et al., Virology, in press). In addition, BKV DNA sequences were amplified by PCR in carcinomas of the uterine cervix, vulva, lips, and tongue (P. Monini, unpublished observations). The average percentage of positive samples in these neoplastic tissues of the urinary and genital tracts and of the oral cavity was similar to that detected in the corresponding normal tissues (61% and 59%, respectively). However, in tumors of the urinary bladder and prostate, two-dimensional gel electrophoresis and Southern hybridization analysis showed either a single integration of BKV DNA sequences associated with disruption of the viral late region or both integrated and extrachromosomal viral sequences (P. Monini et al., Virology, in press). Viral episomes consisted of rearranged oligomers containing cellular DNA sequences whose size was incompatible with assembly in a virus particle. Attempts to rescue these viral sequences by transfection of tumor DNA into permissive cells were unsuccessful, suggesting that in these tumors the process of integration and formation of episomal oligomers produced a rearrangement of viral sequences responsible for the elimination of viral infectivity and potentially leading to stable expression of BKV transforming functions. A PCR analysis for JCV early region in the same urinary tract tumors studied for BKV DNA gave negative results except for 2 of 26 bladder tumors that were positive for JCV DNA sequences, also by Southern hybridization. The overall positivity of urinary tract tumors for JCV was 2 of 52 samples (4%) (P. Monini, unpublished results), a value significantly lower (Fisher's test, $p = 0.025$) than that detected in normal tissues (19%). These data are summarized in Tables I and II. Earlier reports, however, failed to detect polyomavirus footprints in urinary tract tumors, using less sensitive techniques.[181,182]

A PCR analysis for the early and regulatory regions in 25 samples of KS (five classic, 12 African, and eight AIDS-associated) revealed 100% positivity for BKV DNA sequences, whereas JCV sequences were present in one classical KS and in one

TABLE I
BKV and JCV Early Region Sequences Detected by PCR in Urinary Tract Tissues

Organs	BKV		JCV	
	Carcinomas	Nonneoplastic tissues[a]	Carcinomas	Nonneoplastic tissues[a]
Kidney	10/15 (66.6%)	5/6 (83.3%)	0/15	1/6 (16.7%)
Ureter	2/4 (50%)	0/1	0/4	0/1
Bladder	15/26 (57.7%)	5/10 (50%)	2/26 (7.7%)	2/10 (20%)
Urethra	—	0/1	—	0/1
Prostate	4/7 (57.1%)	11/19 (57.9%)	0/7	4/19 (21%)
Total	31/52 (59.6%)	21/37 (56.7%)	2/52 (3.8%)	7/37 (18.9%)

[a]Tissues from nonneoplastic lesions.

TABLE II
BKV Early Region Sequences Detected by PCR in Tissues of the Female Genital Tract

Normal tissues[a]	6/10 (60%)
Genital warts	0/2
Cervical intraepitelial neoplasia II + III	3/5 (60%)
Invasive cervical carcinoma	15/22 (68.2%)
Vulvar carcinoma	2/3 (66.6%)
Total neoplastic	20/32 (62.5%)

[a]Normal cervical and vulvar biopsies as determined by histological examination.

African KS and were negative in AIDS-associated KS. SV40 sequences were not detected in any of the three groups of KS analyzed. Analysis by PCR of four KS cell lines disclosed BKV DNA sequences in all of them, whereas JCV and SV40 sequences were absent (P. Monini, unpublished results). Cytomegalovirus, human papillomavirus, and human herpesvirus 6 have been detected in KS tissue.[183-187] All these viruses, including BKV, are latent in the human host and have oncogenic potential. Therefore, it is likely that they are reactivated by the immunosuppression associated to KS and infect the KS tissue. They could then participate in the development or progression of KS through the production and release of cytokines and growth factors[188,189] induced by their transforming proteins.

7. PRESENCE OF SV40 IN HUMAN TUMORS

SV40 does not normally infect humans. SV40 infection in the human population is restricted to those persons having contacts with monkeys, such as people from Indian villages living close to the jungle and persons attending to monkeys in zoos.[190] However, hundreds of millions of children and adults were treated with inactivated or attenuated oral antipoliovirus vaccines containing infectious SV40 during the period 1955–1962. Since SV40 induces brain tumors in rodents with a high frequency,[191,192] transforms human cells,[193] and is mutagenic in mammalian cells,[194] epidemiologic surveys were initiated in persons who had received the SV40-contaminated polio vaccine. The results of these studies, 30 years after accidental human infection with SV40, do not support clear epidemiologic evidence for the possible involvement of SV40 in human malignancy, although a slightly higher incidence of gliomas, glioblastomas, oligodendrogliomas, medulloblastomas, and spongioblastomas was observed in people exposed to the SV40-contaminated vaccine than in nonexposed people.[195] It was suggested that this epidemiologic evaluation should be continued for at least two more decades in order to permit any firm conclusions to be drawn.

Meanwhile, several studies reported the presence of SV40 DNA or T antigen in human tumors, mainly brain tumors.[196-201] SV40 sequences were cloned directly from two human primary brain tumors without any passage in culture.[202] One of these tumors, a meningioma, yielded DNA sequences indistinguishable from wild-type SV40, and the second tumor, an astrocytoma, contained a rearranged SV40 genome with a defective early region and a tandem duplication of an intact replication origin. Recently, detection of early region SV40 sequences by PCR was reported in 10 of 20 choroid plexus papillomas and in 10 of 11 ependymomas in children not exposed to SV40-contaminated poliovaccine.[203] Further support for an association of SV40 with human tumors was provided by Carbone et al.,[204] who detected early region SV40 sequences by PCR in 60% of human pleural mesotheliomas. This study was stimulated by the induction of mesotheliomas in 100% of Syrian hamsters injected in the pleural space with wild-type SV40.[205] In human mesotheliomas, SV40 T antigen was found by immunocytochemical methods to be specifically associated with the tumor tissue and absent in the surrounding normal pulmonary parenchyma. Interestingly, all the patients analyzed had serum antibodies to SV40 T antigen. Most of the patients had an age compatible with exposure to SV40-contaminated polio vaccine and showed presence of asbestos in their lungs, suggesting that SV40, fortuitously introduced in the human population by contaminated polio vaccines, may cooperate with asbestos as a cocarcinogen in induction of mesotheliomas. A possible mechanism for a potential involvement of SV40 in human oncogenesis was identified in the ability of SV40 T antigen to induce chromosomal aberrations and mutations in normal human cells.[154,155] Other investigations, however, reported negative results for SV40 sequences in human tumors.[174]

8. CONCLUSIONS

The role of polyomaviruses in human malignancy is far from elucidated. Concordant results exclude a relationship of JCV with human tumors, whereas BKV was repeatedly associated with human oncogenesis. This emphasizes the different biological properties of the two viruses, the low transforming activity of JCV T antigen and the better replication of JCV, as compared to BKV, during latency, probably facilitating lytic activity and reducing the JCV transforming potential. Another remarkable difference between BKV and JCV is their detection rate in urinary tract tissues. Both viruses can be revealed only by PCR in nonneoplastic tissues, suggesting that a minor fraction of cells are infected. If the viruses are not etiologically involved in the oncogenic process, because of the clonal nature of neoplasms, the prevalence of viral DNA in tumors should be remarkably lower than that in normal urothelial tissue. As reported in this chapter and shown in other studies,[206] this is the case only for JCV. In contrast, the prevalence of BKV in neoplastic and nonneoplastic specimens is comparable. At least in the urinary tract, BKV seems not to be distributed in a simple random fashion. This might reflect either its involvement in the neoplastic process or a fairly high rate of viral reactivation and reinfection. However, subclinical viruria is by far more frequent for JCV, pointing to a possible specific involvement of BKV in malignancy. Another observation is that BKV DNA sequences can be detected in the urinary tract by Southern blot only in neoplastic specimens. Their organization and arrangement are incompatible with productive infection and viral spreading, so they should be vertically transmitted between cell generations. Their simple restriction pattern indicates that those neoplasms were clonal and originated from an infected cell, suggesting a possible causative role for BKV in the oncogenic process at some early stage of initiation or progression.

The presence of SV40 in human tumors of persons not exposed to SV40-contaminated vaccines requires free circulation of SV40 in the human population. Because SV40 is not a ubiquitous virus in humans, and human cells are only semipermissive for SV40 replication, it is possible that viruses with SV40 sequences now detected in humans are recombinants between human polyomaviruses and SV40 introduced in the general population by vaccination with contaminated polio vaccines. Because the regulatory region of the viral genome, containing the promoter-enhancer for the early region, confers the host specificity and susceptibility to polyomaviruses, a recombinant chimera containing a BKV or JCV regulatory region and SV40 early region sequences would be highly infectious in humans and potentially oncogenic as a result of the remarkable transforming activity of SV40 T antigen for human cells.

As to BKV, the high prevalence of seropositive individuals in the human population and the ubiquity of its DNA sequences in normal human tissues during latency make it difficult to evaluate its involvement in human malignancy. It should be noted, however, that the classical Koch's postulates cannot be applied to latent viruses. New rules should be considered for these viruses in order to establish their oncogenic role.[207] (1) presence and persistence of the virus or its nucleic acid in tumor cells; (2) cell immortalization or neoplastic transformation after transfection of the viral genome or its subgenomic fragments; (3) demonstration that the malignant phenotype of the primary tumor and the modifications induced by transfection of cultured cells depend on specific functions expressed by the viral genome; and (4) epidemiologic and clinical evidence that viral infection represents a risk factor for tumor development. This chapter has reviewed data showing that BKV fulfills the first three criteria, suggesting that BKV may cooperate as a cofactor in the development or progression of human tumors.

Indeed, BKV DNA is present and expressed in human tumors. Furthermore, BKV is oncogenic in rodents and mutagenic in human cells. The BKV T antigen is a transforming protein that induces extensive chromosomal rearrangements and binds the products of tumor suppressor genes p53 and Rb, inactivating their functions. The BKV cooperates with c-*ras* in induction of malignant tumors in hamsters and in neoplastic transformation of rodent cells. Tumor suppressor genes located on chromosomes 6 and 11 suppress BKV tumorigenicity. During prolonged persistent infection and latency, minimal viral replication reduces or eliminates the virus lytic activity. Although BKV does not produce complete transformation of normal human cells, under these conditions BKV oncogenicity may be revealed by cooperating events, such as oncogene activation, loss of tumor suppressor genes, or rearrangements of the viral genome, inducing proliferation of clonal neoplastic cells in a population of BKV latently infected cells. Rare human populations have been described that are not affected by BKV infection.[31] Therefore, assessment of the fourth postulate could be carried out by determining the risk of certain human tumors in these BKV-free human populations as compared to BKV-infected counterparts.

ACKNOWLEDGMENTS. The work of the authors reported in this presentation was supported by grants to G. Barbanti-Brodano and E. Cassai from Associazione Italiana per la Ricerca sul Cancro (A.I.R.C.), from Consiglio Nazionale delle Ricerche (Progetto Finalizzato "Applicazioni Cliniche della Ricerca Oncologica"), and from Ministero dell'Università e della Ricerca Scientifica e Tecnologica (M.U.R.S.T. 60%).

REFERENCES

1. Padgett, B. L., and Walker, D. L., 1976, New human papovaviruses, *Prog. Med. Virol.* **22**:1–35.
2. Walker, D. L., and Frisque, R. J., 1986, The biology and molecular biology of JC virus, in: *The Papoviridae, Vol. 1, The Poliomaviruses* (N. P. Salzman, ed.), Plenum Press, New York, pp. 327–377.
3. Yoshiike, K. and Takemoto, K. K., 1986, Studies with BK virus and monkey lymphotropic papovavirus, in: *The Papoviridae, Vol. 1, The Poliomaviruses* (N. P. Salzman, ed.), Plenum Press, New York, pp. 295–326.
4. Alwine, J. C., 1982, Evidence for simian virus 40 late transcriptional control: Mixed infections of wild-type simian virus 40 and a late leader deletion mutant exhibit trans effects on late viral RNA synthesis, *J. Virol.* **42**:798–803.
5. Hay, N., Skolnick-David, H., and Aloni, Y., 1982, Attenuation in the control of SV40 gene expression, *Cell* **29**:183–193.
6. Ng, S.-C., Mertz, J. E., Sanden-Will, S., and Bina, M., 1985, Simian virus 40 maturation in cells harboring mutants deleted in the agnogene, *J. Biol. Chem.* **260**:1127–1132.
7. Takemoto, K. K., and Mullarkey, M. F., 1973, Human papovavirus, BK strain: Biological studies including antigenic relationship to simian virus 40, *J. Virol.* **12**:625–631.
8. Walker, D. L., Padgett, B. L., zu Rhein, G. M., Albert, A. E., and Marsh, R. F., 1973, Current study of an opportunistic papovavirus, in: *Slow Virus Diseases* (W. Zeman and E. H. Lennette, eds.), Williams & Wilkins, Baltimore, pp. 49–58.
9. Shah, K. V., Ozer, H. L., Ghazey, H. N., and Kelly, T. J., Jr., 1977, Common structural antigen of papovaviruses of the simian virus 40-polyoma subgroup, *J. Virol.* **21**:179–186.
10. Frisque, R. J., Bream, G. L., and Cannella, M. T., 1984, Human polyomavirus JC virus genome, *J. Virol.* **51**:458–469.
11. Yang, R. C. A., and Wu, R., 1979, BK virus DNA: Complete nucleotide sequence of a human tumor virus, *Science* **206**:456–462.
12. Martin, J. D., Padgett, B. L., and Walker, D. L., 1983, Characterization of tissue culture-induced heterogeneity in DNAs of independent isolates of JC virus, *J. Gen. Virol.* **64**:2271–2280.
13. Shinohara, T., Matsuda, M., Yasui, K., and Yoshike, K., 1989, Host range bias of the JC virus mutant enhancer with DNA rearrangement, *Virology* **170**:261–263.
14. Markowitz, R-B., Eaton, B. A., Kubik, M. F., Latorra, D., McGregor, J. A., and Dynan, W. S., 1991, BK virus and JC virus shed during pregnancy have predominantly archetypal regulatory regions, *J. Virol.* **65**:4515–4519.
15. Rubinstein, R., Shoonakker, B. C. A., and Harley, E. H., 1991, Recurring theme of changes in the transcriptional control region of BK virus during adaptation to cell culture, *J. Virol.* **65**:1600–1604.

16. Yogo, Y., Hara, K., Guo, J., Taguchi, F., Nagashima, K., Akatani, K., and Ikegami, N., 1993, DNA-sequence rearrangement required for the adaptation of JC polyomavirus to growth in a human neuroblastoma cell line (IMR-32), *Virology* **197**:793–795.
17. Rubinstein, R., Pare, N., and Harley, E. H., 1987, Structure and function of the transcriptional control region of nonpassaged BK virus, *J. Virol.* **61**:1747–1750.
18. Yogo, Y., Kitamura, T., Sugimoto, C., Ueki, T., Aso, Y., Hara, K., and Taguchi, F., 1990, Isolation of a possible archetypal JC virus DNA sequence from nonimmunocompromised individuals, *J. Virol.* **64**:3139–3143.
19. Flaegstad, T., Sundsfjord, A., Arthur, R. R., Pedersen, M., Traavik, T., and Subramani, S., 1991, Amplification and sequencing of the control regions of BK and JC virus from human urine by polymerase chain reaction, *Virology* **180**:553–560.
20. Negrini, M., Sabbioni, S., Arthur, R. R., Castagnoli, A., and Barbanti-Brodano, G., 1991, Prevalence of the archetypal regulatory region and sequence polymorphism in nonpassaged BK virus variants, *J. Virol.* **65**:5092–5905.
21. Tominaga, T., Yogo, Y., Kitamura, T., and Aso, Y., 1992, Persistence of archetypal JC virus DNA in normal renal tissue derived from tumor-bearing patients, *Virology* **186**:736–741.
22. Loeber, G., and Dörries, K., 1988, DNA rearrangements in organ-specific variants of polyomavirus JC strain GS, *J. Virol.* **62**:1730–1735.
23. Sundsfjord, A., Johansen, T., Flaegstad, T., Moens, U., Villand, P., Subramani, S., and Traavik, T., 1990, At least two types of control regions can be found among naturally occurring BK virus strains, *J. Virol.* **64**:3864–3871.
24. Yogo, Y., Kitamura, T., Sugimoto, C., Hara, K., Iida, T., Taguchi, F., Tajima, A., Kawabe, K., and Aso, Y., 1991, Sequence rearrangement in JC virus DNAs molecularly cloned from immunosuppressed renal transplant patients, *J. Virol.* **65**:2422–2428.
25. Ault, G. S., and Stoner, G. L., 1992, Two major types of JC virus defined in progressive multifocal leukoencephalopathy brain by early and late coding region DNA sequences, *J. Gen. Virol.* **73**:2669–2678.
26. Ault, G. S., and Stoner, G. L., 1993, Human polyomavirus JC promoter/enhancer rearrangement patterns from progressive multifocal leukoencephalopathy brain are unique derivatives of a single archetypal structure. *J. Gen. Virol.* **74**:1499–1507.
27. Negrini, M, Rimessi, P., Mantovani, C., Sabbioni, S., Corallini, A., Gerosa, M. A., and Barbanti-Brodano, G., 1990, Characterization of BK virus variants rescued from human tumors and tumour cell lines. *J. Gen. Virol.* **71**:2731–2736.
28. Gardner, S. D., Field, A. M., Coleman, D. V., and Hulme, B., 1971, New human papovavirus (B.K.), isolated from urine after renal transplantation, *Lancet* **1**:1253–1257.
29. Padgett, B. L., Walker, D. L., zu Rhein, G. M., Eckroade, R. J., and Dessel, B. H., 1971, Cultivation of papova-like virus from human brain with progresive multifocal leucoencephalopathy, *Lancet* **1**:1257–1260.
30. Walker, D. L., and Padgett, B. L., 1983, Progressive multifocal leukoencephalopathy, in: *Comprehensive Virology, Vol. 18* (H. Fraenkel-Conrat, and R. R. Wagner, eds.), Plenum Press, New York, pp. 161–193.
31. Brown, P., Tsai, T., and Gajdusek, D. C., 1975, Seroepidemiology of human papovaviruses: Discovery of virgin populations and some unusual patterns of antibody prevalence among remote peoples of the world, *Am. J. Epidemiol.* **102**:331–340.
32. Gardner, S. D., 1973, Prevalence in England of antibody to human polyomavirus (BK), *Br. Med. J.* **1**:77–78.
33. Mantyjarvi, R. A., Meurman, O. H., Vihma, L., and Berglund, B., 1973, A human papovavirus (BK), biological properties and seroepidemiology, *Ann. Clin. Res.* **5**:283–287.
34. Shah, K. V., Daniel, R. W., and Warszawski, R. M., 1973, High prevalence of antibodies to BK virus, an SV40 related papovavirus, in residents of Maryland. *J. Infect. Dis.* **128**:784–787.
35. Portolani, M., Marzocchi, A., Barbanti-Brodano, G., and La Placa, M., 1974, Prevalence in Italy of antibodies to a new human papovavirus (BK virus), *J. Med. Microbiol.* **7**:543–546.

36. Padgett, B. L., and Walker, D. L., 1973, Prevalence of antibodies in human sera against JC virus, an isolate from a case of progressive multifocal leukoencephalopathy, *J. Infect. Dis.* **127**:467–470.
37. Taguchi, F., Kajioka, J., and Miyamura, T., 1982, Prevalence rate and age of acquisition of antibodies against JC virus and BK virus in human sera, *Microbiol. Immunol.* **26**:1057–1062.
38. Hashida, J., Gaffney, P. C., and Yunis, E. J., 1976, Acute hemorrhagic cystitis of childhood and papovavirus-like particles, *J. Pediatr.* **89**:85–87.
39. Goudsmit, J., Baak, M. L., Slaterus, K. W., and van der Noordaa, J., 1981, Human papovavirus isolated from the urine of a child with acute tonsillitis, *Br. Med. J.* **283**:1363–1364.
40. Goudsmit, J., Wetheim-van Dillen, P., van Strien, A., and van der Noordaa, J., 1982, The role of BK virus in acute respiratory tract disease and the presence of BKV DNA in tonsils, *J. Med. Virol.* **10**:91–99.
41. Mininberg, D. T., Watson, C., and Desquitado, M., 1982, Viral cystitis with transient secondary vesicoureteral reflux, *J. Urol.* **127**:983–985.
42. Padgett, B. L., Walker, D. L., Desquitado, M., and Kim, D. V., 1983, BK virus and nonhemorrhagic cystitis in a child, *Lancet* **1**:770.
43. Blake, K., Pillay, D., Knowles, W., Brown, D. V., Griffiths, P. D., and Taylor, B., 1992, JC virus associated meningoencephalitis in an immunocompetent girl, *Arch. Dis. Child.* **67**:956–957.
44. Coleman, D. V., Gardner, S. D., and Field, A. M., 1973, Human polyomavirus infection in renal allograft recipients, *Br. Med. J.* **3**:371–375.
45. Lecatsas, G., Prozesky, O. W., Van Wyk, J., and Els, H. J., 1973, Papova virus in urine after renal transplantation, *Nature* **241**:343–344.
46. Hogan, T. F., Borden, E. C., McBain, J. A., Padgett, B. L., and Walker, D. L., 1980, Human polyomavirus infections with JC virus and BK virus in renal transplant recipients, *Ann. Intern. Med.* **92**:373–378.
47. Traystman, M. D., Gupta, P. K., Shah, K. V., Reissig, M., Cowles, L. T., Hillis, W. D., and Frost, J. K., 1980, Identification of viruses in the urine of renal transplant recipients by cytomorphology, *Acta Cytol.* **24**:501–510.
48. O'Reilly, R. J., Lee, F. K., Grossbard, E., Kapoor, N., Kirkpatrick, D., Dinsmore, R., Stutzer, C., Shah, K. V., and Nahmias, A. J., 1981, Papovavirus excretion following marrow transplantation. Incidence and association with hepatic disfunction, *Transplant. Proc.* **13**:262–266.
49. Gardner, S. D., MacKenzie, E. F. D., Smith, C., and Porter, A. A., 1984, Prospective study of the human polyomaviruses BK and JC and cytomegalovirus in renal transplant recipients, *J. Clin. Pathol.* **37**:578–586.
50. Rice, S. J., Bishop, J. A., Apperley, J., and Gardner, S. D., 1985, BK virus as a cause of haemorrhagic cystitis after bone marrow transplantation, *Lancet* **2**:844–845.
51. Arthur, R. R., Shah, K. V., Baust, S. J., Santos, G. W., and Saral, R., 1986, Association of BK viruria with hemorrhagic cystitis in recipients of bone marrow transplants, *N. Engl. J. Med.* **315**:230–234.
52. Apperley, J. F., Rice, S. J., Bishop, J. A., Chia, Y. C., Krausz, T., Gardner, S. D., and Goldman, J. M., 1987, Late-onset hemorrhagic cystitis associated with urinary excretion of polyomaviruses after bone marrow transplantation, *Transplantation* **43**:108–112.
53. Arthur, R. R., Shah, K. V., Charache, P., and Saral, R., 1988, BK and JV virus infections in recipients of bone marrow transplants. *J. Infect. Dis.* **158**:563–569.
54. Coleman, D. V., Daniel, R. A., Gardner, S. D., Field, A. M., and Gibson, P. E., 1977, Polyomavirus in urine during pregnancy, *Lancet* **2**:709–710.
55. Coleman, D. V., Wolfendale, M. R., Daniel, R. A., Dhanjal, N. K., Gardner, S. D., Gibson, P. E., and Field, A. M., 1980, A prospective study of human polyomavirus infection in pregnancy. *J. Infect. Dis.* **142**:1–8.
56. Shah, K. V., Daniel, R., Madden, D., and Stagno, S., 1980, Serological investigation of BK papovavirus infection in pregnant women and their offspring, *Infect. Immun.* **30**:29–35.
57. Coleman, D. V., Gardner, S. D., Mulholland, C., Fridiksdottir, V., Portner, A. A., Lilford, R.,

and Valdimarsson, H., 1983, Human polyomavirus in pregnancy. A model for study of defense mechanisms to virus reactivation. *Clin. Exp. Immunol.* **53**:289–296.
58. Takemoto, K. K., Rabson, A. S., Mullarkey, M. R., Blaese, M. F., Garon, C. F., and Nelson, D., 1974, Isolation of papovavirus from brain tumor and urine of a patient with Wiskott–Aldrich syndrome, *J. Natl. Cancer Inst.* **53**:1205–1207.
59. Rhiza, H. J., Belohradsky, B. H., Schneider, V., Schwenk, H. U., Burkamm, G. W., and zur Hausen, H., 1978, BK virus. II. Serological studies in children with congenital disease and patients with malignant tumors and immunodeficiencies, *Med. Microbiol. Immunol.* **165**:83–92.
60. Snider, W. D., Simpson, D. M., Nielsen, S., Gold, J. W., Metroka, C. E., and Posner, J. B., 1983, Neurological complications of the acquired immune deficiency syndrome: Analysis of 50 patients. *Ann. Neurol.* **14**:403–418.
61. Flaegstad, T., Permin, H., Husebekk, A., Husby, G., and Traavik, T., 1988, BK virus infection in patients with AIDS, *Scand. J. Infect. Dis.* **20**:145–150.
62. Wiley, C. A., Grafe, M., Kenney, C., and Nelson, J. A., 1988, Human immunodeficiency virus (HIV) and JC virus in acquired immune deficiency syndrome (AIDS) patients with progressive multifocal leukoencephalopathy, *Acta Neuropathol.* **76**:338–346.
63. Vazeux, R., Cumont, M., Girard, P. M., Nassif, X., Trotot, P., Marche, C., Matthiessen, L., Vedrenne, C., Mikol, J., Henin, D., Katlama, C., and Bolgert, F., 1990, Severe encephalitis resulting from coinfections with HIV and JC virus, *Neurology* **40**:944–948.
64. Gillespie, S. M., Chang, Y., Lemp, G., Arthur, R., Buchbinder, S., Steiml, A., Baumgartner, J., Rando, T., Neal, D., Rutherford, G., Schomberger, L., and Janssen, R., 1991. Progressive multifocal leukoencephalopathy in persons infected with human immunodeficiency virus, S. Francisco, 1981–1989, *Ann. Neurol.* **30**:597–604.
65. Quinlivan, E. B., Norris, M., Bouldin, T. W., Suzuki, K., Meeker, R., Smith, M. S., Hall, C., and Kenney, S., 1992, Subclinical central nervous system infection with JC virus in patients with AIDS. *J. Infect. Dis.* **166**:80–85.
66. Markowitz, R. B., Thompson, H. C., Mueller, J. F., Cohen, J. A., and Dynan, W. S., 1993, Incidence of BK virus and JC virus viruria in human immunodeficiency virus-infected and -uninfected subjects, *J. Infect. Dis.* **167**:13–20.
67. Vallbracht, A., Löhler, J., Gossmann, J., Glück, T., Pertersen, D., Gerth, H.-J., Gencic, M., and Dörries, K., 1993, Disseminated BK type polyomavirus infection in an AIDS patient associated with CNS disease, *Am. J. Pathol.* **143**:1–11.
68. Hogan, T. F., Padgett, B. L., Walker, D. L., Borden, E. C., and Frias, Z., 1983, Survey of human polyomavirus (JCV, BKV) infections in 139 patients with lung cancer, breast cancer, melanoma, or lymphoma, *Prog. Clin. Biol. Res.* **105**:311–324.
69. Taguchi, F., and Nagaki, D., 1978, BK papovavirus in urine of patient with systemic lupus erythematosus, *Acta Virol.* **22**:513.
70. Lecatsas, G., Schoub, B. D., Prozesky, O. W., Pretorius, F., and De Beer, F. C., 1976, Polyomavirus in urine in aplastic anemia, *Lancet* **1**:259–260.
71. Lecatsas, G., Pretorius, F., Crewe-Brown, H., Requadt, E., and Ackthun, I., 1977, Polyomavirus in urine in pernicious anemia, *Lancet* **2**:147.
72. Lecatsas, G., and Bernard, M. M., 1982, BK virus excretion in Fanconi's anemia, *S. Afr. Med. J.* **62**:467.
73. Nagao, S., Iijima, I. S., Suzuki, H., Yokota, T., and Shigeta, S., 1982, BK virus-like particles in the urine of a patient with nephrotic syndrome. An electron microscopic observation, *Fukushima J. Med. Sci.* **29**:45–49.
74. van der Noordaa, J., and Wartheim-van Dillen, P., 1977, Rise in antibodies to human papovavirus BK and clinical disease, *Br. Med. J.* **1**:1471.
75. Taguchi, F., Nagaki, D., Saito, M., Haruyama, C., Iwasaki, K., and Suzuki, T., 1975, Transplacental transmission of BK virus in humans. *Jpn. J. Microbiol.* **19**:395–398.
76. Rhiza, H. J., Belohradsky, B. H., and zur Hausen, H., 1978, BK virus. I. Seroepidemiology and serologic response to viral infection, *Med. Microbiol. Immunol.* **165**:73–81.

77. Borgatti, M., Costanzo, F., Portolani, M., Vullo, C., Osti, L., Masi, M., and Barbanti-Bordano, G., 1979, Evidence for reactivation of persistent infection during pregnancy and lack of congenital transmission of BK virus, a human papovavirus, *Microbiologica* **2**:173–178.
78. Daniel, R., Shah, K., Madden, D., and Stagno, S., 1981, Serological investigation of the possibility of congenital transmission of papovavirus JC, *Infect. Immun.* **33**:319–321.
79. Corallini, A., Barbanti-Brodano, G., Bortoloni, W., Nenci, I., Cassai, E., Tampieri, M., Portolani, M., and Borgatti, M., 1977, High incidence of ependymomas induced by BK virus, a human papovavirus, *J. Natl. Cancer Inst.* **59**:1561–1563.
80. Corallini, A., Altavilla, G., Cecchetti, M. G., Fabris, G., Grossi, M. P., Balboni, P. G., Lanza, G., and Barbanti-Brodano, G., 1978, Ependymomas, malignant tumors of pancreatic islets and osteosarcomas induced in hamsters by BK virus, a human papovavirus, *J. Natl. Cancer Inst.* **61**:875–883.
81. Uchida, S., Watanabe, S., Aizawa, T., Furuno, A., and Muto, T., 1979, Polioncogenicity and insulinoma-inducing ability of BK virus, a human papovavirus, in Syrian golden hamsters, *J. Natl. Cancer Inst.* **63**:119–126.
82. Chenciner, N., Meneguzzi, G., Corallini, A., Grossi, M. P., Grassi, P., Barbanti-Brodano, G., and Milanesi, G., 1980, Integrated and free viral DNA in hamster tumors induced by BK virus, *Proc. Natl. Acad. Sci. USA* **77**:975–979.
83. Corallini, A., Altavilla, G., Carrà, L., Grossi, M. P., Federspil, G., Caputo, A., Negrini, M., and Barbanti-Brodano, G., 1982, Oncogenicity of BK virus for immunosuppressed hamsters, *Acta Virol.* **73**:243–253.
84. Nase, L. M., Karkkaiven, M., and Mantyjarvi, R. A., 1974, Transplantable hamster tumors induced with the BK virus, *Acta Pathol. Microbiol. Scand.* **83**:347–352.
85. Shah, K. V., Daniel, R. W., and Strandberg, J., 1975, Sarcoma in a hamster inoculated with BK virus, a human papovavirus, *J. Natl. Cancer Inst.* **54**:945–949.
86. van der Noordaa, J., 1976, Infectivity, oncogenity and transforming ability of BK virus and BK virus DNA. *J. Gen. Virol.* **30**:371–373.
87. Costa, T., Yee, C., Tralka, T. S., and Rabson, A. S., 1976, Hamster epandymomas produced by intracerebral inoculation of human papovavirus (MMV), *J. Natl. Cancer Inst.* **56**:863–864.
88. Uchida, S., Watanabe, S., Aizawa, T., Kato, F., Furuno, A., and Muto, T., 1976, Induction of papillary ependymomas and insulinomas in the Syrian golden hamster by BK virus, a human papovavirus. *Gann* **67**:857–865.
89. Dougherty, R. M., 1976, Induction of tumors in Syrian hamster by a human renal papovavirus, RF strain, *J. Natl. Cancer Inst.* **57**:395–400.
90. Greenlee, J. E., Narayan, O., Johnson, R. T., and Hernodon, R. M., 1977, Induction of brain tumors in hamsters with BK virus, a human papovavirus, *Lab. Invest.* **36**:636–642.
91. Watanabe, S., Yoshiike, K., Nozawa A., Yuasa, Y., and Uchida, S., 1979, Viable deletion mutant of human papovavirus BK that induces insulinomas in hamsters, *J. Virol.* **32**:934–942.
92. Watanabe, S., Kotake, S., Nozawa, A., Muto, T., and Uchida, S., 1982, Tumorigenicity of human BK papovavirus plaque isolates, wild-type and plaque morphology mutants, in hamsters, *Int. J. Cancer* **29**:583–589.
93. Noss, G., Stauch, G., Mehraein, P., and Georgii, A., 1981, Oncogenic activity of the BK type of human papovavirus in newborn Wistar rats, *Arch. Virol.* **69**:239–251.
94. Noss, G., and Stauch, G., 1984, Oncogenic activity of the BK type of human papova virus in inbred rat strains, *Acta Virol.* **81**:41–50.
95. Caputo, A., Corallini, A., Grossi, M. P., Carrà, L., Balboni, P. G., Negrini, M., Milanesi, G., Federspil, G., and Barbanti-Brodano, G., 1983, Episomal DNA of a BK virus variant in a human insulinoma, *J. Med. Virol.* **12**:37–49.
96. Pagnani, M., Negrini, M., Reschiglian, P., Corallini, A., Balboni, P. G., Scherneck, S., Macino, G., Milanesi, G., and Barbanti-Brodano, G., 1986, Molecular and biological properties of BK virus-IR, a BK virus variant rescued from a human tumor, *J. Virol.* **59**:500–505.
97. Corallini, A., Pagnani, M., Viadana, P., Camellin, P., Caputo, A., Reschiglian, P., Rossi, S.,

Altavilla, G., Selvatici, R., and Barbanti-Brodano, G., 1987, Induction of malignant subcutaneous sarcomas in hamsters by a recombinant DNA containing BK virus early region and the activated human c-Harvey-*ras* oncogene, *Cancer Res.* **47:**6671–6677.
98. Corallini, A., Pagnani, M., Caputo, A., Negrini, M., Altavilla, G., Catozzi, L., and Barbanti-Brodano, G., 1988, Cooperation in oncogenesis between BK virus early region gene and the activated human c-Harvey-*ras* oncogene, *J. Gen. Virol.* **69:**2671–2679.
99. Walker, D. L., Padgett, B. L., zu Rhein, G. M., Albert, A. E., and Marsh, R. F., 1973, Human papovavirus (JC): induction of brain tumors in hamsters, *Science* **181:**674–676.
100. Wold, W. S. M., Green, M., Mackey, J. K., Martin, J. D., Padgett, B. L., and Walker, D. L., 1980, Integration pattern of human JC virus sequences in two clones of a cell line established from a JC virus-induced hamster brain tumor, *J. Virol.* **33:**1225–1228.
101. zu Rhein, G. M., and Varakis, J., 1975, Morphology of brain tumors induced in Syrian hamsters after inoculation with JC virus, a new human papovavirus, in: *Proceedings of the VIIth International Congress of Neuropathology, Budapest*, Volume 1 (S. Kornzey, S. Tariska, and G. Gosztony, eds.), Academic Kiado, Budapest and Excerpta Medica, Amsterdam, pp. 479–481.
102. Padgett, B. L., Walker, D. L., zu Rhein, G. M., and Varakis, J. N., 1977, Differential neuro-oncogenicity of strains of JC virus, a human polyomavirus, in newborn Syrian hamsters, *Cancer Res.* **37:**718–725.
103. zu Rhein, G. M., and Varakis, J., 1979, Perinatal induction of medulloblastomas in Syrian golden hamsters by a human polyoma virus (JC), *Natl. Cancer Inst. Monogr.* **51:**205–208.
104. zu Rhein, G. M., 1983, Studies of JC virus-induced nervous system tumors in the Syrian hamster: A review, in: *Polyomaviruses and Human Neurological Disease* (J. L. Sever and D. L. Madden, eds.), Alan R. Liss, New York, pp. 205–221.
105. Varakis, J. N., zu Rhein, G. M., Padgett, B. L., and Walker, D. L., 1976, Experimental (JC virus-induced) neuroblastomas in the Syrian hamster, *J. Neuropathol. Exp. Neurol.* **35:**314.
106. Ohashi, T., zu Rhein, G. M., Varakis, J., Padgett, B. L., and Walker, D. L., 1978, Experimental (JC virus-induced) intraocular and extraorbital tumors in the Syrian hamster. *J. Neuropathol. Exp. Neurol.* **37:**667.
107. Varakis, J. N., zu Rhein, G. M., Padgett, B. L., and Walker, D. L., 1978, Induction of peripheral neuroblastomas in Syrian hamsters after injection as neonates with JC virus, a human polyoma virus, *Cancer Res.* **38:**1718–1722.
108. Varakis, J. N., and zu Rhein, G. M., 1976, Experiment pineocytoma of the Syrian hamster induced by a human papovavirus (JC), *Acta Neuropathol.* **35:**243–264.
109. Nagashima, K., Yasui, K., Kimura, J., Washizu, M., Yamaguchi, K., and Mori, W., 1984, Induction of brain tumors by a newly isolated JC virus (Tokio-1 strain), *Am. J. Pathol.* **116:**455–463.
110. London, W. T., Houff, S. A., Madden, D. L., Fuccillo, D. A., Gravell, M., Wallen, W. C., Palmer, A. E., Sever, J. L., Padgett, B. L., Walker, D. L., zu Rhein, G. M., and Ohashi, T., 1978, Brain tumors in owl monkeys inoculated with a human polyomavirus (JC virus), *Science* **201:**1246–1249.
111. London, W. T., Houff, S. A., MacKeever, P. E., Wallen, W., Sever, J. L., Padgett, B., and Walker, D., 1983, Viral-induced astrocytomas in squirrel monkeys, in: *Polyomaviruses and Human Neurological Disease* (J. L. Sever and D. L. Madden, eds.), Alan R. Liss, New York, pp. 227–237.
112. Miller, N. R., McKeever, P. E., London, W., Padgett, B. L., Walker, D. L., and Wallen, W. C., 1984, Brain tumors of owl monkeys inoculated with JC virus contain the JC virus genome, *J. Virol.* **49:**848–856.
113. Major, E. O., 1983, JC virus T protein expression in owl monkey tumor cell lines, in: *Polyomaviruses and Human Neurological Disease* (J. L. Sever and D. L. Madden, eds.), Alan R. Liss, New York, pp. 289–298.
114. Major, E. O., Mourrain, P., and Cummins, C., 1984, JC virus-induced owl monkey glioblastoma cells in culture: Biological properties associated with the viral early gene product, *Virology* **136:**359–367.

115. Major, E. O., and Di Mayorca, G., 1973, Malignant transformation of BHK21 clone 13 cells by BK virus—a human papovavirus, *Proc. Natl. Acad. Sci. USA* **70**:3210–3212.
116. Portolani, M., Barbanti-Brodano, G., and La Placa, M., 1975, Malignant transformation of hamster kidney cells by BK virus, *J. Virol.* **15**:420–422.
117. Tanaka, R., Koprowski, H., and Iwasaki, Y., 1976, Malignant transformation of hamster brain cells *in vitro* by human papovavirus BK, *J. Natl. Cancer Inst.* **56**:671–673.
118. Takemoto, K. K., and Martin, M. A., 1976, Transformation of hamster kidney cells by BK papovavirus DNA, *J. Virol.* **17**:247–253.
119. Costa, J., Howley, P. M., Legallais, F., Yee, C., Young, N., and Rabson, A. S., 1977, Oncogenicity of a nude mouse cell line transformed by a human papovavirus, *J. Natl. Cancer Inst.* **58**:1147–1151.
120. Mason, D. H., Jr., and Takemoto, K. K., 1977, Transformation of rabbit kidney cells by BKV (MM) human papovavirus, *Int. J. Cancer* **19**:391–395.
121. Seehafer, J., Salmi, A., and Colter, J. S., 1977, Isolation and characterization of BK virus transformed hamster cells, *Virology* **77**:356–366.
122. Bradley, M. K., and Dougherty, R. M., 1978, Transformation of African green monkey kidney cells with the RF strain of human papovavirus BKV, *Virology* **85**:231–240.
123. Portolani, M., Borgatti, M., Corallini, A., Cassai, E., Grossi, M. P., Barbanti-Brodano, G., and Possati, L., 1978, Stable transformation of mouse, rabbit and monkey cells and abortive transformation of human cells by BK virus, a human papovavirus, *J. Gen. Virol.* **38**:369–374.
124. Seehafer, J., Downer, D. N., Salmi, A., and Colter, J. S., 1979, Isolation and characterization of BK virus-transformed rat and mouse cells, *J. Gen. Virol.* **42**:567–578.
125. van der Noordaa, J., De Jong, W., Pauw, W., Sol, C. J. A., and van Strien, A., 1979, Transformation and T antigen induction by linearized BK virus DNA, *J. Gen. Virol.* **44**:843–847.
126. Grossi, M. P., Caputo, A., Meneguzzi, G., Corallini, A., Carrà, L., Portolani, M., Borgatti, M., Milanesi G., and Barbanti-Brodano, G., 1982, Transformation of human embryonic fibroblasts by BK virus, BK virus DNA and a subgenomic BK virus DNA fragment, *J. Gen. Virol.* **63**:369–403.
127. Grossi, M. P., Corallini, A., Valieri, A., Balboni, P. G., Poli, F., Caputo, A., Milanesi, G., and Barbanti-Brodano, G., 1982, Transformation of hamster kidney cells by fragments of BK virus DNA. *J. Virol.* **41**:319–325.
128. Watanabe, S., and Yoshiike, K., 1982, Change of DNA near the origin of replication enhances the transforming capacity of human papovavirus BK, *J. Virol.* **42**:978–985.
129. Pagnani, M., Corallini, A., Caputo, A., Altavilla, G., Selvatici, R., Cattozzi, L., Possati, L., and Barbanti-Brodano, G., 1988, Co-operation in cell transformation between BK virus and the human c-Harvey-*ras* oncogene, *Int. J. Cancer* **42**:405–413.
130. Haukland, H. H., Vonen, B., and Traavik, T., 1992, Transformed rat pancreatic islet-cell lines established by BK virus infection *in vitro*, *Int. J. Cancer* **51**:79–83.
131. van der Noordaa, J., van Strien, A., and Sol, C. J. A., 1986, Persistence of BK virus in human foetal pancreas cells, *J. Gen. Virol.* **67**:1485–1490.
132. Corallini, A., Pagnani, M., Viadana, P., Silini, E., Mottes, M., Milanesi, G., Gerna, G., Vettor, R., Trapella, G., Silvani, V., Gaist, G., and Barbanti-Brodano, G., 1987, Association of BK virus with human brain tumors and tumors of pancreatic islets, *Int. J. Cancer* **39**:60–67.
133. Shah, K. V., Hudson, J. C., Valis, J., Strandberg, D., 1976, Experimental infection of human foreskin cultures with BK virus, a human papovavirus, *Proc. Soc. Exp. Biol. Med.* **153**:180–186.
134. Purchio, A. F., and Fareed, G. C., 1979, Transformation of human embryonic kidney cells by human papovavirus BK, *J. Virol.* **29**:763–769.
135. Vasavada, R., Eager, K. B., Barbanti-Brodano, G., Caputo, A., and Ricciardi, R. P., 1986, Adenovirus type 12 early region 1A proteins repress class I HLA expression in transformed human cells. *Proc. Natl. Acad. Sci. USA* **83**:5257–5261.
136. Pater, A., and Pater, M. M., 1986, Transformation of primary human embryonic kidney cells to anchorage independence by a combination of BK virus DNA and the Harvey-*ras* oncogene, *J. Virol.* **58**:680–683.

137. Corallini, A., Gianni, M., Mantovani, C., Vandini, A., Rimessi, P., Negrini, M., Giavazzi, R., Bani, M. R., Milanesi, G., Dal Cin, P., van den Berghe, H., and Barbanti-Brodano, G., 1991, Transformation of human cells by recombinant DNA molecules containing BK virus early region and the human activated c-H-*ras* or c-*myc* oncogenes, *Cancer J.* **4:**24–34.
138. Takemoto, K. K., Linke, H., Miyamura, T., and Fareed, G. C., 1979, Persistent BK papovavirus infection of transformed human fetal brain cells, I. Episomal viral DNA in cloned lines deficient in T-antigen expression, *J. Virol.* **29:**1177–1185.
139. Negrini, M., Castagnoli, A., Pavan, J. V., Sabbioni, S., Araujo, D., Corallini, A., Gualandi, F., Rimessi, P., Bonfatti, A., Giunta, C., Sensi, A., Stanbridge, E. J., and Barbanti-Brodano, G., 1992, Suppression of tumorigenicity and anchorage-independent growth of BK virus-transformed mouse cells by human chromosome 11, *Cancer Res.* **52:**1297–1303.
140. Gualandi, F., Morelli, C., Pavan, J. V., Rimessi, P., Sensi, A., Bonfatti, A., Gruppioni, R., Possati, L., Stanbridge, E. J., and Barbanti-Brodano, G., 1994, Induction of senescence and control of tumorigenicity in BK virus transformed mouse cells by human chromosome 6, *Genes Chrom. Cancer* **10:**77–85.
141. Sabbioni, S., Negrini, M., Possati, L., Bonfatti, A., Corallini, A., Sensi, A., Stanbridge, E. J., and Barbanti-Brodano, G., 1994, Multiple loci on human chromosome 11 control tumorigenicity of BK virus transformed cells, *Int. J. Cancer* **57:**185–191.
142. Howley, P. M., and Martin, M. A., 1977, Uniform representation of the human papovavirus BK genome in transformed hamster cells, *J. Virol.* **23:**205–208.
143. Beth, E., Giraldo, G., Schmidt-Ullrich, R., Pater, M. M., Pater, A., and Di Mayorca, G., 1981, BK virus-transformed inbred hamster brain cells. I. Status of the viral DNA and the association of BK virus early antigens with purified plasma membranes, *J. Virol.* **40:**276–284.
144. Meneguzzi, G., Chenciner, N., Corallini, A., Grossi, M. P., Barbanti-Brodano, G., and Milanesi, G., 1981, The arrangement of viral integrated DNA is different in BK virus-transformed mouse and hamster cells, *Virology* **111:**139–153.
145. Grossi, M. P., Corallini, A., Meneguzzi, G., Chenciner, N., Barbanti-Brodano, G., and Milanesi, G., 1982, Tandem integration of complete viral genomes can occur in nonpermissive hamster cells transformed by linear BKV DNA with cohesive ends, *Virology* **120:**500–503.
146. Yogo, Y., Furuno, A., Watanabe, S., and Yoshiike, K., 1980, Occurrence of free, defective viral DNA in a hamster tumor induced by human papovavirus BK, *Virology* **103:**241–244.
147. Nakashatri, H., Pater, M. M., and Pater, A., 1988, Functional role of BK virus tumor antigens in transformation, *J. Virol.* **62:**4613–4621.
148. Dyson, N., Buchkovich, K., Whyte, P., and Harlow, E., 1989, The cellular 107K protein that binds to adenovirus E1A also associates with the large T antigens of SV40 and JC virus, *Cell* **58:**249–255.
149. Kang, S., and Folk, W. R., 1992, Lymphotropic papovavirus transforms hamster cells without altering the amount or stability of p-53, *Virology* **191:**754–764.
150. Galloway, D. A., and McDougall, J. K., 1983, The oncogenic potential or herpes simplex viruses: Evidence for a "hit and run" mechanism, *Nature* **302:**21–24.
151. Yogo, Y., Furuno, A., Nozawa, A., and Uchida, S., 1981, Organization of viral genome in a T antigen-negative hamster tumor induced by human papovavirus BK, *J. Virol.* **38:**556–563.
152. Brunner, M., di Mayorca, G., and Goldman, E., 1989, Absence of BK virus sequences in transformed hamster cells transfected by human tumor DNA, *Virus Res.* **12:**315–330.
153. Theile, M., and Grabowski, G., 1990, Mutagenic activity of BKV and JCV in human and other mammalian cells, *Arch. Virol.* **113:**221–233.
154. Ray, F. A., Peabody, D. S., Cooper, J. L., Cram, L. S., and Kraemer, P. M., 1990, SV40 T antigen alone drives karyotype instability that precedes neoplastic transformation of human diploid fibroblasts, *J. Cell. Biochem.* **42:**13–31.
155. Stewart, N., and Bacchetti, S., 1991, Expression of SV40 large T antigen, but not small t antigen, is required for the induction of chromosomal aberrations in transformed human cells, *Virology* **180:**49–57.

156. Fareed, G. C., Takemoto, K. K., and Gimbrone, M. A., 1978, Interaction of simian virus 40 and human papovavirus, BK and JC, with human vascular endothelial cells, in: *Microbiology* (D. Schlessinger, ed.), American Society for Microbiology, Washington, D. C., pp. 427–431.
157. Walker, D. L., and Padgett, B. L., 1978, Biology of JC virus, a human papovavirus, in: *Microbiology* (D. Schlessinger, ed.), American Society for Microbiology, Washington, D.C., pp. 432–434.
158. Frisque, R. J., Rifkin, D. B., and Walker, D. L., 1980. Transformation of primary hamster brain cells with JC virus and its DNA, *J. Virol.* **35**:265–269.
159. Howley, P. M., Rentier-Delrue, F., Heilman, C. A., Law, M. F., Chowdhury, K., Israel, M. A., and Takemoto, K. K., 1980, Cloned human polyomavirus JC DNA can transform human amnion cells. *J. Virol.* **36**:878–882.
160. Bollag, B., Chuke, W. F., and Frisque, R. J., 1989, Hybrid genomes of the polyomavirus JC virus, BK virus, and simian virus 40: Identification of sequences important for efficient transformation, *J. Virol.* **63**:863–872.
161. Haggerty, S., Walker, D. L., and Frisque, R. J., 1989, JC virus-simian virus 40 genomes containing heterologous regulatory signals and chimeric early regions: Identification of regions restricting transformation by JC virus, *J. Virol.* **63**:2180–2190.
162. Trowbridge, P. W., and Frisque, R. J., 1993, Analysis of G418-selected Rat2 cells containing prototype, variant, mutant, and chimeric JC virus and SV40 genomes, *Virology* **196**:458–474.
163. Heritage, J., Chesters, P. M., and McCance, D. J., 1981, The persistence of papovavirus BK DNA sequences in normal human renal tissue, *J. Med. Virol.* **8**:143–150.
164. Chesters, P. M., Heritage, J., and McCance, D. J., 1983, Persistence of DNA sequences of BK virus and JC virus in normal human tissues and in diseased tissues, *J. Infect. Dis.* **147**:676–684.
165. Grinnel, B. W., Padgett, B. L., and Walker, D. L., 1983, Distribution of nonintegrated DNA from JC papovavirus in organs of patients with progressive multifocal leukoencephalopathy, *J. Infect. Dis.* **147**:669–675.
166. McCance, D. J., 1983, Persistence of animal and human papovaviruses in renal and nervous tissues, in: *Polyomaviruses and Human Neurological Disease* (J. L. Sever and D. L. Madden, eds.), Alan R. Liss, New York, pp. 479–481.
167. Israel, M. A., Martin, M. A., Takemoto, K. K., Howley, P. M., Aaronson, S. A., Solomon, D., and Khoury, G., 1978, Evaluation of normal and neoplastic human tissue for BK virus, *Virology* **90**:187–196.
168. Pater, M. M., Pater, A., Fiori, M., Slota, J., and Di Mayorca, G., 1980, BK virus DNA sequences in human tumors and normal tissues and cell lines, in: *Cold Spring Harbor Conferences on Cell Proliferation, Volume 7, Viruses in Naturally Occurring Cancers, Book A* (M. Essex, G. Todaro and H. zur Hausen, eds.), Cold Spring Harbor Laboratory Press, New York, pp. 329–341.
169. Houff, S. A., Major, E. O., Katz, D. A., Kufta, C. V., Sever, J. L., Pittaluga, S., Roberts, J. R., Gitt, J., Saini, N., and Lux, W., 1988, Involvement of JC virus-infected mononuclear cells from the bone marrow and spleen in the pathogenesis of progressive multifocal leukoencephalopathy, *N. Engl. J. Med.* **318**:301–305.
170. Lecatsas, G., Schoub, B. D., Rabson, A. R., and Joffe, M., 1976, Papovavirus in human lymphocyte cultures, *Lancet* **2**:907–908.
171. Schneider, E. M., and Dörries, K., 1993, High frequency of polyomavirus infected in lymphoid cell preparations after allogenic bone marrow transplantation, *Transplant. Proc.* **25**:1271–1273.
172. Portolani, M., Piani, M., Gazzanelli, G., Borgatti, M., Bortoletti, A., Grossi, M. P., Corallini, A., and Barbanti-Brodano, G., 1985, Restricted replication of BK virus in human lymphocytes, *Microbiologica* **8**:59–66.
173. Elsner, C., and Dörries, K., 1992, Evidence of human polyomavirus BK and JC infection in normal brain tissue, *Virology* **191**:72–80.
174. De Mattei, M., Martini, F., Tognon, M., Serra, M., Baldini, N., and Barbanti-Brodano, G., 1994, Polyomavirus latency and human tumors, *J. Infect. Dis.* **169**:1175–1176.
175. Dörries, K., Vogel, E., Günther, S., and Czub, S., 1994, Infection of human polyomavirus JC

and BK in peripheral blood leukocytes from immunocompetent individuals, *Virology* **198:** 59–70.
176. Fiori, M., and Di Mayorca, G., 1976, Occurrence of BK virus DNA in DNA obtained from certain human tumors, *Proc. Natl. Acad. Sci. USA* **73:**4662–4666.
177. Calos, M. P., and Miller, J. H., 1980, Transposable elements, *Cell* **20:**579–595.
178. Dörries, K., Loeber, G., and Meixenberger, J., 1987, Association of polyomavirus JC, SV40 and BK with human brain tumors, *Virology* **160:**268–270.
179. Barbanti-Brodano, G., Pagnani, M., Viadana, P., Beth-Giraldo, E., Giraldo, E., and Corallini, A., 1987, BK virus in Kaposi's sarcoma, *Antibiot. Chemother.* **38:**113–120.
180. Barbanti-Brodano, G., Pagnani, M., Balboni, P. G., Rotola, A., Cassai, E., Beth-Giraldo, E., Giraldo, E., Giraldo, G., and Corallini, A., 1988, Studies on the association of Kaposi's sarcoma with ubiquitous viruses, in: *AIDS and Associated Cancers in Africa* (G. Giraldo, E. Beth-Giraldo, N. Clumeck, Md-R. Gharbi, S. K. Kyalwazi, and G. de Thè, eds.), S. Karger, Basel, pp. 175–181.
181. Shah, K. V., Daniel, R. W., Stone, K. R., and Elliott, A. Y., 1978, Investigation of human urogenital tract tumors for papovavirus etiology: brief communication, *J. Natl. Cancer Inst.* **60:**579–582.
182. Grossi, M. P., Meneguzzi, G., Chenciner, N., Corallini, A., Poli, F., Altavilla, G., Alberti, S., Milanesi, G., and Barbanti-Brodano, G., 1981, Lack of association between BK virus and ependymomas, malignant tumors of pancreatic islets, osteosarcomas and other human tumors, *Intervirology* **15:**10–18.
183. Giraldo, G., Beth, E., Kourilski, F. M., Henle, W., Mike, V., Huraux, J. M., Andersen, H. K., Gharbi, M. R., Kyalwazi, S. K., and Puissant, A., 1975, Antibody patterns to herpesviruses in Kaposi's sarcoma: Serological association of European Kaposi's sarcoma with cytomegalovirus, *Int. J. Cancer* **15:**839–848.
184. Giraldo, G., Beth, E., Henle, W., Helne, G., Mikè, V., Safai, B., Huraux, J. M., Mchardy, J., and De-Thè, G., 1978, Antibody patterns to herpesviruses in Kaposi's sarcoma. II: Serological association of American Kaposi's sarcoma with cytomegalovirus, *Int. J. Cancer* **22:**126–131.
185. Huang, Y. Q., Li, J. J., Rush, M., Poiesz, B. J., Nicolaides, A., Jacobson, M., Zhang, W. G., Coutavas, E., Abbott, M. A., and Friedman-Kien, A. E., 1992, HPV-16-related DNA sequences in Kaposi's sarcoma, *Lancet* **339:**515–518.
186. Nickoloff, B. J., Huang, Y. Q., Li, J. J., and Friedman-Kien, A. E., 1992, Immunohistochemical detection of papillomavirus antigens in Kaposi's sarcoma, *Lancet* **339:**548–549.
187. Bovenzi, P., Mirandola, P., Secchiero, P., Strumia, R., Cassai, E., and Di Luca, D., 1993, Human herpesvirus 6 (variant A) in Kaposi's sarcoma, *Lancet* **341:**1288–1289.
188. Ensoli, B., Nakamura, S., Salahuddin, S. Z., Biberfeld, P., Larsson, L., Beaver, B., Wong-Staal, F., and Gallo, R. C., 1989, AIDS—Kaposi's sarcoma-derived cells express cytokines with autocrine and paracrine growth effects, *Science* **243:**223–226.
189. Ensoli, B., Nakamura, S., Salahuddin, S. Z., and Gallo, R. C., 1989, AIDS-associated Kaposi's sarcoma: A molecular model for its pathogenesis, *Cancer Cells* **1:**93–96.
190. Shah, K. V., and Nathanson, N., 1976, Human exposure to SV40. Review and comment, *Am. J. Epidemiol.* **103:**1–12.
191. Gerber, P., and Kirschstein, R. L., 1962, SV40-induced ependymomas in newborn hamster. I. Virus-tumor relationship, *Virology* **18:**582–588.
192. Girardi, A. J., Sweet, B. H., Slotnick, V. B., and Hilleman, M. R., 1962, Development of tumors in hamsters inoculated in the neonatal period with vacuolating virus, SV40, *Proc. Soc. Exp. Biol. Med.* **109:**649–660.
193. Shein, H. M., and Enders, J. F., 1962, Transformation induced by simian virus 40 in human renal cell cultures. I. Morphology and growth characteristics, *Proc. Natl. Acad. Sci. USA* **48:**1164–1172.
194. Theyle, M., Strauss, M., and Luebbe, L., 1980, SV40-induced somatic mutations: Possible relevance to viral transformation, *Cold Spring Harbor Symp. Quant. Biol.* **44:**377–382.
195. Geissler, E., 1990, SV40 and human brain tumors, *Prog. Med. Virol.* **37:**211–222.

196. Soriano, F., Shelburne, C. E., and Gökcen, M., 1974, Simian virus 40 in a human cancer, *Nature* **249:**421–424.
197. Weiss, A. F., Portmann, R., Fischer, H., Simon, J., and Zang, K. D., 1975, Simian virus 40-related antigens in three human meningiomas with defined chromosome loss, *Proc. Natl. Acad. Sci. USA* **72:**609–613.
198. Tabuchi, K., Kirsch, W. M., and Low, M., 1978, Screening of human brain tumors for SV40 related T-antigen, *Int. J. Cancer* **21:**12–17.
199. Meinke, W., Goldstein, D. A., and Smith, R. A., 1979, Simian virus 40-related DNA sequences in a human brain tumor, *Neurology* **29:**1590–1594.
200. Scherneck, S., Rudolph, M., and Geissler, E., 1979, Isolation of a SV40 like papovavirus from a human glioblastoma, *Int. J. Cancer* **24:**523–532.
201. Krieg, P., Amtmann, E., Jonas, D., Fischer, H., Zang, K., and Sauer, G., 1981, Episomal simian virus 40 genomes in human brain tumors, *Proc. Natl. Acad. Sci. USA* **78:**6446–6450.
202. Krieg, P., and Scherer, G., 1984, Cloning of SV40 genomes from human brain tumors, *Virology* **138:**336–340.
203. Bergsagel, D. J., Finegold, M. J., Butel, J. S., Kupsky, W. J., and Garcea, R. L., 1992, DNA sequences similar to those of simian virus 40 in ependymomas and choroid plexus tumors of childhood. *N. Engl. J. Med.* **326:**988–993.
204. Carbone, M., Pass, H. I., Rizzo, P., Marinetti, M. R., Di Muzio, M., Mew, D. J. Y., and Procopio, A., 1994, Simian virus 40-like DNA sequences in human pleural mesotheliomas, *Oncogene* **9:**1781–1780.
205. Cicala, C., Pompetti, F., and Carbone, M., 1993, SV40-induced mesotheliomas in hamsters, *Am. J. Pathol.* **142:**1524–1533.
206. Tominaga, T., Yogo, Y., Kitamura, T., and Aso, Y., 1992, Persistence of archetypal JC virus DNA in normal renal tissue derived from tumor-bearing patients, *Virology* **186:**736–741.
207. zur Hausen, H., 1991, Papillomavirus/host cell interactions in the pathogenesis of anogenital cancer, in: *Origins of Human Cancer—A Comprehensive Review* (J. Brugge, T. Curran, E. Harlow, and F. McCormick, eds.), Cold Spring Harbor Laboratory Press, New York, pp. 695–705.

5

Association of Simian Virus 40 with Rodent and Human Mesotheliomas

MICHELE CARBONE, PAOLA RIZZO, and HARVEY I. PASS

1. SIMIAN VIRUS 40

1.1. General Characteristics

Simian virus 40 (SV40) is an oncogenic papovavirus capable of inducing tumors in hamsters and of transforming cells from many species in tissue culture.[1] In the past, although useful for basic studies, the SV40-induced hamster tumor model was considered to have only limited relevance in the elucidation of mechanisms for human cancer development. Recently, however, the discovery of relationships between SV40 proteins and human tumor suppressor gene products suggested that DNA tumor viruses could possibly illuminate molecular mechanisms of human carcinogenesis.[2] Moreover, the very recent finding of SV40-like sequences in human tumors[3,4] underscores the direct relevance of SV40 research to human cancer. The SV40 proteins associated with *in vivo* oncogenesis and *in vitro* cell transformation are encoded by the early region of the virus; these proteins are known as the T (tumor) antigens because animals bearing tumors induced by SV40 have antibodies against these viral proteins. Two SV40 tumor antigens are detected in infected or transformed cells, "large T antigen" (Tag) and "small t antigen" (tag), and these two proteins are produced from a single viral gene by differential splicing of an RNA transcript.[1,2] Unlike the oncogenes of the retroviruses, the T antigen oncogenes of SV40 do not have a homologous cellular protooncogene.[5]

1.2. The SV40 Large T Antigen

The SV40 Tag is a 96-kDa polypeptide principally found in the nuclei of SV40-infected and/or-transformed cells. Although it was known that Tag is the primary and prerequisite gene product responsible for cell transformation by SV40, its mechanism of action remained enigmatic for a long time.[1] The Tag induces cellular DNA replication, but this effect does not in itself seem to account for its ability to act as an oncogenic protein.[6] Of greater relevance to cancer development are the mutagenic capabilities of Tag, which cause structural chromosome aberrations, aneuploidy, and point mutations.[2,7] Thus, it has been proposed that Tag may "hit and run" without obligatory continuous expression for transforming activity.[8] The SV40 Tag may promote transformation through its known interactions with several cellular proteins including the binding and inactivation of the products of a number of tumor suppressor genes and related proteins, such as p53, Rb, p130, p107,[2,9,10] and p300.[11] These tumor suppressor gene products are necessary to prevent the cell from cycling, and in the normal cell they must be inactivated through phosphorylation/dephosphorylation events to permit the resting cell to traverse from G_1 to S.[2] By complexing with the p53, Rb, p107, p130, and p300 proteins, Tag may inactivate these cellular proteins and stimulate the cell to enter the S phase; Tag also induces the expression of insulin-like growth factor 1 (IGF-1) and requires a functional IGF-1 receptor for its growth-stimulating effects.[12-14]

1.3. The SV40 Small t Antigen

The SV40 tag is a 17- to 21-kDa protein found predominantly in the cytoplasm of infected and transformed cells.[1] This protein shares 82 amino acids at its amino terminus with Tag, but the remaining 92 amino acids are unique. Small tag inhibits the cellular phosphatase 2A (PP2A) and indirectly alters the phosphorylation state of some of the tumor suppressor gene proteins mentioned above, contributing to their inactivation.[15,16] Furthermore, tag stimulates the MAP kinase,[17] and AP-1, inducing cell cycle progression.[18]

1.4. Wild-Type SV40 Oncogenicity

Papovaviruses, including SV40, normally infect differentiated cells that are growth-arrested, and these quiescent cells must be stimulated to enter S phase for efficient viral DNA replication to occur. Tag and tag, in combination, will stimulate cell division.[18] Transformation may occur when SV40 infects a cell that is unable to support viral replication, and Tag and tag stimulate continuous cycling.

Wild-type SV40 is highly oncogenic in hamsters. Newborn animals are particularly susceptible, usually developing fibrosarcomas at the injection site following subcutaneous (s.c.) inoculation of a low dose of SV40.[19] When newborn hamsters are inoculated with SV40 intracerebrally, they develop ependymomas.[20] Weanling and adult animals may develop fibrosarcomas if injected s.c. with a high dose of virus, $>10^9$ plaque-forming units (pfu), but only with a low tumor incidence and after prolonged incubation periods.[21] When SV40 ($>10^{8.5}$ pfu) is injected intravenously into weanling hamsters, subjecting many cell types to high concentrations of the virus, lymphocytic leukemia, lymphoma, and osteosarcoma will develop at sites distant from the injection.[22] When SV40 is injected intracardially into 21-day-old hamsters, 60% of the animals develop mesotheliomas (Figs. 1 and 2), while the remaining animals develop true histiocytic lymphomas, osteosarcomas, and rarely B-cell lymphomas.[23] These data indicate that only specific cell types are susceptible to SV40 transformation, i.e., mesothelial cells, osteoblasts/osteocytes, macrophages, and B lymphocytes. When high doses of SV40 are injected subcutaneously, fibrosarcomas develop at the site of virus inoculation. Fibrosarcomas do not develop, however, when the virus is injected directly into the bloodstream. Carcinomas, the most common tumors in humans, never develop following SV40 injection, suggesting that epithelial cells may be resistant to SV40 transformation.

The previous discussion highlights the fact that the specific routes of SV40 inoculation used[23] apparently play a key role in the induction of specific hamster tumors, including mesotheliomas. In fact, in previous studies, mesotheliomas were not observed following s.c.,[24] intracerebral,[20] and intravenous[22] inoculation of SV40. Lipotich and colleagues,[25] however, reported the use of an SV40-induced hamster mesothelioma cell line (800TU) derived accidentally when newborn hamsters were injected between the scapulas with SV40. All the other animals in those experiments developed *in situ* fibrosarcomas. A plausible explanation for such an event could be that in one of those s.c. injections, the needle reached the pleura and induced that single hamster mesothelioma (R. C. Moyer, personal communication). Our results suggested that mesothelial cells are very susceptible to SV40 transformation. Because mesotheliomas did not develop following intravenous

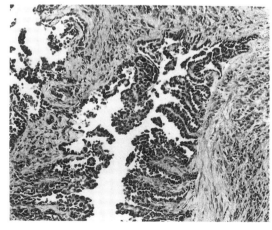

FIGURE 1. Histological appearance of SV40-induced hamster pleural mesothelioma. The morphology and macroscopic, microscopic, and ultramicroscopic appearance are indistinguishable from those of human mesotheliomas (×40).

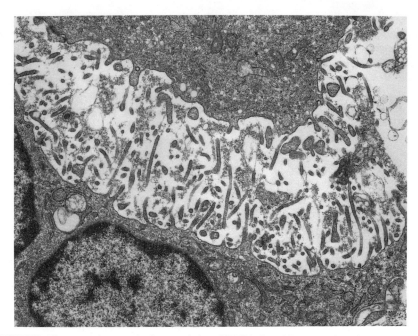

FIGURE 2. Electron microscopic characteristics of SV40-induced hamster pleural mesothelioma. Note the presence of abundant long branching microvilli characteristic of mesothelial cells in both rodents and humans (×24,000).

inoculation of SV40, we speculate that SV40 reached the mesothelial cells of the pleura and pericardium directly through the external surface of the needle used to inject the virus into the left ventricle of the heart. To test the possibility that mesothelial cells were particularly susceptible to SV40 transformation, we injected SV40 directly into the pleural and peritoneal cavities of 21-day-old hamsters. All of the animals injected intrapleurally and 67% of those injected into the peritoneum developed mesotheliomas 3 to 6 months following SV40 inoculation.[23]

1.5. Oncogenicity of SV40 Small t Antigen Deletion Mutants

Studies to determine the oncogenicity of SV40 tag deletion mutants indicate that these mutants are oncogenic when injected s.c. into newborn hamsters, but the tumors develop only after a prolonged latency (1).[8,26,27] The behavior of the tumors induced by tag mutants differs from that of tumors induced by wild-type SV40 in another important way besides latency: A significant fraction of subcutaneously inoculated hamsters develop tumors distant from the site of injection. Characterization of these distant tumors revealed that they were formed by a subpopulation of well-differentiated macrophages expressing the MAC2 surface antigen,[28] and tumor cells derived from these tumors did not produce mitogenic inhibitors.[29] When injected intracardially with SV40 tag mutants, the majority of the injected animals develop true histiocytic lymphomas; a few develop B-cell lymphomas and osteosarcomas.[30] Only a single mesothelioma developed in hamsters injected intracardially with SV40 tag mutants, suggesting that tag plays a key role during the process of mesothelial cell transformation. The involvement of tag in mesothelial cell transformation is intriguing. It is possible that in addition to physical binding between SV40 Tag and the products of the cellular tumor suppressor genes p53, Rb, p107, p130, and p300, alteration of the phosphorylation state of the products of these tumor suppressor genes by tag is also required to inactivate their function completely and allow the cell to progress to the S phase of the cell cycle, during which transformation occurs.[2] Thus, mesothelial cells may have a very low cycling rate making transformation in this system dependent on tag.

2. THE SV40 VIRUS IN HUMANS

2.1. Human Infection by SV40

There is sparse information regarding SV40 multiplication in the human host. *In vitro*, only human cells are truly *semipermissive*, allowing SV40 growth and cellular transformation. Normally monkey cells are totally *permissive*, and rodent cells are *nonpermissive*; that is, they can only be transformed.[1] The SV40 is a monkey virus that infected the human population through contamination of polio vaccines (both oral and inactivated) from 1956 to 1963.[31] These vaccines were produced in 1954, 8 years before SV40 was first isolated,[32] and were prepared by growing poliovirus in monolayers of cell cultures from rhesus monkey kidney cells. Because SV40 is endogenous to the rhesus monkey, produces no cytopathic effects in rhesus monkey kidney cells, and had not yet been reported, its presence in the cultures was not recognized. It was not until 1960 that high titers of virus were discovered in some lots of the vaccines.[31] Killed as well as attenuated vaccines contained infectious SV40. As a result of this contamination, hundreds of millions of children and adults have been injected with SV40-contaminated polio vaccines.[33,34] Another possible transmission of SV40 could have occurred between 1957 and 1960, when parenteral adenovirus vaccines containing SV40 virions were used extensively in the military and, to a limited extent, in the U.S. civilian population.[33] In addition, Shah *et al.*[35,36] demonstrated a high incidence of SV40-neutralizing antibodies in persons handling rhesus monkeys, and Horvath[37] demonstrated SV40-neutralizing antibodies in the sera of laboratory workers handling rhesus monkey kidney cell cultures. As recently as 1980, more than 150 newborn children were treated with a hepatitis A vaccine that was contaminated with SV40.[8] Humans who were injected with SV40-contaminated polio vaccines (Salk) developed SV40-neutralizing antibody that persisted for at least 3 years.[38] No SV40-inactivating antibodies were detected in children who received contaminated live oral attenuated (Sabin) polio vaccines, although some of the children excreted infectious virus in their stools for 3 to 5 weeks after ingestion of the vaccine.[39]

In 1961, Morris *et al.*[40] demonstrated that adult volunteers infected intranasally with SV40 developed subclinical infection as determined by the isolation of the virus from throats swabs and the presence of SV40-neutralizing antibody in convalescent serum specimens. Although these results suggest that people infected with SV40 may become infectious to others, this possibility has never been investigated. Two years after the isolation of SV40 from human polio vaccines, Eddy described the development of tumors in hamsters injected subcutaneously with SV40.[24,32] At the same time, it was recognized that SV40 could cause transformation of human cells in culture,[41-43] and later Jensen et al.,[44] reported that SV40-transformed human tissue culture cells could cause tumors in humans following s.c. injection of these transformed cells. Hence, the agencies responsible for the development and safety of these vaccines were confronted with the problem of millions of people exposed to a tumor virus by inoculation or ingestion of inadvertently contaminated vaccines. The absence of an antibody response in children receiving oral contaminated polio vaccines and the inability to induce tumors in hamsters given SV40 by mouth, suggest that only the parenteral polio vaccines (Salk) may be potentially oncogenic for humans.[33,34]

2.2. Follow-up Studies of Populations Injected with SV40-Contaminated Polio Vaccines

No illnesses attributable to SV40 were detected during the initial studies of patients receiving the contaminated poliovirus or adenovirus vaccines.[33,34] It is surprising that, given the relevance of the matter to the public health, only a few short-term epidemiologic studies have been performed to determine whether recipients of the contaminated vaccines, or the children of pregnant women receiving the vaccine, have developed a higher-than-chance number of neoplasms,[33,34,45] and only one report describes the long-term follow-up in inoculated patients.[46] This report involves only 1073 children and has only a 19-year follow-up (with the known latency period of mesothelioma, for example, being 20–50 years following asbestos exposure). In addition, most of the children studied (920) received SV40 by mouth, and as mentioned above, it is debatable whether oral administration produces a systemic infection.[33,34] Thus, the study of Mortimer et al.[46] must be interpreted with reservations, because an oncogenic rate below 1:1073, or 1:153 if only the children inoculated with the Salk vaccine are considered, would not have been detected. Nevertheless, on the basis of a 19-year follow-up study of 1073 children injected with SV40 that showed no increased cancer incidence, it was concluded that SV40 was not oncogenic for humans.[46] Geissler[8] compared the occurrence of cancer in Germany in persons born in 1959–1961 to that of persons born in 1962–1964, a majority of whom received SV40-free vaccines, and concluded that people treated with polio vaccines presumably contaminated with SV40 did not develop more tumors within 22 years after vaccination than did those who had received SV40-free vaccine. However, although this study would have readily detected a difference in the incidence among the most common human cancers, i.e., carcinomas, it would have probably not have detected differences among rarer tumors such as mesotheliomas, in which additional factors—i.e., exposure to asbestos—play a key role in tumor development.

2.3. SV40-like Sequences in Human Tumors

A number of isolated reports during the past 30 years link SV40 to human diseases, including progressive multifocal leukoencephalopathy[47] and cancer development.[8,48–50] Questions regarding the possible source of SV40 contamination, along with the limitations of molecular biological techniques of 20 years ago [Southern blot and immunofluorescence versus the more sensitive polymerase chain reaction (PCR) and immunohistochemical techniques available presently] prevented scientists from clearly and definitely demonstrating their findings. These limitations resulted in a climate of skepticism in the scientific community regarding the possibility of SV40 as a human pathogen. Moreover, the designation of SV40 as oncogenic for humans, based on the relatively limited data available, could have raised serious concerns for millions of people injected with the contaminated vaccines and implied the requirement for special facilities (P-3 laboratories) in order for scientists to use SV40 and/or molecular reagents containing SV40 genes.

It was not until 1992 that a study published in *The New England Journal of Medicine* by Bergsagel et al.[3] convincingly demonstrated the presence and the expression of SV40-like sequences in human ependymomas and choroid plexus tumors using PCR, sequencing, and immunoperoxidase experiments. More recently, J. Lednicky and collaborators[51] succeeded in isolating an SV40-like virus from one human choroid plexus tumor and demonstrated that it was a wild-type strain of SV40 not previously isolated.

When Bergsagel et al.[3] reported SV40-like sequences in human brain tumors, we had just discovered that SV40 could induce mesotheliomas in 100% of hamsters injected intrapleurally and in more than 60% of those injected intracardially or intraperitoneally.[23] Viruses had not been previously associated with the development of mesotheliomas in mammals, and the epidemiology of mesotheliomas suggested the existence of as yet unknown cocarcinogens acting in concert with asbestos in mesothelioma carcinogenesis. Could SV40 or a similar virus be related to the enormous increase in incidence in mesotheliomas (from virtually no cases to 2000–3000 cases per year in the United States alone in the last 50 years)?

3. MESOTHELIOMAS

3.1. General Characteristics

Mesotheliomas are tumors originating from the serosal lining of the pleural, pericardial, and peritoneal cavities.[52–54] Mesotheliomas are among the most aggressive human tumors. Survival after diagnosis is usually less than 1 year, and none of the current therapeutic approaches has been shown to alter the natural history of this disease.[54–58] The entire field of pathology as it was known in the 1930s was reviewed in 30 volumes of the *Handbuch der Speziellen Patologischen Anatomie und Histologie*. Three volumes of the *Handbuch* were devoted to lung diseases. It is of note that if mesotheliomas existed prior to 1930, the cases were so sporadic that the authors of the *Handbuch* could not state definitively whether the disease did or did not exist.[59] Physicians can diagnose a disease only when they know it exists, so one explanation for the rise in incidence is more widespread appreciation of the clinical and pathological features of mesothelioma. To address this possibility, Mark and Yokoi[60] reviewed the autopsy records of the Massachusetts General Hospital from 1896 to 1991 (47,000 autopsies). Their study revealed that of the mesotheliomas recorded at autopsy during that 95-year period, all occurred in the last 45 years, and none occurred in the first 50 years. Thus, it has been suggested that mesotheliomas might be considered a new disease [H. L. Stewart, personal communication, and (60)]. The continued rise in incidence of mesotheliomas, to 2000–3000 cases per year in the United States,[53,56] has been related to the widespread commercial use of asbestos during the last 50 years.[60–62] The exact biochemical mechanism whereby asbestos can induce mesothelioma is unclear.[53,63–67] In tissue culture, asbestos fibers can cause mutagenic events, including DNA strand breaks and deletion mutations, through the production of hydroxyl radicals and superoxide anions and alter chromosome morphology and ploidy by mechanically interfering with their segregation during mitosis. Furthermore, macrophages will produce DNA-damaging oxyradicals following phagocytosis of asbestos fibers and elaborate lymphokines, which may depress the host immune response. However, asbestos is not a complete carcinogen for induction of lung carcinomas—it has a multiplicative effect with cigarette smoking, and smoking is not related to mesothelioma development.[68,69] Thus, either asbestos has different effects on different cell types (i.e., complete carcinogen for mesothelial cells, incomplete carcinogen for lung cells) or additional carcinogens must act in concert with asbestos in the process of mesothelial cell transformation.

Although the association between asbestos and mesothelioma is indisputable, fewer than 10% of the people exposed to high doses of asbestos develop mesotheliomas, and approximately half of the reported cases of mesothelioma have no documented exposure to asbestos.[67] In areas where asbestos mines exist, more than 70% of mesothelioma patients have a positive exposure history, but in areas where no substantial asbestos-using industry exists, as few as 10% of patients with mesothelioma have a positive exposure history.[67] These findings, and reports of mesothelioma developing in childhood,[70] or even in utero,[71] suggest that there are probably other unknown factors involved in the pathogenesis of mesothelioma.[72] Thus, researchers have long sought additional carcinogens that might be responsible for mesotheliomas in non-asbestos-exposed individuals or that could render particular individuals more susceptible to the carcinogenic effect of asbestos.[67] SV40 immortalizes human mesothelial cells, but these cells are not oncogenic unless additional DNA alterations are induced.[73,74] Asbestos induces DNA alterations and chromosomal abnormalities and facilitates transformation of mouse cells by plasmid DNA and by SV40 in tissue cultures.[75,76] The importance of existing DNA alterations in the process of SV40 transformation of human cells is underscored by the observation that cell lines established from skin biopsies of patients with Down's syndrome (thus containing chromosomal abnormalities) can be transformed by SV40 with a much higher frequency than control human skin cultures.[77] These reports, together with our own finding that SV40 induces mesotheliomas in hamsters, prompted us to search for SV40 sequences in human mesotheliomas.

3.2. SV40-like Sequences in Human Mesotheliomas

Using an approach similar to the one described by Bergsagel et al.,[3] which demonstrated SV40-like sequences in human ependymomas and choroid plexus tumors, we analyzed the DNAs extracted from 48 human pleural mesotheliomas.[4] In 28 of these mesothelioma cases, DNA was also extracted from lung tissue apparently not infiltrated by the tumor. Additional controls included 53 samples of varying histology. The DNAs were amplified by PCR reactions using the same set of primers used by Bergsagel, which amplify a 172-base-pair (bp) conserved region of Tag containing the Rb-family binding domain. Twenty-nine of the mesothelioma samples contained SV40-like sequences (60%). Control DNAs were essentially negative. Further sequence analyses of the amplified sequences revealed that they were 97% homologous to SV40 and that the amplified region did not contain a 9-bp sequence that is present in other papovaviruses.

Following the publication of these data, we have been able to sequence up to 576 bp of viral DNA corresponding to SV40 Tag from specimens obtained from human mesotheliomas and osteosarcomas. These analyses revealed the presence of different deletions within the intron of Tag in specimens from different tumors (P. Rizzo, unpublished observations). Immunohistochemical analyses revealed that mesothelioma frozen specimens contained a nuclear protein that specifically reacted with an anti-Tag monoclonal antibody (Ab-1, Oncogene Science) (Fig. 3). This same antibody precipitated a 90-kDa protein (the expected molecular weight of Tag) from extracts obtained from frozen mesothelioma specimens. Finally, we demonstrated that mesothelioma patients have anti-Tag antibodies in their serum. We also performed light and electron microscopic studies on the available frozen lung tissues of 26 mesotheliomas. All contained asbestos fibers. Moreover, ten additional patients had a history of heavy exposure to asbestos. Thus, at least 36 of these 48 mesothelioma patients had been exposed to asbestos. Of the 36 patients exposed to asbestos, 22 had viral sequences in their tumors.[4]

These findings suggest the possibility that asbestos and an SV40-like virus might act independently or in some instances as cocarcinogens during mesothelioma development, and experiments are in progress to test this hypothesis.

4. SHOULD SV40 BE CONSIDERED A POTENTIAL HEALTH HAZARD TO HUMANS?

As previously discussed, SV40 is a DNA virus that is capable of transforming rodent cells in tissue culture, and when these cells are injected into animals, they induce tumors. Compared to other DNA-virus-transformed cells, SV40 is highly oncogenic. SV40-transformed cells induce tumors in both adult syngeneic and allogeneic animals, compared to adenovirus (Ad) type 5-transformed cells, which induce tumors only in immunocompromised animals, and Ad12-transformed cells which induce tumors in adult immunocompetent syngeneic animals but not in allogeneic animals.[78] In addition, the direct injection of SV40 virus particles into hamsters induces tumors of different histology depending on the route of injection

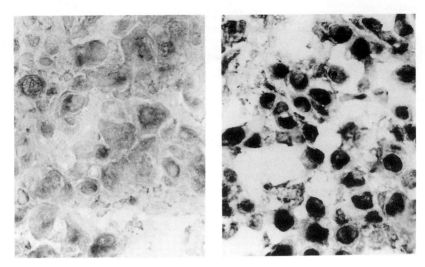

FIGURE 3. Histochemical reactivity of Tag in human mesothelioma biopsy. **Left**, control (×63) immunoperoxidase staining using the monoclonal antibody Trp E (control for the Ab-1 series, Oncogene Science); right, immunoperoxidase staining (×63) with the monoclonal anti-Tag Ab-1 (Oncogene Science), showing nuclear staining of mesothelioma cells.

used (see Section 1.4.). SV40 is capable of infecting human cells: These cells may become transformed, especially if they contain preexisting DNA alterations.[77] Human cells transformed by SV40 may produce tumors when injected into men[44] or may require additional DNA alterations to become oncogenic.[74] These data suggest that human infection by SV40, or injection of SV40 into humans, should raise some health concerns. In addition, the recent finding of wild-type SV40 from one choroid plexus tumor[51] suggests that SV40 should be considered potentially oncogenic for humans.

It should be noted that in rodents SV40 induces only specific tumor types, ependymomas and choroid plexus tumors,[20] mesotheliomas,[23] bone tumors, true histiocytic lymphomas, and, rarely, B-cell lymphomas,[22,28,30] depending on the route of inoculation. Fibrosarcomas are induced only with high titers of virus delivered subcutaneously.[19,24] In humans SV40-like sequences have been demonstrated in the same tumor types, i.e., ependymomas, choroid plexus tumors,[3] and mesotheliomas.[4] In addition, we recently analyzed 349 different human tumor specimens and found that, in addition to mesotheliomas, ependymomas, and choroid plexus tumors, only human bone tumors contained SV40-like sequences.[79] This study has also revealed a different incidence of SV40-like sequences in osteosarcoma patients from different countries.

We suggest that, for reasons as yet unclear, only particular cell types are susceptible to SV40 transformation in both hamsters and humans. Because these tumors are rare in men, it is not surprising that most studies for possible SV40 footprints were negative. The presence of viral footprints does not *per se* establish a relationship of cause and effect. In other words, SV40-like sequences or SV40 virus could be accidentally associated with some human cancers without having a causative role. We find it hard to believe, however, that the presence and the expression of SV40 or SV40-like Tag sequences in human cells can be totally harmless. The effects of Tag and tag on a number of cellular genes that regulate cell growth are discussed in Sections 1.1 and 1.2, and it is generally accepted that human cancer is a multistep process in which several factors play a role: preexisting genetic alterations, smoke, alcohol, asbestos, radiation, etc.

We suggest that the problem of SV40 oncogenicity in humans cannot be studied as an all-or-nothing phenomenon. Just as a minority of people exposed to asbestos develop cancer, the corollary may apply with SV40 in that only a fraction of people exposed to the virus might be at risk. By studying the overall increase of cancer incidence in people injected or otherwise exposed to SV40, we may have taken an excessively simplistic approach. Instead, people exposed to SV40 could have a higher risk of developing specific rare malignancies, such as mesotheliomas, when exposed to asbestos or other carcinogens, or these malignancies may develop in exposed individuals who are also carriers of some as yet undescribed genetic abnormality. It may also be that the Tag as a result of SV40 infection sets off a series of events altering the normal balance of cell cycle events at the tumor suppressor gene level. In support of this hypothesis, we recently found that mesotheliomas expressing Tag also express high levels of the cellular p53 oncogene.[80] We found that p53 was wild-type in these mesothelioma specimens; however, it was possible that the p53 was inactive because it was physically associated with Tag. Previous studies had reported the surprising finding of a low incidence of mutations of p53 in mesotheliomas. Mutations in p53 were detected in fewer than 30% of the mesothelioma studied.[81] This was an unexpected finding because p53 mutations have been repeatedly associated with aggressive tumor phenotypes,[82] and mesotheliomas are among the most aggressive human cancers. Also, mesotheliomas are very resistant to radio- and chemotherapy,[53,55] and tissue culture experiments suggest that cells containing wild-type p53 should be sensitive to radiochemotherapy and that cells containing mutated p53 should be resistant.[83] Our findings that Tag binds and inactivates p53 in approximately 60% of human mesotheliomas provides a rationale for the results discussed above, i.e., that p53 is wild-type in most mesotheliomas but because it is physically associated with Tag, it may be functionally inactive.[80]

Finally, is it true that the finding of viral footprints in a human tumor does not establish a cause-and-effect relationship? It is probably just as provocative to say that the inability to find viral footprints in a tumor does not exclude a viral etiology! Most scientists, however, will refuse to consider the possibility that a given virus contributes to cancer development unless footprints of the virus are found in the tumor. The implications of SV40 as an etiological agent in human cancer may be such a difficult-to-digest phenomenon, overshadowed by the specter of monumental public health considerations, as well as implications for the molecular biological technology industry, that it will take years to sort out its relevance. The study of this phenomenon is self-limited in a way, also, because the tumors associated with SV40-like DNA are fairly rare, and only certain centers will have the availability of tissue necessary for study of the occurrence.

5. CONCLUSIONS

The data discussed suggest to us that SV40 should be considered potentially oncogenic for humans and that precautions should be taken by investigators who handle SV40. The detection of SV40-like sequences in a portion of human mesotheliomas, osteosarcomas, ependymomas, and choroid plexus tumors opens the possibility for designing new therapeutic approaches for these malignancies. Specifically, immunotherapeutic approaches aimed at producing T lymphocytes directed against viral antigens expressed on the surface of the malignant cells, such as Tag,[84] and/or genetic approaches aimed at interfering with the transforming function of Tag. Experiments are in progress to test the effects of these experimental therapeutic approaches in rodents. It is hoped to be able to use similar immunogenetic approaches in the future for human cancer patients.

ACKNOWLEDGMENTS. This work was partly supported by The University of Chicago Cancer Research Center grant 7-50386 (to M. C.); by Pfizer-Groton, CT; and by Pfizer-Italy with a fellowship to Dr. Paola Rizzo.

REFERENCES

1. Topp, W. C., Lane, D., and Pollack, R., 1981, Transformation by SV40 and polyoma virus, in: *DNA Tumor Viruses, part 2, revised* (J. Tooze, ed.), Cold Spring Harbor Press, New York, pp. 205–296.
2. Fanning, E., and Knippers, R., 1992, Structure and function of simian virus 40 large tumor antigen, *Annu. Rev. Biochem.* **61**:55–85.
3. Bergsagel, D. J., Finegold, M. J., Butel, J. S., Kupsky, W. J., and Garcea, R., 1992, DNA sequences similar to those of simian virus 40 in ependymomas and choroid plexus tumors of childhood, *N. Engl. J. Med.* **36**:988–993.
4. Carbone, M., Pass, H. I., Rizzo, P., Marinetti, M. R., Di Muzio, M., Mew, D. J. Y., Levine, A. S., and Procopio, A., 1994, Simian virus 40-like DNA sequences in human pleural mesothelioma, *Oncogene* **9**:1781–1790.
5. Carbone, M., and Levine, A. S., 1990, Oncogenes, antioncogenes and the regulation of cell growth, *Trends Endocrinol. Metab.* **1**:248–253.
6. Butel, J. S., 1986, SV40 large T antigen: Dual oncogene, *Cancer Surv.* **5**:343–365.
7. Geissler, E., and Theile, M., 1983, Virus-induced gene mutations of eukaryotic cells. *Hum. Genet.* **63**:1–12.
8. Geissler, E., 1990, SV40 and human brain tumors, *Prog. Med. Virol.* **37**:211–222.
9. Tiemann, F., and Deppert, W., 1994, Immortalization of Balb/c mouse embryo fibroblasts alters SV40 large T-antigen interactions with the tumor suppressor p53 and results in a reduced SV40-transformation-efficiency, *Oncogene* **9**:1990–1915.
10. Claudio, P. P., Howard, C. M., Baldi, A, De Luca, A., Fu, Y., Condorelli, G., Sun, Y., Colburn, N., Calabretta, B., and Giordano, A., 1994, p130/pRb2 has growth suppressive properties similar to

yet distinctive from those of retinoblastoma family members pRb and p107, *Cancer Res.* **54:** 5556–5560.
11. Avantaggiati, M. L., Carbone, M., Graessmann, A., Howard, B., and Levine, A. S., The SV40 large T antigen and adenovirus E1a interact with different isoforms of the transcriptional coactivator p300, *EMBO J.*, in press.
12. Porcu, P., Ferber, A., Pietrzkowski, Z., Roberts, C. T., Adamo, M., LeRoith, D., and Baserga, R., 1992, The growth-stimulatory effect of simian virus 40 T antigen requires the interaction of insulin-like growth factor 1 with its receptor, *Mol. Cell. Biol.* **12:**5069–5077.
13. Sell, C., Rubini, M., Rubin, R., Liu, J. P., Efstradiatis, A., and Baserga, R., 1993, Simian virus 40 large tumor antigen is unable to transform mouse embryonic fibroblasts lacking type 1 insulin-like growth factor receptor, *Proc. Natl. Acad. Sci. USA* **90:**11217–11221.
14. Valentinis, B., Porcu, P., Quinn, K., and Baserga, R., 1994, The role of the insulin-like growth factor I receptor in the transformation by simian virus 40 T antigen, *Oncogene* **9:**825–831.
15. Scheidtmann, K. H., Mumby, M. C., Rundell, K., and Walter, G., 1991, Dephosphorylation of simian virus 40 large-T antigen and p53 protein by protein phosphatase 2A: Inhibition by small-t antigen, *Mol. Cell. Biol.* **11:**1996–2003.
16. Carbone, M., Hauser, J., Carty, M. C., Rundell, K., Dixon, K., and Levine, A. S., 1992, Simian virus 40 (SV40) small t antigen inhibits SV40 DNA replication in vitro. *J. Virol.* **66:**1804–1808.
17. Sontag, E., Fedorov, S., Kmibayashi, C., Robbins, D., and Mumby, M., 1993, The interaction of the SV40 small t tumor antigen with protein phosphatase 2A stimulates the Map kinase pathway and induces cell proliferation, *Cell* **75:**887–897.
18. Cicala, C., Avantaggiati, M. L., Graessmann, A., Rundell, K., Levine, A. S., and Carbone, M., 1994, Simian virus 40 small t antigen stimulates viral DNA replication in permissive monkey cells, *J. Virol.* **68:**3138–3144.
19. Lewis, A. M., Jr., and Martin, R. G., 1979, Oncogenicity of simian virus 40 deletion mutants that induce altered 17-kilodalton t-proteins, *Proc. Natl. Acad. Sci. USA* **76:**4299–4302.
20. Gerber, P., and Kirschstein, R. L., 1962, SV40-induced ependymomas in newborn hamsters, *Virology* **18:**582–588.
21. Allison, A. C., Chesterman, F. C., and Baron, S., 1967, Induction of tumors in adult hamsters with simian virus 40, *J. Natl. Cancer Inst.* **38:**567–577.
22. Diamandopoulous, G. T., 1972, Leukemia, lymphoma and osteosarcoma induced in the Syrian golden hamster by simian virus 40, *Science* **176:**73–75.
23. Cicala, C., Pompetti, F., and Carbone, M., 1993, SV40 induces mesotheliomas in hamsters, *Am. J. Pathol.* **142:**1524–1533.
24. Eddy, B. E., Simian virus 40: An oncogenic virus, 1964, *Prog. Exp. Tumor Res.* **4:**1–26.
25. Lipotich, G., Moyer, M. P., and Moyer, R. C., 1982, Rescue of SV40 following transfection of TC7 cells with cellular DNAs containing complete and partial SV40 genomes, *Mol. Gen. Genet.* **186:**78–81.
26. Carbone, M., Pompetti, F., Cicala, C., Nguyen, P., Dixon, K., and Levine, A. S., 1991, The role of small t antigen in SV40 oncogenesis, in: *Molecular Basis of Human Cancer* (C. Nicolini, ed.), Plenum Press, New York, pp. 191–206.
27. Dixon, K., Ryder, B. J., and Burche-Jaffe, E., 1982, Enhanced metastasis of tumors induced by a SV40 small t deletion mutant, 1982, *Nature* **296:**672–675.
28. Carbone, M., Lewis, A. M., Jr., Matthews, B. J., Levine, A. S., and Dixon, K., 1989, Characterization of hamster tumors induced by simian virus 40 small t deletion mutants as true histiocytic lymphomas, *Cancer Res.* **49:**1565–1571.
29. Carbone, M., Kajiwara, E., Patch, C. T., Lewis, A. M., Levine, A. S., and Dixon, K., 1989, Biochemical properties of media conditioned by simian virus 40-induced hamster tumor cells: Correlation with distinct cell phenotypes but not with oncogenicity, *Cancer Res.* **49:**6809–6812.
30. Cicala, C., Pompetti, F., Nguyen, P., Dixon, K., Levine, A. S., and Carbone, M., 1992, SV40 small t deletion mutants preferentially transform mononuclear phagocytes and B-lymphocytes *in vivo*, *Virology* **190:**475–479.

31. Sweet, B. H., and Hilleman, R. M., 1960, The vacuolating virus SV40, *Proc. Soc. Exp. Biol. Med.* **105**:420–427.
32. Eddy, B. E., 1962, Identification of the oncogenic substance in rhesus monkey kidney cell cultures as SV40, *Virology* **17**:65–75.
33. Lewis, A. M., Jr., 1973, Experience with SV40 and adenovirus-SV40 hybrids, in: *Biohazards in Biological Research* (A. Hellman, M. Oxman, and R. Pollack, eds.), Cold Spring Harbor Laboratory Press, New York, pp. 96–113.
34. Shah, K. V., and Nathanson, N., 1976, Human exposure to SV40: Review and comment, *Am. J. Epidemiol.* **103**:1–12.
35. Shah, K. V., Goverdhan, M. K., and Ozer, H. L., 1970, Neutralizing antibodies to SV40 in human sera from south India: Search for additional hosts of SV40, *Am. J. Epidemiol.* **93**:291–297.
36. Shah, K. V., McCrumb, F. R., Jr., Daniel, R. W., and Ozer, H. L., 1972, Serologic evidence for a SV40-like infection in man. *J. Natl. Cancer Inst.* **48**:557–561.
37. Horvath, L. B., 1965, Incidence of SV40 virus neutralizing antibodies in sera of laboratory workers, 1965, *Acta Microbiolol. Acad. Sci. Hung.* **12**:201–211.
38. Gerber, P., 1967, Patterns of antibodies to SV40 in children following the last booster with inactivated poliomyelitis vaccine, *Proc. Soc. Exp. Biol. Med.* **125**:1284–1287.
39. Melnick, J. L., and Stinebaugh, S., 1962, Excretion of vacuolating SV40 virus (papova virus group) after ingestion as a containment of oral poliovaccine, *Proc. Soc. Exp. Biol. Med.* **109**:965–968.
40. Morris, J. A., Johnson, K. M., Aulisio, G. C., Chanock, R. M., and Knight, V., 1961, Clinical and serologic responses in volunteers given vacuolating virus (SV40) by respiratory route, *Proc. Soc. Exp. Biol. Med.* **108**:56–64.
41. Shein, H. M., and Enders, J. F., 1962, Transformation induced by simian virus 40 in human renal cell cultures. Morphology and growth characteristics, *Proc. Natl. Acad. Sci. USA* **48**:1164–1169.
42. Koprowski, H., Ponten, J. A., Jensen, F., Ravdin, R. G., Moorhead, P., and Saksela, E., 1962, Transformation of cultures of human tissue infected with SV40, *J. Cell. Comp. Physiol.* **59**:281–292.
43. Rabson, A. S., Malmgren, R. A., O'Conor, G. T., and Kirschtein, R. L., 1962, Simian vacuolating virus (SV40) infection in cell cultures derived from adult human thyroid tissue, *J. Natl. Cancer Inst.* **29**:1123–1145.
44. Jensen, F., Koprowski, H., Pagano, J. S., Ponten, J., and Ravdin, R. G., 1964, Autologous and homologous implantation of human cells transformed *in vitro* by SV40, *J. Natl. Cancer Inst.* **32**:917–925.
45. Fraumeni, J., Ederer, F., and Miller, R. W., 1963, An evaluation of the carcinogenicity of SV40 in man, *J.A.M.A.* **185**:713–718.
46. Mortimer, E. A., Lepow, M. L., Gold, E., Robbins, F. C., Burton, G. J., and Fraumeni, J. F., 1981, Long-term follow up of persons inadvertently inoculated with SV40 as neonates, *N. Engl. J. Med.* **305**:1517–1518.
47. Weiner, L. P., Herndon, R. M., Narayan, O., Johnson, R. T., Shah, K., Rubinestein, L. J., Preziosi, T. J., and Conley, F. K., 1972, Isolation of virus related to SV40 from patients with progressive multifocal leukoencephalopathy, *N. Engl. J. Med.* **286**:385–390.
48. Soriano, F., Shelburne, C. E., and Gokcen, M., 1974, Simian virus 40 in a human cancer, *Nature* **249**:421–424.
49. Krieg, P., Amtmann, E., Jonas, D., Fisher, H., Zang, K., and Sauer, G., 1981, Episomal simian virus 40 in human brain tumors, *Proc. Natl. Acad. Sci. USA* **78**:6446–6450.
50. Walsh, J. W., Zimmer, S. G., and Perdue, M. L., 1982, Role of viruses in the induction of primary intracranial tumors, *Neurosurgery* **10**:643–662.
51. Lednicky, J., Garcea, R. L., Bergsagel, D. J., and Butel, J., Natural simian virus 40 in human choroid plexus tumors and ependymomas, *Virology*, in press.
52. Henderson, D. W., Shilkin, K. B., Whitaker, D., Attwood, H. D., Constance, T. J., Steele, R. H.,

and Leppard, P. J., 1992, The pathology of malignant mesothelioma, including immunohistology and ultrastructure, in *Malignant Mesothelioma* (D. W. Henderson, K. B. Shilkin, S. P. Le Langlois, and D. Whitaker, eds.), The Cancer Series, Hemisphere Press, New York, pp. 69–139.
53. Pass, H. I., 1994, Contemporary approaches in the investigation and treatment of malignant pleural mesothelioma, *Chest Surg. Clin. North Am.* **4:**497–515.
54. Hammar, S. P., 1994, Pleural diseases, in: *Pulmonary Pathology, 2d ed.* (D. H. Dail and S. P. Hammar, eds.), Springer Verlag, Berlin, pp. 1463–1579.
55. Vogelzang, N. J., Weissman, L. B., Herndon, J. E., Antman, K. H., Cooper, R. M., Corson, J. M., and Green, M. R., 1994, Trimetrexate in malignant mesothelioma: A cancer and leukemia group B phase study, *J. Clin. Oncol.* **12:**1436–1442.
56. Antman, K. H., Pass, H. I., DeLaney, T., and Recht, A., 1993, Benign and malignant mesothelioma, in: *Cancer: Principles and Practice of Oncology, 4th ed.* (V. T. De Vita, S. Hellman, and S. A. Rosenberg, eds.), J. B. Lippincott, Philadelphia, pp. 1489–1508.
57. Mew, D., and Pass, H. I., 1993, Malignant mesotheliomas: A clinical challenge, *Contemp. Oncol.* **3:**50–67.
58. Pass, H. I., and Pogrebniak, H. W., Malignant pleural mesothelioma, 1993, *Curr. Probl. Surg.* **30:**921–1020.
59. Fisher, W., 1931, Die Gewachse der Lunge und des Brustfells, in: *Handbuch der speziellen pathologischen Anatomie und Hisotlogie. Vol. III/3* (F. Henke and O. Lubarsch, eds.), Springer Verlag, Berlin, pp. 509–539.
60. Mark, E. J., and Yokoi, T., 1991, Absence of evidence for a significant background incidence of diffuse malignant mesothelioma apart from asbestos exposure, in: *The Third Wave of Asbestos Disease: Exposure of Asbestos in Place* (P. J. Landrigan and H. Kazemi, eds.), *Ann. N.Y. Acad. Sci.* **643:**196–204.
61. Wagner, J. C., Sleggs, C. A., and Marchand, P. E., 1960, Diffuse pleural mesothelioma and asbestos exposure in the Northwestern Cape Province, *Br. J. Ind. Med.* **17:**260–271.
62. Selikoff, I. J., Hammond, E. C., and Seidman, H., 1980, Latency of asbestos disease among insulation workers in the United States and Canada, *Cancer* **46:**2736–2740.
63. Weitzman, S. A., and Graceffa, P., 1984, Asbestos catalyzed hydroxyl and superoxide radical generation from hydrogen peroxide, *Arch. Biochem. Biophys.* **228:**373–376.
64. Jaurand, M. C., Fleury, J., Monchaux, G., Nebut, M., and Bignon, J., 1987, Pleural carcinogenic potency of mineral fibers (asbestos attapulgite) and their toxicity on cultured cells, *J. Natl. Cancer Inst.* **79:**797–804.
65. Weissman, L. B., and Antman, K. H., 1989, Incidence, presentation and promising new treatments for malignant mesothelioma, *Oncology* **3:**67–72.
66. Mossman, B. T., Bignon, J., Corn, M., Seaton, A., and Gee, J. B. L., Asbestos: Scientific developments and implications for public policy, 1990, *Science* **247:**294–301.
67. Roggli, V. L., Sanfilippo, F., and Shelburne, J., 1992, Mesothelioma, in: *Pathology of Asbestos and Associated Diseases* (V. L. Roggli, S. D. Greenberg, and P. C. Pratt, eds.), Little, Brown, Boston, pp. 383–391.
68. Muscat, J. E., and Wynder, E. L., 1991, Cigarette smoking asbestos exposure and malignant mesothelioma, *Cancer Res.* **51:**2263–2267.
69. Churg, A., 1993, Asbestos, asbestosis and lung cancer, *Mod. Pathol.* **6:**509–511.
70. Cooper, S. P., Fraire, A. E., Buffler, P. A., Greenberg, S. D., and Lanston, C., 1989, Epidemiologic aspects of childhood mesothelioma, *Pathol. Immunopathol. Res.* **8:**276–286.
71. Nishioka, H., Furusho, K., Yasunaga, T., Tanaka, K., Yamanouchi, A., Yokota, T., Ishihara, T., and Nakashima, Y., 1988, Congenital malignant mesothelioma: A case report and electron-microscopic study, *Eur. J. Pediatr.* **147:**428–430.
72. Peterson, J. T., Greenberg, S. D., and Bufler, P. A., Non-asbestos related malignant mesothelioma, 1984, *Cancer* **54:**951–960.
73. Ke, Y., Reddel, R. R., Gerwin, B. I., Reddel, H. K., Somers, A. N. A., McMenamin, M. G., La Veck, M. A., Stahel, R. A., Lechner, J. F., and Harris, C. C., 1989, Establishment of a human in vitro

mesothelial cell model system for investigating mechanisms of asbestos-induced mesothelioma, *Am. J. Pathol.* **134:**979–991.
74. Reddel, R. R., Malan-Shibley, L., Gerwin, B. I., Metcalf, R. A., and Harris, C. C., 1989, Tumorigenicity of human mesothelial cell line transfected with EJ-*ras* oncogene, *J. Natl. Cancer Inst.* **81:**945–948.
75. Appel, J. D., Fasy, T. M., Kohtz, D. S., and Johnson, E. M., 1988, Asbestos fibers mediate transformation of monkey cells by exogenous plasmid DNA, *Proc. Natl. Acad. Sci. USA* **85:**7670–7674.
76. Dubes, G. R., 1993, Asbestos mediates viral DNA transformation of mouse cells to produce multilayered foci. *Proc. Am. Assoc. Cancer Res.* **34:**185.
77. Todaro, G. J., and Martin, G. M., 1967, Increased susceptibility of Down's syndrome fibroblasts to transformation by SV40, *Proc. Soc. Exp. Biol. Med.* **124:**232–1237.
78. Lewis, A. M., Jr., and Cook, J. L., 1985, A new role for DNA virus early proteins in viral carcinogenesis, *Science* **227:**15–20.
79. Carbone, M., Rizzo, P., Giuliano, M. T., Procopio, A., Gephardt, M., Hansen, M., Malkin, D., Pompetti, F., Bushart, G., Levine, A. S., Pass, H. I., and Garcea, R., 1995, Incidence of SV40-like sequences in human tumors. *Proc. Am. Assoc. Cancer Res.* **36:**201.
80. Carbone, M., Rizzo, P., Procopio, A., Mew, D. J. Y., Giuliano, M. T., Steinberg, S. M., Levine, A. S., Grimly, P., and Pass, H. I., 1995, SV40 large T antigen and p53 in human pleural mesothelioma, *J. Cell. Biochem.* **19A**(suppl.):321.
81. Metcalf, R., Welsh, J., Bennet, W., Seddon, M., Lehman, T., Pelin, K., Limainmaa, K., Tammilehto, L., Mattson, K., Gerwin, B., and Harris, C. C., 1992, p53 and Kirsten-*ras* mutations in human mesothelioma cell lines, *Cancer Res.* **52:**2610–2615.
82. Greenblatt, M. S., Bennett, W. P., and Harris, C. C., 1994, Mutations in the p53 tumor suppressor gene: Clues to cancer etiology and molecular pathogenesis, *Cancer Res.* **54:**4855–4878.
83. Loewe, S. W., Ruley, H. E., Jacks, T., and Housman, D. E., 1993, p53 dependent apoptosis modulates the cytotoxicity of anticancer agents, *Cell* **74:**957–967.
84. Bright, R. K., Shearer, M. H., and Kennedy, R. C., 1994, Immunization of BALB/c mice with recombinant SV40 large T tumor antigen induces antibody-dependent cell-mediated citotoxicity against simian virus 40-transformed cells, *J. Immunol.* **153:**2064–2072.

6

Immortalization of Keratinocytes by Human Papillomaviruses

CRAIG D. WOODWORTH and JOSEPH A. DiPAOLO

1. INTRODUCTION

Human papillomaviruses (HPVs) have a unique position in viral carcinogenesis studies. They are ubiquitous, cause chronic infection, and are well adapted to the host immune system. In 1907, Ciuffo demonstrated that the infectious agent for the common wart persisted in filtered homogenates and, thus, clearly established that it was not either a bacterium or protozoan.[1] Subsequently, it was determined that common warts are associated with HPV-1 and -2. Analysis of the genetic drift of HPV-16, the most common HPV associated with cervical cancer, suggests that the viruses evolved slowly.[2] Presently, there are more than 100 types of HPV, including more than 65 that have been characterized and partially cloned.[3] The uniqueness of HPV and nonhuman papillomaviruses (PV) is that animal viruses do not infect humans, and vice versa; thus, there is species specificity. In addition, tissue tropism exists. Thus, a specific HPV type is associated with a specific anatomic site of infection. The cell biology and pathogenesis of different HPV infections has recently been reviewed.[4,5]

The history of animal PVs is perhaps even older than that of the HPVs; both equine and canine PVs followed by the cottontail rabbit papillomavirus (CRPV) and bovine PV (BPV) contributed important animal models for studying PV infections. Shope and Hurst isolated and characterized the first oncogenic DNA virus now referred to as CRPV.[6] Subsequently, Rous and associates demonstrated that

---

CRAIG D. WOODWORTH and JOSEPH A. DiPAOLO • National Cancer Institute, Laboratory of Biology, Bethesda, Maryland 20892.

DNA Tumor Viruses: Oncogenic Mechanisms, edited by Giuseppe Barbanti-Brodano *et al.* Plenum Press, New York, 1995.

Shope PV injected into domestic rabbits often progresses to cancer.[7] Furthermore, this process could be accelerated by the application of either coal tar or a well-characterized carcinogenic polycylic aromatic hydrocarbon, 3-methylcholanthrene. With this model rabbits injected with CRPV also responded to X irradiation, which was shown to contribute to the progression of papillomas to carcinomas. Inadvertently, in the 1950s, therapy of juvenile laryngeal papillomas by X irradiation resulted in carcinomas in some cases. The first epidemiologic report on cancer of the uterus was published by Rigoni-Stern, who concluded that carcinomas were rare in virgins and nuns, thus making the first suggestion that the disease was sexually transmitted. Today it is acknowledged that HPV infections are transmitted primarily by person-to-person contact, and the genital types most frequently by sexual contact. The HPVs have been associated worldwide with 15% of the human cancers, most notably cancer of the uterine cervix, penis, vulva, and anal region; HPV-16, -18, -31, -33, -35, and -39 have been associated with over 90% of high-grade cervical neoplasias. Other genetically distinguishable types of HPVs, such as HPV-6 and -11, have been identified in benign condylomas of the genital region; still other HPV types are found in common cutaneous warts and plantar warts. A subset of HPVs, such as HPV-5 and -8, are associated with skin cancers in patients with epidermodysplasia verruciformis.

Historically, the various HPV types have been classified on sequence homology based on liquid hybridization and reassociation kinetics. Related viruses with less than 50% cross-hybridization to a previously typed HPV were considered a unique type and given a new number in the order of acceptance by the Nomenclature Committee established at the German Cancer Research Center, Heidelberg. When the nucleotide sequence homology was more than 50%, a subtype was assigned, and it was classified with a given HPV prototype. In 1991, the Papillomavirus Nomenclature Committee stated that for an HPV to be recognized as a new type, the entire genome and the nucleotide sequence of the upstream regulatory region, E6, and L1 should demonstrate less than 90% nucleotide sequence homology with established PVs.

2. GENOME ORGANIZATION AND BIOLOGY

The PVs were originally classified as papovaviruses along with polyoma, simian vacuolating virus 40 (SV40), human BK, and JC, because they all have a closed, circular double-stranded DNA genome and are encapsulated in an icosahedral virion. Currently, there is a question whether this taxonomic classification is correct. Papillomaviruses can constitute a distinct group of viruses because there are fundamental differences in the genomic organization and biology that differentiate them from the papovaviruses. For example, the SV40 and polyomavirus group have a capsid of 45 nm, whereas the PV capsid has a diameter of 55 nm. Thus, it is possible to accommodate a 50% larger DNA, approximately 7900–8000 kbp compared to 5250 for SV40 and polyoma, respectively. Furthermore, the genomic organization and pattern of transcription in PVs are different than other papovaviruses. For example, the papovaviruses, polyoma and SV40, transcribe both

DNA strands, whereas in PVs all the open reading frames are encoded and transcribed along only one of the two DNA strands, and the messages for both the early and late functions require interdispersed transcription and processing signals.

The PV genome can be divided into three major regions (Fig. 1). The early region contains several genes that regulate virus replication and transcription. The E1 gene product, in concert with the E2 protein, binds to the origin of replication and promotes replication of the episomal virus DNA.[8–10] The E2 gene encodes several regulatory proteins that transactivate or repress transcription of other PV early genes.[9] The proteins encoded by the E5, E6, and E7 genes facilitate host cell DNA replication and are implicated in cell transformation and immortalization.[12] Because PVs do not encode their own replication enzymes, their ability to stimulate DNA synthesis of the host cells plays a critical role in their life cycle. The late region contains two genes, L1 and L2, that encode the capsid proteins. In addition, the E4 gene is also expressed late in the virus life cycle and alters the cytokeratin network,[13] possibly facilitating release of mature virions from the cell.

The third major region of the PV genome is the noncoding region (NCR), which is comprised of numerous binding sites for host and virus transcription factors.[14,15] This region presumably holds the key to understanding why these

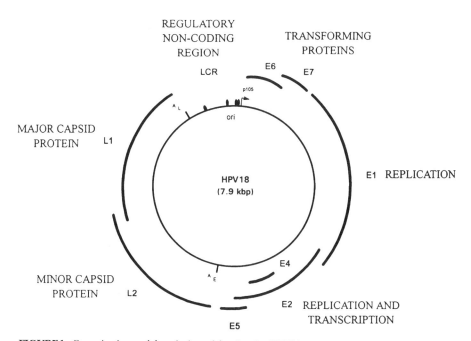

FIGURE 1. Organization and description of the circular HPV-18 genome. Each of the early (E) and late (L) open reading frames is identified by heavy black lines, and major biological functions are indicated. LCR, long control region; p105, early promoter; AE, early polyadenylation site; AL, late polyadenylation site; ori, origin of replication.

viruses exhibit strong species and tissue specificity. One particularly interesting feature of PVs is that the life cycle is intimately regulated by the state of differentiation of the host cell. The viruses appear to require terminally differentiated keratinocytes for vegetative replication, and virions are undetectable in the basal layer. It is presumed, but unproven, that the replication cycle of the PV is synchronized to epithelial cell differentiation. As cells differentiate and migrate from the basal to the spinous and granular stages, episomal DNA replication occurs in the cell nucleus. Only differentiated cells contain complete virus with protein coat; thus, the completed virus is found in the stratum corneum.

The strict requirement for host cell differentiation may contribute to the inability to propagate HPVs efficiently in tissue culture. Although HPV virions have been produced experimentally by transplanting virus-infected cells beneath the renal capsule or within chambers on the dorsal skin of nude mice,[16,17] these methods work with only certain HPV types or yield extremely low levels of virus. More recently, a technique employing organotypic culture has been used to produce low levels of purified virions *in vitro*.[18,19] Refinement of this method may yield higher levels of virus particles and may allow the virus life cycle to be examined by molecular methods.

At one time it was believed that one HPV was responsible for all manifestations of the disease in humans. However, the development of molecular cloning methods enabled isolation of multiple HPV types from human tissues. All HPVs are similar in that they are strictly epitheliotropic; however, HPVs differ widely in their predilection to infect specific anatomic sites. The HPVs are also differentiated according to whether infection results in benign tumors (warts) or progression to dysplasia and invasive cancer. For example, a subset of HPVs are considered high-risk viruses, including HPV-16, -18, -31, and -33, because these specific types are reproducibly associated with cervical cancers. Some HPVs such as HPV-6 and -11 are termed low risk because they are most frequently associated with condylomas that rarely progress to malignancy.

The HPV genome persists as an episomal molecule in the benign and premalignant infections; however, conversion to malignancy usually is accompanied by integration of the virus DNA into the host cell genome.[20] Molecular analyses of integrated virus sequences in cervical carcinoma cell lines demonstrated retention and expression of the early genes, E6 and E7.[21] This provides strong circumstantial evidence that these genes provide a growth advantage. For example, HeLa cells, one of the older HPV-containing cell lines, which has been subpassaged for years, still express the HPV-18 E6/E7 oncoproteins. Furthermore, experiments employing antisense oligonucleotides to E7 reaffirm the necessity of continuous high-level expression of E7 for continuous cell proliferation and tumorigenicity.[224-24] Southern analysis of cervical carcinoma cell lines demonstrates that viral integration often occurs within the HPV E1 or E2 genes. Interruption or loss of the E2 gene removes transcriptional repression by the E2 protein and, therefore, results in enhanced expression of E6 and E7.[21] Thus, clones containing integration within E1 and E2 acquire a selective growth advantage.

It has been appropriate and relevant to use recombinant HPV DNA for transfection into human keratinocytes to determine the role of specific viral genes in the process of aberrant differentiation that leads to carcinomas. Recombinant HPV-16 DNA has been used most widely and has been found in most cervical cancer in either integrated or episomal form.

3. IMMORTALIZATON

The transforming or immortalizing ability of PVs has been studied in different *in vitro* models, and, in general, the results differ significantly depending on the specific cell type.[12,25] The natural target for HPV infection is the keratinocyte; therefore, these cells represent a particularly appropriate bioassay to examine the transforming activity of HPVs.

Rheinwald and Green cultured human keratinocytes on an irradiated feeder layer of NIH3T3 cells in the presence of medium containing calf serum.[26] More recently, Boyce and Ham developed a serum-free medium supplemented with bovine pituitary extract that supported clonal growth of keratinocytes.[27] Normal cultures of keratinocytes derived from foreskin or cervical tissue proliferated rapidly for 40 to 60 population doublings when placed in serum-free medium.[28] Eventually, the keratinocytes senesced.

Pirisi *et al.* first demonstrated that transfection of cultured human keratinocytes in a modified serum-free medium (MCDB-153LB) with recombinant HPV-16 DNA resulted in an indefinite life-span and subsequently, immortalization.[28] Dürst *et al.* simultaneously reported similar results.[29] In both experiments, HPV gene expression was regulated by the natural virus promoter and enhancer in the recombinant HPV construct. These initial observations were rapidly extended to HPV-18[30] and, subsequently, to cultures of keratinocytes derived from both ectocervical and endocervical epithelia.[31–33] A particularly interesting observation was that only HPV DNAs associated with anogenital carcinomas (HPV-16, -18, -31,and -31) induced immortality, whereas those associated with benign condylomas did not.[34,35] These observations imply that the virus immortalization function has an important role in development of anogenital cancer.[34] A number of different studies demonstrated that HPV-16-immortalized cells were nontumorigenic when injected into nude mice; however, two laboratories have reported that keratinocytes immortalized with HPV-18 may eventually become tumorigenic when maintained in culture for extended periods.[36,37]

The HPV-immortalized cell lines resembled normal keratinocytes in both growth rate and morphology. However, they exhibited an aberrant response to induction of terminal differentiation by agents such as serum or 1.0 mM calcium.[30,31,35] Southern analysis of HPV-immortalized cell lines detected either single or multiple rearranged copies of the HPV DNA integrated into the host genome.[28,29,31] In this respect, immortal cells resembled cervical carcinoma lines, which retained integrated and transcriptionally active E6 and E7 genes.[21] Northern analysis demonstrated that the cells actively expressed RNAs encoding the HPV early but not late genes. In fact, one cell line immortalized by a recombinant HPV-18 construct that has a cloning site in the E1 gene expressed only E6 and E7 RNAs, indicating the importance of one or both of these genes for immortalization.[34]

Kaur et al.[38] also immortalized normal keratinocytes using a cloned cervical carcinoma DNA containing HPV-16 E6 and E7 open reading frames.

Transfection studies using subgenomic fragments of HPV DNA under the Control of strong heterologous promoters directly proved that both E6 and E7 were necessary for efficient immortalization.[39–41] E7 alone induced an extended life-span, but eventually the cells senesced; E6 alone was not effective on foreskin keratinocytes. Constructs expressing alternately spliced portions of E6 (E6*) could not substitute for E6.[40] Furthermore, the E6/E7 region of HPV-18 was more efficient than the corresponding region of HPV-16,[41] possibly because of increased expression from the HPV-18 long control region.[42] The frequency of immortalization by recombinant HPV-16 was increased by introducing a point mutation within the E2 gene to prevent E2 protein expression.[43] Recently, E6 transfection of mammary epithelia has been shown to be sufficient to immortalize.[44]

Retrovirus-mediated gene transfer of HPV E6 and E7 cDNAs has advantages for studying the role of these genes in immortalization; high-titer recombinant retroviruses efficiently infect keratinocytes, the HPV genes integrate intact, and expression occurs at a high level. With this approach, introduction of HPV-16 or -18 E7 alone results in immortalization of keratinocytes, although at a reduced efficiency compared to the combination of E6 and E7.[45,46] The E6 cDNA alone was not effective.

Retroviruses encoding low-risk HPV-6 constructs (E7 or both E6 and E7) failed to immortalize. However, either the E6 or E7 cDNA of low-risk HPV-6 cooperated with the complementary cDNA (E7 and E6) from high-risk HPV-16 to induce immortality in this assay.[47] Thus, the E6 and E7 oncoproteins from low-risk HPVs exhibited limited activity in keratinocyte immortalization assays. Together, the data indicate that the ability of HPV oncogenes to immortalize keratinocytes depends on the virus type, the level of oncoprotein expression, and to a certain extent the anatomic source of the target cells.

The molecular mechanisms underlying immortalization by HPV oncoproteins involve direct interaction with cell cycle regulatory proteins [reviewed in (48)]. Important clues to the mechanisms of immortalization by HPV were obtained from studying two other DNA tumor viruses, SV40 and adenovirus. The HPV-16 E7 protein contains sequence domains that are conserved within the SV40 T antigen and the adenovirus Ela genes.[49] The SV40 large T antigen and adenovirus Ela protein bind to the product of the tumor suppressor gene Rb via these conserved sequences. Because the accumulation of unphosphorated Rb protein within the nucleus may prevent cells from progressing into S phase,[50] binding and sequestration of Rb by viral oncoproteins may represent an important mechanism for deregulation of cell growth. Münger et al.[51] showed that the HPV E7 protein bound to Rb both *in vitro* and in intact cells. Furthermore, the level of Rb binding by E7 from high-risk HPVs (HPV-16, -18) was greater than that of low-risk types (HPV-6); mutations in E7 that inhibited binding to Rb *in vitro* were inactive in immortalization of keratinocytes.[52] Recent studies have provided insight into the molecular mechanisms through which Rb may regulate cell proliferation. An important consequence of Rb binding to viral oncoproteins such as HPV E7 is the release of transcription faction E2F, previously sequestered by Rb.[53,54] The released E2F directly activates a number of genes involved in growth regulation.

Although E2F-1 is unable to cooperate with E6 to immortalize keratinocytes, it did, in conjunction with E6, complement an E7 mutant that is defective in immortalization and binding to Rb.[55] These results define a molecular sequence of events through which E7 might directly stimulate cell proliferation.

Although the association between the E7 protein and Rb may be important for keratinocyte immortalization, other studies have suggested that E7 may have additional properties critical for this process. Jewers et al.[56] demonstrated that a mutation in the E7 gene that prevents Rb binding does not abrogate the ability by E7 to immortalize keratinocytes. Similarly, CRPV DNA containing E7 mutations that abolish Rb binding are still effective in inducing papilloma formation on rabbit skin.[57] Whether these mutant E7 proteins retain the ability to bind to other Rb-related proteins is unclear.

The product of the tumor suppressor gene p53 can suppress growth and inhibit transformation of cultured cells.[58] In fact, p53 is the most commonly mutated gene that has been detected in a variety of human tumors. Unlike mutant forms of other tumor suppressor genes, many p53 mutations act dominantly in the presence of wild-type p53 and induce transformation.[58] Both the SV40 T antigen and adenovirus E1b oncoprotein bind to the product of the p53 gene.[59] The interaction stabilizes p53, resulting in significantly increased steady-state levels of the protein within cells.[60] Werness et al.[61] identified a related function for HPV-16 E6 protein in mixing assays of E6 and p53 made in reticulocyte lysates. In contrast to SV40 T antigen and adenovirus E1b, which stabilize p53, the HPV E6 protein led to degradation of wild-type p53 protein via a ubiquitin-mediated pathway.[62] Further studies led to identification of E6 sequences that regulated binding, degradation,[63] and cloning of an E6-associated protein that mediated this effect.[64] Subsequently, it was demonstrated that the E6 protein also mediated degradation of p53 within cultured keratinocytes.[65,66] Specifically, expression of the HPV-16 E6 protein was associated with reduction in the half-life of p53 in either HPV-16-immortalized keratinocytes or normal keratinocytes transiently expressing E6. The ability of E6 to catalyze p53 degradation in cells is biologically important because it inhibits p53-mediated transactivation[67] or repression of gene expression.[66] Paradoxically, keratinocytes expressing E6 do not always exhibit a dramatic reduction in steady-state levels of the p53 protein.[65,68] Thus, E6 might target only specific intracellular pools of p53. Furthermore, reconstitution of steady-state levels of p53 protein via retrovirus transfer of wild-type p53 to HPV-immortalized cell lines did not reverse aberrant differentiation or the immortal phenotype.[69] These results suggest that down-regulation of p53 protein levels might be an important early event in the immortalization process, but continued low levels of steady-state p53 protein may not be necessary to maintain the immortal phenotype.

Recently, Kastan et al.[70] reported that p53-mediated growth arrest serves as a cellular checkpoint for repair of DNA damage. DNA-damaging agents induce elevated intracellular levels of p53 posttranscriptionally, which in turn arrests cells in the G_1 stage of the cell cycle, allowing the damaged DNA to be repaired. Kessis et al.[71] have shown that cells transfected with the HPV-16 E6 gene lose this p53-mediated checkpoint and continue to proliferate after exposure to DNA-damaging agents such as ultraviolet light. Recent studies have shown that p53-dependent G_1 arrest involves pRb-related proteins and is also disrupted by the human papillomavirus-16 E7 oncoprotein.[72,73] These observations provide a novel mechanism to explain how overexpression of HPV E6 and E7 might result in cumulative DNA damage.

Although both HPV E6 and E7 oncoproteins are required for efficient immortalization of genital keratinocytes, the role of other HPV early genes in immortalization or transformation is unclear. So far, no role for the HPV E5 oncoprotein in keratinocyte immortalization has been identified; however, E5 is a transforming gene of BPV[74] and binds to a component of the vacuolar ATPases[75] and the receptor for platelet-derived growth factor.[76] The BPV E5 also cooperates with epidermal growth factor (EGF) receptors to induce transformation in rodent cells.[77] High-level expression of HPV-16 E5 by retrovirus-mediated gene transfer was recently shown to induce malignant transformation of mouse keratinocytes.[78] E5 alters expression and recycling of the EGF receptor in human keratinocytes.[79] Furthermore, both HPV-immortalized cells[80] and papillomas[81] often demonstrate overexpression of EGF receptors. However, the role of E5 in cervical cancer is unclear, and the gene is often lost when the HPV is integrated into cervical carcinoma cell lines.

4. DIFFERENTIATION OF HPV-CONTAINING CELLS

Papillomavirus infections are characterized by altered cellular differentiation manifested as papillomatosis, koilocytosis, or dysplasia. The cellular and molecular mechanisms responsible for regulation of differentiation in normal stratified squamous epithelium are not completely understood. Likewise, it is unclear how expression of specific HPV genes results in altered differentiation in infected keratinocytes. The presence of high-risk HPV DNAs in most cervical dysplasias provides strong circumstantial evidence that the virus alters differentiation. The low-risk HPV DNAs associated with benign condylomas do not induce cellular atypia or disorganization.

Kreider *et al.*[16] developed an experimental model demonstrating that HPV-11 infection directly induces condylomatous changes *in vivo*. In these studies small "chips" of cervical epithelium were infected with a filtrate of condylomas and transplanted beneath the renal capsule of nude mice. Several months later the transplanted tissue developed papillomatosis and produced HPV virions. Unfortunately, HPVs associated with cervical cancer cannot be propagated using this model.

McCance *et al.*[82] used an organotypic culture model in which cells grow at the air–liquid interface on collagen rafts to study epithelial differentiation in HPV-immortalized keratinocytes. Normal foreskin keratinocytes formed stratified squamous epithelia resembling normal epidermis *in vivo*. In contrast, HPV-immortalized cells formed dysplastic disorganized epithelia resembling high-grade dysplasia. In addition, dysplastic epithelial cells invaded the underlying layer of collagen.[32] In this system the degree of differentiation of immortalized foreskin or cervical cultures varied between cell lines,[83] and work by Hudson *et al.*[84] suggested that overexpression of E6 might promote dysplastic differentiation.

To assess whether HPV-immortalized cell lines expressed a dysplastic phenotype *in vivo*, a variety of foreskin and cervical cells transfected with high-risk HPVs (HPV-16, -18, -31, -33) were transplanted beneath a skin–muscle flap in nude mice.[85] Most cell lines immortalized by HPV-18 produced high-level dysplastic changes, whereas lines immortalized by other oncogenic HPVs formed a flattened, abnormal epithelium, at least at early passages. However, late-passage immortal cell lines (150 population doublings) formed dysplastic epithelia. Thus, HPV leads to dysplastic differentiation *in vitro* or *in vivo*, with HPV-18 causing dysplastic transformation earlier than other HPV types. This observation is correlated with clinical observations indicating that HPV-18 infection is associated with rapidly growing and aggressive cervical cancer in younger women.[86]

Dürst *et al.*[87] used a slightly different approach to examine differentiation *in vivo*. A suspension of HPV-16-immortalized cells was injected subcutaneously in nude mice. The two cell lines that were analyzed grew transiently and, subsequently, formed small cysts that demonstrated well-differentiated, stratified squamous epithelia. However, *in situ* analysis of HPV-16 gene expression indicated an inverse correlation between cell differentiation and virus gene expression.[87] In particular, HPV genes were not expressed in the well-differentiated cells but were detected at low levels in the proliferating basal cells. Thus, the *in vivo* and *in vitro* observations suggest that keratinocytes often develop aberrant differentiation after immortalization with high-risk HPV DNAs. However, aberrant differentiation and HPV gene transcription are repressed by factors present in the animal host.

An important question emerging from this work is whether specific HPV oncoproteins directly interfere with normal keratinocyte differentiation or whether aberrant differentiation results secondarily from immortalization and continued subpassaging of cells. High-titer recombinant retroviruses encoding HPV E6 and E7 genes, either alone or in combination, were used to answer this question. Retrovirus infection allowed the HPV genes to be expressed rapidly at high levels in the majority of cells and, thus, minimized secondary alterations that might have occurred in previous experiments as a result of subcloning and expanding colonies after transfection with HPV DNA. The immediate effect of high-level expression of HPV-16 or -18 E6 and E7 was to induce keratinocyte proliferation[46] and sustain expression of proliferating cellular nuclear antigen in cells that had begun terminal differentiation.[88] When infected keratinocytes were maintained in organotypic culture, the stratified squamous epithelia exhibited basal cell hyperplasia and contained more epithelial layers[46]; however, there was minimal evidence of aberrant or disorganized differentiation.[46,47,88] Thus, direct expression of HPV oncoproteins led to alterations in cell growth control but were not sufficient to promote dysplastic differentiation. This conclusion is logical because host cell differentiation is required to support HPV late gene expression and virus maturation. Therefore, it is unlikely that HPV oncoproteins would evolve a function that directly inhibits epithelial differentiation.

5. CYTOGENETIC ALTERATIONS IN CERVICAL NEOPLASIA

Transfection of human keratinocytes with HPV-containing plasmids results in acute changes that are also reflected in the chromosome analysis. Metaphases are observed with chromosome breaks, pulverization, endoreduplication, and dicentric and double minutes.[89] The development of chromosomal rearrangements and the establishment of marker chromosomes are indicative of a cell population undergoing progressive changes often associated with either immortality or malignancy. Normal genital epithelial cells transfected with recombinant HPV became permanent cell lines only with oncogenic HPVs that exhibited transcriptionally active viral sequences with the HPV integrated into the cellular genome.[90,91] The most frequent alteration involved chromosome 1.[92-94] The majority of these HPV-immortalized cell lines were nontumorigenic, independent of the origin of the target cells. The evidence suggests that alterations of chromosome 1 are important to escape from terminal cell differentiation and result in selective growth advantage. Further evidence for the involvement of chromosome 1 in the process of immortality has been provided by the analysis of colorectal adenomas.[95] Alteration of chromosome 1, particularly the long arm, has been suggested to modify a gene associated with cell senescence at the region 1q22-31.[96]

Although nonrandom chromosome alterations may represent pathologically relevant lesions and have been demonstrated in hematologic malignancies as well as in a number of solid tumors, studies of invasive carcinoma of the cervix have been limited. In 1986, Atkin[97] reported results of 32 carcinomas of the cervix and found nonrandom involvement of chromosome 1, 11, and either 4 or 5. Again, chromosome 1 was the most frequently involved in structural and numerical alterations.[98] More recently, Sreekantaiah et al.[99] found a correlation between nonrandom changes and histological types, with alterations of chromosome 1 the most predominant. The suggestion was raised that chromosome 1 may represent the primary alteration in cervical neoplasia. However, the death of cytogenetic and molecular data on preinvasive lesions is a major obstacle in focusing on primary alterations and in evaluating chromosome evolution in cervical cancer.

Without exception, HPV-immortalized human foreskin keratinocytes and cervical cells are aneuploid and frequently exhibit alteration of chromosomes.[92-94] Analyses of several foreskin and cervical keratinocyte lines hybridized with radiolabeled HPV-16 DNA indicated that one or two integration sites were present in each cell line with the viral sequences at the junction of chromosome translocations within duplicated chromosomal regions, HSR, or diffusely stained regions. Furthermore, a comparison of the HPV data with those of literature regarding other DNA viruses has led to the conclusion that there is a clustering of viral integration sites within fragile chromosomal sites strongly indicating the existence of nonrandom viral integration in a human genome. The latter provides evidence for chromosome site specificity of viral integration. With the exception of one HPV cell line, all HPV integration sites regionally mapped by *in situ* hybridization in cervical carcinoma correspond to a fragile site and also coincide with the location of a protooncogene.[100,101] This alteration in protooncogene expression caused by the integration of HPV results in *myc* amplification and in increased gene expression.[102,103] Subsequently, additional HPV integrations have been observed in the same region, namely 8q24.1; thus far, 47% of the HPV-16 and -18 integration sites on genital neoplasia examined have mapped to the same position. Thus, this region appears to be a highly preferred target for HPV in genital cancer. The genetic basis for HPV integration and recombination is supported by analysis of peripheral lymphocytes from women with cervical lesions harboring HPV. A significant difference in induction of chromosome alterations compared to control cultures derived from women without cervical lesions occurred after treatment with aphidicolin (Apc).[104] A significant increase in chromosome fragility was observed in peripheral leukocytes from women infected with HPV-16 or -18 associated with premalignant invasive cancer as opposed to those women carrying HPV-6 or -11, which are associated with benign warts.[104] It can be concluded that the data suggest a higher risk and poorer prognosis of the disease associated with high-risk HPVs. In fact, chromosome fragility may represent a sensitive cytological assay for predicting the prognosis of HPV-infected women. The nonrandom association among HPV integration, fragile sites, and protooncogenes suggests a strategy for further molecular studies of anogenital malignancies.

6. RISK FACTORS FOR CERVICAL NEOPLASIA

The two most important factors for cervical cancer are the age of onset of sexual activity and a woman's lifetime number of sex partners. The sum of epidemiologic, experimental, and clinical evidence indicates that HPV is a causal factor of cervical cancer. However, the need for cofactors is appreciated because only a small proportion of the women positive with an oncogenic strain of HPV develop cervical intraepithelial neoplasia or squamous carcinoma. The promoting factors that increase the risk of infection are poorly understood. Host factors and exogenous agents may influence expression of the virus. Depending on the geographic region examined, smoking may have a role in some stage of cervical carcinogenesis.[105,106] A similar conclusion can be stated for oral contraceptives, which appear to be weakly involved in HPV detection and correlate with increased cervical HPV in some studies.[107] The identification of steroid-responsive elements within the viral promoter region and the effect of steroids on growth-regulatory viral genes suggest that steroids may be important in carcinogenesis, although cervical squamous cell carcinomas are not known to be hormone-dependent cancers. An increased risk of cancer from the use of oral contraceptives could also be associated with the presence of estradiol and progesterone receptors in premalignant cells and may be involved in predisposition to further transformation.[108] Nutritional deficiency and immunologic depression are currently being considered as factors important to the development of cervical cancer. Folate deficiency may occur as a result of the use of oral contraceptive hormones; it has been suggested that low red blood cell folate levels may enhance HPV-16 infection.[109]

The U.S. Centers for Disease Control and Prevention has now indicated that invasive cervical cancer is an AIDS-indicator disease.[110] There is evidence that HIV-positive women with invasive cervical cancer have higher recurrence and death rates with shorter intervals of recurrence and death than HIV-negative controls.[111] In a recent report investigating viral coinfection in males with HPV-associated anogenital lesions, it was found that high-grade lesions were predominant in HPV-positive men who were also positive for HIV.[112]

At one time, a model for genital cancer was proposed that involved herpes simplex virus type 2 (HSV-2). Whether HSV-2 has a role in cervical cancer has been debated for at least 20 years; there is evidence on both sides. The problem is that past reports indicate that HSV DNA is not consistently retained in tumor tissue. An interesting epidemiologic report has recently concluded that women positive for HPV have a ninefold increased risk for cervical cancer with evidence of the presence of HSV-2.[113]

Attempts to determine the importance of cofactors utilizing *in vitro* models with human cells and recombinant HPV have been frustrating. In general, there is a marked contrast between the effects of a carcinogen on rodent and human cells.[114] Whereas a number of chemical and physical viral agents are effective in transforming normal rodent cells to the malignant state, they are for the most part ineffective when applied to human cells. The same can be said for cells that have been immortalized with HPV-16 or -18. The addition of diverse chemical carcinogens to HPV-16-immortalized cells has consistently failed to induce malignancy; however, evidence of effects has been obtained as reflected by new chromosomal rearrangements. It should be pointed out that there is one report that an HPV-18-immortalized cell line became tumorigenic following a chemical carcinogen and 12-O-tetradecanoylphorbol-13-acetate.[115]

HIV-1 has been isolated from cervical biopsy material of HIV-1-positive women with cervicitis.[116,117] Attempts to infect cervical epithelial cells that lack CD4 receptors with a variety of HIVs have failed (unpublished data). The virus attached to the cell surface and produced p24 capsid protein, which was lost on trypsinization. Human herpesvirus-6 (HHV-6) is a ubiquitous virus in the general population, and there is preliminary evidence correlating active HHV-6 infection with progression of HIV infection.[118,119] Furthermore, epithelial cells from bronchial mucosa and salivary glands were found to be infected by HHV-6 *in vivo*. Recently, HHV-6 was shown to productively infect genital epithelial cells, transactivate HPV gene expression, and accelerate tumor growth in nude mice.[120] These observations raise the possibility that HHV-6 may also interact with HPV- and HIV-infected women and possibly contribute to the development of squamous cervical carcinoma.

Tissue from patients with cervical dysplasia was examined for the presence of both HHV-6 and HPV DNA sequences (unpublished data). HHV-6 DNA sequences were detected in 6 of 72 cases of squamous cell carcinoma and cervical intraepithelial neoplasia; HPV-16 was also found in four of these HHV-6-positive cases. Simultaneously, normal cervices and biopsies of patients with cervicitis were found to be negative for HHV-6 DNA. *In situ* hybridization on fixed tissue indicated that HPV-positive cervical cancers may have HHV-6 DNA in epithelial cell nuclei, thus supporting a possible relationship between HHV-6 and HPV-16 *in vivo*. Because HHV-6 increases expression of the HPV genome, it may be involved in progression of cervical cancer. The latter suggests that HHV-6 may be a cofactor with HPV-16 in the genesis of some cervical cancers.

Analysis of clinical material has suggested that the activation of Ha-*ras* gene is associated with malignancy or with metastatic lesions.[121] When the v-Ha-*ras* gene was transfected into immortalized HPV-16-containing cervical cells, the transfected cells gene rapidly became malignant.[92] Thus, this was the first *in vitro* model commencing with normal cervical cells that demonstrates progression to squamous cell carcinomas. Histological analyses of the tumors were consistent with their derivation from the squamous columnar junction that represent the target for malignant transformation. Furthermore, the transformation and expression of the *ras* gene induced amplification and rearrangement of HPV-16 sequences; *ras*-transfected cervical cells demonstrated up-regulated expression of the *ras* protein (p21) compared to the parental line.

Because of the confusion concerning the role of HSV-2 in cervical cancer, HSV-2 is being investigated *in vitro*. The lytic nature of HSV-2 precludes use of the intact virus genome in cell culture transformation assays. A recombinant expression vector containing a morphological transformation region from HSV-2 (*Bgl* II N restriction fragment) can induce malignant transformation in hamster and mouse cells but is ineffective on normal human cells, causing neither transformation nor extension of the life-span.[122] However, this recombinant form transfected into HPV-immortalized cells induced frank malignancies. The *Bgl* II N fragment is approximately 8.2 kb and can be divided into three different *Xho* restriction fragments (unpublished data). Slowly growing tumors obtained with a *Xho* II fragment exhibit integrated HSV-2. The relevance of this integration needs to be explored.

Thus, *in vitro* studies confirm the need for other factors in addition to specific HPV types commonly found in cervical dysplasia or carcinomas in order to obtain tumors. By using *in vitro* models, it is now possible to identify the underlying causes associated with advanced cervical intraepithelial neoplasia in cancer.

REFERENCES

1. Ciuffo, G., 1907, Innesto positivo con filtrato di verruca volgare, *G. Ital. Mal. Venereol.* **48**:12–18.
2. Ho, L., Chan, S.-Y., Burk, R. D., Das, B. C., Fujinaga, K., Icenogle, J. P., Kahn, T., Kiviat, N., Lancaster, W., Mavromara-Nazos, P., Labropoulou, V., Mitrani-Rosenbaum, S., Norrild, B., Pillai, M. R., Stoerker, J., Syrjaenen, K., Syrjaenen, S., Tay, S.-K., Villa, L. L., Wheeler, C. M., Williamson, A.-L., and Bernard, H.-U., 1993, The genetic drift of human papillomavirus type 16 is a means of reconstructing prehistoric viral spread and the movement of ancient human populations. *J. Virol.* **67**:6413–6423.
3. van Ranst, M., Tachezy, R., Delius, H., and Burk, R. D., 1993, Taxonomy of human papillomaviruses, *Papillomavirus Rep.* **4**:61–65.
4. zur Hausen, H., and de Villiers, E.-M., 1994, Human papillomaviruses, *Ann. Rev. Microbiol.* **48**:427–447.
5. Lowy, D. R., Kirnbauer, R., and Schiller, J. T., 1994, Genital human papillomavirus infection, *Proc. Natl. Acad. Sci. USA* **91**:2436–2440.
6. Shope, R. E., and Hurst, E. W., 1933, Infectious papillomatosis of rabbits; with a note on the histopathology, *J. Exp. Med.* **58**:607–624.
7. Kreider, J. W., and Bartlett, G. L., 1981, The Shope papilloma–carcinoma complex of rabbits: A model system of neoplastic progression and spontaneous regression, *Adv. Cancer Res.* **35**:81–110.
8. Ustav, M., and Stenlund, A., 1991, Transient replication of BPV-1 requires two viral polypeptides encoded by the E1 and E2 open reading frames, *EMBO J.* **10**:449–457.
9. Mohr, I. J., Clark, R., Sun, S., Androphy, E. J., MacPherson, P., and Botchan, M. R., 1990,

Targeting the E1 replication protein to the papillomavirus origin of replication by complex formation with the E2 transactivator, *Science* **250**:1694–1699.
10. Lambert, P. F., 1991, Papillomavirus DNA replication, *J. Virol.* **65**:3417–3420.
11. McBride, A. A., Romanczuk, H., and Howley, P. M., 1991, The papillomavirus E2 regulatory proteins, *J. Biol. Chem.* **266**:18411–18414.
12. DiPaolo, J. A., Popescu, N. C., Alvarez, L., and Woodworth, C. D., 1993, Cellular and molecular alterations in human epithelial cells transformed by recombinant human papillomaviruses, *Crit. Rev. Oncogen.* **4**:337–360.
13. Doorbar, J., Ely, S., Sterling, J., McLean, C., and Crawford, L., 1991, Specific interaction between HPV-16 E1-E4 and cytokeratins results in collapse of the epithelial cell intermediate filament network. *Nature* **352**:824–827.
14. Sousa, R., Dostatni, N., and Yaniv, M., 1990, Control of papillomavirus gene expression, *Biochim. Biophys. Acta* **1032**:19–37.
15. Turek, L. P., 1994, The structure, function and regulation of papillomaviral genes in infection and cervical cancer, *Adv. Cancer Res.* **44**:305–356.
16. Kreider, J. W., Howett, M. K., Wolfe, S. A., Bartlett, G. L., Zaino, R. J., Sedlacek, T. V., and Mortel, R., 1985, Morphological transformation *in vivo* of human uterine cervix with papillomavirus from condylomata acuminata, *Nature* **317**:639–641.
17. Sterling, J., Stanley, M., Gatward, G., and Minson, T., 1990, Production of human papillomavirus type 16 virions in a keratinocyte cell line, *J. Virol.* **64**:6305–6397.
18. Meyers, C., Frattini, M. G., Hudson, J. B., and Laimins, L. A., 1992, Biosynthesis of human papillomavirus from a continuous cell line upon epithelial differentiation, *Science* **257**:971–973.
19. Dollard, S. C., Wilson, J. L., Demeter, L. M., Bonnez, W., Reichman, R. C., Broker, T. R., and Chow, L. T., 1992, Production of human papillomavirus and modulation of the infectious program in epithelial raft cultures, *Genes Dev.* **6**:1131–1142.
20. Dürst, M., Kleinheinz, A., Hotz, M., and Gissmann, L., 1985, The physical state of human papillomavirus type 16 DNA in benign and malignant genital tumors, *J. Gen. Virol.* **66**:1515–1522.
21. Schneider-Gädicke, A., and Schwarz, E., 1986, Different human cervical carcinoma cell lines show similar transcription patterns of human papillomavirus type 18 early genes, *EMBO J.* **5**:2285–2292.
22. von Knebel-Doeberitz, M., Ottersdorf, T., Schwarz, E., and Gissmann, L., 1988, Correlation of modified human papillomavirus early gene expression with altered growth properties in C4-1 cervical carcinoma cells, *Cancer Res.* **48**:3780–3786.
23. Steele, C., Cowsert, L. M., and Shillitoe, E. J., 1993, Effects of human papillomavirus type 18-specific antisense oligonucleotides on the transformed phenotype of human carcinoma cell lines, *Cancer Res.* **53**:2330–2337.
24. von Knebel-Doeberitz, M., Rittmüller, C., and zur Hausen, H., 1992, Inhibition of tumorigenicity of cervical cancer cells in nude mice by HPV E6-E7 antisense RNA, *Int. J. Cancer* **51**:831–834.
25. Mansur, C. P., and Androphy, E. J., 1993, Cellular transformation by papillomavirus oncoproteins, *Biochim. Biophys. Acta* **1155**:323–345.
26. Rheinwald, J. G., and Green, H., 1975, Serial cultivation of strains of human epidermal keratinocytes: The formation of keratinizing colonies from single cells, *Cell* **6**:331–344.
27. Boyce, S. T., and Ham, R. G., 1983, Calcium regulated differentiation of normal human epidermal keratinocytes in chemically defined clonal culture and serum-free serial culture. *J. Invest. Dermatol.* **81**:33s–40s.
28. Pirisi, L., Yasumoto, S., Feller, M., Doniger, J., and DiPaolo, J. A., 1987, Transformation of human fibroblasts and keratinocytes with human papillomavirus type 16 DNA, *J. Virol.* **61**:1061–1066.
29. Dürst, M., Dzarlieva-Petrusevska, R. T., Boukamp, P., Fusenig, N. E., and Gissmann, L., 1987,

Molecular and cytogenic analysis of immortalized human primary keratinocytes obtained after transfection with human papillomavirus type 16 DNA, *Oncogene* **1**:251–256.
30. Kaur, P., and McDougall, J. K., 1988, Characterization of primary human keratinocytes transformed by human papillomavirus type 18, *J. Virol.* **62**:1917–1924.
31. Woodworth, C. D., Bowden, P. E., Doniger, J., Pirisi, L., Barnes, W., Lancaster, W. D., and DiPaolo, J. A., 1988, Characterization of normal human exocervical epithelial cells immortalized *in vitro* by papillomavirus types 16 and 18 DNA, *Cancer Res.* **48**:4620–4628.
32. Pecoraro, G., Morgan, D., and Defendi, V., 1989, Differential effects of human papillomavirus type 6, 16, and 18 DNAs on immortalization and transformation of human cervical epithelial cells, *Proc. Natl. Acad. Sci. USA* **86**:563–567.
33. Tsutsomi, K., Belaguli, N. Q. S., Michalak, T. I., Gulliver, W. P., Pater, A., and Pater, M. M., 1992, Human papillomavirus 16 DNA immortalizes two types of normal human epithelial cells of the uterine cervix, *Am. J. Pathol.* **140**:255–261.
34. Woodworth, C. D., Doniger, J., and DiPaolo, J. A., 1989, Immortalization of human foreskin keratinocytes by various human papillomavirus DNAs corresponds to their association with cervical carcinoma, *J. Virol.* **63**:159–164.
35. Schlegel, R., Phelps, W. C., Zhang, Y. L., and Barbosa, M., 1988, Quantitative keratinocyte assay detects two biological activities of human papillomavirus DNA and identifies viral types associated with cervical carcinoma, *EMBO J.* **7**:3181–3187.
36. Hurlin, P. J., Kaur, P., Smith, P. P., Perez-Reyes, N., Blanton, R. A., and McDougall, J. K., 1991, Progression of human papillomavirus type 18-immortalized human keratinocytes to a malignant phenotype, *Proc. Natl. Acad. Sci. USA* **88**:570–574.
37. Pecoraro, G., Lee, M., Morgan, D., and Defendi, V., 1991, Evolution of *in vitro* transformation and tumorigenesis of HPV16 and HPV18 immortalized primary cervical epithelial cells, *Am. J. Pathol.* **138**:1–8.
38. Kaur, P, McDougall, J. K., and Cone, R., 1989, Immortalization of primary human epithelial cells by cloned cervical carcinoma DNA containing human papillomavirus type 16 E6/E7 open reading frames, *J. Gen. Virol.* **70**:1261–1266.
39. Hawley-Nelson, P., Vousden, K. H., Hubbert, N. L., Lowy, D. R., and Schiller, J. T., 1989, HPV16 E6 and E7 proteins cooperate to immortalize human foreskin keratinocytes. *EMBO J.* **8**:3905–3910.
40. Münger, K., Phelps, W. C., Bubb, V., Howley, P. M., and Schlegel, R., 1989, The E6 and E7 genes of the human papillomavirus type 16 together are necessary and sufficient for transformation of primary human keratinocytes, *J. Virol.* **63**:4417–4421.
41. Barbosa, M. S., and Schlegel, R., 1989, The E6 and E7 genes of HPV-18 are sufficient for inducing two stage *in vitro* transformation of human keratinocytes, *Oncogene* **4**:1529–1532.
42. Romanczuk, H., Villa, L. L., Schlegel, R., and Howley, P. M., 1991, The viral transcriptional regulatory region upstream of the E6 and E7 genes is a major determinant of the differential immortalization activities of human papillomavirus types 16 and 18, *J. Virol.* **65**:2739–2744.
43. Romanczuk, H., and Howley, P. M., 1992, Disruption of either the E1 and E2 regulatory gene of human papillomavirus type 16 increases viral immortalization capacity, *Proc. Natl. Acad. Sci. USA* **89**:3159–3164.
44. Band, V., DeCaprio, J. A., Delmolino, L., Kulesa, V., and Sager, R., 1991, Loss of p53 protein in human papillomavirus type 16 E6-immortalized human mammary epithelial cells, *J. Virol.* **65**:6671–6676.
45. Halbert, C. L., Demers, G. W., and Galloway, D. A., 1991, The E7 gene of human papillomavirus type 16 is sufficient for immortalization of human epithelial cells, *J. Virol.* **65**:473–478.
46. Woodworth, C. D., Cheng, S., Simpson, S., Hamacher, L., Chow, L., Broker, T. R., and DiPaolo, J. A., 1992, Recombinant retroviruses encoding human papillomavirus type 18 E6 and E7 genes stimulate proliferation and delay differentiation of human keratinocytes early after infection, *Oncogene* **7**:619–626.
47. Halbert, C. L., Demers, G. W., and Galloway, D. A., 1992, The E6 and E7 genes of human

papillomavirus type 6 have weak immortalizing activity in human epithelial cells, *J. Virol.* **66**:2125–2134.
48. Vousden, K. H., 1994, Interactions between papillomavirus proteins and tumor suppressor gene products, *Adv. Cancer Res.* **64**:1–24.
49. Phelps, W. C., Yee, C. L., Münger, K., and Howley, P. M., 1988, The human papillomavirus type 16 E7 gene encodes transactivation and transformation functions similar to those of adenovirus Ela. *Cell* **53**:539–547.
50. Mihara, K., Cao, X.-R., Yen, A., Chandler, S., Driscoll, B., Murphree, A. L., T'Ang, A., and Fung, Y.-K., 1989, Cell cycle-dependent regulation of phosphorylation of the human retinoblastoma gene product, *Science* **246**:1300–1303.
51. Münger, K., Werness, B. A., Dyson, N., Phelps, W. C., Harlow, E., and Howley, P. M., 1989, Complex formation of human papillomavirus E7 proteins with the retinoblastoma tumor suppressor gene product *EMBO J.* **8**:4099–4105.
52. Münger, K., and Phelps, W. C., 1993, The human papillomavirus E7 protein as a transforming and transactivating factor, *Biochim. Biophys. Acta* **1155**:111–123.
53. Chellappan, S., Kraus, V. B., Kroger, B., Münger, K., Howley, P. M., Phelps, W. C., and Nevins, J. R., 1992, Adenovirus E1A, simian virus 40 tumor antigen, and human papillomavirus E7 protein share the capacity to disrupt the interaction between transcription factor E2F and the retinoblastoma gene product, *Proc. Natl. Acad. Sci. USA* **89**:4549–4553.
54. Nevins, J. R., 1992, E2F: A link between the Rb tumor suppressor protein and viral oncoproteins, *Science* **258**:424–429.
55. Melillo, R. M., Helin, K., Lowy, D. R., and Schiller, J. T., 1994, Positive and negative regulation of cell proliferation by E2F-1: Influence of protein level and human papillomavirus oncoproteins, *Mol. Cell. Biol.* **14**:8241–8249.
56. Jewers, R. J., Hildebrandt, P., Ludlow, J. W., Kell, B., and McCance, D. J., 1992, Regions of human papillomavirus type 16 E7 oncoprotein required for immortalization of human keratinocytes, *J. Virol.* **66**:1329–1335.
57. Defeo-Jones, D., Vuocolo, G. A., Haskell, K. M., Hanobik, M. G., Kiefer, D. M., McAvoy, E. M., Ivey-Hoyle, M., Brandsma, J. L., Oliff, A., and Jones, R. E., 1983, Papillomavirus E7 protein binding to the retinoblastoma protein is not required for viral induction of warts, *J. Virol.* **67**:716–725.
58. Vogelstein, B., and Kinzler, K., 1992, p53 function and disfunction, *Cell* **70**:523–526.
59. Sarnow, P., Ho, Y. S., Williams, J., and Levine, A. J., 1982, Adenovirus E1b-58 Kd tumor antigen and SV40 large tumor antigen are physically associated with the same 54 Kd cellular protein in transformed cells, *Cell* **28**:387–394.
60. Lane, D. P., and Crawford, L. V., 1979, T antigen is bound to a host protein in SV40-transformed cells, *Nature* **278**:261–263.
61. Werness, B. A., Levine, A. J., and Howley, P. M., 1990, Association of human papillomavirus types 16 and 18 E6 proteins with p53, *Science* **248**:76–79.
62. Scheffner, M., Werness, B. A., Huibregtse, J. M., Levine, A. J., and Howley, P. M., 1990, The E6 oncoprotein encoded by the human papillomavirus types 16 and 18 promotes degradation of p53, *Cell* **63**:1129–1136.
63. Crook, T., Tidy, J. A., and Vousden, K. H., 1991, Degradation of p53 can be targeted by HPV E6 sequences distinct from those required for p53 binding and transactivation, *Cell* **67**:547–556.
64. Huibregtse, J. M., Scheffner, M., and Howley, P. M., 1993, Cloning and expression of the cDNA for E6-AP, a protein that mediates the interaction of the human papillomavirus E6 oncoprotein with p53, *Mol. Cell. Biol.* **13**:775–784.
65. Hubbert, N. L., Sedman, S. A., and Schiller, J. T., 1992, Human papillomavirus type 16 E6 increases the degradation rate of p53 in human keratinocytes, *J. Virol.* **66**:6237–6241.
66. Lechner, M. S., Mack, D. H., Finicle, A. B., Crook, T., Vousden, K. H., and Laimins, L. A., 1992, Human papillomavirus E6 proteins bind p53 *in vivo* and abrogate p53-mediated repression of transcription, *EMBO J.* **11**:3045–3052.

67. Mietz, J. A., Unger, T., Huibregtse, J. M., and Howley, P. M., 1992, The transcriptional transactivation function of wild type p53 is inhibited by SV40 large T antigen and by HPV-16 E6 oncoprotein, *EMBO J.* **11**:5013–5020.
68. Scheffner, M., Münger, K., Byrne, J., and Howley, P. M., 1991, The state of the p53 and retinoblastoma genes in human cervical carcinoma lines, *Proc. Natl. Acad. Sci. USA* **88**:5523–5527.
69. Woodworth, C. D., Wang, H., Simpson, S., Alvarez-Salas, L. M., and Notario, V., 1993, Overexpression of wild type p53 alters growth and differentiation of normal human keratinocytes but not human papillomavirus-expressing cell lines, *Cell Growth Differ.* **4**:367–376.
70. Kastan, M. B., Onyekwere, O., Sidransky, D., Vogelstein, B., and Craig, R., 1991, Participation of p53 protein in the cellular response to DNA damage, *Cancer Res.* **51**:6304–6311.
71. Kessis, T. D., Slebos, R. J., Nelson, W. G., Kastan, M. B., Plunkett, B. S., Han, S. M., Lorincz, A. T., Hedrick, L., and Cho, K. R., 1993, Human papillomavirus 16 E6 expression disrupts the p53-mediated cellular response to DNA damage, *Proc. Natl. Acad. Sci. USA* **90**:3988–3992.
72. Demers, G. W., Foster, S. A., Halbert, C. L., and Galloway, D. A., 1994, Growth arrest by induction of p53 in DNA damaged keratinocytes is by-passed by human papillomavirus 16 E7, *Proc. Natl. Acad. Sci. USA* **91**:4382–4386.
73. Slebos, R. J. C., Lee, M. H., Plunkett, B. S., Kessis, T. D., Williams, B. O., Jacks, T., Hedrick, L., Kastan, M. B., and Cho, K. R., 1994, p53-dependent G1 arrest involves pRb-related proteins and is disrupted by the human papillomavirus 16 E7 oncoprotein. *Proc. Natl. Acad. Sci. USA* **91**:5320–5324.
74. Schiller, J. T., Vass, W. C., Vousden, K. H., and Lowy, D. R., 1986, E5 open reading frame of bovine papillomavirus type I encodes a transforming gene, *J. Virol.* **57**:1–6.
75. Goldstein, D. J., Finbow, M. E., Andresson, T., McLean, P., Smith, K., Bubb, V., and Schlegel, R., 1991, Bovine papillomavirus E5 oncoprotein binds to the 16K component of vacuolar H^+ ATPases, *Nature* **352**:347–349.
76. Petti, L., and DiMaio, D., 1992, Stable association between the bovine papillomavirus E5 transforming protein and activated platelet-derived growth factor receptor in transformed mouse cells, *Proc. Natl. Acad. Sci. USA* **89**:6736–6740.
77. Martin, P., Vass, W. C., Schiller, J. T., Lowy, D. R., and Velu, T. J., 1989, The bovine papillomavirus E5 transforming protein can stimulate the transforming activity of EGF and CSF-1 receptors, *Cell* **59**:21–32.
78. Leptak, C., Ramon, Y., Cajal, S., Kulke, R., Horwitz, B. H., Riese, D. J., Dotto, G. P., and DiMaio, D., 1991, Tumorigenic transformation of murine keratinocytes by the E5 genes of bovine papillomavirus type 1 and human papillomavirus type 16, *J. Virol.* **65**:7078–7083.
79. Straight, S. W., Hinkle, P. M., Jewers, R. J., and McCance, D. J., 1993, The E5 oncoprotein of human papillomavirus type 16 transforms fibroblasts and effects the downregulation of the epidermal growth factor receptor in keratinocytes, *J. Virol.* **67**:4521–4532.
80. Sizemore, N., and Rourke, E. A., 1993, Human papillomavirus 16 immortalization of normal human ectocervical cells alters retinoic acid regulation of cell growth and epidermal growth factor receptor expression, *Cancer Res.* **53**:4511–4517.
81. Vambutas, A., DiLorenzo, T. P., and Steinberg, B. M., 1993, Laryngeal papilloma cells have high levels of epidermal growth factor receptor and respond to EGF by a decrease in epithelial differentiation, *Cancer Res.* **53**:910–914.
82. McCance, D. J., Kopan, R., Fuchs, E., and Laimins, L. A., 1988, Human papillomavirus type 16 alters human epithelial cell differentiation *in vitro*, *Proc. Natl. Acad. Sci. USA* **85**:7169–7173.
83. Blanton, R. A., Perez-Reyes, N., Merrick, D. T., and McDougall, J. K., 1991, Epithelial cells immortalized by human papillomaviruses have premalignant characteristics in organotypic culture, *Am. J. Pathol.* **138**:673–685.
84. Hudson, J. B., Bedell, M. A., McCance, D. J., and Laimins, L. A., 1990, Immortalization and altered differentiation of human keratinocytes *in vitro* by the E6 and E7 open reading frames of human papillomavirus type 18, *J. Virol.* **64**:519–526.
85. Woodworth, C. D., Waggoner, S., Barnes, W., Stoler, M. H., and DiPaolo, J. A., 1990, Human

cervical and foreskin epithelial cells immortalized by human papillomavirus DNAs exhibit hysplastic differentiation *in vivo, Cancer Res.* **50:**3709–3715.
86. Barnes, W., Delgado, G., Kurman, R. J., Petrilli, E. S., Smith, D. M., Ahmed, S., Lorincz, A. T., Temple, G. F., Jenson, A. B., and Lancaster, W. D., 1988, Possible prognostic significance of human papillomavirus type in cervical cancer, *Gynecol. Oncol.* **29:**267–273.
87. Dürst, M., Bosch, F. X., Glitz, D., Schneider, A., and zur Hausen, H., 1991, Inverse relationship between human papillomavirus (HPV) type 16 early gene expression and cell differentiation in nude mouse epithelial cysts and tumors induced by HPV-positive human cell lines, *J. Virol.* **65:**796–804.
88. Blanton, R. A., Coltrera, M. D., Gown, A. M., Halbert, C. L., and McDougall, J. K., 1992, Expression of the HPV16 E7 gene generates proliferation in stratified squamous cell cultures which is independent of endogenous p53 levels, *Cell Growth Differ.* **3:**791–802.
89. DiPaolo, J. A., Pirisi, L., Popescu, N. C., Yasumoto, S., and Doniger, J., 1987, Progressive changes induced in human and mouse cells by human papillomavirus type-16 DNA, in: *Papillomavirus* (B. N. Steinberg, J. L. Brandsma, and L. B. Taichman, eds.), Cold Spring Harbor Laboratory, New York, pp. 253–257.
90. Pirisi, L., Yasumoto, S., Feller, M., Doniger, J., and DiPaolo, J. A., 1987, Transformation of human fibroblasts and keratinocytes with human papillomavirus type 16 DNA, *Carcinogenesis* **61:**1061–1066.
91. Popescu, N. C., and DiPaolo, J. A., 1990, Integration of human papillomavirus 16 DNA and genomic rearrangements in immortalized human keratinocyte lines. *Cancer Res.* **50:**1316–1323.
92. DiPaolo, J. A., Woodworth, C. D., Popescu, N. C., Notario, V., and Doniger, J., 1989, Induction of human cervical squamous cell carcinoma by sequential transfection with human papillomavirus 16 DNA and viral *ras*, *Oncogene* **4:**395–399.
93. Smith, P. P., Bryant, E. M., Kaur, P., and McDougall, J. K., 1989, Cytogenetic analysis of eight human papillomavirus immortalized human keratinocyte cell lines, *Int. J. Cancer* **44:**1124–1131.
94. Debiec-Rychter, M., Zukowski, M. K., Wang, C. Y., and Wen, W. N., 1991, Chromosomal characterizations of human nasal and nasopharyngeal cells immortalized by human papillomavirus type 16 DNA. *Cancer Genet. Cytogenet.* **52:**51–56.
95. Paraskeva, C., Harvey, A., Finerty, S., and Powell, S. C.,1989, Possible involvement of chromosome 1 in *in vitro* immortalization: Evidence from progression of a human adenoma-derived cell line *in vitro, Int. J. Cancer* **43:**743–746.
96. Sugarawa, O., Oshimura, M., Koi, M., Annab, L. A., and Barrett, J. C., 1990, Induction of cellular senescence in immortalized cells by human chromosome 1, *Science* **247:**707–710.
97. Atkin, B. N., 1986, Chromosome changes in preneoplastic and genital lesions, *Branbury Rep.* **21:**303–310.
98. Atkin, B. N., 1986, Chromosome 1 aberrations in cancer, *Cancer Genet. Cytogenet.* **21:**279–285.
99. Sreekantaiah, C., De Braekeleer, M., and Hass, O., 1991, Cytogenetic findings in cervical carcinoma. A statistical approach, *Cancer Genet. Cytogenet.* **53:**75–81.
100. Popescu, N. C., and DiPaolo, J. A., 1989, Preferential sites for viral integration on mammalian genome, *Cancer Genet. Cytogenet.* **42:**157–173.
101. Popescu, N. C., Zimonjic, D., and DiPaolo, J. A., 1989, Viral integration, fragile sites, and proto-oncogenes in human neoplasia, *Hum. Genet.* **84:**383–386.
102. Lazo, P., DiPaolo, J. A., and Popescu, N. C., 1989, Amplification of viral transforming genes of human papillomavirus-18 and its 5' flanking sequences located near *myc* proto-oncogene in HeLa cells, *Cancer Res.* **49:**4305–4310.
103. Dürst, M., Croce, C. M., Gissman, L., Schwarz, E., and Huebner, K., 1987, Papillomavirus sequences integrate near cellular oncogenes in some cervical carcinomas, *Proc. Natl. Acad. Sci. USA* **84:**1070–1074.
104. Paz-y-Mino, C., Ocampo, L., Narvaez, R., and Narvaez, L., 1992, Chromosome fragility in

lymphocytes of women with cervical uterine lesions produced by human papillomavirus, *Cancer Genet. Cytogenet.* **59**:173–176.
105. Winkelstein, W., 1977, Smoking and cancer of the uterine cervix: Hypothesis, *Am. J. Epidemiol.* **106**:257–259.
106. Brinton, L. A., Schairer, C., Haenszel, W., Stolley, P., Lehman, H. F., Levine, R., and Savitz, D. A., 1986, Cigarette smoking and invasive cervical cancer, *J.A.M.A.* **255**:3265–3269.
107. Vessey, M. P., Lawless, M., McPherson, K., and Yeates, D., 1983, Neoplasia of the cervix uteri and contraception: A possible adverse effect of the pill, *Lancet* **2**:930–934.
108. Monsonego, J., Magdelenat, H., Catalan, F., Coscas, Y., Zerat, L., and Sastre, X., 1991, Estrogen and progesterone receptors in cervical human papillomavirus related lesions, *Int. J. Cancer* **48**:533–539.
109. Butterworth, C. E., Jr., Hatch, K. D., Macaluso, M., Cole, P., Sauberlich, H. E., Soong, S.-J., Borst, M., and Baker, V., 1992, Folate deficiency and cervical dysplasia. *J.A.M.A.* **267**:528–533.
110. Centers for Disease Control and Prevention, 1992, 1993 revised classification system for HIV infection and expanded surveillance case definition for AIDS among adolescents and adults, *Morbid, Mortal. Week. Rep.* **41(RR17)**:1–19.
111. Vermund, S. H., Kelley, K. F., Klein, R. S., and Feingold, A. R., 1991, High risk of human papillomavirus infection and cervical squamous intraepithelial lesions among women with symptomatic human immunodeficiency virus infection, *Am. J. Obstet. Gynecol.* **162**:392–400.
112. Bernard, C., Mougin, C., Madoz, L., Drobacheff, C., van Landuyt, H., Laurent, R., and Lab, M., 1992, Viral co-infections in human papillomavirus-associated anogenital lesions according to the serostatus for the human immunodeficiency virus, *Int. J. Cancer* **52**:731–737.
113. Hildesheim, A., Mann, V., Brinton, L., Szklo, M., Reeves, W. C., and Rawls, W. E., 1991, Herpes simplex virus type 2: A possible interaction with human papillomavirus types 16/18 in the development of invasive cervical cancer, *Int. J. Cancer* **49**:335–340.
114. DiPaolo, J. A., 1983, Relative difficulties in transforming human and animal cells *in vitro*, *Cancer Res.* **70**:3–8.
115. Garrett, L. R., Perez-Reyes, N., Smith, P. P., and McDougall, J. K., 1993, Interactions of HPV-18 and nitrosomethylurea in the induction of squamous cell carcinoma, *Carcinogenesis* **14**: 329–332.
116. Feingold, A. R., Vermund, S. H., Burk, R. D., Kelley, K. F., Schrager, L. K., Schreiber, K., Munk, G., Friedland, G. H., and Klein, R. S., 1990, Cervical cytologic abnormalities and papillomavirus in women infected with human immunodeficiency virus. *J. AIDS* **3**:896–903.
117. Pomerantz, R. J., de la Monte, S. M., Donnegan, S. P., Rota, T. R., Vogt, M. W., Craven, D. E., and Hirsch, M. S., 1988, Human immunodeficiency virus infection of the uterine cervix, *Ann. Intern. Med.* **108**:321–327.
118. Huemer, P. H., Larcher, C., Wachter, H., and Dierich, M. P., 1989, Prevalence of antibodies to human herpesvirus 6 in human immunodeficiency virus 1-seropositive and -negative intravenous drug addicts, *J. Infect. Dis.* **160**:549–550.
119. Levy, J. A., Greenspan, D., Ferro, F., and Lennette, E. R., 1990, Frequent isolation of HHV-6 from saliva and high seroprevalence of the virus in the population, *Lancet* **335**:1047–1050.
120. Chen, M., Popescu, N., Woodworth, C., Berneman, Z., Corbellino, M., Lusso, P., Ablashi, D. V., and DiPaolo, J. A., 1994, Human herpesvirus 6 infects cervical epithelial cells and transactivates human papillomavirus gene expression, *J. Virol.* **68**:1173–1178.
121. Riou, G. F., Barrois, M., Sheng, Z., Duvillard, P., and L'Homme, C., 1988, Somatic deletions and mutations of x-Ha-*ras* gene in human cervical cancers, *Oncogene* **3**:329–333.
122. DiPaolo, J. A., Woodworth, C. D., Popescu, N. C., Koval, D. L., Lopez, J. V., and Dongier, J., 1990, HSV-2-induced tumorigenicity in HPV16-immortalized human genital keratinocytes, *Virology* **177**:777–779.

7

Cooperation between Bovine Papillomaviruses and Dietary Carcinogens in Cancers of Cattle

MARIA E. JACKSON and M. SAVERIA CAMPO

1. INTRODUCTION

In recent years it has become increasingly apparent that carcinogenesis is a multistage process requiring the cooperation of several cofactors. The predictions of Nordling[1] and Armitage and Doll[2] that six or seven successive changes occur during progression to carcinoma have now been confirmed, most notably in the instance of colorectal cancer with the demonstration that mutations of several defined protooncogenes and tumor suppressor genes accumulate during the transition from normal colonic epithelium to carcinoma.[3] Studies in *in vitro* transformation have also demonstrated that the introduction of more than one oncogene into normal diploid cells is required for the acquisition of a malignant phenotype.[4-6] The identification of both genetic predispositions and environmental factors that contribute to the process of carcinogenesis is a major goal of cancer research. In some cases the agents that cooperate in the induction of naturally occurring cancer have been identified. Thus, papillomavirus infection and the presence of bracken fern in the diet have been demonstrated to elicit both alimentary canal and urinary bladder cancers in cattle. This chapter reviews the evidence

---

MARIA E. JACKSON and M. SAVERIA CAMPO • The Beatson Institute for Cancer Research, CRC Beatson Laboratories, Bearsden, Glasgow G61 1BD, Scotland.

DNA Tumor Viruses: Oncogenic Mechanisms, edited by Giuseppe Barbanti-Brodano *et al.* Plenum Press, New York, 1995.

for cooperation between viruses and chemicals in the bovine system and our current understanding of how these factors interact in oncogenesis.

2. THE IDENTIFICATION OF PAPILLOMAVIRUSES AND BRACKEN FERN AS COFACTORS IN TUMORIGENESIS

A high incidence of alimentary tract and urinary bladder cancer in cattle was observed in localized areas of Scotland, Northern England, and the Nasampolai Valley of Kenya, where the animals were grazing on bracken-infested land.[7-10] Urinary bladder cancer could be induced experimentally by feeding cattle on bracken[11-13] or by injecting extracts of bovine cutaneous papillomas into the urinary bladder of cattle.[14] Furthermore, suspensions of naturally occurring, bracken-associated bladder tumors produced papillomas of the skin and vagina as well as polyps and fibromas in the bladder.[15] These experiments suggested the involvement of both bracken and a transmissible virus in naturally occurring bladder cancers. However, the virus type was not identified because at this time the bovine papillomaviruses (BPVs) had not been characterized, and the existence of multiple BPV types was unknown. Six BPVs have now been described,[16-20] which fall into two groups, the fibropapillomaviruses BPV1, BPV2, and BPV5 and the epitheliotropic viruses BPV3, BPV4, and BPV6.[20] More recent studies have shown an association between BPV2 and urinary bladder cancer in bracken-fed cattle. Thus, cattle inoculated on the skin with BPV2 and kept on a hay diet developed cutaneous warts at the inoculation site but not bladder cancers, whereas cattle inoculated on the skin with BPV2 and fed on bracken developed cancers of the urinary bladder as well as cutaneous warts at the site of injection. However, cattle that were fed bracken but not injected with BPV2 also developed urinary bladder cancers. Nevertheless, BPV DNA was found in the cancers from both groups (W. F. H. Jarrett and M. S. Campo, unpublished data), strongly suggesting that latent asymptomatic virus had been activated by the bracken diet,[21] possibly as a result of the chronic immunosuppression caused by bracken feeding.[22] This result was confirmed in an experiment primarily designed to test for synergism between BPV4 and bracken in alimentary tract cancer (see below), in which none of the cattle were experimentally exposed to BPV2, but bladder cancers occurred only in bracken-fed animals (Table I), and in 60% of the animals these cancers were positive for BPV2 DNA.[23,24] Of the naturally occurring bladder cancers, 46% were also found to be positive for BPV2 DNA.[24] A group of cattle that were kept on a hay diet but immunosuppressed with azathioprine developed BPV2-positive hemangiomas of the bladder, again suggesting that immunosuppression leads to activation of latent virus.[21]

A study of naturally occurring alimentary canal cancers in cattle feeding on bracken found that 96% of the cattle had coexisting papillomas at the site of the carcinoma and that a direct transformation could be observed histologically between the benign papilloma and the carcinoma.[9,25] BPV4 was isolated from these papillomas and was thus a strong candidate for the etiologic agent of alimentary canal cancer.[18] In addition, cattle grazing on bracken-infested land often had

TABLE I
Synergism between Viral and Chemical Cofactors in Cattle[a]

Group treatment		Immuno-suppression	Alimentary papillomas	Alimentary carcinomas	Bladder lesions
1(6)	Bracken	+	0	0	5/6 transitional cell carcinoma
					2/6 hemangiosarcoma
					2/6 hemangioendothelioma
					1/6 hemangioma
2(6)	BPV4	−	6/6 +	0	0
3(6)	Bracken BPV4	+	6/6 ++	2/6 esophageal carcinoma	3/6 transitional cell carcinoma
					3/6 hemangiosarcoma
					1/6 hemangioendothelioma
					1/6 hemangioma
					1/6 polyps
4(6)	Azathoprine BPV4	+	6/6 ++	0	6/6 hemangiomas
5(4)	Azathioprine	+	0	0	4/4 hemangiomas
6(2)	Quercetin BPV4	−	2/2 +	0	0
7(4)	Control	—	0	0	0

[a]Figures in parentheses indicate the number of animals in each group; the degree of alimentary papillomatosis is indicated by + and ++. Data are from references 23, 26, and 27.

widespread papillomatosis of the alimentary tract, whereas single papillomas were usual in animals from noncancer areas,[9,25] probably reflecting the immunosuppressed status of cattle feeding on bracken.[22]

As a direct test of the hypothesis that infection by BPV4 and a bracken diet were cofactors in alimentary canal carcinogenesis, cattle were injected in the palate with BPV4 and fed on bracken[23,26] (Table I). Two of six animals developed cancer of the alimentary canal, whereas no such cancers were observed in animals exposed to virus only or to bracken only, strongly suggesting that both virus and bracken were required for malignant progression.[26,27] Furthermore, a group of animals that were inoculated with BPV4 and immunosuppressed with azathioprine developed widespread papillomatosis but not cancer, indicating that increased papillomatosis results from immunosuppressive components of bracken but that additional compounds present in bracken induce the transition from papilloma to carcinoma.

3. CARCINOGENIC AND MUTAGENIC COMPONENTS OF BRACKEN

In addition to the association with urinary bladder and alimentary canal cancer described above, bracken fern ingestion by cattle is known to cause acute toxicity with symptoms similar to those resulting from exposure to ionizing radiation.[28,29] Cattle feeding on bracken have also been shown to have a relatively high incidence of chromosomal abnormalities.[30] The immunosuppressant constituents of bracken may be the sesquiterpene pterosins and pterosides[22] because sesquiterpenes are known to have immunomodulatory effects.[31–33] A wide variety of compounds with carcinogenic and mutagenic properties have been isolated from bracken,[34] including shikimic acid, tannin, ptaquiloside, α-ecdysone, and quercetin. Ptaquiloside, a norsesquiterpene glucoside, induces tumors and hematuria in rats[35,36] and produces chromosomal aberrations in Chinese hamster cells,[37] and breakdown products of ptaquiloside exhibit mutagenic activity in the Ames test.[38] Egyptian toads fed a diet containing α-ecdysone develop neoplastic lesions.[39] The flavonoid quercetin (3,3',4',5,7-pentahydroxyflavone) has been reported to be carcinogenic in rats,[40] is mutagenic in both prokaryotic[41] and eukaryotic cells,[42–45] can act as an initiator *in vitro* in a two-stage transformation assay,[46] and induces chromosomal aberrations *in vitro*.[47] In addition, quercetin activates protein tyrosine kinases and inhibits phosphatidylinositol-3-kinase and phosphatases.[48,49] Quercetin was tested as a cocarcinogen in alimentary canal carcinogenesis by inoculating cattle with BPV4 and administering quercetin orally.[23,26] No alimentary canal cancers were observed; however, these animals were not immunosuppressed, and there is increasing evidence to suggest that immunosuppression is an important factor in some papillomavirus-associated cancers.[50] Nevertheless, *in vitro* experiments indicate that quercetin is able to cooperate with BPV4 in the malignant transformation of primary bovine fibroblasts (see below).

The cellular targets of bracken-derived chemicals in the progression to cancer are not yet known. However, one probable target is the c-Ha-*ras* protooncogene, which is activated in bovine upper alimentary canal carcinomas.[51] The precise nature of the activating mutation has not yet been defined, although it lies outside exons 1 and 2 of the gene,[51] where activating mutations are commonly found at codons 11–13 and 61 in other systems.[52] The finding of an increased cancer risk for carriers of rare mutant alleles of the c-Ha-*ras* minisatellite suggests that mutations outside the coding sequence may be associated with cancer in some cases.[53] A second potential target of bracken fern mutagens is the p53 gene, in which mutations have been detected in BPV4-associated alimentary cancers of cattle (L. Scobie, M. E. Jackson, and M. S. Campo, unpublished data). The multistage nature of carcinogenesis and the multiplicity of bracken chemicals that can contribute to this process make it probable that more than one step in the production of alimentary canal and bladder cancers are elicited by components of the bracken fern.

4. GENOMIC ORGANIZATION OF BPV4

The BPV4 genome is a double-stranded closed circle of 7265 base pairs that can be divided into three regions: a long control region, which contains elements through which viral gene expression and replication are regulated, and regions encoding the early and late gene products[54] (Fig. 1). The gene products of the major open reading frames (ORFs) have been well characterized for BPV1 and the

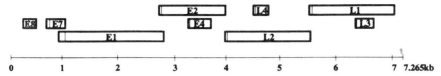

FIGURE 1. Genomic organization of BPV4. The viral genome is represented as linear, and the boxes represent ORFs in the three forward-reading frames. The early ORFs are designated E, and the late ORFs L. The first ATG codon in each ORF is indicated by a vertical line.

human papillomaviruses (HPVs). Thus, L1 and L2 encode the major and minor capsid proteins, respectively; the E1 protein functions in viral DNA replication[55,56]; and the E2 ORF is involved in control of both replication[56] and transcription, encoding both a full length transcriptional transactivator and N-terminally truncated transcriptional repressors.[57–59] The E4 ORF appears to be expressed predominantly as an E1–E4 fusion containing the first five residues of E1[60]; this protein may function by interfering with cytokeratin assembly, possibly disturbing the keratinocyte differentiation program and favoring the production of virion progeny.[61] Three ORFs, E5, E6, and E7, have been shown to play major roles in cell transformation by papillomaviruses, although the relative importance of each of these ORFs in *in vitro* transformation varies between virus types [reviewed by Campo (62)]. An unusual feature of the epitheliotropic B subgroup of BPVs is their lack of the E6 ORF, which is present in all other papillomaviruses characterized to date.[63] The E6 proteins of the oncogenic HPVs have been demonstrated to bind and promote degradation of the p53 tumor suppressor gene product.[64,65] However, it appears that for the epitheliotrophic BPVs, an E6 protein is not required for successful infection, and papillomas induced by BPV4 can progress to carcinoma despite the lack of an E6 ORF. Mutations of p53 detected in alimentary carcinomas (L. Scobie, M. E. Jackson, and M. S. Campo, unpublished data) may reflect the lack of an E6-encoded p53 degradation function in the BPV4 genome. The E8 ORF of the B subgroup BPVs occupies the genomic position where E6 is present in other papillomaviruses, but the E8 protein strongly resembles the papillomavirus E5 proteins that are normally encoded in the distal part of the early region.

5. THE VIRAL CONTRIBUTION TO CARCINOGENESIS

Analysis of alimentary canal carcinomas revealed that although infection by BPV4 is a prerequisite for carcinogenesis (see above), the viral DNA is normally undetectable in cancers.[66] Loss of viral DNA during progression has also been demonstrated to occur both in *in vitro* transformation systems[67] and in carcinoma induction in xenografts of BPV4-infected bovine palate tissue.[68] Thus, BPV4 appears to act by a "hit-and-run" mechanism in carcinogenesis of the upper alimentary canal of cattle; i.e., the virus is required for the initiation but not the maintenance of the malignant phenotype.

BPV4 DNA alone is capable of transforming immortalized murine cell lines to the fully malignant phenotype[67] and induces cellular gene amplification.[69,70] However, for transformation of primary bovine fibroblasts (PalF), BPV4 requires cooperation with an activated *ras* oncogene.[71] The transformed PalF cells are capable of anchorage-independent growth in Methocel but are not tumorigenic in nude mice[72] (Table II). Morphological transformation appears to be a function of the E7 ORF,[71] and the E8 ORF is responsible for anchorage-independent growth.[73] The similarity of BPV4 E7 to the E7 proteins of other papillomaviruses suggest that it acts in the same way in transformation and that one of its functions is to bind the p105 Rb tumor suppressor gene product[71]; indeed, mutations in the putative p105 Rb binding domain and Cys-X-X-Cys zinc-binding motifs of BPV4 E7 abrogate its transformation function (G. J. Grindlay and M. S. Campo, unpublished data). The E8 protein, like BPV1 E5,[74] binds the 16 kDa ductin protein, which is a component of vacuolar ATPase and intercellular gap junctions[75] (A. M. Faccini, M. Cairney, R. A. Anderson, M. E. Finbow, M. S. Campo, and J. D. Pitts, unpublished data). E8-expressing cells have a diminished capacity for gap-junctional intercellular communication, suggesting that E8 may interact with the gap-junctional form of 16 kDa ductin (A. M. Faccini, M. Cairney, R. A. Anderson, M. E. Finbow, M. S. Campo, and J. D. Pitts, unpublished data). Thus, at least two BPV4-encoded proteins are capable of contributing to the process of carcinogenesis.

6. COOPERATION BETWEEN BPV4 AND CHEMICALS IN EXPERIMENTAL SYSTEMS

PalF cells transformed by BPV4 plus *ras* have an extended life-span in culture but are not immortal and are anchorage independent but not tumorigenic in nude mice, indicating that additional events are required for full malignant transforma-

TABLE II
Synergy between BPV4 Gene Products and Quercetin (Q) in the Transformation of Primary Bovine Fibroblasts (PalF)[a]

Treatment (plus *ras* DNA)	Morphological transformation	Anchorage independence	Immortality	Tumorigenicity
BPV4	Yes	Yes	No	No
E7	Yes	No	No	No
E7 + E8	Yes	Yes	No	No
Q + BPV4	Yes	Yes	Yes	Yes
Q + E7	Yes	Yes	Yes	Yes/no[b]
Q + E7 + E8	Yes	No	n.d.	No
Q	No[c]	n.d.	n.d.	n.d.

[a]Data from references 54 and 71–73.
[b]When quercetin treatment is immediately before transfection, Q + E7 cells are nontumorigenic; however, if quercetin treatment is immediately after transfection, Q + E7 cells are tumorigenic (76).
[c]Cells treated with quercetin and transfected with *ras* alone produced only small contact-inhibited colonies that could not be expanded; n.d., not done.

tion of primary cells.[72] However, if the PalF cells are exposed to quercetin, which is present in bracken (see above), immediately before transfection with BPV4 plus *ras*, the cells become more aggressively transformed and are immortal and tumorigenic in nude mice[72] (Table II). Quercetin treatment immediately after transfection is even more effective in the induction of an aggressive phenotype, indicating that the timing of exposure to viral products and quercetin is crucial. In fact, when PalF cells are treated with quercetin immediately before transfection with E7 plus *ras*, the transformed clones are nontumorigenic, whereas if quercetin treatment is immediately after transfection, the clones exhibit tumorigenicity[76] (Table II). Increasing the time interval between quercetin treatment and transfection weakens the observed synergy for all permutations, suggesting that BPV4 DNA and quercetin are required to be present simultaneously for maximal effect.[76] Furthermore, PalF cells transformed by the combination of quercetin, BPV4 E7, and *ras* show immortality and anchorage independence in the absence of the E8 ORF, and it appears that in some circumstances E8 and quercetin act antagonistically.[76] Contrary to previous reports for established cell lines, cytogenetic analysis of quercetin-treated PalF cells and screening with a panel of minisatellite probes have not detected any gross chromosomal changes (M. Cairney, M. S. Campo, and R. Frier, unpublished data). In regard to epigenetic effects of quercetin treatment, preliminary results indicate increased protein tyrosine phosphorylation and up-regulation of BPV4 promoter activity in PalF cells following exposure to quercetin. In view of the importance of phosphorylation in signal transduction, transcription factor activity, cell growth, and transformation, this may well contribute to the observed effects of quercetin (J. A. Connolly, M. Cairney, and M. S. Campo, unpublished data).

The interaction of BPV4 with chemicals has also been investigated in the nude mouse xenograft system.[77] Fetal bovine palate biopsies that are infected with BPV4 and implanted either beneath the renal capsule or subcutaneously in nude mice develop virus-producing papillomas.[78] In the absence of other agents, these papillomas normally remain benign, and spontaneous transition to carcinoma has been observed only once.[68] However, when the mice are also implanted with slow-release pellets of either the initiator 7,12-dimethylbenz(a)anthracene (DMBA) or the tumor promoter 12-O-tetradecanoylphorbol-13-acetate (TPA), the frequency of neoplastic conversion is dramatically increased[79] (Table III). The DMBA induces both activating mutations in *ras*[80] and amplification of the epidermal growth factor receptor (EGF-R) gene.[81] As described above, *ras* is activated in alimentary canal carcinomas,[51] and elevated levels of EGF-R are present in cells explanted from an alimentary canal cancer.[82] The TPA treatment leads to BPV4 DNA amplification and enhanced viral transcription,[67,82] and increased BPV4 transcription results in more efficient cell transformation.[71] The TPA also perturbs gap-junctional communication between cells, causing inhibition of cell–cell communication in cultured cells[83,84] and inappropriate cell–cell communication in mouse skin.[85] The results using the nude mouse xenograft system indicate that BPV4 can cooperate with both an initiator and a tumor promoter in carcinogenesis, suggesting that BPV4-associated cancers may be the result of different series of events depending on the nature of the environmental carcinogen(s) in each case. Although bracken fern is the only known cofactor cooperating with BPV4 in alimen-

TABLE III
Synergism between BPV4 and TPA or DMBA in the Generation of Neoplastic Changes in Bovine Xenografts[a]

Treatment	Papillomas (%)	Neoplasia (%)
BPV4	0/9	0/9
BPV4 + TPA	11/33 (33%)	4/33 (12%)
TPA	0/25	0/25
BPV4 + DMBA	10/20 (50%)	13/20 (65%)
DMBA	0/10	0/10

[a]The term neoplasia includes premalignant changes in TPA-treated implants and carcinomas in DMBA-treated implants. Data are from reference 79.

tary canal carcinogenesis, a number of different mutagenic and carcinogenic compounds are present in bracken, and several of these may be involved with BPV4 in the tumorigenic process.

7. DISCUSSION

The interactions of BPV and chemicals derived from bracken fern in carcinogenesis appear to be complex. The BPV-induced papilloma itself represents an increased target of proliferating cells for chemicals to act on, and the immunosuppression caused by bracken fern ingestion further increases the pool of target cells by permitting widespread and persistent papillomatosis. Immunosuppression also allows the activation of latent BPV2, leading to nonproductive infections of the urinary bladder that may progress to cancer. Cells proliferating under the influence of viral oncoproteins may then sequentially undergo additional events such as *ras* activation, p53 mutation, and EGF-R increase as a result of cofactors provided by the bracken, with the eventual emergence of a fully malignant clone. In the case of BPV4, the viral oncoproteins are no longer required at this stage, and viral DNA is lost in the absence of selective pressure for its maintenance.

REFERENCES

1. Nordling, C. O., 1953, A new theory on the cancer-inducing mechanism, *Br. J. Cancer* **7**:68–72.
2. Armitage, P., and Doll, R., 1954, The age distribution of cancer and a multi-stage theory of carcinogenesis, *Br. J. Cancer* **8**:1–12.
3. Fearon, E. R., and Vogelstein, B., 1990, A genetic model for colorectal tumorigenesis, *Cell* **61**:759–767.
4. Land, H., Parada, L. F., and Weinberg, R. A., 1983, Tumorigenic conversion of primary embryo fibroblasts requires at least two cooperating oncogenes, *Nature* **304**:596–602.

5. Newbold, R. F., and Overell, R. W., 1983, Fibroblast immortality is a prerequisite for transformation by EJ c-Ha-*ras* oncogene, *Nature* **304**:648–651.
6. Ruley, H. E., 1983, Adenovirus early region 1A enables viral and cellular transforming genes to transform primary cells in culture, *Nature* **304**:602–606.
7. Plowright, W., Linsell, C. A., and Peers, F. G., 1971, A focus of rumenal cancer in Kenyan cattle, *Br. J. Cancer* **25**:72–80.
8. Jarrett, W. F. H., 1973, Oesophageal and stomach cancer in cattle; a candidate viral and carcinogen model system and its possible relevance to man, *Br. J. Cancer* **28**:93.
9. Jarrett, W. F. H., 1982, Bracken and cancer, *Proc. R. Soc. Edin.* **81**:79–83.
10. Bertone, A. L., 1990, Neoplasms of the bovine intestinal tract, *Vet. Clin. North Am. Food Anim. Pract.* **6**:515–524.
11. Pamukcu, A. M., Goskoy, S. K., and Price, J. M., 1967, Urinary bladder neoplasms induced by feeding bracken fern to cows, *Cancer Res.* **27**:917–924.
12. Price, J. M., and Pamukcu, A. M., 1968, The induction of neoplasms of the urinary bladder of the cow and the small intestine of the rat by feeding bracken fern, *Cancer Res.* **28**:2247–2251.
13. Pamukcu, A. M., Price, J. M., and Bryan, G. T., 1976, Naturally-occurring and bracken fern-induced bovine urinary bladder tumors, *Vet. Pathol.* **13**:110–122.
14. Olson, C., Pamukcu, A. M., Brobst, D. F., Kowalczyk, T., Satter, E. J., and Price, J. M., 1959, A urinary bladder tumor induced by a bovine cutaneous papilloma agent, *Cancer Res.* **19**: 779–783.
15. Olson, C., Pamukcu, A. M., and Brobst, D. F., 1965, Papilloma-like virus from bovine urinary bladder tumors, *Cancer Res.* **25**:840–849.
16. Lancaster, W. D., and Olson, C., 1978, Demonstration of two distinct classes of bovine papillomavirus, *Virology* **89**:372–379.
17. Pfister, H., Linz, U., Gissmann, L., Huchthausen, B., Hoffmann, D., and zur Hausen, H., 1979, Partial characterisation of a new type of bovine papillomaviruses, *Virology* **96**:1–8.
18. Campo, M. S., Moar, M. H., Jarrett, W. F. H., and Laird, H. M., 1980, A new papillomavirus associated with alimentary cancer in cattle, *Nature* **286**:180–182.
19. Campo, M. S., Moar, M. H., Laird, H. M., and Jarrett, W. F. H., 1981, Molecular heterogeneity and lesion site specificity of cutaneous bovine papillomaviruses, *Virology* **113**:323–335.
20. Jarrett, W. F. H., Campo, M. S., O'Neil, B. W., Laird, H. M., and Coggins, L. W., 1984, A novel bovine papillomavirus (BPV-6) causing true epithelial papillomas of the mammary gland skin: A member of a proposed new BPV subgroup, *Virology* **136**:255–264.
21. Campo, M. S., Jarrett, W. F. H., O'Neil, B. W., and Barron, R. J., 1994, Latent papillomavirus infection in cattle, *Res. Vet. Sci.* **56**:151–157.
22. Evans, W. C., Patel, M. C., and Koohy, Y., 1982, Acute bracken poisoning in homogastric and ruminant animals, *Proc. R. Soc. Edin.* **81**:29–64.
23. Campo, M. S., and Jarrett, W. F. H., 1986, Papillomavirus infection in cattle: Viral and chemical cofactors in naturally occurring and experimentally induced tumours, *CIBA Found. Symp.* **120**:117–130.
24. Campo, M. S., Jarrett, W. F. H., Barron, R., O'Neil, B. W., and Smith, K. T., 1992, Association of bovine papillomavirus type 2 and bracken fern with bladder cancer in cattle, *Cancer Res.* **52**:6898–6904.
25. Jarrett, W. F. H., McNeil, P. E., Grimshaw, W. T. R., Selman, I. E., and McIntyre, W. I. M., 1978, High incidence area of cattle cancer with a possible interaction between an environmental carcinogen and a papilloma virus, *Nature* **274**:215–217.
26. Campo, M. S., O'Neil, B. W., Barron, R. J., and Jarrett, W. F. H., 1994, Experimental reproduction of the papilloma-carcinoma complex of the alimentary canal in cattle, *Carcinogenesis* **15**:1597–1601.
27. Campo, M. S, 1987, Papillomaviruses and cancer in cattle, *Cancer Surv.* **6**:39–54.
28. Evans, W. C., Evans, I. A., Thomas, A. J., Watkin, J. E., and Chamberlain, A. T., 1958, Studies on bracken poisoning in cattle, Pt IV, *Br. Vet. J.* **114**:180–267.

29. Heath, G. B. S., and Wood, B., 1958, Bracken poisoning in cattle, *J. Comp. Pathol.* **68:**201–212.
30. Moura, J. W., Stocco dos Santos, R. C., Dagli, M. L. Z., D'Angelino, J. L., Birgel, E. H., and Becak, W., 1988, Chromosome aberrations in cattle raised on bracken fern pasture, *Experientia* **44:**785–788.
31. Tawfik, A. F., Bishop, S. J., Ayalp, A., and el-Feraly, F. S., 1990, Effects of artemisinin, dihydroartemisinin and arteether on immune responses of normal mice, *Int. J. Immunopharmacol.* **12:**385–389.
32. Bourget-Kondracki, M. L., Longeon, A., Morel, E., and Guyot, M., 1991, Sesquiterpene quinones as immunomodulating agents, *Int. J. Immunopharmacol.* **13:**393–399.
33. Toki, J., Miyazaki, W., Inaba, M., Saigo, S., Nishino, T., Fukuba, Y., Good, R. A., and Ikehara, S., 1992, Effects of K-76cooh (MX-1) on immune response: Induction of suppressor cells by MX-1, *Int. J. Immunopharmacol.* **14:**1093–1098.
34. Evans, I. A., Prorok, J. H., Cole, R. C., Al-Samani, M. H., Al-Samarrai, A. M., Patel, M. C., and Smith, R. M. N., 1982, The carcinogenic, mutagenic and teratogenic toxicity of bracken, *Proc. R. Soc. Edin.* **81:**65–77.
35. Hirono, I., 1986, Carcinogenic principles isolated from bracken fern, *CRC Crit. Rev. Toxicol.* **17:**1–22.
36. Hirono, I., Ogino, H., Fujimoto, M., Yamada, K., Yoshida, Y., Ikagawa, M., and Okumura, M., 1987, Induction of tumours in ACI rats given a diet containing ptaquiloside, a bracken carcinogen, *J. Natl. Cancer Inst.* **79:**1143–1149.
37. Matsuoka, A., Hirosawa, A., Natori, S., Iwasaki, S., Sofuni, T., and Ishidate, M., 1989, Mutagenicity of ptaquiloside, the carcinogen in bracken, and its related illudane-type sesquiterpenes. II. Chromosomal aberration tests with cultured mammalian cells. *Mut. Res.* **215:** 179–185.
38. Matoba, M., Saito, E., Saito, K., Koyama, K., Natori, S., Matsushima, T., and Takimoto, M., 1987, Assay of ptaquiloside, the carcinogenic principle of bracken, *Pteridium aquilinum*, by mutagenicity testing in *Salmonella typhimurium*. *Mutagenesis* **2:**419–423.
39. El-Mofty, M., Sadek, I., Soliman, A., Mohamed, A., and Sakre, S., 1987, α-Ecdysone, a new bracken fern factor responsible for neoplasm induction in the Egyptian toad (*Bufo regularis*). *Nutr. Cancer* **9:**103–107.
40. Pamukco, A. M., Yalciner, S., Hatcher, J. F., and Bryan, G. T., 1980, Quercetin, a rat intestinal and bladder carcinogen present in bracken fern, *Cancer Res.* **40:**3468–3472.
41. Bjieldanes, L. F., and Chang, G. W., 1977, Mutagenic activity of quercetin and related compounds, *Science* **197:**577–578.
42. Maruta, A., Enaka, K., and Umeda, M., 1979, Mutagenicity of quercetin and kaempferol on cultured mammalian cells, *GANN* **70:**273–276.
43. Amacher, D. E., Paillet, S., and Ray, V. A., 1979, Point mutations at the thymidine kinase locus in L5178Y mouse lymphoma cells, *Mut.. Res.* **64:**391–406.
44. Nakayasu, M., Sakamoto, H., Terada, M., Nagao, M., and Sugimura, T., 1986, Mutagenicity of quercetin in Chinese hamster lung cells in culture, *Mut.. Res.* **174:**79–83.
45. Ishikawa, M., Okada, F., Hamada, J., Hosokawa, M., and Kobayashi, A., 1987, Changes in the tumorigenic and metastatic properties of tumour cells treated with quercetin or 5-azacytidine, *Int. J. Cancer* **39:**338–342.
46. Sakai, A., Sasaki, K., Mizusawa, H., and Ishidate, M., 1990, Effects of quercetin, a plant flavonol, on the two stage transformation *in vitro*, *Terat. Carc. Mutagen.* **10:**333–340.
47. Ishidate, M., 1988, *Data Book on Chromosomal Aberration Tests in Vitro*, LIC, Tokyo, Elsevier, Amsterdam.
48. Van Wart-Hood, J. E., Linder, M. E., and Burr, J. G., 1989, TPCK and quercetin act synergistically with vanadate to increase protein-tyrosine phosphorylation in avian cells, *Oncogene* **4:**1267–1271.
49. Matter, W. F., Brown, R. F., and Vlahos, C. J., 1992, The inhibition of phosphatidyl-inositol 3-kinase by quercetin and analogs, *Biochem. Biophys. Res. Commun.* **186:**624–631.

50. Jackson, M. E., Campo, M. S., and Gaukroger, J. M., 1993, Cooperation between papillomavirus and chemical cofactors in oncogenesis, *CRC Crit. Rev. Oncogen.* **4:**277–291.
51. Campo, M. S., McCaffery, R. E., Doherty, I., Kennedy, I. M., and Jarrett, W. F. H., 1990, The Harvey *ras* 1 gene is activated in papillomavirus-associated carcinomas of the upper alimentary canal in cattle, *Oncogene* **5:**303–308.
52. Barbacid, M., 1987, *ras* genes, *Annu. Rev. Biochem.* **56:**779–827.
53. Krontiris, T. G., Devlin, B., Karp, D. D., Robert, N. J., and Risch, N., 1993, An association between the risk of cancer and mutations in the HRAS1 minisatellite locus, *N. Engl. J. Med.* **329:**517–523.
54. Campo, M. S., Anderson, R. A., Cairney, M., and Jackson, M. E., 1994, Bovine papillomavirus type 4: From transcriptional control to control of disease, in: *128th SGM Symposium—Viruses and Cancer* (A. C. Minson, ed.), Cambridge University Press, Cambridge, pp. 47–70.
55. Lambert, P. F., 1991, Papillomavirus DNA replication, *J. Virol.* **65:**3417–3420.
56. Yang, L., Li, R., Mohr, I. J., Clark, R., and Botchan, M. R., 1991, Activation of BPV1 replication *in vitro* by the transcription factor E2, *Nature* **353:**628–632.
57. Hirochika, H., Broker, T. R., and Chow, L. T., 1987, Enhancers and transacting E2 transcriptional factors of papillomaviruses, *J. Virol.* **61:**2599–2606.
58. Lambert, P. F., Spalholz, B. A., and Howley, P. M., 1987, A transcriptional repressor encoded by BPV1 shares a common C-terminus with the E2 transactivator, *Cell* **50:**69–78.
59. Choe, J., Vaillancourt, P., Stenlund, A., and Botchan, M., 1989, Bovine papillomavirus type 1 encodes two forms of a transcriptional repressor: Structural and functional analysis of new viral cDNAs, *J. Virol.* **63:**1743–1755.
60. Doorbar, J., Evans, H. S., Coneron, I., Crawford, L. V., and Gallimore, P. H., 1988, Analysis of HPV-1 E4 gene expression using epitope-defined antibodies, *EMBO J.* **7:**825–833.
61. Doorbar, J., Ely, S., Sterling, J., McLean, C., and Crawford, L., 1991, Specific interaction between HPV-16 E1^E4 and cytokeratins results in collapse of the epithelial cell intermediate filament network, *Nature* **352:**824–827.
62. Campo, M. S., 1992, Cell transformation by animal papillomaviruses, *J. Gen. Virol.* **73:**217–222.
63. Jackson, M. E., Pennie, W. D., McCaffery, R. E., Smith, K. T., Grindlay, G. J., and Campo, M. S., 1991, The B subgroup bovine papillomaviruses lack an identifiable E6 open reading frame, *Mol. Carcinogen.* **4:**382–387.
64. Werness, B. A., Levine, A. J., and Howley, P. M., 1990, Association of human papillomavirus type 16 and 18 E6 proteins with p53, *Science* **248:**76–79.
65. Scheffner, M., Werness, B. A., Huibregtse, J. M., Levine, A. J., and Howley, P. M., 1990, The E6 oncoprotein encoded by human papillomavirus types 16 and 18 promotes the degradation of p53, *Cell* **63:**1129–1136.
66. Campo, M. S., Moar, M. H., Sartirana, M. L., Kennedy, I. M., and Jarrett, W. F. H., 1985, The presence of bovine papillomavirus type 4 DNA is not required for the progression to, or the maintenance of, the malignant state in cancers of the alimentary canal in cattle, *EMBO J.* **4:**1819–1825.
67. Smith, K. T., and Campo, M. S., 1988, "Hit and run" transformation of mouse C127 cells by bovine papillomavirus type 4: The viral DNA is required for the progression to but not for the maintenance of the transformed phenotype, *Virology* **164:**39–47.
68. Gaukroger, J. M., Chandrachud, L., Jarrett, W. F. H., McGarvie, G. E., Yeudall, W. A., McCaffery, R. E., Smith, K. T., and Campo, M. S., 1991, Malignant transformation of a papilloma induced by bovine papillomavirus type 4 in the nude mouse renal capsule, *J. Gen. Virol.* **72:**1165–1168.
69. Smith, K. T., and Campo, M. S., 1989, Amplification of specific DNA sequences in C127 mouse cells transformed by bovine papillomavirus type 4, *Oncogene* **4:**409–413.
70. Smith, K. T., Coggins, L. W., Doherty, I., Pennie, W. D., Cairney, M., and Campo, M. S., 1993, BPV4 induces amplification and activation of "silent" BPV1 in a subline of C127 cells, *Oncogene* **8:**151–156.
71. Jaggar, R. T., Pennie, W. D., Smith, K. T., Jackson, M. E., and Campo, M. S., 1990, Cooperation

between bovine papillomavirus type 4 and *ras* in the morphological transformation of primary bovine fibroblasts, *J. Gen. Virol.* **71**:3041–3046.
72. Pennie, W. D., and Campo, M. S., 1992, Synergism between bovine papillomavirus type 4 and the flavonoid quercetin in cell transformation *in vitro*, *Virology* **190**:861–865.
73. Pennie, W. D., Grindlay, G. J., Cairney, M., and Campo, M. S., 1993, Analysis of the transforming functions of bovine papillomavirus type 4, *Virology* **193**:614–620.
74. Goldstein, D. J., Finbow, M. E., Andresson, T., McLean, P., Smith, K. T., Bubb, V., and Schlegel, R., 1991, Bovine papillomavirus E5 oncoprotein binds to the 16k component of vacuolar H^+ ATPases, *Nature* **352**:347–349.
75. Holzenburg, A., Jones, P. C., Franklin, T., Pali, T., Heimburg, T., Marsh, D., Findlay, J. B. C., and Finbow, M. E., 1993, Evidence for a common structure for a class of membrane channels, *Eur. J. Biochem.* **213**:21–30.
76. Cairney, M., and Campo, M. S., 1995, The synergism between bovine papillomavirus type 4 and quercetin is dependent on the timing of exposure, *Carcinogenesis* **16**:1997–2001.
77. Kreider, J. W., Howett, M. K., Lill, N. L., Bartlett, G. L., Zaino, R. J., Sedlacek, T. V., and Mortel, R., 1986, *In vivo* transformation of human skin with human papillomavirus type 11 from condylomata acuminata, *J. Virol.* **59**:369–376.
78. Gaukroger, J., Bradley, A., O'Neil, B., Smith, K., Campo, S., and Jarrett, W., 1989, Induction of virus producing tumours in athymic nude mice by bovine papillomavirus type 4, *Vet. Rec.* **125**:391–392.
79. Gaukroger, J. M., Bradley, A., Chandrachud, L., Jarrett, W. F. H., and Campo, M. S., 1993, Interaction between bovine papillomavirus type 4 and cocarcinogens in the production of malignant tumours, *J. Gen. Virol.* **74**:2275–2280.
80. Quintanilla, M., Brown, K., Ramsden, M., and Balmain, A., 1986, Carcinogen-specific mutation and amplification of Ha-*ras* during mouse skin carcinogenesis, *Nature* **322**:78–80.
81. Wong, D. T . W., 1987, Amplification of the c-*erb*-B1 oncogene in chemically induced oral carcinomas, *Carcinogenesis* **8**:1963–1965.
82. Smith, K. T., Campo, M. S., Bradley, J., Gaukroger, J., and Jarrett, W. F. H., 1987, Cell transformation by bovine papillomavirus: Cofactors and cellular responses, *Cancer Cells* **5**: 267–274.
83. Yotti, L. P., Chang, C. C., and Trosko, J. E., 1979, Elimination of metabolic cooperation in Chinese hamster cells by a tumor promoter, *Science* **206**:1089–1091.
84. Newbold, R. F., and Amos, J., 1982, Inhibition of metabolic cooperation between mammalian cells in culture by tumour promoters, *Carcinogenesis* **2**:243–249.
85. Kam, E., and Pitts, J. D., 1988, Effects of the tumor promoter 12-O-tetradecanoylphorbol-13-acetate on junctional communication in intact mouse skin: Persistence of homologous communication and increase of epidermal–dermal coupling, *Carcinogenesis* **9**:1389–1394.

8

Papillomaviruses as Promoting Agents in Human Epithelial Tumors

CHRISTA CERNI and CHRISTIAN SEELOS

1. INTRODUCTION

The papillomaviruses (PVs) are a large family of DNA viruses indigenous to many, if not all, vertebrate species. The unique feature of these viruses is that they all induce primarily benign proliferations of epithelial cells in their natural hosts. Individual members of this family show a high degree of both species specificity and tissue specificity. Squamous or mucosal epithelial cells are the targets of infection. The life cycle of the virus is intimately coupled to the differentiation program of infected tissue. Conversely, expression of the early viral genes interferes with the normal pattern of keratin expression, cell cycle regulation, and terminal differentiation, which ultimately gives rise to aberrant cells. With some viral types, however, these benign proliferations progress into aggressive malignancies.

In recent years, much attention has been focused on this progression or transition to invasive carcinoma because development of human cervical cancer, the second most common cancer in women worldwide, is definitely associated with infection by several types of human PVs (HPVs). Human PV DNA is frequently found in biopsies obtained from patients with cervical cancer when assayed by nucleic acid hybridization techniques. The development of more sensitive methods based on the polymerase chain reaction (PCR) revealed the presence of HPV DNA in almost 100% of cervical cancer biopsies. An overwhelming amount of evidence links the action of viral proteins to cellular transformation, and causality of HPV infection in the development of malignancies is scarcely doubted.

CHRISTA CERNI and CHRISTIAN SEELOS • Institute of Tumor Biology—Cancer Research, University of Vienna, 1090 Vienna, Austria.

DNA Tumor Viruses: Oncogenic Mechanisms, edited by Giuseppe Barbanti-Brodano *et al.* Plenum Press, New York, 1995.

In view of the broad spectrum of PV-induced proliferations, the two most obvious questions concern the molecular basis for the tissue specificity of the vast majority of PVs and for the different oncogenic potential of individual PV types. Although little progress has been achieved regarding the former aspect, undoubtedly because of the previous lack of adequate *in vitro* systems to propagate HPVs, much progress has been attained regarding the latter in recent years. Findings that oncogenic proteins of certain HPV types are able to form complexes with cellular tumor suppressor gene products have eventually provided the long-sought approach to link cellular transformation to cell cycle deregulation. Because of the paramount importance of this topic in extending the initial aims of clinically applied research, we have focused this review on recent data on the biological and molecular properties of HPVs that have contributed to a better understanding of how a small genome can govern a giant such as a cell.

It has become increasingly clear that the final outcome of viral–cellular interaction is determined by five variables: (1) the activities of the noncoding region of the viral DNA with a complex array of target sequences for cellular regulatory proteins; (2) the properties of individual viral proteins; (3) the interaction between virally encoded proteins and the genetic makeup of infected cells; (4) the tissue context of infected cells with continuous exposure to a multitude of external signals; and (5) the accessibility of viral proteins to the immune system of the host.

2. HUMAN PAPILLOMAVIRUSES

2.1. Classification of HPVs

The family *Papovaviridae* consists of the *Polyomavirus* genus and the *Papillomavirus* genus. Yet, several differences in the organization, size, and homology of the genomes as well as in the replicative cycles are at variance with maintaining the two virus groups in the same family.

In 1979, at one of the first international workshops on PVs, general directions for characterizing new types of HPVs were laid down, and the genome organization of HPVs was defined on a comparative basis with that of the well-characterized bovine PV type 1 (BPV-1). The types are differentiated by comparison of DNA homology. If a new PV has 50% or greater homology (percentage cross-hybridization in liquid phase) to a known virus, it will be placed in that typing group.[1] For some types, stable mutations exist that differ in some restriction endonuclease sites, although they show extensive cross-hybridization. These subtypes are indicated by a letter (e.g., -1a, -1b, -1c). To date, 70 distinct types of HPVs have been isolated and characterized.[2] About half of them are associated with anogenital or mucosal lesions, and the other half with nongenital cutaneous lesions.

2.1.1. Mucosal HPV Types

The mucosal HPVs infect the anogenital tract. These genital HPVs can be subdivided according to their oncogenic potential. "Low-risk"HPVs, with HPV-6 and HPV-11 as the classical representatives, induce benign proliferations such as condylomata acuminata, which only rarely develop into malignancy. Conversely, "high-risk" HPVs are found associated with the vast majority of cervical and anogenital malignancies. Only four viral types, HPV-16, HPV-18, HPV-31, and HPV-33, are associated with over 90% of cervical carcinomas.

As judged by histopathological criteria, the early steps of transformation of clearly normal cells into definite tumor cells occur in a rather blurred manner. Moreover, the individual evolution of many slightly distinct cell clones within a cancer-prone tissue renders precise diagnosis at any given time more difficult. These precancerous lesions of cervical cancer are named cervical intraepithelial neoplasias (CIN) and are graded into three categories (I–III) according to the extent of cellular dedifferentiation. In general, low-risk HPVs, but also HPV-18, are detected in CIN I or CIN II dysplasias.[3] In CIN I, viral DNA is found in episomal form; CIN III lesions, not always clearly marked off from carcinoma *in situ*, are suggestive of high-risk HPV infection, and viral DNA sequences are found integrated at high frequency.

HPV-16 is preferentially associated with invasive squamous lesions, whereas HPV-18 is related mainly to adenocarcinomas and is rarely found in squamous cervical cancers (SCC).[4] When HPV-18 DNA is present in SCC, the tumor cells are usually poorly differentiated and have already migrated to the lymph nodes at the time of surgery. In addition, the mean age of women with HPV-18-related SCC is lower than that of women with HPV-16-related cancers. This supports the notion that HPV-18-induced tumors have a worse prognosis than cancers induced by other HPV types.[5]

The fact that viral DNAs of several of the so-called genital HPVs have also been found occasionally in mucosal proliferations of the respiratory and digestive tracts has raised some doubt about the stringency of tissue specificity and the expediency of a tissue-related classification, the more so because a small number of HPV types such as type 57 and the widespread type 2 infect both squamous and mucosal epithelia, and this at rather distinct localizations. In addition, genital HPVs might also play a role in the genesis of Kaposi's sarcoma,[6] a collective noun for malignancies thought to arise from endothelial cells.

2.1.2. Cutaneous HPV Types

The nongenital cutaneous HPVs of at least nine genotypes have been detected in benign warts of the skin of otherwise healthy individuals. A larger number of cutaneous subtypes have been isolated from patients suffering from epidermodysplasia verruciformis (EV),[7] which is an autosomal recessive disease characterized by disseminated wart-like proliferations of the skin that progress or transit to SCC in up to half of the affected patients, usually at sun-exposed sites. The main HPV subtypes found in malignant EV-associated lesions are HPV-5 and HPV-8, which can also be detected in skin carcinomas of therapeutically immunosuppressed patients.[8]

2.2. Organization of the HPV Genome

The genome of HPVs consists of a double-stranded, supercoiled, circular DNA molecule. The length of the genome is about 8000 base pairs, and all HPVs exhibit a similar genome organization.[9] All of the potential open reading frames (ORFs) are found on one strand. This is in contrast to the polyomavirus genus, where the early proteins are coded by half the genome of one strand and the late proteins by the other strand in the opposite direction.[10] Although minor variations do occur, HPVs have six different ORFs in the early region, designated E1, E2, E4, E5, E6, and E7, and two ORFs in the late region, L1 and L2 (Fig. 1). Whereas the late genes code for the structural proteins of the virus particle and mediate virus uptake by putative cellular receptors, the early proteins are responsible for the biological effects exerted on target cells.

2.3. Upstream Regulatory Region of HPVs

Immediately upstream of the early ORFs, a region of approximately 500 to 1000 base pairs of DNA is located that has no coding potential. Instead, several *cis*-acting elements have been defined in this area by comparison to known binding sites for *trans*-acting factors and have been confirmed by biological assays. This region is thus referred to as the upstream regulatory region (URR) or long control region (LCR). In Fig. 2, a comparison of the URRs of two members each of highly oncogenic HPVs (HPV-16 and -18) and HPVs that cause mainly benign condylomatous warts (HPV-6 and -11) is depicted.

Despite a common genomic organization, large differences exist in the consequences of infection by different HPV types, particularly with regard to oncogenic potential. As we shall see, this is partly explained by genomic variance, especially within the early genes E6 and E7, and also by the regulation of expression of these genes. Actually, the fact that roughly 10% of the HPV genome is devoted to transcriptional regulation of the early (onco)genes points to an essential role of the URRs in mediating the biological effects of these viruses.

2.3.1. Regulation of General Transcription

Transcriptional regulation of gene expression is primarily exerted by a combination of *cis*-acting sequences and *trans*-acting factors.[11] Three types of *trans*-acting

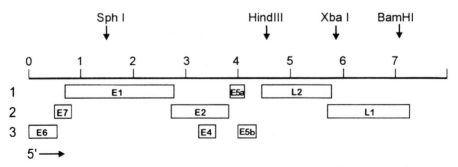

FIGURE 1. Genomic organization of HPV-11. Viral genes in the three reading frames are depicted, and single-cut restriction enzyme sites are indicated.

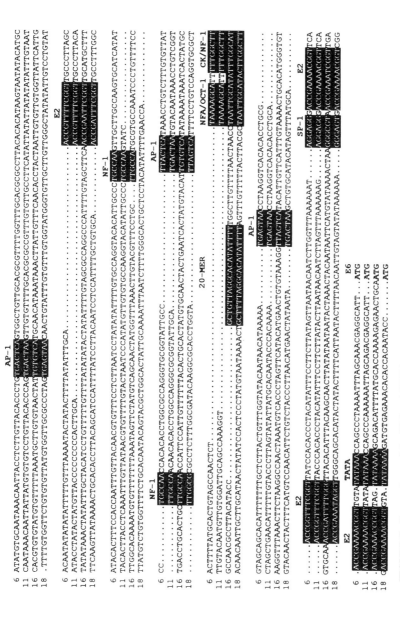

FIGURE 2. Comparison of the URRs of HPV-6, -11, -16, and -18 with regard to conserved binding sites for transcription factors. No attempt was made to align sequences between conserved binding sites for the indicated factors. The start of the E6 ORF is indicated.

factors are known: (1) the basal transcription factors, (2) gene-specific transcription factors with an activation domain, and (3) gene-specific transcription factors without activation domains. The latter group has also been termed architectural transcription factors.[12] They function by changing the structure of DNA and thereby promote the positioning of upstream factors close to the transcription initiation site.

With regard to HPV gene expression, much is known about the effects of *cis*-sequences in the URRs that are responsible for tissue-specific or drug-inducible transcription of the early genes.[13–20] Two important qualities of HPVs are thought to be dependent mainly on the structure of the URR, namely, (1) the rate of expression of early oncogenes and (2) the strict epithelial tropism of HPV expression. Many of the *cis*-elements in the URR are conserved between HPV family members. Nevertheless, differences exist between HPV types, and, as stated earlier, this may be at least one of the reasons for the different tumorigenicity of different HPV types.

2.3.2. *Transcriptional Regulation by Viral Gene Products*

Transcriptional analysis of PVs initially has focused on functions of proteins expressed from the E2 ORF. The full-length 48-kDa E2 protein has been shown to act in an enhancer-like fashion on promoters linked to the short palindromic sequence $5'$-$ACCN_6GGT$-$3'$.[21–24] The smaller 38-kDa protein expressed from the $3'$ part of the E2 ORF binds to the same sequence but represses the activity of linked promoters.[19,25,26] Most of the E2 results have been obtained with the BPV model; however, all the PVs that have been sequenced so far were revealed to contain the E2 consensus motif several times in their control regions. Although the homology of E2 proteins of unrelated PVs is only about 35%, E2 proteins of a number of different PVs have demonstrated their ability to bind to this sequence, even in an heterologous manner.[19,24,27–29]

Although the BPV-1 URR contains 12 E2 binding sites, the HPVs contain at most four, with two sites located close to the TATA box and one further upstream (Fig. 2). The single most distal site does not seem to have any effect on early gene expression[24,29,30]; however, binding of either full-length or $3'$-E2 to the two proximal sites lowers promoter strength.[13,31,32] Because the proximal sites are located close to the TATA box, steric hindrance of access to basal transcription factors of this region exerted by the E2 proteins is thought to be responsible for this effect. Most likely, this mechanism will help to regulate the expression of early oncogenes during the episomal part of the HPV life cycle. Integration into the host genome is site-specific with regard to the viral genome and in many cases leads to destruction of the E1-E2 genes.[33,34] Therefore, repressor activity is lost at this point, possibly resulting in enhanced expression of E6/E7 oncogenes.[35,36] This might explain how the viral genes gain enhanced carcinogenic potential following integration as compared to the episomal state.

2.3.3. *Transcriptional Regulation by Cellular Factors*

In addition to the viral E2 protein, several host cell intrinsic *trans*-acting factors interact with HPV URRs. In Fig. 2, conserved binding sites in some HPV URRs for the factors AP-1, NF-1, NFA, and SP-1 are indicated. By DNase I footprinting, binding sites for these and several other factors have been mapped in the regulatory regions in various numbers and locations.[14,37–40] Although certain sites have been claimed to be essential for the expression of early genes, a scenario can be envisioned in which the interplay of all factors mediates the cellular environment that determines the biological effects of the presence of viral genes in the host cell. Indeed mutation of any of various single binding sites within the HPV-16 enhancer has been shown to eliminate its activity.[41] This kind of synergism is thought to coordinate and allow the response to small changes in transcriptional activator concentrations.[42]

2.4. The Viral Proteins

2.4.1. Properties of HPV E7

a. Biochemical and Structural Properties. The E7 protein is well conserved among the various HPV types (Fig. 3). The small acidic phosphoprotein consists of about 100 amino acids with a predicted molecular mass of 11 kDa. In SDS-PAGE, some E7 proteins migrate at 14 to 21.5 kDa.[43,44] The difference in calculated and observed mobility results from substantial negative charge on the protein. It was shown that substitution of one or two basic amino acids for acidic residues in the N-terminus of HPV-16 E7 restored the normal electrophoretic migration.[43,45] The protein contains in its C-terminal half two conserved Cys-X-X-Cys amino acid motifs typical of zinc finger structures of the type found in the glucocorticoid receptor.[46] Indeed, E7 has been shown to bind zinc; however, no DNA-binding activity of E7 has been reported. Obviously it does not function as a conventional zinc finger, which might

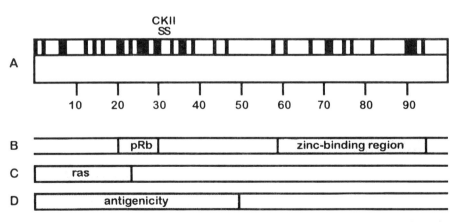

FIGURE 3. Properties of the HPV E7 oncoprotein. (A) Amino acids that are conserved throughout all genital HPVs are indicated by black bars; serines 31 and 32, which are targets for phosphorylation by casein kinase II (CKII), are also shown. Shaded areas represent the pRb and the zinc-binding regions (B), the region responsible for cooperation with *ras* (C), and the antigenic region (D) of HPV-16 E7.

be because of the large spacing of 29 amino acids between the two motifs. Recent evidence suggests that these regions contribute to the stabilization and dimerization of the protein.[47,48] Initially the protein appeared to be mainly localized in the cytoplasm. The small size of the molecule, which diffuses rapidly from nuclei during fractionation, and, moreover, its association with several nuclear proteins had obviously masked its predominantly nuclear localization.[49] The E7 protein has no known enzymatic activities.

b. *Binding to pRb.* As stated earlier, viral oncoproteins are able to functionally eliminate cellular proteins controlling cell cycle progression. The E7 protein can form a complex with the retinoblastoma tumor suppressor protein (pRb), which mimics the loss of cellular pRb.[50] The Rb-1 gene was first identified through its association with an inherited predisposition to retinoblastomas. Although the consequences of this protein–protein interaction remain poorly understood, the fact that pRb is mutated in a variety of human malignancies and targeted for inactivation by several tumor viruses points to a central role of pRb in tumor suppression and cell cycle regulation.[51]

The N-terminal half of E7 shows similarities with the conserved region of the adenovirus E1A protein. The E1A genes of different adenovirus serotypes have three highly conserved regions (CR1,2,3) in common. Almost all the CR2 and the N-terminal part of CR1 share striking amino acid similarities with part of E7 and SV40 large T antigen. This conserved structure is responsible for pRb binding. It was shown for HPV-16 E7 that the pRb binding site is contained within a region likely to have high conformational flexibility that might be of importance in allowing promiscuous interactions with other proteins.[48] The pRb is bound to E7 proteins of high-risk HPVs with a higher affinity than to E7 proteins of the low-risk group or of cutaneous HPVs.[43,44,52] Interestingly, a single amino acid residue within the pRb binding domains, consistently different in the high-risk and low-risk HPV-derived E7 proteins (aspartic acid versus glycine), is primarily responsible for the different pRb binding capabilities.[43] Mutations in this region that cause an impairment of E7 proteins to bind to pRb also result in a reduced cooperation with *ras*,[43,44] indicating that binding to pRb is a major determinant of cellular transformation. However, the consideration that mere binding of pRb protein alone might not be sufficient for efficient transformation is supported by the fact that mutations in the N- and C-terminal regions of the E7 protein, outside of the pRb binding site and casein kinase II (CKII) recognition sequence (see below), can also affect the transformation capability.[53]

c. *Phosphorylation of E7 Proteins.* The amino-terminal halves of high-risk and low-risk HPVs contain consensus recognition sites for CKII.[52] It has been shown that the E7 proteins of high-risk HPVs are phosphorylated *in vitro* at a higher rate than the E7 proteins of low-risk HPVs. In HPV-16, E7 protein is phosphorylated *in vivo* by CKII at two adjacent serines at positions 31 and 32, just C-terminal to the pRb-binding region. Mutation analysis revealed that phosphorylation of either or both serines is important for the ability of E7 to immortalize and transform rodent cells. However, a negative charge at this position is not required for pRb-binding and *trans*-activation of the adenovirus E2 promoter.[52,54] The efficiency of pRb binding and phosphorylation correlates well with the relative oncogenic potential of the different virus types and suggests that quantitative rather than qualitative biochemical differences might account for the comparatively weak oncogenicity of low-risk and cutaneous HPVs.

d. *Trans-Activation.* The E7 protein of HPV-16 can *trans*-activate heterologous promoters, which was first demonstrated by using the adenovirus E2 early promoter.[55] The adenoviruses are extensively studied model systems for cellular transformation. E1A proteins encoding the CR1 and CR2 regions, although obviously dispensable for lytic infection, are critical for cellular transformation through the activation of cellular and viral genes with E2F-binding sites.[56,57]

Mutations within the adenovirus CR1 and CR2 domains affect both transcription activation and transformation. This is at variance with high-risk E7 proteins, where mutants with impaired *trans*-activation potential are still able to transform.[43,53,54] Moreover, transcriptional transactivation of the adenovirus 2 promoter is a property shared by E7 proteins of both low-risk and high-risk HPVs.[58]

e. *Biological Properties.* The first indication that certain HPV types encode efficient oncoproteins was derived from cell culture experiments in which transfection of HPV-16 DNA together with a resistance gene resulted in the appearance of transformed, drug-resistant colonies of the otherwise nontransformed murine cell lines NIH3T3. These transformants were anchorage independent and tumorigenic in nude mice.[59] In subsequent experiments it was shown that the expression of ORF E7 in the absence of E6 is capable of transforming rodent fibroblast lines.[55,59–61] This activity could be assigned to the N-terminal part of E7. In contrast to the transforming activities of high-risk HPV E7, E7s of low-risk HPVs and cutaneous HPVs such as types 5 and 8 showed no *in vitro* transformation of established rodent cells.[44,58] The apparent stringency of transformation of rodent cell lines is, however, quite questionable, as DNA of the "no-risk" HPV-1, which generates definitely harmless palmar warts, induced transformed foci in the established murine C127 line.[62]

With regard to association between biochemical and biological properties, cooperative action with a mutated *ras* oncogene appears to be a more appropriate criterion. Again, high-risk HPV E7 proteins cooperate very efficiently with *ras* in primary rodent cells leading to anchorage-independent and tumorigenic lines.[57] By mutations in the E7 ORF and construction of chimeric E7 genes, a region at the N-terminus of the E7 protein was found responsible for this activity.[58,63] Cooperation of low-risk HPVs or cutaneous HPVs with *ras*, however, is not abolished but reduced in comparison to the high-risk HPV E7s. In fact, tumorigenic transformed clones of rodent cells had been readily obtained on transfer of the genital HPV-6 and -11[64,65] and cutaneous HPV-5, although HPV-8 was ineffective in this assay.[44] The extent of cooperative ability correlates quite well with the clinical observation on tumorigenicity of the respective HPV types. This in turn seems to correlate well with their respective pRb binding capacities.[43,44,52]

An important characteristic of nuclear oncoproteins is whether they are able to induce long-term growth of primary cells with emphasis for normal human keratinocytes. The full-length genomes of high-risk HPVs can efficiently immortalize epithelial cells and fibroblasts of human or rodent origin.[55,66–68] This activity is provided mainly by the E7 ORF. However, the concomitant expression of the corresponding E6 proteins enhances the E7-mediated effects.[69,70] Recent improvement of tissue culture techniques has further substantiated the contribution of high-risk HPVs to cellular alterations. In the organotypic raft cultures human keratinocytes are cultured on top of collagen fibers and a feeder layer of nondividing cells. Keratinocytes are exposed to the liquid–air interface, somehow similar to the *in vivo* situation. Transfection with high-risk HPV sequences can alter the morphological differentiation of human keratinocytes in a manner similar to that found in cervical lesions *in vivo*.[71]

2.4.2. Properties of HPV E6

a. Structure and Biochemical Properties. The E6 gene of cancer-associated HPVs is the second major oncogene encoded by the viruses. The E6 protein is well conserved among the various HPV types (Fig. 4). The E6 proteins of PVs are approximately 150 amino acids long and contain multiple regularly spaced Cys-X-X-Cys sequences, which are characteristic of zinc finger domains. Indeed, it was shown that the E6 protein of HPV-18 is able to bind zinc *in vitro*.[72] From studies with antisera against overlapping synthetic peptides, it was deduced that the putative zinc finger regions, which are separated by a flexible linker region, are obviously hidden in the E6 protein and apparently take part in higher-ordered protein structures.[48]

It is somehow surprising that little consensus exists about the subcellular localization of E6. In previous studies E6 proteins were found associated with the nuclear matrix and nonnuclear membranes[73]; however, a recent immunohistochemical analysis of widely used cervical cancer cell lines, such as HeLa, CaSki, and SiHa, demonstrated a definitely cytoplasmic localization of oncogenic E6 proteins.[74] A perinuclear signal that colocalized with the Golgi body and a punctate nuclear signal were also described by Keen *et al.*[75]

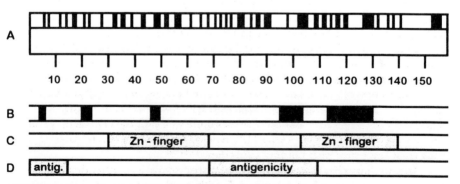

FIGURE 4. Properties of the HPV E6 oncoprotein. (A) Amino acids that are conserved throughout all genital HPVs are indicated by black bars. (B) Black bars indicate regions that mediate degradation of p53 by HPV-18 E6. (C) Zinc-binding regions and (D) regions of antigenicity of HPV-16 E6 are shaded.

b. *Transcription.* In high-risk HPVs, the ORFs E6 and E7 are separated by a few base pairs, which allows transcription of a polycistronic mRNA from the unique promoter for the transcription of viral early genes. Despite the adjacent localization of the E6 ORF to promoter sequences, only a minor transcript constituting a few percent of early viral transcripts encodes a full-length E6 protein.[76] The predominant E6–E7 colinear transcripts in cervical cancer and cancer-derived cell lines are generated by splicing out introns within the E6 ORF. These transcripts encode a full-length E7 protein as well as two different short E6 proteins, designated E6* and E6**. They have no obvious detectable biological activity but rather serve to facilitate the translation of E7 transcripts.[77] In accord with their transcription levels, the relative abundance of E7 proteins is opposed to low or even undetectable levels of full-length E6 proteins. In contrast with genome organization in high-risk HPVs, in low-risk HPVs the ORFs E6 and E7 overlap by a short stretch of nucleotides. This demands an additional promoter within the E6 ORF for E7 transcripts and, in turn, omits the need for splicing sequences.[78] Indeed, with the exception of HPV-43, low-risk HPVs cannot express E6* proteins. Nevertheless, at the mRNA level, the ratio between E6- and E7-encoding transcripts is similar to that observed with high-risk HPVs.[78]

c. *Binding to p53.* Cellular wild-type (wt) p53 is a tumor suppressor protein that binds as a homotetramer to site-specific DNA sequences.[79] The p53 has transcriptional activity and is a cellular target for several viral oncoproteins including oncogenic E6. Several studies have shown that p53 sequence-specific *trans*-activation is efficiently inhibited by E6 oncoproteins but not by low-risk HPV E6 or HPV-16 E6 mutants that do not bind p53.[80,81]

In contrast to SV40 large T antigen, where tight binding of p53 increases the half-life of p53 severalfold, binding of p53 to E6 induces subsequent specific degradation of the p53 protein. This is mediated by the ubiquitin-dependent protease system[82] and the presence of an additional protein called E6-AP, which promotes the degradation process through its function as a ubiquitin ligase.[83,84] The gene for E6-AP was recently cloned,[85] and this will definitely allow insights into its role in the context of normal cells in the near future. Efficient abolition of p53 functions furthermore appears guaranteed by the ability of E6 to target all quaternary forms of p53, including monomers and dimers,[86] and to firmly keep p53 from entering the nucleus as evidenced by the exclusively cytoplasmatic colocalization of the protein complex.[74] In any case, the presence of E6 oncogenes can significantly reduce the level of available active p53 protein as observed in many but not all cervical tumors and tumor cells.[74,82,87,88]

Several efforts were undertaken to localize the p53-binding domains within the E6 proteins. From a series of mutated E6 genes, assayed in various *in vitro* and *in vivo* systems, a complex picture emerged. In contrast to E7, where the binding site to other cellular proteins has been localized rather precisely, the corresponding region in the E6 for p53 is apparently spread over the entire protein including the second zinc finger motif.[81,89] In a recently reported analysis with a series of mutation and deletion mutants of HPV-18 E6, it was demonstrated that at least three domains of the E6 protein are essential for the ability to degrade p53, suggesting a complex pattern of interactions among E6, p53, and the ubiquitine pathway.[81] In the case of HPV-18 E6, there are more extensive regions involved in the interaction with p53 than in the case of HPV-16 E6,[90] where binding appears to be confined to a short stretch in the N-terminus.[80] Moreover and most notably, E6 mutants had been described that did not stimulate p53 degradation; nevertheless, they abolished p53 *trans*-activation.[81,87] From these data it was concluded that p53 binding alone is sufficient for the abolition of transcriptional activity and that degradation is an independent event.

E6–p53 protein complexes have not yet been detected *in vivo* using E6-specific antibodies. By using glutathione-S-transferase–E6 chimeric proteins, Keen *et al.*[75] recently succeeded in identifying seven cellular proteins that interact with HPV-16 E6, which suggests that the E6 protein may participate in large complexes of cellular proteins. At present, only two proteins, of 100 kDa and 33 kDa, respectively, were found to interact directly with E6. The 100-kDa protein was assumed to be E6-AP, but the nature and function of the smaller protein remain unknown. In addition, a strong histone H1 kinase activity was specifically associated with E6 of high-risk and low-risk HPVs, although it was absent when E6 proteins of the cutaneous types HPV-5 and -8 were used. This kinase phosphorylates a protein, p182, also found in the E6 complexes.

d. *Trans-Activation.* As already discussed, the adenovirus E2 promoter is responsive to *trans*-activation by several viral oncogenes. It was of interest to investigate, in analogy to E7, whether E6 proteins also have transcriptional activities, especially because HPV-18 E6 protein appear to have nonspecific DNA-binding properties.[73,91] Cotransfection experiments with various HPV-16 E6 constructs and plasmids that contained the adenovirus E2 promoter 5' to the CAT reporter gene revealed a severalfold activation of transcription by the full-length E6 protein and a largely reduced activity of the short E6*. The extent of transcriptional activation of E6 even exceeded that obtained with a corresponding E7 construct. Interestingly, both E6 and E7 of the low-risk HPV-6 were also transcriptionally active in the same experimental system.[77] In an extension of these observations, Desaintes, *et al.*[76] investigated the transcriptional activity of HPV-16 E6 on different viral promoters and mutations therein. From the data the authors concluded that E6 might interact with the transcription initiation complex to anchor it properly to the promoter and to potentiate its activation. This activity of E6 is restricted to TATA-containing promoters and does not require enhancer elements. Thus, the repertoire of genes that might be potential targets for E6, at least theoretically, is clearly at variance with E7-mediated *trans*-activation, which is confined to E2F responsive sequences.

In search of the various domains within the E6 oncoproteins, Pim *et al.*[8 l] analyzed a series of mutated HPV-18 E6 proteins with regard to transcriptional potency. Interestingly, none of the altered E6 proteins showed a reduction of *trans*-activating activity on the adenovirus E2 promoter.

e. *Biological Activities.* Compared to the plethora of data on biological effects of E7, the transforming potential of E6 proteins has attracted attention only recently. In fact, the ORFs E6 of individual oncogenic HPVs exert a similar spectrum of activities in various cell systems, albeit at lower efficiency, as the corresponding E7 proteins. The E6 genes of high-risk HPVs and several cutaneous HPVs were able to transform established rodent cells.[77,92] In addition, E6 oncoproteins of genital

HPVs can cooperate with a mutated *ras* gene to transform epithelial rodent cells in culture, although this cooperation appears to be dependent on the genetic background of the cells.[81,93,94] At the genetic level, the cooperative activity of the E6 molecule could be assigned to a large stretch of its carboxy-terminal half. Several mutations and small deletions within this region were shown to abolish binding and degradation of p53, yet they did not affect cellular transformation in cooperation with *ras*.[81] These data together with recent observations made with HPV-16 E6-immortalized human mammary epithelial cells[88] indicate that immortalization and/or cooperation with *ras* do not crucially depend on the functional inactivation of cellular p53 and that another function of E6, probably binding to other cellular proteins, is required that is different from its association with p53.

In contrast with E7, several efforts to immortalize human keratinocytes by transfection of E6-expressing plasmids have failed so far. Although this tissue obviously appears resistant to E6-mediated immortalization, at least *in vitro*, human mammary epithelial cells can be readily immortalized by several different E6 genes including those of HPV-6 and BPV-1.[88,95] Expression of these proteins was sufficient for human mammary epithelial cells to bypass the M_1 stage of cellular senescence followed by a period of cell crisis, after which eventually immortal lines evolve. This postulated two-step model of immortalization was originally deduced from the growth behavior of SV40-infected human cells.[96]

In vivo, HPV-18-associated cancers in general appear to be more aggressive than HPV-16-associated malignancies.[97] It is reported that the immortalization of human foreskin keratinocytes is 10–50 times more efficient than HPV-16. This difference might either result from distinct transcriptional regulation caused by different URRs or be the result of intrinsic properties of the respective E6–E7 proteins.[16]

It is clear that high-risk HPVs are contributing to the early steps of cellular transformation. However, HPV oncoproteins can influence cells at more than one step in their progression toward malignancy. It was shown that the introduction of HPV-16 E6 and E7 genes into nontumorigenic or nonmetastatic cell lines converted them into metastasizing lines. The HPV-16 E7 gene had a greater effect on the induction of metastasis than that of the E6 gene. Again, the low-risk viruses differed rather quantitatively than qualitatively: HPV-6b E6 or E7 genes produced a few nodules when injected intravenously into nude mice, but not when tumor cells were injected intramuscularly.[98]

2.4.3. Properties of HPV E5

The E5 represents a third potential oncogene of HPVs. Although the E5 ORF is lost in the majority of cervical cancers in the course of viral integration, transcripts encompassing the ORF E5 and E4 are the predominant mRNAs found in premalignant lesions.[99] This suggests that these proteins play a role in early stages of tumor development, at a time when HPV DNA is in general episomal.

What is common to E5 proteins of PVs is that they are hydrophobic, are approximately 10 kDa in size, are localized in cellular endomembranes including the Golgi apparatus, and are complexed with a 16-kDa pore-forming component of the vacuolar ATPase.[100] They also share some structural similarities: a Cys-containing sequence near the C terminus and a glutamine residue within the hydrophobic domain.

In both low-risk and high-risk HPVs, the E5 proteins exert a low oncogenic potential when assayed in established rodent cells,[101,102] indicating that the E5 ORF might encode a third oncogene.[103] Indeed, the transforming ability of the HPV-16 E5 proteins is qualitatively similar to the effect of BPV-1 E5, which is almost exclusively responsible for the efficient focus-forming activity of the entire BPV-1.[103] The BPV-1 E5 binds to and activates growth factor receptors including epidermal growth factor (EGF)-R, β-type platelet-derived growth factor (PDGF)-R, and colony-stimulating factor (CSF)-1-R. Interestingly, this receptor-binding activity of BPV-1 E5 is shared by the HPV-6 E5 protein but not by the HPV-16 E5.[104] In view of the different biological effects of HPVs *in vivo*, it is assumed that the transforming activities of the various HPV E5s involve distinct mechanisms. The direct activation of growth receptors might contribute to the biological effects of low-risk HPVs, but this activity appears to be either dispensable or circumvented in the case of high-risk HPVs. Evidence for the latter possibility was presented recently by Bouvard *et al.*[105] who showed that HPV-16 E5 enhances the expression and activity of E7.

2.4.4. Properties of HPV E1 and E2

As in other instances, BPV-1 has served as prototype for the genetic analysis. The BPV E1 has been studied extensively in recent years and was shown to be a multifunctional phosphoprotein that is an important factor in PV replication [reviewed by Lambert[106]]. Its activities include oribinding,[107,108] DNA helicase/ATPase,[109,110] and E2 binding.[111,112] Only recently have similar functions been demonstrated for HPV E1.[113,114]

The E2 proteins function as *trans*-acting transcriptional modulators. E2 proteins consist of three functionally different domains, and their arrangement is common to all PVs. The amino-terminal and carboxy-terminal regions are highly conserved among all the E2 proteins, but the hinge region in between is variable with respect to length and amino acid sequences. The N-terminal part of E2 proteins, which is unique to the full-length E2 protein, constitutes the activator domain, and the C-terminal part mediates dimerization and binding to DNA. All forms of E2 are able to bind as dimers to the highly conserved consensus sequences in the URRs. A recent crystallographic analysis of BPV-1 E2 revealed an unusually folded structure of the DNA-binding domain with a multitude of mutual contacts.[115]

The full-length E2 protein of BPV-1 comprises 410 amino acids and is an activator or repressor of transcription, depending on the promoter context of a target gene. In BPV-1, the E2 ORF additionally encodes two proteins that lack the N-terminal part and therefore function as transcriptional repressors by binding to the same enhancer elements. One of these repressors, E2-TR, is translated from an internal methionine downstream in the E2 gene. This is the most abundant E2 transcript in BPV-transformants, a finding that correlates well with the observed low level of viral transcription in these cells. In HPV-16-transfected cervical keratinocytes, the analogue of the major E2 repressor of BPV-1 could not be synthesized, but two transcripts with presumably repressing capacities have been identified in HPV-11, one of which is similar to the minor repressive E2 protein in BPV-1.[32]

With regard to transcriptional regulation of early viral genes, the gene dosage (i.e., number of viral copies in a cell) seems to play only a minor role for overall transcription level in a cell. This is obviously balanced by the E2 protein that controls its own transcription as well as that of the E6 and E7 ORFs by transcriptional stimulation at low levels (small numbers of viral copies) and inhibition of transcription at high levels (large numbers of viral copies).[116] Integration accompanied by disruption of the E2 gene alters this balance in favor of more abundant oncoproteins. In fact, E1 and E2 can negatively regulate one of the most apparent biological effects to be observed in cell systems, cellular immortalization.[36]

2.4.5. Properties of HPV E4

E4 proteins are found predominantly in the cytoplasm of suprabasal cells. Localization to the intermediate filament network might indicate that the proteins interfere with the cross-linked cytokeratin matrix *in vivo*. However, E4-induced collapse of cytokeratin filament is not observed regularly *in vitro*.[117] Thus, the consequence of this protein's associations *in vivo* is unclear. It was shown recently that the E4 protein of HPV-1 binds zinc readily and that this chelation is mediated by three histidine residues and an unidentified fourth ligand, although the arrangement of the residues did not conform to the classical metal-binding motif. Zinc binding was essential neither for E4 dimerization nor for binding to the cytokeratin network.[118] By comparison with other zinc-binding proteins, it is speculated that the zinc ions might mediate protein interactions. Whether this feature is a common characteristic of all E4 proteins is not yet known. In fact, all of the known E4 protein sequences contain cysteine and histidine conformations with potential metal-binding motifs.

2.4.6. Properties of HPV L1 and L2

The coding capacity of the various HPV L1 ORFs is about 500 amino acids, yielding proteins with an approximate M_r value of 55 kDa. There is a broad genetic diversity in the L1 ORF of the various HPVs. Indeed, a short stretch of residues within the L1 protein was used by Chan *et al.*[119] to construct a PV phylogenetic tree. Comparative analysis of this sequence, which comprises only a small segment of the genome, shows the relatively close evolutionary relationship between HPV-11 and -6 and the distance to HPV-16 and -18, which themselves diverge rather extensively.

A specific feature of the HPV capsid proteins has recently drawn the attention of several investigators. It was shown that recombinant L1 major capsid proteins, expressed in eukaryotic or insect cells, assemble spontaneously into virus-like particles (VLP), which consist of spherical particles with a regular array of capsomers morphologically similar to native virions.

The ability of the capsid proteins to assemble spontaneously into VLPs is a general feature of PVs and not restricted to certain human or animal types. The VLPs retain conformational neutralizing epitopes and are very antigenic as demonstrated by the high titers of neutralizing antibodies elicited on injection into animals, without need for additional adjuvants.[120–122]

The predominant structure of L2 is a 90-kDa protein, and it represents about 10% of the total protein content of HPV particles. Its exact localization in viral particles has not been determined. Recent studies have suggested that L2 may be required for the encapsidation of DNA,[123] probably mediated by the relatively conserved N-terminal region of the L2 protein as shown by Zhou *et al.*[124] for HPV-16 L2. Although the L2 proteins contain a number of type-specific epitopes, their contribution to eliciting an immune response is unclear. In a recent study HPV-1 VLP, consisting of both L1 and L2, was used to detect capsid-specific antibodies in sera from individuals with plantar warts. Interestingly, humoral immune response was directed only against L1 epitopes, with no detectable serum reactivity toward L2 protein.[125] This might suggest that L2 is either not accessible or poorly antigenic in the virions. The only effect of L2 observed so far is that concomitant expression of recombinant L1 and L2 increased the yield of VLP by severalfold.[121]

3. INTERACTION OF VIRAL PROTEINS WITH CELLULAR FACTORS

3.1. Binding of E7 to pRb and p107

The viral oncoproteins encoded by adenovirus E1A, SV40 large T-antigen, and HPV E7 are able to dissociate a protein complex formed by E2F-1 and pRb. The transcription factor E2F is in the midst of regulatory mechanisms for cell cycle control and is contained in several multimeric protein complexes. E2F consists of two groups of proteins that form heterodimers and are at present mainly defined by their molecular masses.[126] E2F-1, one species of E2F, is a 437-amino-acid nuclear protein that is synthesized in a cell-cycle-dependent manner. A region in its N-terminal half mediates specific DNA binding, and the 60-amino-acid *trans*-activation domain was localized to the C-terminal part and overlaps with the 18-amino-acid pRb binding site. E2F-1 binds pRb in G_1 when pRb is underphosphorylated.[127] pRb becomes gradually phosphorylated on serine and threonine residues at middle to late G_1 and is increasingly phosphorylated in S and G_2 phase before it undergoes dephosphorylation in mitosis. The waves of phosphorylation events are regulated by cyclin-dependent kinases (cdks),[128,129] and dephosphorylation is induced by a type-1 phosphatase.[30]

High-risk HPV E7 has been shown to bind pRb, and this is thought to cause disruption of pRb/E2F complexes. Within the pRb-binding regions of the viral proteins, the sequence Leu-X-Cys-X-Glu is invariant.[131,132] In E7, individual point mutations in this region abolish or severely reduce binding of E7 to pRb in *in vitro* assays, reduce *trans*-activation of the adenovirus 2 promotor, and decrease transforming activities.[52,55] It was shown that E7 and E2F bind to separate sites on the pRb and that additional C-terminal sequences in the E7 protein corresponding to the zinc-finger-like region are required for the efficient interruption of pRb/E2F complexes.[133] The release of free E2F-1 molecules from interference with HPV E7 could lead to deregulated expression of E2F-responsive genes. The number of genes with identified E2F-responsive promoter sequences is increasing. Some of them encode enzymes important in nucleotide metabolism (thymidine kinase, thymidylate reductase, dihydrofolate reductase, ribonucleotide reductase, and DNA polymerase α). In addition, cyclin A and the cellular transcription factors *myc* and *myb* are activated via their E2F sites before onset of DNA synthesis.[51,134,135] Conversely, in the case of pRb and cdc2, E2F is apparently involved in down-regulation of these genes.[51] Obviously, modulation of expression of E2F-sensitive genes is a prerequisite for both immediate early and late G_1 transcriptional events and transversion of the cell cycle. That deregulated expression of E2F can indeed lead to transformation of primary cells has recently been demonstrated by Singh *et al.*[136] Overexpression of E2F-1 in rat embryo fibroblasts was clearly associated with tumorigenic transformation of these cells.

Interestingly, the Leu-X-Cys-X-Glu sequence is also found in the cyclin D proteins. Recent reports demonstrated that each of the three G_1 cyclins, D1, D2, and D3, can physically interact with pRb and two specific kinases, cdk4 and cdk6, which in turn mediate pRb phosphorylation.[137–139] Because of the complexity of cell cycle regulation, a huge number of feedback loops must definitely exist. First examples are at hand: cyclin D1 expression is positively regulated by pRb at the transcriptional level. Cyclin D/pRb complex formation could be impaired by interference by HPV E7. Binding of pRb to E7 might either lower the expression of G_1 cyclins D, as E7-bound pRb probably does not function like the genuine cyclin D/pRb complex in transcriptionally activating cyclin D genes, or E7 might interfere with complex formation itself. In fact, HPV-containing cervical carcinoma cells or HPV-immortalized keratinocytes have almost undetectable levels of cyclin D/pRb complexes,[129,132,140] a situation that is functionally analogous to mutation or loss of pRb gene.

The cellular protein p107 shares many structural and biochemical features with the Rb protein, and both are potent inhibitors of E2F-mediated transcription. In addition, p107 is phosphorylated in a cell-cycle-dependent manner similar to that observed with pRb. However, pRb-induced growth arrest of sensitive cells can be rescued by coexpression of cyclin A or E, presumably by mediating hyperphosphorylation of pRb, but the same experimental approach is ineffective when the cell cycle is arrested by overexpressed p107. In addition, E2F-1 can rescue pRb growth arrest but not p107-mediated arrest, which is suggestive of distinct underlying mechanisms. Whereas pRb associates with E2F mainly in late G_1 and S phase, p107 forms two independent complexes with E2F: one includes cdk2 and cyclin A and is found mainly in S phase; the other includes cdk2 and cyclin E and is found in G_1 phase. It has been shown that E7 binds p107,[141,142] and targeting of complexes of p107/E2F and of others that include pRb, cyclin A plus cdk2,[142–144] and yet unidentified factors may have consequences for the ability of E2F to modulate transcription. However, and this in contrast to E1A, neither high-risk nor low-risk HPV E7 seems to dissociate the various components.[140,144] Hence, the cellular consequences of this event are unclear. Recently, Lam *et al.*[134] have shown that cell-cycle-dependent expression of B-*myb* is deregulated by targeting of a p107/E2F complex by HPV-16 E7, and this is the first example of a cellular gene whose expression is directly regulated by interactions between HPV E7 and E2F containing protein complexes.

Apart from the different capacities of E7 proteins to bind cellular proteins, there is also an apparently distinct affinity for cellular RNA. In contrast with HPV-6b E7, HPV-16 E7 protein was found tightly associated with RNA molecules.[145] The region in the protein responsible for RNA binding and the functional significance of this association are unknown.

3.2. Binding of E6 to p53

As stated earlier, E6 proteins of high-risk HPVs bind more efficiently to wt p53 than those of low-risk HPVs, and efficient binding leads to abolition of p53 function.[89,90,147] There is an emerging consensus that p53, if required, exerts crucial functions in the regulation of cell cycle progression: p53 is a transcription modulator capable of repressing promoters, particularly those with TATA box sequences,[148] and of stimulating cellular or synthetic promoters containing p53 consensus sequences[79,149]; it is a phosphoprotein that is spatially regulated during the cell cycle.[150]

In deliberating possible consequences for cells in which p53 is bound to E6, it might be advantageous to look at the most extreme situation in which p53 is completely lacking in all cells of an entire organism. In fact, the successful creation of p53-knockout mice has provided a surprise. These animals developed normally without any obvious defects, but after a few months postpartum a series of tumors with quite distinct pathologies arose in the p53-deficient animals. The main conclusion from these experience was that wt p53 is definitely not needed for normal development, but its absence predisposes most, if not all, tissues to segregate malignant derivatives.[151] In recent years mutations and loss of p53 alleles have been extensively reported to occur in about half of malignant human tumors, a finding that *a priori* pointed to a central role of wt p53 in opposing aberrant cell growth [reviewed by Vogelstein and Kinzler[152]]. In addition, p53 mutations are associated with the inherited cancer-susceptibility Li–Fraumeni syndrome in which one mutated p53 gene is inherited, and additional loss or mutation of the normal allele is found in the tumors of affected people.

The functions of normal p53 were quickly discovered.[153] It acts as a cellular checkpoint controller for the transition from late G_1 to S phase and is able to mediate growth arrest in G_1 when the cell or cellular DNA is damaged by a variety of (but not all) distinct factors such as energy-rich radiation, several genotoxic agents, or exposure to suboptimal conditions. Blocking the start of DNA synthesis provides sufficient time for repair of damaged DNA. It follows that growth arrest followed by DNA repair demands coordinated activation and repression of genes involved in these processes. Indeed, the number of identified genes sensitive to p53 activities is continuously growing.[154] For example, one is another tumor suppressor gene, p21 (initially named WAF1/CIP1), which is a potent inhibitor of cdks; p53-induced p21 is abundantly associated with cyclin-E-containing complexes obviously leading to a decrease in associated cdk activity.[155,156] If, however, repair fails, wt p53 can induce apoptosis, thereby eliminating cells that might otherwise replicate their damaged DNA and, furthermore, pass the various lesions on to their progeny. The involvement of wt p53 in apoptosis is best documented with thymocytes, where p53-null cells can tolerate a severalfold higher irradiation dosage than their normal counterparts.[157]

Abrogation of p53 function by efficient binding to E6 should have consequences on most or all p53 activities identified so far. Indeed, Mietz et al.[80] demonstrated that the transcriptional function of p53 could be inhibited by the SV40 large T antigen and the E6 of HPV-16. Mutants of both oncoproteins that could not bind p53 and, similarly, E6 genes of the low-risk types were unable to block transcriptional activation. In an extension of these observations, Crook et al.[89] analyzed the extent of repression of p53 transcriptional activity in the presence of mutant and chimeric E6 proteins and found a good correlation between the decrease in p53-mediated transactivation and oncogenic potential of E6 proteins. It was recently shown that E6 of HPV-18 inhibits p53 trans-activation following genotoxic DNA damage with UV irradiation.[158] In addition, cells expressing HPV-16 E6 were found unable to accumulate high levels of p53 in response to actinomycin D and failed to arrest in G_1, as is normally observed in wt-p53-expressing cells in response to such damage.[159]

An example of reduced apoptosis of E6-expressing cells was recently published by Pan and Griep,[160] who studied cell cycle regulation in the lens of HPV-16 E6 or E7 transgenic mice. At least a part of the observed morphogenic alterations of lens development could be attributed to the lack of apoptotic-like DNA degradation in E6-transgenic animals.

One consequence of p53 abolition deserves special notion. Cells that have no p53 protein are at least a million times more likely to permit DNA amplification than cells with normal amounts of p53 protein.[161,162] The cause for this enormous genome rearrangement is not understood. Whatever the underlying mechanisms might be, it is conceivable that enhanced genomic instability can easily accelerate the occurrence of genetic changes that allow cells to overcome the normal controls over excessive multiplication. In agreement with this idea, it was found recently that embryonic fibroblasts from p53-knockout mice, in addition to growing faster than hemizygotic or wt populations and showing karyotypic abnormalities, have acquired the ability to grow indefinitely.[163] Although immortalization of p53-null cells did not appear to be a direct consequence of lacking p53, it points to the importance of normal p53 functions in a cell.

To interfere with wt p53 activities is probably not the unique feature of E6 oncoproteins. The ability of E6 proteins to trans-activate TATA-sequence-containing promoter[76,77] suggests that E6, like E7, may exert its effects through a number of different cellular targets and mechanisms, almost all of which are still awaiting elucidation.

4. EPITHELIAL CELLS

4.1. Differentiation of Epithelial Cells

Epidermis and mucosa are highly and strictly organized tissues comprising multiple layers of specialized cells at different stages of differentiation. Proliferative cells are restricted to the basal layer, which is composed of cells committed to differentiation. The adherence of basal cells to the basement membrane is mediated by hemidesmosomes. These are calcium-activated adhesion plaques that consist of many different proteins including $\alpha_6\beta_4$ integrins. Basal cells make contract with each other and to the suprabasal cells via desmosomes, which consist of a different set of proteins. Hemidesmosomes and desmosomes function as anchor structures for the intracellular keratin filaments.

The capacity for mitosis is immediately lost when cells leave this layer. Differentiating cells migrate toward the upper layers and synthesize structural proteins characteristic of each type of epithelium. For instance, adult basal keratinocytes synthesize keratins 5 (K5) and 14 (K14). In suprabasal cells, synthesis of K5 and K14 progressively declines, and synthesis of K1 and K10 takes over. In the cell layers closer to the surface, the keratins make up up to 85% of the total cellular protein. Terminal differentiation of keratinocytes is associated with profound changes in the synthesis of macromolecules, leading eventually to the constitution of the cornified scaling layers of dead skin. This is at variance with multilayered mucosal tissue, where the upper layer is composed of living, nucleated cells.[164,165]

The expression pattern of various keratins depends on the tissue. Other stratified nonskin squamous epithelia express K4 and K13. Migration of basal cells to the upper layers of epidermis is regulated by complex mechanisms that are not yet elucidated. The pattern of keratin expression also depends on physiological processes such as tissue repair. In the course of wound healing, suprabasal cells transiently switch on the expression of K6 and K16.

4.2. Effects of HPV Infection on Cellular Response to Growth Factors

Growth factors are unique regulators of cellular proliferation and differentiation, and loss of sensitivity of this regulative action might be a fundamental step in the process of carcinogenesis.[166] Human PV infection interferes with a cell's ability to react to growth factors, and this could be a main determinator for tumorigenicity.

Transforming growth factor-β (TGF-β) is a family of structurally related polypeptides that are regulators of cellular proliferation and differentiation. In epithelial cells, TGF-β has been shown to inhibit cell proliferation.[167] In HPV-infected keratinocytes, the degree of malignancy seems to be correlated with sensitivity to the actions of TGF-β. Several investigators have shown that HPV-16- or HPV-18-immortalized keratinocytes were nearly as responsive as normal cells to TGF-β,[168–170] and growth inhibition was accompanied by repression of HPV E6/E7 expression. However, several cell lines, particularly after prolonged cultivation or malignant transformation, exhibited decreased or no sensitivity to TGF-β.[168,169,171,172]

Repression of early gene expression in HPV-immortalized keratinocytes has also been achieved by treatment with EGF, and EGF was necessary for optimum growth. An EGF-responsive silencer element has been mapped in the proximal 124-bp of the HPV-16 URR.[173] Another group postulated a role for EGF in HPV-16-immortalized keratinocytes in stimulating cell growth by reducing the levels of inhibitory insulin-like growth factor (IGF)-binding proteins, thereby potentiating the effects of IGF.[174]

Recently, it has been indicated that resistance to tumor necrosis factor-α (TNF-α) is associated with the tumorigenicity of keratinocytes containing HPV-16. Escape from TNF-α-mediated growth inhibition was accompanied by lower numbers of TNF-α receptors in highly tumorigenic cell lines.[175]

Retinoids are potent regulators of epithelial cell growth and differentiation and have been shown to reverse preneoplastic cervical lesions caused by infection with HPVs. Several mechanisms have been suggested to mediate this effect. The expression of HPV-18 E6 and E7 genes in HeLa cells was down-regulated by retinoic acid.[176] A similar effect of retinoic acid as well as retinol on E6/E7 expression has been observed in human keratinocytes immortalized by transfection with HPV-16 DNA.[171] Sizemore and Rorke[177] found that retinoids reduced EGF receptor levels that have been found elevated in HPV-16-immortalized ectocervical epithelial cells. Hence, retinoic acid can possibly attenuate the increased responsiveness to EGF in these cells. However, altered epithelial cell differentiation of HPV-immortalized keratinocytes is resistant to high levels of retinoic acid as evidenced by maintenance of elevated expression of differentiation markers. Blockage of terminal differentiation in these cells required 10- to 30-fold higher concentrations of retinoic acid than in control cells.[178]

It has been suggested that retinoid-induced secretion of TGF-β in human as well as murine keratinocytes[179,180] might account at least for some of the effects of retinoids on normal as well as HPV-immortalized keratinocytes.

4.3. Aberration of Cellular Genes

Although integration of viral genes into the host genome is definitely important for release of the E6 and E7 ORFs from regulation by E1 and/or E2, another aspect closely associated with integration should not be neglected namely, the alteration of normal cellular genes. The frequency of finding a common chromosomal alteration in cervical carcinomas is in the range of 10–20%. Although only a small number of cases have been analyzed so far, it is a striking observation that both HPV-16 and HPV-18 sequences are integrated in several cases in two chromosomal regions, 8q24 and 12q13. These regions have been found altered in several other tumors with quite distinct pathologies. This suggests that eventual transformation of normal cells into tumor cells might share common pathways at some stage of their development. A variety of potential oncogenes involved in signal transduction and cell cycle regulation have been located to these two integration sites, most notably c-*myc*. Yet no clear correlation was found between chromosal integration of HPV DNA and deregulation of specific genes.[181]

4.4. HPV-Negative Cervical Carcinomas

A minority of cervical cancer do not harbor any detectable HPV sequences. Because of the importance of efficient inactivation of the tumor suppressor proteins pRb and p53, the question of a substitute mechanism arose. Spontaneous mutations in either one or both of these genes would be an obvious possibility. Indeed, several studies suggested that p53 mutations occur with a high frequency in

HPV-negative carcinoma cell lines and with low frequency in HPV-positive ones.[182–184] However, a recent analysis of the p53 status in 257 cervical cancer samples clearly showed that the rate of p53 mutations is much lower than initially reported.[185] Park et al.[185] concluded from compiled data that the frequency of p53 mutations in HPV-positive cervical carcinoma is 3%, and in HPV-negative ones 15%. This suggests that other mechanisms than those deregulated by HPVs can be involved in development of cervical cancer.

5. IMMUNE RESPONSE

5.1. To Oncogenic HPV Proteins

Human serological immune response to HPV-16 E6 and E7 proteins have been reported to occur frequently in women with cervical malignancies, but with disappointingly insufficient HPV type specificity. Thus, the clinical relevance of circulating antibodies directed toward oncogenic HPV proteins is unclear. However, the HPV-16 E6 and E7 proteins that are abundantly expressed in precancerous and malignant cervical lesion have been shown to be tumor rejection antigens in murine and rat systems.[186–188] Several human T-cell epitopes have been identified in these oncoproteins, yet no clear correlation was found between the presence and/or extent of cellular immune response and the clinical course of HPV-mediated proliferations.

5.2. To Capsid Proteins L1 and L2

The results of serological tests using bacterially derived proteins or synthetic peptides have not correlated well with other measures of HPV infections. Significant antivirion immune response might have escaped detection, because the majority of epitopes recognized after experimental inoculation are conformation dependent. A major breakthrough might now be the use of noninfectious VLP for vaccination. The efficiently self-assembled capsid proteins L1 and L2 of all PVs investigated so far and derived from various recombinant expression systems retained the correct conformational neutralizing epitopes, are very antigenic, and elicit high titers of neutralizing antibodies when experimentally inoculated into animals.[120–122] Considering the cross-reactivity of capsid proteins among the various HPVs, it is conceivable to apply a composite vaccine of several VLPs that provides cross-protection to the most common HPV types.

6. CONCLUSION

Many efforts have been undertaken in recent years to get detailed insights into the actions of PVs with overt emphasis on oncogenic HPVs. Without doubt, enormous scientific progress has been made. Apart from a large number of details on biochemical and biological properties of HPV oncoproteins, much information on various facets of the intricate network of cell cycle regulation has been obtained. In fact, HPV oncoproteins have turned out to represent indispensable tools in various model systems applied to studies of cell cycle regulation and differentiation. However, despite this unequivocal progress and its benefit for future research and clinical applications, three limitations should be kept in mind.

First, in order to circumvent the complexity of viral–host interactions experimental systems had to be chosen where interference of viral with host proteins is restricted to a few measurable parameters. Thus, almost all of the biochemical data had necessarily been obtained with synthetically produced proteins, and their interaction was studied in highly artificial *in vitro* systems or at least in rather peculiar cell lines. Extrapolation of these pure *in vitro* data to the human *in vivo* situation must be done with caution.

Second, despite advantages in tissue culture techniques, especially those related to the development of raft cultures for keratinocytes, the cells studied in the laboratory are clearly not the natural targets for HPVs *in vivo*. The majority of HPV-containing malignancies, such as cervical and anogenital cancer, develop in areas of transitional epithelia. The accurate regulation of transition from multilayered unpolarized squamous cells to monolayered polarized cylindrical cells must represent a rather difficult local problem. The fact that in women this transition zone is shifted twice with age underlies the evident complexity of local epithelial regulation. Because no experimental systems have been developed for culturing mucosal cells or even such transition epithelia, conceptions of the very early interaction between virus and host remain largely speculative.

Finally, it should be emphasized that HPVs by themselves are in no instance tumorigenic. They rather may act as promoter-like agents in synergy with carcinogenic initiators. Indications for this decisive cooperation are at present mainly deduced from epidemiologic studies rather than from systematic analysis. However, HPVs contribute to malignant transformation of cells by inducing profound and comprehensive deregulation of the cell cycle machinery and circumventing cell cycle checkpoints, which eventually render infected cells sensitive for subsequent genetic and epigenetic alterations.

ACKNOWLEDGMENTS. The authors are supported by grants from Jubiläumsfonds der Österreichische Nationalbank (Project No. 4366 to C.C.) and Hochschuljubilämsstiftung der Stadt Wien (Project No. H-00135/93 to C.S.).

REFERENCES

1. Coggin, J. R., and zur Hausen, H., 1979, Workshop on papillomaviruses and cancer, *Cancer Res.* **39:**545–546.
2. De Villiers, E.-M., 1994, Human pathogenic papillomavirus types: An update, *Curr. Top. Microbiol. Immunol.* **186:**1–12.
3. McLachlin, C. M., Tate, J. E., Zitz, J. C., Sheets, E. E., and Crum, C. P., 1994, Human papillomavirus type 18 and intraepithelial lesions of the cervix, *Am. J. Pathol.* **144:**141–147.

4. Lörincz, A. T., Reid, R., Jenson, A. B., Greenberg, M. D., Lancaster, W., and Kurman, R. J., 1992, Human papillomavirus infection on the cervix: Relative risk association of 15 common anogenital types, *Obstet. Gynecol.* **79**:328–337.
5. Czeglédy, J., Evander, M., Hernádi, Z., Gergely, L., and Wadell, G., 1994, Human papillomavirus type 18 E6* mRNA in primary tumors and pelvic lymph nodes of Hungarian patients with squamous cervical cancer, *Int. J. Cancer* **56**:182–186.
6. Scinicariello, F., Dolan, M. J., Nedelcu, I., Tyring, S., and Hilliard, J. K., 1994, Occurrence of human papillomavirus and p53 gene mutations in Kaposi's sarcoma, *Virology* **203**:153–157.
7. De Villiers, E.-M., 1989, Heterogeneity of the human papillomavirus group, *J. Virol.* **63**:4898–4903.
8. Barr, B. B. B., Beriton, E. C., McLaren, K., Bunney, M. H., Smith, J. W., Blessing, K., and Hunter, J. A., 1989, Human papillomavirus infection and skin cancer in renal allograft recipients, *Lancet* **1**:124–129.
9. Baker, C. C., 1987, Sequence analysis of papillomavirus genomes, in: *The Papovaviridae*, Vol. 2, *The Papillomaviruses* (N. P. Salzman and P. M. Howley, eds.), Plenum Press, New York, pp. 321–385.
10. Griffin, B. E., 1980, Structure and genomic organization of SV40 and polyoma virus, in: *DNA Tumor Viruses*, Part 2 (J. Tooze, ed.), Cold Spring Harbor Laboratory, New York, pp. 61–123.
11. Mitchell, P. J., and Tjian, R., 1989, Transcriptional regulation in mammalian cells by sequence-specific DNA binding proteins, *Science* **245**:371–378.
12. Wolffe, A. P., 1994, Architectural transcription factors, *Science* **264**:1100–1101.
13. Dollard, S. C., Broker, T. R., and Chow, L. T., 1993, Regulation of the human papillomavirus type 11 E6 promoter by viral and host transcription factors in primary human keratinocytes, *J. Virol.* **67**:1721–1726.
14. Chong, T., Apt, D., Gloss, B., Isa, M., and Bernard, H.-U., 1991, The enhancer of human papillomavirus type 16: Binding sites for the ubiquitous transcription factors oct-1, NFA, TEF-2, NF1, and AP-1 participate in epithelial cell-specific transcription, *J. Virol.* **65**:5933–5943.
15. Taniguchi, A., Kikuchi, K., Nagata, K., and Yasumoto, S., 1993, A cell-type-specific transcription enhancer of type 16 human papillomavirus (HPV 16)-P_{97} promoter is defined with HPV-associated cellular events in human epithelial cell lines, *Virology* **195**:500–510.
16. Romanczuk, H., Villa, L. L., Schlegel, R., and Howley, P. M., 1991, The viral transcriptional regulatory region upstream of the E6 and E7 genes is a major determinant of the differential immortalization activities of human papillomavirus types 16 and 18, *J. Virol.* **65**:2739–2744.
17. Mack, D. H., and Laimins, L. A., 1991, A keratinocyte-specific transcription factor, KRF-1, interacts with AP-1 to activate expression of human papillomavirus type 18 in squamous epithelial cells, *Proc. Natl. Acad. Sci. USA* **88**:9102–9106.
18. Chan, W.-K., Chong, T., Bernhard, H.-U., and Klock, G., 1990, Transcription of the transforming genes of the oncogenic human papillomavirus-16 is stimulated by tumor promotors through AP1 binding sites, *Nucleic Acids Res.* **18**:763–769.
19. Cripe, T. P., Haugen, T. H., Turk, J. P., Tabatabai, F., Schmid III, P. G., Dürst, M., Gissmann, L., Roman, A., and Turek, L. P., 1987, Transcriptional regulation of the human papillomavirus-16 E6-E7 promoter by a keratinocyte-dependent enhancer, and by viral E2 transactivator and repressor gene products: Implications for cervical carcinogenesis, *EMBO J.* **6**:3745–3753.
20. Ishiji, T., Lace, M. J., Parkkinen, S., Anderson, R. D., Haugen, T. H., Cripe, T. P., Xiao, J.-H., Davidson, I., Chambon, P., and Turek, L. P., 1992, Transcriptional enhancer factor (TEF-1) and its cell-specific co-activator activate human papillomavirus 16 E6 and E7 oncogene transcription in keratinocytes and cervical carcinoma cells, *EMBO J.* **11**:2271–2281.
21. Androphy, E. J., Lowy, D. R., and Schiller, J. T., 1987, Bovine papillomavirus E2 *trans*-activating gene product binds to specific sites in papillomavirus DNA, *Nature* **325**:70–73.
22. Haugen, T. H., Cripe, T. P., Ginder, G. D., Karin, M., and Turek, L. P., 1987, *Trans*-activation of an upstream early gene promoter of bovine papillomavirus-1 by a gene product of the viral E2 gene, *EMBO J.* **6**:145–152.

23. Moskaluk, C., and Bastia, B, 1987, The E2 "gene" of bovine papillomavirus encodes an enhancer-binding protein, *Proc. Natl. Acad. Sci. USA* **84:**1215–1218.
24. Phelps, W. C., and Howley, P. M., 1987, Transcriptional *trans*-activation by the human papillomavirus type 16 E2 gene product, *J. Virol.* **61:**1630–1638.
25. Lambert, P. F., Spalholz, B. A., and Howley, P. M., 1987, A transcriptional repressor encoded by BPV-1 shares a common carboxy-terminal domain with the E2 transactivator, *Cell* **50:**69–78.
26. Hirochika, H., Broker, T. R., and Chow, L. T., 1987, Enhancers and *trans*-acting E2 transcriptional factors of papillomaviruses, *J. Virol.* **61:**2599–2606.
27. Giri, I., and Yaniv, M., 1988, Study of the E2 gene product of the cottontail rabbit papillomavirus reveals a common mechanism of transactivation among papillomaviruses, *J. Virol.* **62:** 1573–1581.
28. Chin, M. T., Hirochika, R., Hirochika, H., Broker, T. R., and Chow, L. T., 1988, Regulation of human papillomavirus type 11 enhancer and E6 promoter by activating and repressing proteins from the E2 open reading frame: Functional and biochemical studies, *J. Virol.* **62:** 2994–3002.
29. Hirochika, H., Hirochika, R., Broker, T. R., and Chow, L. T., 1988, Functional mapping of the human papillomavirus type 11 transcriptional enhancer and its interaction with the *trans*-acting E2 proteins, *Genes Dev.* **2:**54–67.
30. Gius, D., Grossmann, S., Bedell, M. A., and Laimins, L. A., 1988, Inducible and constitutive enhancer domains in the noncoding region of human papillomavirus type 18, *J. Virol.* **62:**665–672.
31. Thierry, F., and Yaniv, M., 1987, The BPV1-E2 transactivating protein can be either an activator or a repressor of the HPV18 regulatory region. *EMBO J.* **6:**3391–3397.
32. Chin, M. T., Broker, T. R., and Chow, L. T., 1989, Identification of a novel constitutive enhancer element and an associated binding protein: Implications for human papillomavirus type 11 enhancer regulation, *J. Virol.* **53:**2967–2977.
33. Baker, C. C., Phelps, W. C., Lindgren, V., Braun, M. J., Gonda, M. A., and Howley, P. M., 1987, Structural and transcriptional analysis of human papillomavirus type 16 sequences in cervical carcinoma cell lines, *J. Virol.* **61:**962–971.
34. Lehn, H., Krieg, P., and Sauer, G., 1985, Papillomavirus genomes in human cervical tumors: Analysis of their transcriptional activity. *Proc. Natl. Acad. Sci. USA* **82:**5540–5544.
35. Sousa, R., Dostatni, N., and Yaniv, M., 1990, Control of papillomavirus gene expression, *Biochim. Biophys. Acta* **1032:**19–37.
36. Romanczuk, H., and Howley, P. M., 1992, Disruption of either the E1 or the E2 regulatory gene of human papillomavirus type 16 increases viral immortalization capacity, *Proc. Natl. Acad. Sci. USA* **89:**3159–3163.
37. Garcia-Carranca, A., Thierry, F., and Yaniv, M., 1988, Interplay of viral and cellular proteins along the long control region of human papillomavirus type 18, *J. Virol.* **62:**4321–4330.
38. Gloss, B., Chong, T., and Bernard, H.-U., 1989, Numerous nuclear proteins bind the long control region of human papillomavirus type 16: A subset of 6 of 23 DNaseI-protected segments coincides with the location of the cell-type-specific enhancer, *J. Virol.* **63:**1142–1152.
39. Royer, H. D., Freyaldenhoven, M. P., Napierski, I., Spitkovsky, D. D., Bauknecht, T., and Dathan, N., 1991, Delineation of human papillomavirus type 18 enhancer binding proteins: The intracellular distribution of a novel octamer binding protein p92 is cell cycle regulated, *Nucleic Acids Res.* **19:**2363–2371.
40. Sibbet, G. J., and Campo, M. S., 1990, Multiple interactions between cellular factors and the non-coding region of human papillomavirus type 16, *J. Gen. Virol.* **71:**2699–2707.
41. Chong, T., Chan, W.-K., and Bernard, H.-U., 1990, Transcriptional activation of human papillomavirus 16 by nuclear factor I, AP1, steroid receptors and a possibly novel transcription factor, PVF: A model for the composition of genital papillomavirus enhancers, *Nucleic Acids Res.* **18:**465–470.
42. Herschlag, D., and Johnson, F. B., 1993, Synergism in transcriptional activation: A kinetic view, *Genes Dev.* **7:**173–179.

43. Heck, D. V., Yee, C. L., Howley, P., and Münger, K., 1992, Efficiency of binding the retinoblastoma protein correlates with the transforming capacity of the E7 oncoproteins of the human papillomaviruses, *Proc. Natl. Acad. Sci. USA* **89:**4442–4446.
44. Yamashita, T., Segawa, K., Fujinaga, Y., Nishikawa, T., and Fujinaga, K., 1993, Biological and biochemical activity of E7 genes of the cutaneous human papillomavirus type 5 and 8, *Oncogene* **8:**2433–2441.
45. Armstrong, D. J., and Roman, A., 1993, The anomalous electrophoretic behavior of the human papillomavirus type 16 E7 protein is due to the high content of acidic amino acid residues, *Biochem. Biophys. Res. Commun.* **192:**1380–1387.
46. Vallee, B. L., Coleman, J. E., and Auld, D. S., 1991, Zinc fingers, zinc clusters, and zinc twists in DNA-binding protein domains, *Proc. Natl. Acad. Sci. USA* **88:**999–1003.
47. McIntyre, M. C., Frattini, M. G., Grossman, S. R., and Laimins, L. A., 1993, Human papillomavirus type 18 E7 protein requires intact Cys-X-X-Cys motifs for zinc binding, dimerization, and transformation but not for Rb binding, *J. Virol.* **67:**3142–3150.
48. Stacey, S. N., Eklund, C., Jordan, D., Smith, N. K., Stern, P. L., Dillner, J., and Arrand, J. R., 1994, Scanning the structure and antigenicity of HPV-16 E6 and E7 oncoproteins using antipeptide antibodies, *Oncogene* **9:**636–645.
49. Greenfield, I., Nickerson, J., Penman, S., and Stanley, M., 1991, Human papillomavirus 16 E7 protein is associated with the nuclear matrix, *Proc. Natl. Acad. Sci. USA* **88:**11217–11221.
50. Dyson, N., Howley, P. M., Münger, K., and Harlow, E., 1989, The human papilloma virus-16 E7 oncoprotein is able to bind to the retinoblastoma gene product, *Science* **243:**934–936.
51. Ewen, M. E., 1994, The cell cycle and the retinoblastoma protein family, *Cancer Metastasis Rev.* **13:**45–66.
52. Barbosa, M. S., Edmonds, C., Fisher, C., Schiller, J. T., Lowy, D. R., and Vousden, K. H., 1990, The region of the HPV E7 oncoprotein homologous to adenovirus E1a and SV40 large T antigen contains separate domains for Rb binding and casein kinase II phosphorylation, *EMBO J.* **9:**153–160.
53. Edmonds, C., and Vousden, K. H., 1989, A point mutational analysis of human papillomavirus type 16 E7 protein, *J. Virol.* **63:**2650–2656.
54. Firzlaff, J. M., Lüscher, B., and Eisenman, R. N., 1991, Negative charge at the casein kinase II phosphorylation site is important for transformation but not for Rb protein binding by the E7 protein of human papillomavirus type 16, *Proc. Natl. Acad. Sci. USA* **88:**5187–5191.
55. Phelps, W. C., Yee, C. L., Münger, K., and Howley, P. M., 1988, The human papillomavirus type 16 E7 gene encodes transactivation and transformation functions similar to those of adenovirus E1A, *Cell* **53:**539–547.
56. Nevins, J. R., 1993, Transcriptional activation by the adenovirus E1A proteins, *Semin. Virol.* **4:**25–31.
57. Münger, K., and Phelps, W. C., 1993, The human papillomavirus E7 protein as a transforming and transactivating factor. *Biochim. Biophys. Acta* **1155:**111–123.
58. Münger, K., Yee, C. L., Phelps, W. C., Pietenpol, J. A., Moses, H. L., and Howley, P. M., 1991, Biochemical and biological differences between E7 oncoproteins of the high- and low-risk human papillomavirus types are determined by amino-terminal sequences, *J. Virol.* **65:**3943–3948.
59. Yasumoto, S., Burkhardt, A. L., Doninger, J., and DiPaolo, J., 1986, Human papillomavirus type 16 DNA induced malignant transformation of NIH3T3 cells, *J. Virol.* **57:**572–577.
60. Bedell, M. A., Jones, K. H., Grossman, S. R., and Laimins, L. A., 1989, Identification of human papillomavirus type 18 transforming genes in immortalized and primary cells, *J. Virol.* **63:**1247–1255.
61. Tanaka, A., Noda, T., Yajima, H., Hatanaka, M., and Ito, Y., 1989, Identification of a transforming gene of human papillomavirus type 16, *J. Virol.* **63:**1465–1469.
62. Watts, S. L., Phelps, W. C., Ostrow, R. S., Zachow, K. R., and Faras, A. J., 1984, Cellular transformation by human papillomavirus DNA *in vitro*, *Science* **225:**634–636.

63. Banks, L., Edmonds, C., and Vousden, K. H., 1990, Ability of the HPV16 E7 protein to bind RB and induce DNA synthesis is not sufficient for efficient transforming activity in HIH3T3 cells, *Oncogene* **5**:1383–1389.
64. Cerni, C., Patocka, K., and Meneguzzi, G., 1990, Immortalization of primary rat embryo cells by human papillomavirus type 11 DNA is enhanced upon cotransfer of *ras*, *Virology* **177**:427–436.
65. Chester, P. M., and McCance, D. J., 1989, Human papillomavirus types 6 and 16 in cooperation with Ha-*ras* transform secondary rat embryo fibroblasts, *J. Gen. Virol.* **70**:353–365.
66. Cerni, C., Binetruy, B., Schiller, J. T., Lowy, D. R., Meneguzzi, G., and Cuzin, F., 1989, Successive steps in the process of immortalization identified by transfer of separate bovine papillomavirus genes into rat fibroblasts, *Proc. Natl. Acad. Sci. USA* **86**:3266–3270.
67. Kaur, P., and McDougall, J. K., 1989, HPV-18 immortalization of human keratinocytes, *Virology* **173**:302–310.
68. Pirisi, L., Yasumoto, S., Feller, M., Doninger, J., and DiPaolo, J. A., 1987, Transformation of human fibroblasts and keratinocytes with human papillomavirus type 16 DNA, *J. Virol.* **61**:1061–1066.
69. Hawley-Nelson, P., Vousden, K. H., Hubbert, N. L., Lowy, D. R., and Schiller, J. T., 1989, HPV16 E6 and E7 proteins cooperate to immortalize human foreskin keratinocytes, *EMBO J.* **8**:3905–3910.
70. Münger, K., Phelps, W. C., Bubb, V., Howley, P. M., and Schlegel, R., 1989, The E6 and E7 genes of the human papillomavirus type 16 together are necessary and sufficient for transformation of primary human keratinocytes, *J. Virol.* **63**:4417–4421.
71. McCance, D. J., Kopan, R., Fuchs, E., and Laimins, L. A., 1988, Human papillomavirus type 16 alters human epithelial cell differentiation in vitro. *Proc. Natl. Acad. Sci. USA* **85**:7169–7173.
72. Grossman, S. R., and Laimins, L. A., 1989, E6 protein of human papillomavirus type 18 binds zinc, *Oncogene* **4**:1089–1093.
73. Grossman, S. R., Mora, S., and Laimins, L. A., 1989, Intracellular localization and DNA-binding properties of human papillomavirus type 18 E6 protein expressed with a baculovirus vector, *J. Virol.* **63**:366–374.
74. Liang, X. H., Volkmann, M., Klein, R., Herman, B., and Lockett, S. J., 1993, Co-localization of the tumor-suppressor protein p53 and human papillomavirus E6 protein in human cervical carcinoma cell lines, *Oncogene* **8**:2645–2652.
75. Keen, N., Elston, R., and Crawford, L., 1994, Interaction of the E6 protein of human papillomavirus with cellular proteins, *Oncogene* **9**:1493–1499.
76. Desaintes, C., Hallez, S., van Alphen, P., and Burny, A., 1992, Transcriptional activation of several heterologous promoters by the E6 protein of human papillomavirus type 16, *J. Virol.* **66**:325–333.
77. Sedman, S. A., Barbosa, M. S., Vass, W. C., Hubbert, N. L., Haas, J. A., Lowy, D. R., and Schiller, J. T., 1991, The full-length E6 protein of human papillomavirus type 16 has transforming and *trans*-activating activities and cooperates with E7 to immortalize keratinocytes in culture, *J. Virol.* **65**:4860–4866.
78. Smotkin, D., Prokoph, H., and Wettstein, F. O., 1989, Oncogenic and nononcogenic human genital papillomaviruses generate the E7 mRNA by different mechanisms, *J. Virol.* **63**:1441–1447.
79. Kern, S. E., Kinzler, K. W., Bruskin, A., Jarosz, D., Friedman, P., Prives, C., and Vogelstein, B., 1991, Identification of p53 as a sequence-specific DNA-binding protein, *Science* **252**:1708–1711.
80. Mietz, J. A., Unger, T., Huibregtse, J. M., and Howley, P. M., 1992, The transcriptional transactivation function of wild type p53 is inhibited by SV40 large T-antigen and by HPV-16 E6 oncoprotein, *EMBO J.* **11**:5013–5020.
81. Pim, D., Storey, A., Thomas, M., Massimi, A., and Banks, L., 1994, Mutational analysis of HPV-18 E6 identifies domains required for p53 degradation *in vitro*, abolition of p53 transactivation *in vivo* and immortalization of primary BMK cells, *Oncogene* **9**:1869–1876.
82. Scheffner, M., Werness, B. A., Huibregtse, J. M., Levine, A. J., and Howley, P. M., 1990, The E6

oncoprotein encoded by human papillomavirus type 16 and 18 promotes the degradation of p53, *Cell* **63**:1129–1136.
83. Huibregtse, J. M., Scheffner, M., and Howley, P. M., 1991, A cellular protein mediates association of p53 with the 6 oncoprotein of human papillomavirus types 16 or 18, *EMBO J.* **10**:4129–4135.
84. Scheffner, M., Huibregtse, J. M., Vierstra, R. D., and Howley, P. M., 1993, The HPV-16 E6 and E6-AP complex functions as a ubiquitin–protein ligase in the ubiquitination of p53, *Cell* **75**:495–505.
85. Huibregtse, J. M., Scheffner, M., and Howley, P. M., 1993, Cloning and expression of the cDNA for E6-AP, a protein that mediates the interaction of the human papillomavirus E6 oncoprotein with p53, *Mol. Cell. Biol.* **13**:775–784.
86. Medcalf, E. A., and Milner, J., 1993, Targeting and degradation of p53 by E6 of human papillomavirus type 16 is preferentially for the 1620+ conformation, *Oncogene* **8**:2847–2851.
87. Lechner, M. S., Mack, D. H., Finicle, A. B., Crook, T., Vousden, K. H., and Laimins, L. A., 1992, Human papillomavirus E6 proteins bind p53 *in vivo* and abrogate p53-mediated repression of transcription, *EMBO J.* **11**:3045–3052.
88. Band, V., Dalal, S., Delmolino, L., and Androphy, E. J., 1993, Enhanced degradation of p53 protein in HPV-6 and BPV-1 E6-immortalized human mammary epithelial cells, *EMBO J.* **12**:1847–1852.
89. Crook, T., Fisher, C., Masterson, P. J., and Vousden, K. H., 1994, Modulation of transcriptional regulatory properties of p53 by HPV E6, *Oncogene* **9**:1225–1230.
90. Crook, S. T., Tidy, J. A., and Vousden, K. H., 1991, Degradation of p53 can be targeted by HPV E6 sequences distinct from those required for p53 binding and transactivation, *Cell* **67**:547–556.
91. Lamberti, C., Morrissey, L. C., Grossman, S. R., and Androphy, E. J., 1990, Transcriptional activation by the papillomavirus E6 zinc finger oncoprotein, *EMBO J.* **9**:1907–1913.
92. Koyono, T., Hiraiwa, A., and Ishibashi, M., 1992, Differences in transforming activity and coded amino acid sequence among E6 genes of several papillomaviruses associated with epidermodysplasia verruciformis, *Virology* **186**:628–639.
93. Storey, A., and Banks, L., 1993, Human papillomavirus type 16 E6 gene cooperates with EJ-*ras* to immortalize primary mouse cells, *Oncogene* **8**:919–924.
94. Liu, Z., Ghai, J., Ostrow, R. S., McGlennen, R. C., and Faras, A. J., 1994, The E6 gene of human papillomavirus type 16 is sufficient for transformation of baby rat kidney cells in cotransfection with activated Ha-*ras Virology* **201**:388–396.
95. Shay, J. W., Wright, W. E., Brasiskyte, D., and Van der Haegen, B. A., 1993, E6 of human papillomavirus type 16 can overcome the M_1 stage of immortalization in human mammary epithelial cells but not human fibroblasts, *Oncogene* **8**:1407–1413.
96. Shay, J. W., Wright, W. E., and Werbin, H., 1991, Defining the molecular mechanisms of human cell immortalization, *Biochim. Biophys. Acta* **1072**:1–7.
97. Cullen, A. P., Reid, R., Campion, M., and Lorincz, A. T., 1991, Analysis of the physical state of different human papillomavirus DNAs in intraepithelial and invasive cervical neoplasm, *J. Virol.* **65**:606–612.
98. Chen, L., Ashe, S., Singhal, M. C., Galloway, D. A., Hellström, I., and Hellström, K. E., 1993, Metastatic conversion of cells by expression of human papillomavirus type 16 E6 and E7 genes. *Proc. Natl. Acad. Sci. USA* **90**:6523–6527.
99. Böhm, S., Wilczynski, S. P., Pfister, H., and Iftner, T., 1993, The predominant mRNA class in HPV16-infected genital neoplasias does not encode the E6 or the E7 protein, *Int. J. Cancer* **55**:791–798.
100. Conrad, M., Bubb, V. J., and Schlegel, R., 1993, The human papillomavirus type 6 and 16 E5 proteins are membrane-associated proteins which associate with the 16-kilodalton pore-forming protein, *J. Virol.* **67**:6170–6178.

101. Chen, S.-L., and Mounts, P., 1990, Transforming activity of E5a protein of human papillomavirus type 6 in NIH 3T3 and C127 cells, *J. Virol.* **64**:3226–3233.
102. Pim, D., Collins, M., and Banks, L., 1992, Human papillomavirus type 16 E5 gene stimulates the transforming activity of the epidermal growth factor receptor, *Oncogene* **7**:27–32.
103. Leptak, C., Ramon, S., Cajal, Y., Kulke, R., Horowitz, B. H., Riese, D. J., Dotto, G. P., and DiMaio, D., 1991, Tumorigenic transformation of murine keratinocytes by the E5 gene of bovine papillomavirus type 1 and human papillomavirus type 16, *J. Virol.* **65**:7078–7083.
104. Conrad, M., Goldstein, D., Andersson, T., and Schlegel, R., 1994, The E5 protein of HPV-6, but not of HPV-16, associates efficiently with cellular growth factor receptors, *Virology* **200**:796–800.
105. Bouvard, V., Matlashewski, G., Gu, Z.-M., Storey, A., and Banks, L., 1994, The human papillomavirus type 16 E5 gene cooperates with the E7 gene to stimulate proliferation of primary cells and increase viral gene expression, *Virology* **203**:73–80.
106. Lambert, P. F., 1991, Papillomavirus DNA replication, *J. Virol.* **65**:3417–3420.
107. Ustav, M., and Stenlund, A., 1991, Transient replication of BPV-1 requires two viral polypeptides encoded by the E1 and E2 open reading frames, *EMBO J.* **10**:449–458.
108. Wilson, V. G., and Ludes-Meyers, J., 1991, A bovine papillomavirus E1-related protein binds specifically to bovine papillomavirus DNA, *J. Virol.* **65**:5314–5322.
109. Seo, Y.-S., Müller, F. Lusky, M., and Hurwitz, J., 1993, Bovine papilloma virus (BPV)-encoded E1 protein contains multiple activities required for BPV DNA replication, *Proc. Natl. Acad. Sci. USA* **90**:702–706.
110. Yang, L., Mohr, I., Fouts, E., Lim, D. A., Nohaile, M., and Botchan, M, 1993, The E1 protein of bovine papilloma virus 1 is an ATP-dependent DNA helicase, *Proc. Natl. Acad. Sci. USA* **90**:5086–6090.
111. Mohr, I. J., Clark, R., Sun, S., Androphy, E. J., MacPherson, P., and Botchan, M. R., 1990, Targeting the E1 replication protein to the papillomavirus origin of replication by complex formation with the E2 transactivator, *Science* **250**:1694–1699.
112. Lusky, M., and Fontane, E., 1991, Formation of the complex of bovine papillomavirus E1 and E2 proteins is modulated by E2 phosphorylation and depends upon sequences within the carboxyl terminus of E1, *Proc. Natl. Acad. Sci. USA* **88**:6363–6367.
113. Bream, G. L., Ohmstede, C.-A., and Phelps, W. C., 1993, Characterization of human papillomavirus type E1 and E2 proteins expressed in insect cells, *J. Virol.* **67**:2655–2663.
114. Hughes, F. J., and Romanos, M. A., 1993, E1 protein of human papillomavirus is a DNA helicase/ATPase, *Nucleic Acids Res.* **25**:5817–5823.
115. Hedge, R. S., Grosman, S. R., Laimins, L. A., and Sigler, P. B., 1992, Crystal structure at 1.7 A of the bovine papillomavirus-1 E2 DNA-binding domain bound to its DNA target, *Nature* **359**:505–512.
116. Nasseri, M, Gage, J. R., Lorincz, A., and Wettstein, F. O., 1991, Human papillomavirus type 16 immortalized cervical keratinocytes contain transcripts encoding E6, E7, and E2 initiated at the p97 promoter and express high levels of E7, *Virology* **184**:131–140.
117. Sterling, J. C., Skeeper, J. N., and Stanley, M. A., 1993, Immuno-electronmicroscopical localization of human papillomavirus type 16 L1 and E4 proteins in cervical keratinocytes cultured *in vitro*, *J. Invest. Dermatol.* **100**:154–158.
118. Roberts, S., Ashmole, I., Sheehan, T. M. T., Davies, A. H., and Gallimore, P. H., 1994, Human papillomavirus type 1 E4 protein is a zinc-binding protein, *Virology* **202**:865–874.
119. Chan, S.-Y., Bernard, H.-U., Ong, C.-K., Chan, S.-P., Hofmann, B., and Delius, H., 1992, Phylogenetic analysis of 48 papillomavirus types and 28 subtypes and variants: A showcase for the molecular evolution of DNA viruses, *J. Virol.* **66**:5714–5725.
120. Hagensee, M. E., Yaegashi, N., and Galloway, D., 1993, Self-assembly of human papillomavirus type 1 capsids by expression of the L1 protein alone or by coexpression of the L1 and L2 capsid proteins, *J. Virol.* **67**:315–322.
121. Kirnbauer, R., Taub, J., Greenstone, H., Roden, R., Dürst, M., Gissmann, L., Lowy, D. R., and

Schiller, J. T., 1993, Efficient self-assembly of human papillomavirus type 16 L1 and L1–L2 into virus-like particles, *J. Virol.* **67:**6929–6936.
122. Christensen, N. D., Höpfl, R., DiAngelo, S. L., Cladel, N. M., Patrick, S. D., Welsh, P. A., Budgeon, L. R., Reed, C. A., and Kreider, J. W., 1994, Assembled baculovirus-expressed human papillomavirus type 11 L1 capsid protein virus like particles are recognized by neutralizing monoclonal antibodies and induce high titres of neutralizing antibodies, *J. Gen. Virol.* **75:**2271–2276.
123. Zhou, J., Stenzel, D. J., Sun, X.-Y., and Frazer, I. H., 1993, Synthesis and assembly of infectious bovine papillomavirus particles *in vitro*, *J. Gen. Virol.* **71:**2185–2190.
124. Zhou, J., Sun, X.-Y., Louis, K., and Frazer, I. H., 1994, Interaction of human papillomavirus (HPV) type 16 capsid proteins with HPV DNA requires an intact L2 N-terminal sequence, *J. Virol.* **68:**619–625.
125. Carter, J. J., Hagensee, M. B., Lee, S. K., McKnight, B., Koutsky, L. A., and Galloway, D. A., 1994, Use of HPV1 capsids produced by recombinant vaccinia viruses in an ELISA to detect serum antibodies in people with foot warts, *Virology* **199:**284–491.
126. Huber, H. E., Edwars, G., Goddhart, P. J., Patrick, D. R., Huang, P. S., Ivey-Hoyle, M., Barnett, S. F., Oliff, A., and Heimbrook, D. C., 1993, Transcription factor E2F binds DNA as a heterodimer, *Proc. Natl. Acad. Sci. USA* **90:**3525–3529.
127. Chellappan, S. P., Hiebert, S., Mudryj, M., Horowitz, J. M., and Nevins, J. R., 1991, The E2F transcription factor is a cellular target for the RB protein, *Cell* **65:**1053–1061.
128. Hu, Q., Lees, J. A., Buchkovich, K. J., and Harlow, E., 1992, The retinoblastoma protein physically associates with the human cdc2 kinase, *Mol. Cell. Biol.* **12:**971–980.
129. Bates, S., Parry, D., Bonetta, L., Vousden, K. H., and Dickson, C., 1994, Absence of cyclin D/cdk complexes in cells lacking functional retinoblastoma protein, *Oncogene* **9:**1633–1640.
130. Huang, P. S., Patrick, D. R., Edwards, G., Goodhart, P. J., Huber, H., Miles, L., Durfee, T., Becherer, K., Chen, P.-L., Yeh, S.-H., Yang, Y., Killburn, A. E., Lee, W.-H., and Elledge, S. J., 1993, The retinoblastoma protein associates with the protein phosphatase type 1 catalytic subunit, *Genes Dev.* **7:**555–569.
131. Whyte, P., Williamson, N. M., and Harlow, E., 1989, Cellular targets for transformation by the adenovirus E1A proteins, *Cell* **56:**67–75.
132. Chellappan, S., Kraus, V. B., Kroger, B., Munger, K., Howley, P. M., Phelps, W. C., and Nevins, J. R., 1992, Adenovirus E1A, simian virus 40 tumor antigen, and human papillomavirus E7 protein share the capacity to disrupt the interaction between transcription factor E2F and the retinoblastoma gene product, *Proc. Natl. Acad. Sci. USA* **89:**4549–4553.
133. Wu, E. W., Clemens, K. E., Heck, V., and Münger, K., 1993, The human papillomavirus E7 oncoprotein and the cellular transcription factor E2F bind to separate sites on the retinoblastoma tumor suppressor protein, *J. Virol.* **67:**2402–2407.
134. Lam, E. W.-F., Morris, J. D. H., Davies, R., Crook, T., Watson, R. J., and Vousden, K., 1994, HPV16 E7 oncoprotein deregulates B-*myb* expression: Correlation with targeting of p107/E2F complexes, *EMBO J.* **13:**871–878.
135. Oswald, F., Lovec, H., Möröy, T., and Lipp, M., 1994, E2F-dependent regulation of human *MYC*: Transactivation by cyclins D1 and A overrides tumor suppressor protein functions, *Oncogene* **9:**2029–2036.
136. Singh, P., Wong, S. H., and Hong, W., 1994, Overexpression of E2F-1 in rat embryo fibroblasts leads to neoplastic transformation, *EMBO J.* **13:**3329–3338.
137. Kaelin, W. G., Pallas, D. C., DeCaprio, J. A., Kaye, F. J., and Livingston, D. M., 1991, Identification of cellular proteins that can interact specifically with the T/E1A-binding region of the retinoblastoma gene product, *Cell* **64:**521–532.
138. Ewen, M. E., Sluss, H. K., Sherr, C. J., Matsushime, H., Kato, J.-Y., and Livingston, D. M., 1993, Functional interactions of the retinoblastoma protein with mammalian D-type cyclins, *Cell* **73:**487–497.

139. Dowdy, S. F., Hinds, P. W., Louie, K., Reed, S., Arnold, A., and Weinberg, R. A., 1993, Physical interaction of the retinoblastoma protein with human D cyclins, *Cell* **3:**499–511.
140. Pagano, M, Dürst, M., Joswig, S., Draetta, G., and Jansen-Dürr, P., 1992, Binding of the human E2F transcription factor to the retinoblastoma protein but not to cyclin A is abolished in HPV-16-immortalized cells, *Oncogene* **7:**1681–1986.
141. Davies, R., Hicks, R., Crook, T., Morris, J., and Vousden, K., 1993, Human papillomavirus type 16 E7 associates with a histone H1 kinase and with p107 through sequences necessary for transformation, *J. Virol.* **67:**2521–2528.
142. Dyson, N., Guida, P., Munger, K., and Harlow, E., 1992, Homologous sequences in adenovirus E1A and human papillomavirus E7 proteins mediate interaction with the same set of cellular proteins, *J. Virol.* **66:**6893–6902.
143. Tommasino, M., Adamczewski, J. P., Carlotti, F., Barth, C. F., Manetti, R., Contorni, M., Cavalieri, F., Hunt, T., and Crawford, L., 1993, HPV16 E7 protein associates with the protein kinase p33 cdk2 and cyclin A, *Oncogene* **8:**195–202.
144. Arroyo, M, Bagchi, S., and Raychaudhuri, P., 1993, Association of the human papillomavirus type 16 E7 protein with the S-phase-specific E2F–cyclin A complex, *Mol. Cell. Biol.* **13:**6537–6546.
145. Chinami, M., Moriyama, K., Fukumaki, Y., Terada, M., and Shingu, M., 1993, Association of RNA with human papillomavirus E7 protein of type 16 but not type 6b, *Biochem. Biophys. Res. Commun.* **197:**1609–1614.
146. Werness, B. A., Levine, A. J., and Howley, P. M., 1990, Association of human papillomavirus types 16 and 18 E6 proteins with p53, *Science* **248:**76–79.
147. Lechner, M. S., and Laimins, L. A., 1994, Inhibition of p53 DNA binding by human papillomavirus E6 proteins, *J. Virol.* **68:**4262–4273.
148. Seto, E., Usheva, A., Zambetti, G. P., Momand, J., Horikoshi, N., Weinmann, R., Levine, A. J., and Shenk, T., 1992, Wild-type p53 binds to the TATA-binding protein and represses transcription, *Proc. Natl. Acad. Sci. USA* **89:**12028–12032.
149. Foord, O., Navot, N., and Rotter, V., 1993, Isolation and characterization of DNA sequences that are specifically bound by wild-type p53 protein, *Mol. Cell. Biol.* **13:**1378–1384.
150. Slingerland, J. M., Jenkins, J. R., and Benchimol, S., 1993, The transforming and suppressor functions of p53 alleles: Effects of mutations that disrupt phosphorylation, oligomerization and nuclear translocation, *EMBO J.* **12:**1029–1027.
151. Donehower, L. A., Harvey, M., Siagle, B. L., McArthus, M. J., Montgomery, C. A., Butel, J. S., and Bradley, A., 1992, Mice deficient for p53 are developmentally normal but susceptible to spontaneous tumours, *Nature* **356:**215–221.
152. Vogelstein, B., and Kinzler, K. W., 1992, p53 function and dysfunction, *Cell* **70:**523–526.
153. Deppert, W., 1994, The yin and yang of p53 in cellular proliferation, *Semin. Cancer Biol.* **5:** 187–202.
154. Selter, H., and Montearh, M., 1994, The emerging picture of p53, *Int. J. Biochem.* **26:**145–154.
155. Dulic, V., Kaufmann, W. K., Wilson, S. J., Tlsty, T. D., Lees, E., Harper, J. W., Elledge, S. J., and Reed, S. I., 1993, p53-dependent inhibition of cyclin-dependent kinase activities in human fibroblasts during radiation-induced G1 arrest, *Cell* **76:**1013–1023.
156. El-Deiry, W. S., Harper, J. W., O'Connor, P. M., Velculescu, V. E., Canman, C. E., Jackman, J., Pietenpol, J. A., Burell, M., Hill, D. E., Wang, Y., Wiman, K. G., Mercer, W. E., Kastan, M. B., Kohn, K. W., Elledge, S. J., Kinzler, K. W., and Vogelstein, B., 1994, WAF/CIP1 is induced in p53-mediated G1 arrest and apoptosis, *Cancer Res.* **54:**1169–1174.
157. Clarke, A. R., Purdie, C. A., Harrison, D. J., Morris, R. G., Bird, C. C., Hooper, M. L., and Willie, A. H., 1993, Thymocyte apoptosis induced by p53-dependent and independent pathways, *Nature* **362:**849–852.
158. Zhengming, G., Pim, D., Labrecque, S., Banks, L., and Matlashewski, G., 1994, DNA damage induced p53 mediated transcription is inhibited by human papillomavirus type 18 E6, *Oncogene* **9:**629–633.

159. Kessis, T. D., Slebos, R. J., Nelson, W. G., Kastan, M., Plunkett, B. S., Han, S. M., Lorincz, A. T., Hedrick, L., and Cho, K. R., 1993, Human papillomavirus 16 E6 expression disrupts the p53-mediated cellular response to DNA damage, *Proc. Natl. Acad. Sci. USA* **90**:3988–3992.
160. Pan, H., and Griep, A. E., 1994, Altered cell cycle regulation in the lens of HPV-16 E6 or E7 transgenic mice: Implications for tumor suppressor function in the development, *Genes Dev.* **8**:1285–1299.
161. Livingston, L. R., White, A., Sprouse, J., Livanos, E., Jacks, T., and Tlsty, T., 1992, Altered cell cycle arrest and gene amplification potential accompany loss of wild-type p53, *Cell* **70**:923–935.
162. Yin, Y., Tainsky, M. A., Bischoff, F. Z., Strong, L. C., and Wahl, G. M., 1992, Wild-type p53 restores cell cycle control and inhibits gene amplification in cells with mutant p53 alleles, *Cell* **70**:937–948.
163. Harvey, M., Sands, A. T., Weiss, R. S., Hegi, M. E., Wiseman, R. W., Pantazis, P., Giovanella, B. C., Tainsky, M. A., Bradley, A., and Donehower, L. A., 1993, In vitro growth characteristics of embryo fibroblasts isolated from p53-deficient mice, *Oncogene* **8**:2457–2467.
164. Fuchs, E., 1993, Epidermal differentiation and keratin gene expression, *J. Cell Sci. [Suppl.]* **17**:197–208.
165. Yuspa, S. H., 1994, The pathogenesis of squamous cell cancer: Lessons learned from studies of skin carcinogenesis—Thirty-third G. H. A. Clowes memorial award lecture, *Cancer Res.* **54**:1178–1189.
166. Goustin, A. S., Leof, E. B., Shipley, G. D., and Moses, H. L., 1986, Growth factors and cancer, *Cancer Res.* **46**:1015–1029.
167. Roberts, A. B., Thompson, N. L., Heine, U., Flanders, C., and Sporn, M. B., 1988, Transforming growth factor-β: Possible roles in carcinogenesis, *Br. J. Cancer* **57**:594–600.
168. Braun, L., Dürst, M., Mikumo, R., and Gruppuso, P., 1990, Differential response of nontumorigenic and tumorigenic human papillomavirus type 16-positive epithelial cells to transforming growth factor β1, *Cancer Res.* **50**:7324–7332.
169. Woodworth, C. D., Notario, V., and DiPaolo, J. A., 1992, Transforming growth factors beta 1 and 2 transcriptionally regulate human papillomavirus (HPV) type 16 early gene expression in HPV-immortalized human genital epithelial cells, *J. Virol.* **64**:4767–4775.
170. Braun, L., Durst, M., Mikumo, R., Crowley, A., and Robinson, M., 1992, Regulation of growth and gene expression in human papillomavirus-transformed keratinocytes by transforming growth factor-beta: Implications for the control of papillomavirus infection, *Mol. Carcinogen.* **6**:100–111.
171. Pirisi, L., Batova, A., Jenkins, G. R., Hodam, J. R., and Creek, K. E., 1992, Increased sensitivity of human keratinocytes immortalized by human papillomavirus type 16 DNA to growth control by retinoids, *Cancer Res.* **52**:187–193.
172. Moses, H. L., 1992, TGF-beta regulation of epithelial cell proliferation, *Mol. Reprod. Dev.* **32**:179–184.
173. Yasumoto, S., Taniguchi, A., and Sohma, K., 1991, Epidermal growth factor (EGF) elicits down-regulation of human papillomavirus type 16 (HPV-16) E6/E7 mRNA at the transcriptional level in an EGF-stimulated human keratinocyte cell line: Functional role of EGF-responsive silencer in the HPV-16 long control region, *J. Virol.* **65**:2000–2009.
174. Hembree, J. R., Agarwal, C., and Eckert, R. L., 1994, Epidermal growth factor suppresses insulin-like growth factor binding protein 3 levels in human papillomavirus type 16-immortalized cervical epithelial cells and thereby potentiates the effects of insulin-like growth factor 1, *Cancer Res.* **54**:3160–3166.
175. Malejczyk, J, Malejczyk, M., Majewski, S., Breitburd, F., Luger, T. A., Jablonska, S., and Orth, G., 1994, Increased tumorigenicity of human keratinocytes harboring human papillomavirus type 16 is associated with resistance to endogenous tumor necrosis factor-alpha-mediated growth limitation, *Int. J. Cancer* **56**:593–598.
176. Bartsch, D., Boye, B., Baust, C., zur-Hausen, H., and Schwarz, E., 1992, Retinoic acid-mediated

repression of human papillomavirus 18 transcription and different ligand regulation of the retinoic acid receptor beta gene in nontumorigenic and tumorigenic HeLa hybrid cells, *EMBO J.* **11**:2283–2291.
177. Sizemore, N., and Rorke, E. A., 1993, Human papillomavirus 16 immortalization of normal human ectocervical epithelial cells alters retinoic acid regulation of cell growth and epidermal growth factor receptor expression, *Cancer Res.* **53**:4511–4517.
178. Merrick, D. T., Gown, A. M., Halbert, C. L., Blanton, R. A., and McDougall, J. K., 1993, Human papillomavirus-immortalized keratinocytes are resistant to the effects of retinoic acid on terminal differentiation, *Cell Growth Differ.* **4**:831–840.
179. Batova, A., Danielpour, D., Pirisi, L., and Creek, K. E., 1992 Retinoic acid induces secretion of latent transforming growth factor beta 1 and beta 2 in normal and human papillomavirus type 16-immortalized human keratinocytes, *Cell Growth Differ.* **3**:763–772.
180. Glick, A. B., Flanders, K. C., Danielpour, D., Yuspa, S. H., and Sporn, M. B., 1989, Retinoic acid induces transforming growth factor-β2 in cultured keratinocytes and mouse epidermis, *Cell Regul.* **1**:87–97.
181. Lazo, P. A., Gallego, M. I., Ballester, S., and Feduchi, E, 1992, Genetic alterations by human papillomaviruses in oncogenesis, *FEBS Lett.* **300**:109–113.
182. Crook, T., Wrede, D., and Vousden, K. H., 1991, p53 point mutation in HPV negative human cervical carcinoma cell lines, *Oncogene* **6**:873–875.
183. Wrede, D., Tidy, J. A., Crook, T., Lane, D., and Vousden, K. H., 1991, Expression of RB and p53 proteins in HPV-positive and HPV-negative cervical carcinoma cell lines, *Mol. Carcinogen.* **4**:171–175.
184. Srivastava, S., Tong, Y. A., Devadas, K., Zou, Z. Q., Chen, Y., Pirollo, F. K., and Chang, E. H., 1992, The status of the p53 gene in human papilloma virus positive and negative cervical carcinoma cell lines, *Carcinogenesis* **13**:1273–1275.
185. Park, D. J., Sharon, P., Wilczynski, R. L., Paquette, R. L., and Miller, C. W., 1994, p53 mutations in HPV-negative cervical carcinoma, *Oncogene* **9**:205–210.
186. Chen, L., Thomas, E. K., Hu, S., Hellström, I., and Hellström, K. E., 1991, Human papillomavirus type 16 nucleoprotein E7 is a tumor rejection antigen, *Proc. Natl. Acad. Sci. USA* **88**:110–114.
187. Chen, L., Mizuno, M. T., Singhal, M. C., Hu, S.-L., Galloway, D. A., Hellström, I., and Hellström, K. E., 1992, Induction of cytotoxic T lymphocytes specific for a syngeneic tumor expressing the E6 oncoprotein of human papillomavirus type 16, *J. Immunol.* **148**:2617–2621.
188. Meneguzzi, G., Cerni, C., Kieny, M. P., and Lathe, R., 1991, Immunization against human papillomavirus type 16 tumor cells with recombinant vaccinia viruses expressing E6 and E7, *Virology* **181**:62–69.

9

Vaccines against Human Papillomaviruses and Associated Tumors

LIONEL CRAWFORD

1. INTRODUCTION

In principle there is no reason why vaccination against human papillomavirus (HPV) infection and against HPV-associated tumors should not already be in progress. Systems for production of the necessary antigens have been available for some time, and the problems of producing papillomavirus-coded proteins in adequate amounts and of acceptable quality have largely been overcome. For a long time the fact that HPVs could not be grown *in vitro*, and that virus particles from *in vivo* lesions were available at only very low levels, ruled out vaccination against the initial virus infection. By the use of recombinant DNA technology, components of the virus particle and, more recently, synthetic empty virus particles with many of the properties of authentic virus particles have been produced. These are discussed under prophylactic vaccines.

Because there is a long interval between the initial infection with HPV and the generation of malignant cells, there is a second possible point of intervention against HPV-associated tumors, that is, against the premalignant cells carrying the HPV genome in an integrated or nonintegrated state. This therapeutic approach has to focus on proteins other than those of the virus particle, i.e., the early proteins coded by the virus and responsible for transforming the host cell. This therefore defines the problem and the two possible points of intervention, first against the infecting virus and second against the premalignant cell.

LIONEL CRAWFORD • ICRF Tumour Virus Group, Department of Pathology, University of Cambridge, Cambridge CB2 1QP, England.

DNA Tumor Viruses: Oncogenic Mechanisms, edited by Giuseppe Barbanti-Brodano *et al.* Plenum Press, New York, 1995.

For a vaccination strategy to be successful, it must normally be based on augmentation of natural immunity, although there are some circumstances in which there is no natural immunity but vaccination can succeed. In the case of papillomaviruses, the evidence for natural immunity is rather conflicting. On one hand, repeated infections with the same or closely related viruses occur, but on the other hand, immunosuppression appears to increase the risk of cervical cancer substantially, as discussed in Section 4. Even if natural immunity is not easy to demonstrate, it is still possible to generate immunity in some circumstances. These are where the failure to generate immunity results from the way in which antigens are presented to the immune system, generating no response or tolerance rather than a positive response.

2. INFECTION BY HPV

Anogenital tumors, in particular carcinoma of the uterine cervix, are strongly associated with infection by a subgroup of the genital HPV. The types most commonly found in these tumors are 16, 18, 31, 33, and 45 in descending order of frequency. In different parts of the world the relative importance of these types varies, but type 16 always seems to account for at least half the cases. Infection with these malignant types is mostly by sexual contact, although there may also be occasional transmission by perinatal infection. The infection itself has little immediate impact, and the lesions produced in the genital tract frequently regress without treatment. A significant percentage of normal women at any time have an active HPV lesion on their cervices, by the criterion of cells containing HPV DNA detectable by PCR. The virus DNA is present in the epithelial layer only. The cells in the basal layer contain HPV DNA, usually as a free plasmid at this stage, and only the virus-coded early proteins are being expressed.

A map of the genome of HPV-16 is shown in Fig. 1, and a listing of the known functions of virus-coded proteins is given in Table 1. The early functions of the virus, those expressed soon after infection, alter the metabolic state of the cell, changing it from a resetting state to a more active one in which virus replication can take place. Transcription of the virus genes takes place in one direction, clockwise on the map as drawn, and there is a good deal of overlap among the various messenger RNAs and a complex relationship among the proteins produced. Some of them share sequences and are coded in the same region of the genome but in different reading frames, for example, E2 and E4.

By interacting with cellular control proteins such as Rb1 and p107, E6 and E7 activate the transcription machinery of the cell to make the array of enzymes that are needed for the synthesis of nucleic acid precursors and DNA. The virus proteins concerned with control of virus gene transcription and DNA synthesis are E1 and E2, and these take up most of the early region. Cells containing virus DNA in free or integrated form can persist for long periods without progressing to the later stages of the virus life cycle in which E4, L1, and L2 proteins are expressed. Expression of the virus particle (late) proteins and assembly of virus particles take place as the cells move up through the epithelium, differentiate, and die before being shed from the surface as squames.

HUMAN PAPILLOMAVIRUS VACCINES

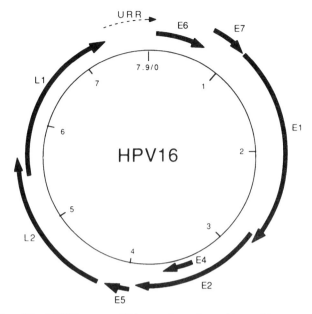

FIGURE 1. Map of the HPV-16 genome. This map shows the positions of the open reading frames of HPV-16.

In a small fraction of infected women, and in an extremely small fraction of infected men, additional changes occur in the cells containing the HPV DNA before they are committed to virus production, differentiation, and death. The accumulation of these changes over an extended period is what generates malignancy. The interval between the initial infection and the production of premalignant cells detectable by cervical smear test may be as long as 10 to 20 years. During

TABLE I
Sizes and Functions of HPV16 Proteins

Predicted size (amino acids)		Functions
Early proteins		
E1	649	Virus DNA replication
E2	365	Transcription and replication control
E4	92	Binds cytokeratins preceding virus assembly
E5	83	Affects growth factor receptors
E6	151	Binds p53, trans-activates, contributes to immortalization
E7	98	Main transforming protein, binds Rb1 and cyclin A/$p33^{CDK2}$
Late proteins		
L1	505	Major virus coat protein (glycoprotein)
L2	473	Minor virus coat protein

this time the virus DNA persists and may become integrated into the genome of the infected cell while undergoing deletion of substantial parts of its own genome, particularly of the late region. What is left includes the genes for the two smallest virus-coded early proteins, E6 and E7. These are therefore the most obvious targets for therapeutic vaccines.[1,2]

3. TRANSFORMATION AND TUMORIGENESIS

Infection with HPV is clearly not sufficient for development of a tumor because most women who become infected do not develop cervical cancer, even though the HPV type involved is a malignant one. This is even more true for men: cancer of the penis is very rare in developed countries in spite of the fact that comparable numbers of men and women are probably infected with malignant HPV types. However, the association of cervical cancer (and penile cancer) with HPV infection is very close and implies that the infection is the essential first step in a series of events in which genetic changes accumulate in the progeny of the epithelial cells initially infected. These finally render them malignant and invasive. Some HPV-negative cervical tumors are found, although these are relatively uncommon. It is not clear how many of these are caused by HPV types that have not yet been characterized and therefore cannot be detected with existing probes. Alternatively they may have been HPV-positive initially and subsequently lost the HPV genome by chance or as a result of immune selection at a stage where the malignant phenotype is no longer dependent on the continued expression of HPV genes. The third alternative, that they have a different etiology and relate to carcinogens other than HPV, cannot be ruled out, but this must apply to only a small percentage of all cervical tumors because the association of cervical cancer with HPV is so strong.

Maintenance of the transformed state of cells containing at least part of the HPV genome is related to interference with cell cycle control by the virus-coded early proteins E7 and E6. E7 associates with protein coded for by the retinoblastoma suppressor gene, Rb1. This protein normally prevents cells entering the cell cycle by sequestering essential transcription factors such as E2F. When Rb1 is phosphorylated or complexed with E7, the transcription factors are released and DNA replication and cell division can go ahead. At the end of one round of replication, Rb1 normally reexerts its control, but the presence of E7 in the cell prevents this. The cell continues to divide in an uncontrolled way but at this stage is still otherwise normal. Other Rb1-related proteins, such as p107 and p300, are also affected in a similar way by complexing with E7, probably with similar consequences.

The role of E6 is less clear but involves another tumor suppressor gene, p53. This exerts its effect on the cell cycle in a conditional way, preventing the cell from entering DNA replication when its DNA contains unrepaired damage caused by irradiation, exposure to mutagens, virus infection, or other insults. This gives time for the DNA repair machinery of the cell to correct the lesions(s) before DNA replication makes them permanent. If this fails, p53 is also capable of triggering programmed cell death, apoptosis, thereby avoiding the production of an altered and potentially malignant cell. In a majority of most common human tumors, p53 is found to be mutant, losing these protective activities and at the same time acquiring the properties of an active oncogene. The E6 proteins of malignant HPV types associate with p53 and cause its degradation via the ubiquitin pathway. Loss of p53 leaves the cell open to accumulation of mutation in all sorts of other genes, particularly in other tumor suppressor genes, of which the cellular genome contains a significant number. Loss of E6 and restoration of p53 activity at a later stage would not affect these other lesions and, if the cell was already malignant, achieve little. Elimination of cells expressing E6- and E7-derived peptides on their cell surfaces would, on the other hand, be very effective in preventing development of tumors.

4. IMMUNE RESPONSE TO HPV INFECTION

Striking evidence that an immune response can be effective against HPV-associated tumors comes from studies of allograft patients who have been immunosuppressed over an extended period to avoid rejection of their transplants. These patients have substantially increased rates of cervical cancer, implying that in normal women many of the potentially malignant events that occur are dealt with by the immune system before they become detectable. The means by which this is achieved are not entirely clear but involve cell-mediated immunity rather than a humoral response (antibodies). This is apparent from the fact that inherited immunodeficiency diseases that affect this arm of the immune response also show increased cervical cancer rates. On the other hand, defects in antibody production do not affect the rate significantly.[3] Work with animals[4-6] strongly supports the role of cell-mediated immunity in rejection of papillomavirus-induced tumors. Spontaneous rejection is usually accompanied by massive lymphocyte infiltration into the lesions.

The role of antibodies in HPV immunity is less clear although antibodies against a variety of virus-coded proteins are certainly produced. These are active against a whole variety of virus-coded proteins, both early proteins and particle proteins. The highest levels of antibodies against the major transforming protein, E7, are found in patients with cervical cancer[7] and seem to be more an indication of advanced disease than a protection against it. The best candidates for protective antibodies would be mucosal IgA antibodies in cervical mucus directed against the virus particle proteins. There is little evidence for production of antibodies of this type at the high levels that would be likely to be needed for inactivation of incoming virus. This is consistent with the observation that prior infection with HPV has little protective effect against subsequent infection. However, in some cases, the lack of antibodies may be more apparent than real, related to the use of fusion proteins produced in bacterial systems as antigens. Recently antibodies have been detected in a significant number of women using better quality antigens in the tests, i.e., L1 in the form of synthetic virus-like (empty) particles produced in baculovirus systems.[8] The way in which the viral antigens are presented to the immune system may account for the seemingly poor antibody response to HPV infection. The superficial nature of the virus infection, presentation of antigens by keratinocytes, and the fact that assembled virus particles are present only in the outermost layers of the mucosa may all contribute to this poor response. Good-quality antigens presented in a more effective way may nevertheless be capable of generating a protective immune response.

5. REQUIREMENTS FOR GENERATION OF CELL-MEDIATED IMMUNITY

The first step in this process is the presentation of virus-coded peptides complexed with HLA molecules, either class I or class II, on the surface of antigen-presenting cells. Exogenous proteins are taken up and processed by antigen-presenting cells, and the peptides derived from them are displayed on the cell surface complexed with HLA class II. This generates the $CD4^+$ helper response, which is important for both antibody and CTL responses. For presentation on HLA class I, the peptides are normally derived from proteins synthesized within the cell and processed by specific proteases. The peptides are transported to the lumen of the endoplasmic reticulum by specific transporters and associate there with HLA class I heavy chains and β_2-microglobulin. This complex is then expressed on the cell surface and generates $CD8^+$ cytotoxic T cells. Failure at any stage of antigen processing or peptide transport can prevent proper expression of the peptide. Insertion of exogenously derived peptide into this pathway is difficult because the association with HLA normally takes place inside the cell. Proteins presented outside the cell are normally broken down by a different pathway and presented on HLA class II to $CD4^+$ T cells. This class II presentation may also be important here for stimulation of helper T cells and production of cytokines early in the immune response, as mentioned above.

The next step involves the association of the antigen-presenting cell carrying the peptide with T cells bearing specific receptors and additional accessory molecules such as CD28 to interact with B7 on the antigen-presenting cell. So long as all these requirements are satisfied, cytotoxic T cells (CTLs) may be generated, usually $CD8^+$, although $CD4^+$ CTLs have been reported.

The recognition of target cells by the CTLs involves many of the same components as are involved in their generation. The cells are recognized by virtue of their displaying on their surfaces the same virus-coded peptide associated with HLA class I. Evasion of the CTL response can be achieved in a variety of ways by virus-transformed cells. Allele loss and down-regulation of HLA expression have been postulated as a means of escape from this type of immune surveillance. This and related topics are discussed in detail in a recent review.[9] Its importance is clear in that no amount of immunization would be effective if the target cells are devoid of the elements such as HLA class I needed to render them susceptible or if they can shed the virus-coded elements against which the immunity is directed. Fortunately, there is no indication that this occurs as a rule.

6. PROPHYLACTIC VACCINES

In the long term it is likely that control of HPV infections will be achieved by the use of prophylactic vaccines. This is based on analogy with other virus diseases that have been controlled or, in the case of smallpox, eliminated by the use of attenuated live vaccines or subunit vaccines. The use of attenuated live vaccines seems unlikely at present with HPV, and other alternatives must therefore be used.

Virus-like (empty) particles appear to be the antigen that mimics most closely the properties of the infecting virus particle and should therefore elicit the production of antibodies against epitopes of the major capsid protein, L1, and perhaps also the minor capsid protein, L2. It is essential that these include the configurational determinants that are involved in neutralization of virus infectivity The initial infection with HPV may be followed by several rounds of self-infection in which virus produced by the first infected cell is released and infects surrounding cells to enlarge the initial lesion. This phase should also be inhibited by the presence of antibodies in the cervical mucus.

Cell-mediated immunity may also be involved even at this stage. Generation of CTLs against targets producing L1 and L2 would limit the infection very effectively. However, the administration of L1 or L2 as free protein is unlikely to generate this response unless accompanied by special adjuvants, which are now being developed. Expression of L1 from a live recombinant vaccinia vector has been shown to generate CTL that are active against syngeneic mouse cells expressing L1.[10] Restrictions on the use of vaccinia, now that smallpox has been eliminated, mean that this vector is unlikely to be approved for human use. However, the use of other poxvirus live vectors remains an attractive possibility, bearing in mind the economy, stability, and ease of use of these vectors in developing countries where most of the world's cervical cancer occurs. Several systems now exist for the production of virus-like (empty) particles.[11-15] Initially it seemed that HPV-16 virus-like particles were either more unstable or more difficult to produce in both baculovirus and vaccinia systems than other papillomaviruses. This now seems to be a consequence of the DNA sequence[16] of the original isolate of HPV-16, which came from a cervical carcinoma rather than from a virus-producing lesion.[17] Recent isolates from earlier-stage lesions have sequences that differ at several positions. One of these, at residue 202 in L1, is seen in almost all isolates and appears to be critical. Constructs with the corrected sequence produce much higher levels of virus-like particles *in vitro*.[14,18]

Virus proteins other than L1 may be considered targets for prophylactic vaccines in the sense of prophylactic against the infected cell rather than against the infecting virus. Any of the virus-coded early proteins, E1, E2, E5, E6, or E7, would be a potential target and could ensure the elimination of the first infected cell before production of virus progeny.

The prime requirement for any effective prophylactic vaccine would be that it would have to be administered early in life, preferably to both boys and girls. So long as solid immunity was established before the onset of sexual activity, prevalence of HPV in the population would gradually be reduced.

7. THERAPEUTIC VACCINES

The first vaccines to be tested on a large scale are likely to be therapeutic rather than prophylactic for ethical reasons, because they would be given to patients who already have cervical cancer rather than to healthy individuals. Because E6 and E7 genes are uniformly retained in cervical cancers and other HPV genes are often lost, attention has focused on these two small proteins. This retention could have been for one or both of two reasons. First, only those cells in which control of cell division remained abnormal because of the interaction of the two viral proteins with cell cycle control proteins would continue to divide and therefore be potentially malignant. Second, the small size of the virus proteins and selection over a long period had ensured that their sequences contained no T-cell epitopes. Cells expressing them would therefore not be potential targets for CTLs and would escape immune surveillance. Initially the difficulty of defining T-cell epitopes gave some support to this idea, but more recently T-cell epitopes for mice have been defined. Surprisingly, although the sequence of E7 contained what seemed a reasonable consensus T-cell epitope in the N-terminal half, the main response to E7 was directed against a sequence in the C-terminal half that had little similarity to the consensus.[19] Injection of synthetic peptides covering these and other sequences generated tumor protection in mice, confirming their importance *in vivo*.[20] Peptides coupled to a lipid moiety are highly immunogenic and produce good CTL responses in experimental animals. However, because the response to peptides is MHC restricted, there are serious difficulties in extending the results obtained in inbred mice to outbred human populations. In addition, the cost of producing peptides and ensuring that a course of several injections is completed would be barriers to the widespread use of peptide immunization. These studies do eliminate the concern that E7 contains no T-cell epitopes and support the usefulness of targeting E7.

8. DELIVERY

The mode of delivery of the chosen immunogen(s) is critical for their efficacy and must be considered at the same time as the form of the immunogen. Protein immunogens under normal conditions, even with the addition of Freund's adjuvant, tend to favor the production of antibodies rather than the induction of cytotoxic T cells, which is probably what is required in this context. Other adjuvants are therefore needed, although alum is the only adjuvant approved for general use in humans at present. Adjuvants based on squalene or other oils and muramyl peptides should soon be approved for human use and prove much better than alum. One of these new adjuvants, MF 59, has already been shown to reduce the number of tumors in mice when used with HPV-16 E7 protein to protect against syngeneic tumor cells transformed by HPV-16 plus *myc*.[21] Emulsifying immunogen with adjuvant may alter the form of complex antigens such as virus-like (empty) particles. It is not clear whether synthetic empty particles are stable enough to withstand this procedure and whether configurational determinants would be retained if the particles were broken down to free capsomeres.

Both protein and peptide immunogens are expensive to produce, and it is likely that second-generation vaccines will be live, using either viral or bacterial vectors. These have the additional advantage that HPV proteins produced in cells infected with recombinant viruses can be processed by the normal pathway and presented on the cell surface bound to HLA. As this mimics more closely the sequence of events in a normal HPV infection it should increase the chances of a good cell-mediated response being produced. The use of vaccinia and other poxvirus vectors introduces additional complications because of the peculiarities of their transcription machinery and their impact on the immune system. This necessitates the removal of virus-specific transcription termination signals to avoid premature termination of the HPV mRNA and the use of early rather than late promoters for expression of the HPV protein(s).[9]

Particulate antigens may have the advantage of being less dependent on the addition of adjuvants for their effectiveness, especially in the induction of mucosal immunity. In addition to synthetic empty particles consisting entirely of HPV-coded protein(s), other particles have been generated with epitopes from HPV proteins inserted into other particle-forming proteins. These include picornaviruses, yeast Tu particles, hepatitis B pre-S/S particles, and hepatitis B core antigen particles. Hepatitis B pre-S/S particles produced in yeast have been widely used as a vaccine against hepatitis and are administered to children at an early age. This would be ideal for HPV as well, setting up immunity before sexual maturity. Because hepatitis B is strongly associated with liver cancer, with dietary aflatoxin acting as a cofactor in many cases, the hepatitis immunization program should also reduce liver cancer in due course. The extent to which it does so should provide strong encouragement for this approach to prevention of virus-associated cancers and thus for a parallel approach to HPV and cervical cancer.

Bacterial vectors have received relatively little attention in spite of the advantages they offer of cheapness, stability, and ease of propagation in comparison with viral vectors. Attenuated pathogens such as BCG have been widely used against tuberculosis and are attractive vehicles for foreign antigens,[22] although studies with HPV proteins have not been reported. Attenuated *Salmonella* are also possible vectors for HPV proteins,[23] with the advantage that these can be administered orally and then penetrate the cells of the mucosal lining of the gut. From this location they can present antigens in such a way as to induce mucosal cell-mediated immunity. It is not clear whether mucosal immunity in the intestine is also accompanied by immunity in the genital tract.

Rather than using attenuated pathogens with the possibility of their reverting to virulence, there is also the option of using commensal bacteria that are part of the normal flora of the gut or genital tract. One construct that has been shown to be capable of generating an antibody response to an HPV protein is HPV-16 E7 inserted into the M protein of *Streptococcus gordonii*, a normal human commensal.[24] The E7 protein displaces the immunodominant determinant of the M protein and is displayed on the outer membrane of the bacteria. Initial studies have been done with E7 in mice, inducing serum IgA and IgG antibodies (B. Chain and L. Gao, personal communication), and it would be desirable to extend this work to cell-mediated immunity and to other HPV antigens. If HPV-16 L1 and L2 could be inserted into the same vector and shown to generate cell-mediated and humoral responses in the genital tract of mice, this would be a good indication that this was a feasible approach to prophylactic vaccination.

9. VALIDATION OF VACCINE EFFICACY

One of the most serious problems that remains to be solved is the devising of satisfactory criteria for evaluating vaccine effectiveness that are also practical and can be achieved in a reasonable time span. Safety testing of candidate vaccines is relatively straightforward, but the problems are much worse with efficacy. Some circumstantial evidence can be generated along with the safety trials, the generation of antibodies and activated T cells for example, but this is a long way from showing that the vaccines would be effective in preventing tumors from the natural infection. The narrow host range of HPV means that animal systems are of limited use, and no system exists in which infection with HPV virus particles results in tumors analogous to cervical tumors. The use of other primate papillomaviruses to produce tumors in their natural hosts has been suggested as a test system for vaccine strategies, but apart from the cost and length of time needed for generation of tumors it is uncertain how close the analogy to human tumors would be. At best the results would be only suggestive. Bovine systems[4] have been used to show that both therapeutic and prophylactic vaccines work to some extent, and, although this is very encouraging, the extent to which the BPV tumors are analogous to HPV tumors is unclear. Murine systems in which the endpoint is the prevention of tumor production by syngeneic transformed cells are cheaper and quicker but are artificial to the extent that the transformed cells have to be produced *in vitro* and then injected into the test animal. Transformation of epithelial cells requires both HPV-16 and an activated oncogene, and it is essential to show that any tumor prevention or rejection is directed against the HPV component and not against the oncogene. This type of system can be used to test therapeutic vaccines and has shown that peptides containing T-cell epitopes from HPV-16[20] and E7 protein produced in fission yeast[21] can be effective in reducing tumor production. Recombinant vaccinia virus expressing E7 and/or E6 has also been shown to be effective in rodent test systems.[1,2]

In many ways the most realistic approach to validation of vaccines against malignant types would be via benign HPV types. Genital warts produced by HPV types such as 6 and 11 are a serious clinical problem in their own right. Although the tumors they produce are in general benign, the progress of the disease is otherwise very similar to that of cervical cancer. The advantage of genital warts as a test system would be that results would be seen more quickly and be easier to assess because the lesions are external. The patients most at risk for genital warts are also those at risk for cervical cancer, and the methods developed for measurement of the immune response to the benign HPV proteins would be easily adapted for the malignant types, based on practical experience.

10. CURRENT VACCINE TRIALS

One of the most encouraging developments in the last year has been the approval and initiation of two small-scale trials. These are at the level of safety and efficacy testing and are being carried out in patients who have had cervical cancer and have limited further treatment options available. Their success will be judged by the level of immune response that is generated, and it is unreasonable to expect that they will produce the cure or remission of what is already advanced disease.

The trial being carried out in Wales (L. K. Borysiewicz, personal communication) is using a live vaccinia recombinant expressing mutant proteins combining the E6 and E7 genes of HPV-16 and -18. The vaccinia recombinant was constructed by Cantab Pharmaceuticals and is intended to retain the antigenic activity of all four proteins while eliminating their other undesirable properties. In any case, patients with cervical lesions have clearly been exposed to significant amounts of intact E6 and E7 in the normal course of their HPV infection, and further exposure would be unlikely to have any deleterious effect.

The other trial being carried out in Australia (I. H. Frazer, personal communication) is also using E7, but in the form of pure protein rather than as a live virus. The protein is produced in *E. coli* as a fusion protein with glutathione-S-transferase, purified, and then combined with adjuvant for injection into the patients.

Both of these trials will involve only small numbers of patients but should pave the way for much larger and more ambitious trials, and these are indeed at the planning stage.

11. CLOSING REMARKS

Studies on HPV have now reached the exciting stage where there is a sound basis for development of strategies for intervention against both the initial virus infection and the malignant or premalignant cell containing HPV DNA and expressing virus-coded proteins. More information would be desirable on the differences in immune response between women who are infected with malignant HPV types and do not get cervical carcinoma and those who suffer from the tumor. This would highlight the areas in which alteration of the response would be likely to have the greatest impact on the prevalence of the tumor. However, this information is difficult to acquire, and it is more likely that in the short term advances will come from a more pragmatic approach, testing vaccines that have a reasonable chance of being effective and studying the response to these in detail. Vaccines based on the small transforming proteins of HPV-16 and -18, E6 and E7, are already in trial on a small scale. Proof of their safety should allow larger-scale trials to be carried out soon. In parallel with trials on vaccines against malignant HPV types, it would be very desirable to have similar trials of vaccines against the benign types that cause genital warts. Benign, in this context, is very much a relative term because tumors containing HPV-6 and -11 do occur, and even so-called benign tumors can be life-threatening, as seen with laryngeal tumors. The direct benefits of an effective vaccine against these benign types would be accompanied by a greatly improved understanding of what is required for effective intervention against the malignant types that are the prime cause of so many deaths worldwide.

REFERENCES

This is not a complete reference list for vaccine-related work, and more extensive references may be found in the reviews given as references.[3,6,9,15,25]

1. Chen, L., Thomas, E. K., Hu, S.-L., Hellstrom, I., and Hellstrom, K. E., 1991, Human papillomavirus type 16 nucleoprotein E7 is a tumor rejection antigen, *Proc. Natl. Acad. Sci. USA* **88:** 110–114.
2. Meneguzzi, G., Cerni, C., Kieny, M. P., and Lathe, R., 1991, Immunization against human papillomavirus type 16 tumor cells with recombinant vaccinia viruses expressing E6 and E7, *Virology* **181:**62–69.
3. Frazer, I. H., and Tindle, R. W., 1992, Cell-mediated immunity to papillomaviruses, *Papillomavirus Rep.* **3:**53–58.
4. Lin, Y. L., Borenstein, L. A., Selvakumar, R., Ahmed, R., and Wettstein, F. O., 1992, Effective vaccination against papilloma development by immunization with L1 or L2 structural proteins of cottontail rabbit papillomavirus, *Virology* **187:**612–619.
5. Campo, M. S., Grindlay, G. J., O'Neil, B. W., Chandrachud, L. M., McGarvie, G. M., and Jarrett, W. F. H., 1993, Effective vaccination against a mucosal papillomavirus, *J. Gen. Virol.* **74:**945–953.
6. Jochmus, I., and Altmann, A., 1993, Immune response to papillomaviruses: Prospects of an anti-HPV vaccine, *Papillomavirus Rep.* **4:**147–151.
7. Jochmus-Kudielka, I., Schneider, A., Braun, R., Kimmig, R., Koldovsky, U., Schneweiss, K. E., Seedorf, K., and Gissmann, L., 1989, Antibodies against the human papillomavirus type 16 early proteins in human sera: Correlation of anti-E7 reactivity with cervical cancer, *J. Natl. Cancer Inst.* **81:**1689–1704.
8. Kirnbauer, R., Hubbert, N. L., Wheeler, C. M., Becker, T. M., Lowy, D. R., and Schiller, J. T., 1994, A virus-like particle enzyme-linked immunosorbent assay detects serum antibodies in a majority of women infected with human papillomavirus type 16, *J. Natl. Cancer Inst.* **86:**484–498.
9. Duggan-Keen, M., Keating, P. J., Comme, F. V., Walboomers, J. M. M., and Stern, P. L., 1994, Alterations in major histocompatibility complex expression in cervical cancer: Possible consequences for immunotherapy, *Papillomavirus Rep.* **5:**3–9.
10. Zhou, J., McIndoe, A., Davies, H., Sun, X.-Y., and Crawford, L., 1991, The induction of cytotoxic T lymphocyte precursor cells by recombinant vaccinia virus expressing human papillomavirus type 16 L1, *Virology* **181:**203–210.
11. Zhou, J., Sun, X.-Y., Stenzel, J., and Frazer, I. H., 1991, Expression of vaccinia recombinant HPV 16 L1 and L2 ORF proteins in epithelial cells is sufficient for assembly of HPV virion-like particles, *J. Virol.* **185:**251–257.
12. Xi, S.-Z., and Banks, L. M., 1991, Baculovirus expression of human papillomavirus type 16 capsid proteins: Detection of L1–L2 protein complexes, *J. Gen. Virol.* **72:**2981–2988.
13. Kirnbauer, R., Booy, F., Cheng, N., Lowy, D., and Schiller, J. T., 1992, Papillomavirus L1 major capsid protein self-assembles into virus-like particles that are highly immunogenic, *Proc. Natl. Acad. Sci. USA* **89:**12180–12184.
14. Kirnbauer, R., Taub, J., Greenstone, H., Roden, R., Durst, M., Gissmann, L., Lowy, D. R., and Schiller, J. T., 1993, Efficient assembly of human papillomavirus type 16 L1 and L1–L2 into virus-like particles, *J. Virol.* **67:**6929–6936.
15. Hagensee, M. E., and Galloway, D. A., 1993, Growing human papillomaviruses and virus-like particles in the laboratory, *Papillomavirus Rep.* **4:**121–124.
16. Seedorf, K., Krammer, G., Durst, M., Suhai, S., and Rowekamp, W. G., 1985, Human papillomavirus type 16 DNA sequence, *Virology* **145:**181–185.
17. Durst, M., Gissmann, L., Ikenberg, H., and zur Hausen, H., 1983, A papillomavirus DNA from a cervical carcinoma and its prevalence in cancer biopsy samples from different geographical regions, *Proc. Natl. Acad. Sci. USA* **80:**3812–3815.

18. Pushko, P., Sasagawa, T., Cuzick, J., and Crawford, L., 1994, Sequence variation in the capsid protein genes of human papillomavirus type 16, *J. Gen. Virol.* **75**:911–916.
19. Sadovnikova, E., Zhu, Z., Collins, S. M., Zhou, J., Vousden, K., Crawford, L., Beverley, P., and Strauss, H. J., 1994, Limitations of predictive motifs revealed by cytotoxic T lymphocyte epitope mapping of the human papilloma virus E7 protein, *Int. Immun.* **6**:289–296.
20. Feltkamp, M. C. W., Smits, H. L., Vierboom, M. P. M., Minnaar, R. P., de Jongh, B. M., Drijhout, J. M. W., ter Schegget, J., Melief, C. J. M., and Kast, W. M., 1993, Vaccination with cytotoxic T lymphocyte epitope containing peptide protects against a tumor induced by human papillomavirus type 16-transformed cells, *Eur. J. Immunol.* **23**:2242–2249.
21. Hibma, M. H., Tommasino, M., Van Nest, G., Ely, S. J., Contorni, M., and Crawford, L., 1993, Immune responses to HPV 16 E7, in: *Immunology of Human Papillomaviruses* (M. A. Stanley, ed.), Plenum Press, New York, pp. 291–297.
22. Strover, C. K., de la Cruz, V. F., Fuerst, T. R., Burlein, J. E., Benson, L. A., Bennett, L. T., Bansal, G. P., Young, J. F., Lee, M. H., Hatfull, G. F., Snapper, S. B., Barletta, R. G., Jacobs, W. R., and Bloom, B. R., 1991, New use of BCG for recombinant vaccines, *Nature* **351**:456–460.
23. Londono, P., Tindle, R., Frazer, I., Chatfield, S., and Dougan, G., 1993, Use of double *aro* *Salmonella* mutants to stably express HPV 16 E7 protein epitopes carried by HBV core antigen, in: *Immunology of Human Papillomaviruses*, (M. A. Stanley, ed.), Plenum Press, New York, pp. 299–303.
24. Pozzi, G., Contorni, M., Oggioni, M. R., Manganelli, R., Tommasino, M., Cavalieri, F., and Fischetti, V. V., 1992, The delivery and expression of a heterologous antigen on the surface of streptococci, *Infect. Immun.* **60**:1902–1907.
25. Crawford, L., 1993, Prospects for cervical cancer vaccines, *Cancer Surv.* **16**:215–229.

10

The Complex Role of Hepatitis B Virus in Human Hepatocarcinogenesis

MARIE ANNICK BUENDIA and PASCAL PINEAU

1. INTRODUCTION

Viral hepatitis in humans has been causally related to a variety of agents, which belong to different virus groups and differ markedly in their structural and biological properties. As many as five viruses associated with human hepatitis (the hepatitis A, B, C, D, and E viruses) have been isolated and fully characterized,[1-5] and the question of whether additional, uncharacterized viruses may be responsible for liver disease in some cases is still debated. Among the known human hepatitis viruses, two—the hepatitis B and C viruses—appear to be more pathogenic for humans because of their ability to induce persistent infections in the infected host. Chronic, long-lasting infections with both agents frequently result in the development of liver cirrhosis and ultimately evolve to primary liver cancer in a significant proportion of virus carriers.

Evidence that chronic infection with the hepatitis B virus (HBV) is tumorigenic for humans was first provided by a strong epidemiologic association between long-lasting HBV carriage and the development of hepatocellular carcinoma (HCC).[6-9] This association is particularly obvious in endemic countries, in which seropositive males incur a 30- to 100-fold increased risk of developing HCC.[10-12] It has been estimated that five out of 1000 HBV carrier patients worldwide will eventually succumb to primary liver cancer. Moreover, HBV DNA has been detected

---

MARIE ANNICK BUENDIA and PASCAL PINEAU • Unité de Recombinaison et Expression Génétique, INSERM U163, Département des Rétrovirus, Institut Pasteur, 75724 Paris Cedex 15, France.

DNA Tumor Viruses: Oncogenic Mechanisms, edited by Giuseppe Barbanti-Brodano *et al.* Plenum Press, New York, 1995.

by sensitive techniques in about one-half of liver tumors from serologically recovered or seronegative patients, further extending the epidemiologic association.[13,14] Another argument linking HBV and HCC comes from studies of related viruses of the *Hepadnaviridae* group, showing that the woodchuck hepatitis virus (WHV) and the ground squirrel hepatitis virus (GSHV) induce liver disease and HCC in their respective hosts.[15,16]

Despite extensive studies during the last decade, the precise role of HBV in malignant transformation of infected hepatocytes remains elusive. It is also striking that liver cancer remains one of the few human neoplasms in which oncogenic events are poorly defined at the molecular level. None of the previously identified tumorigenic pathways operating in other malignancies has been demonstrated in the case of HCC, although a number of etiological factors have been identified, including the above-cited viral agents, environmental factors such as dietary aflatoxins and alcohol, and inherited metabolic diseases such as hemochromatosis and α_1-antitrypsin deficiency. Even the ubiquitous tumor suppressor gene p53, implicated in an altered form in more than one-half of human neoplasms of various kinds, appears to be mutated in only 20% of HCCs worldwide and more frequently in advanced-stage tumors.[17]

We present in this chapter an overview of different possible mechanisms by which persistent HBV infection may lead to hepatocarcinogenesis after a long latency period. We also report on studies of naturally occurring as well as experimental animal models, supporting some current hypotheses on the role of HBV in human liver cancer.

2. INDIRECT MECHANISMS: IMPORTANCE OF THE IMMUNE RESPONSE IN NECROINFLAMMATORY LIVER DISEASE

It is generally admitted that HBV has no direct oncogenic effect on the infected hepatocytes but, rather, that malignant transformation results from decades of hepatocellular injury and subsequent mitogenic stimulation of liver cells, allowing accumulation of genomic mutations and finally transformation. First, a quick overview on the phenomena paving the way to liver cancer is necessary.

2.1. Physiopathological Features

The natural history of chronic hepatitis B usually begins insidiously, following a mild or clinically inapparent acute hepatitis. It involves both inflammation of the liver and hepatocyte damage. Regeneration accompanies continuous cell destruction, with appearance of foci of new cells. Changes in parenchymal cells are associated with cellular infiltrates rich in lymphocytes and containing polymorphonuclear leukocytes and Kupffer cells. The clinical and histological classification of chronic hepatitis distinguishes cases with significant parenchymal damage, named chronic active hepatitis (CAH), from those with inflammatory infiltrates but minimal cell damage, named chronic persistent hepatitis (CPH). During CAH, HBV antigens are readily detected in liver tissue, and many liver cells contain both the core (HBcAg) and surface (HBsAg) antigens. On the contrary, during CPH, only few cells with HBcAg are detected among many HBsAg-positive cells.[18,19]

It is mainly during CAH that cirrhosis may develop when collagen fibers are laid down in the portal tracts and when nodular regeneration becomes prominent.[20] Thus, the initiating event for liver cirrhosis is an important hepatocellular injury followed by overreactivity of the regeneration process: cell division continues beyond simple replacement, giving rise to hyperplastic nodules. The hepatitis B virus is usually associated with macronodular cirrhosis. Each nodule seems to have its own pattern of HBV antigen distribution: some are free of viral markers, but other cells stain intensely by immunofluorescent labeling. This heterogeneity may reflect different degrees of differentiation among regenerative nodules.[19] Toxins usually inactivated by cells situated near the portal triad of the hepatic lobule may accumulate in peculiarly susceptible hepatocytes as a result of the alteration of parenchymal organization and blood flows.[21]

About 60% to 80% of all HCCs arise on an underlying cirrhosis. It is not clear whether cirrhosis by itself is a predisposing factor or whether the underlying causes of cirrhosis are actually carcinogenic factors.[22] Indeed, patients with cirrhosis of any etiology—but particularly those infected with HBV (or HCV)—are at increased risk of developing HCC.[19] However, a substantial percentage of HCCs (around 25%) arise in a noncirrhotic liver, providing evidence that the key precursor lesion to HCC is hepatocellular injury rather than cirrhosis.[23] Recent studies of patients with HCC diagnosed at an early stage indicate that most or all of these tumors begin as macroregenerative (also called adenomatous hyperplastic) nodules.[24] Viral antigens are differently distributed in the cancerous and noncancerous parts of the liver. Both surface and core antigens are less frequently observed in tumor than in nontumor tissue. The surface antigen is more frequently seen in well-differentiated than in poorly differentiated tumors.[19]

2.2. Immunopathogenesis

There is a general consensus that hepatocellular damage in hepatitis B is caused by the host immune response and not by the virus itself. The chronicity of infection my result from the inability of the immune response to destroy all virus-infected cells.[25] The liver parenchyma is infiltrated by a mixed population of inflammatory cells including mainly MHC class-I-restricted cytotoxic T lymphocytes (CTL).[26] The remaining populations include polynuclear neutrophils, granulocytes, and predominant mononuclear phagocytic cells of the organ: the Kupffer cells.[27] The immune response mediated by polyclonal CTL is the principal effector of cell injury in human hepatitis B.[28] The clearance of infected hepatocytes is achieved by CTL only if viral antigens are appropriately processed and displayed on the cell surface in association with products of the major histocompatibility complex (MHC).[26] In fact, it has been shown that HBc/HBe antigens are the major targets of the cell-mediated immune response and play an important part in liver damage during acute and chronic hepatitis.[29] In chronic hepatitis B, the HLA class-I- and class-II-restricted HBV-specific T cells are much less abundant than in patients with acute hepatitis who successfully clear the virus.[28] It has been demonstrated in transgenic mice that CTL directly bind and induce cytological changes characteristic of apoptosis in HBsAg-positive hepatocytes.

The severity of the liver damage is not determined by the direct cytotoxic potential of CTLs but by their capacity to recruit antigen-nonspecific inflammatory cells and by amplification mechanisms triggered by interferon-γ (IFN-γ) production.[30] Several other inflammatory cytokines secreted by different cellular types present in the parenchymal tissue have been implicated in immune response during chronic hepatitis B: IFN-α, IFN-β, interleukin-2 (IL-2), and tumor necrosis factor-α (TNF-α).[27] Cytokines play an important part in the local immune response by recruiting activated inflammatory cells and allowing adequate expression of MHC class I molecules on the hepatocyte membranes.[25] The production of IFN-α, IFN-β, and IL-2 has been shown to be deficient in patients with chronic HBV infection.[31] However, recent studies have shown that these cytokines play a more complex role: IFN-α, IFN-β, IL-2, and TNF-α are able to profoundly down-regulate HBV gene expression. Thus, they may contribute either to viral clearance or to evasion from the host immune response and progression to viral persistence.[28] It would be interesting to know whether other inflammatory cytokines such as IL-1 or IL-6 play a part in viral persistence or immune-mediated liver damage.

Activated expression of growth factors has also been associated with chronic hepatitis B: recent reports have revealed the implication of tumor growth factor-α (TGF-α) in nodule regeneration during cirrhosis, and elevated levels of TGF-α can be detected in the urine of patients diagnosed with HCC. Tumor growth factor β1 may have an important role in the pathogenesis of cirrhosis; TGF-β1 is secreted by lipocytes (or Ito cells) and other nonparenchymal cells and has been shown to stimulate collagen type-I synthesis in the liver of chronically infected patients.[20]

2.3. Molecular Aspects

Because any dividing cell is much more at risk for mutations, liver regeneration following liver injury expands the pool of cells prone to genetic lesions.[32] The rare cells sustaining the appropriate genetic hits can then undergo clonal expansion. Hepatocyte DNA lesions during persistent HBV infection may result from chronic cellular dysfunction characterized by exposure to mutagenic products secreted by inflammatory cells, impaired detoxification pathways, or DNA repair mechanisms. The activity of a variety of microsomal monooxygenases involved in the inactivation of chemical carcinogens is known to be enhanced in liver preparations from HBV-infected individuals.[33] Cytochrome P-450 isoforms involved in the hepatic activation of aflatoxin B1 (AFB1), the most widely spread human hepatocarcinogen, are induced in HBV-infected cells, making them more susceptible to AFB1-induced genotoxicity.[34]

Another mechanism whereby chronic liver injury as well as inflammation may lead to hepatocarcinogenesis is through the endogenous production of mutagens such as oxygen free radicals and nitrosamines.[35] The phagocytic cells scavenge the necrotic debris, but they also contribute to hepatocyte damage by directly releasing products of oxidative metabolism such as H_2O_2 and hydroxyl radicals.[19] Oxygen free radicals known to be produced during liver inflammation by activated Kupffer cells may lead to the formation of 8-hydroxyguanosine adducts, which are promutagenic DNA lesions inducing G-to-T transversion.[36] Reactive oxygen species are able to enhance lipid peroxidation, producing malondialdehyde, a carcinogenic compound leading to the formation of malondialdehyde–deoxyguanosine adducts.[37] It should be noted that the oxidative stress (hydrogen peroxide commonly produced by phagocytic cells) markedly reduces HBV gene expression in cultured cells, possibly contributing to viral escape from immune response.[38] This phenomenon may be compared with the action of some cytokines on viral gene expression in transgenic mice.[39] The potential for endogenous production of nitrosamines has been demonstrated to be mediated *in vitro* by nitrosating agents derived from nitric oxide (NO). It has been shown that, under stimulation by lipopolysaccharides and cytokines, human hepatocytes are able to produce NO through amino acid oxidation catalyzed by nitric oxide synthetase.[40] An increased nitrosation potential has been detected during cirrhosis and liver infestation by trematodes.[41]

The complex immunoinflammatory reactions involved in HBV-associated liver disease and eventually in hepatocarcinogenesis are schematically represented in Fig. 1. It is well known that mitogenic agents are proper carcinogens and are important in human cancer.[32] However, regarding the molecular aspect developed in the following sections of this chapter, the contribution of HBV to liver carcinogenesis may not be restricted to the sole indirect liver injury followed by cell regeneration.

3. DOES THE HBV GENOME CONTAIN ANY CONCLUSIVELY TRANSFORMING GENE?

3.1. The Genetic Organization of HBV

The HBV is a small enveloped DNA virus that displays a unique genomic structure and replication pathway. The HBV virion consists of an eicosahedral nucleocapsid containing the viral genome, surrounded by an envelope made of viral glycoproteins and host-derived lipids. During HBV infections, abundant viral subparticles made of empty envelopes (HBsAg particles) are produced in the circulating blood. As illustrated in Fig. 2, the HBV genome is a partially double-stranded DNA molecule of 3.2 kb, maintained in circular configuration by base pairing of the 5' ends of the two DNA strands. Four viral genes have been mapped on the HBV genome; two of them encode structural proteins of the virion. The pre-C/C region encodes the capsid protein carrying the core antigen (HBcAg) and a secreted polypeptide, the HBe antigen. The pre-S/S region codes for three proteins of the viral envelope carrying the surface antigen determinants (HBsAg): the large surface protein (LHBs), the middle protein (MHBs), and the small or major protein (SHBs). The longest HBV gene (P), covering two-thirds of the genome, encodes a polypeptide with DNA polymerase and reverse transcriptase activity involved in viral replication, and the last gene (X) codes for a transcriptional *trans*-activator.

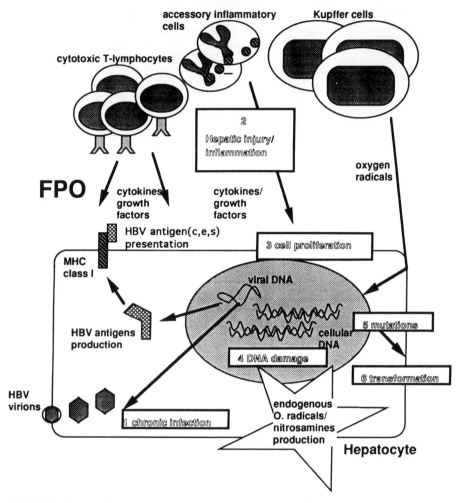

FIGURE 1. Nonspecific immunoinflammatory processes suspected to favor hepatocarcinogenesis during chronic HBV infection.

How might the viral replicative forms and the viral gene products interfere with the normal growth control of infected hepatocytes? As represented in Fig. 3, the HBV life cycle is entirely extrachromosomal: after the first steps of viral entry, which remain largely unknown, the viral genome migrates to the nucleus, and the formation of covalently closed circular DNA (cccDNA) precedes the transcription step. Although integration into the host genome is not required for viral replication, integrated HBV sequences are frequently observed in chronic infections and in HCC. Whether HBV DNA integration plays a part in the tumorigenic process is

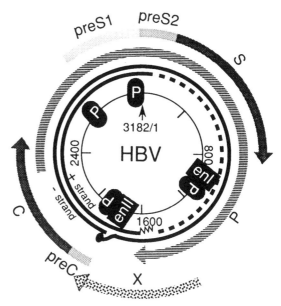

FIGURE 2. Structure and genetic organization of the HBV genome. The inner circles represent the long (−) and short (+) DNA strands, and the terminal protein linked to the 5' end of the long strand is shown by a black spot. Nucleotide numbering from the unique EcoRI site in HBV ayw.[1] Large arrows depict the functional HBV open reading frames. The regulatory elements, four promoters and two enhancers, are positioned.

discussed later (Section 4). The viral transcripts made from cccDNA serve different functions: they are translated into seven proteins, and the RNA pregenome serves as an intermediate for viral replication. After synthesis of the two DNA strands inside the nucleocapsid, the viral assembly occurs in the endoplasmic reticulum (ER), and the virions, together with a large excess of empty envelope particles, are secreted through the Golgi apparatus.

Different HBV gene products have been suspected to contribute to the oncogenic potential of the virus. They are the viral X *trans*-activator, in native or modified form, and surface proteins including the large protein and truncated versions of the middle protein.

3.2. Potential Oncogenicity of the Viral X *Trans*-Activator

Similarities in the organization and mode of replication between retroviruses and HBV have suggested that the HBX gene might encode a *trans*-activator analogous to the HTLV-1 *tax* and the HIV *tat* genes.[42] Indeed, like *tat* and *tax*, HBX can *trans*-activate a variety of eukaryotic cellular and viral promoters. It has recently been shown that the X gene of the closely related hepadnavirus WHV is necessary for viral infection *in vivo*,[43,44] although HBX is dispensable for the viral life cycle *in vitro*.[45] *Trans*-activation by the X protein is mediated by a number of transcription factors, including AP-1,[46] AP-2,[47] C/EBP,[48] CREB/ATF,[49] and NF-KB.[50] In the absence of a DNA-binding domain in HBX, the question of whether this protein is present in DNA-binding protein complexes in the cell nucleus is still debated. The X protein might also operate from a cytoplasmic compartment through various

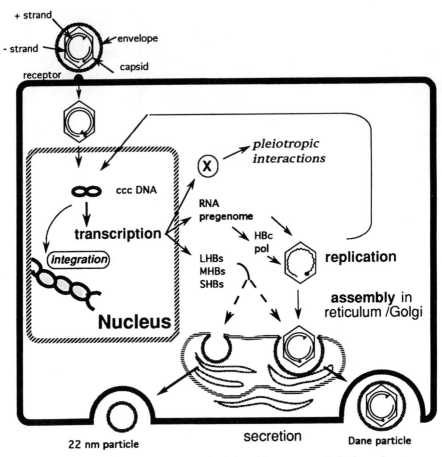

FIGURE 3. Hepatitis B virus replication in the infected hepatocytes. Only the major steps are indicated: viral entry through attachment to a still unknown receptor; migration of the genome to the nucleus and formation of covalently closed circular DNA (cccDNA), which serves as a template for viral transcription; synthesis of the viral proteins; encapsidation of pregenomic RNA; assembly of the virions in the ER, and secretion. Detailed replication mechanisms may be found in recent reviews.[133,134]

intracellular signal transduction pathways, such as the signaling cascade upstream of protein kinase C (PKC)[51] and of Ras, Raf, and MAP kinases,[52] leading to the stimulation of cell proliferation. In addition, the X gene product may interact with the tumor suppressor p53[53,54] and with cellular serine proteases,[55] two associations that might be relevant to explain its potential role in the pathogenesis of human hepatocellular carcinoma. The pleiotropic interactions of X with cellular proteins are reported in Fig. 4.

Arguments favoring the oncogenicity of HBX have been provided in experiments showing that HBX can render certain NIH3T3 cells tumorigenic in nude

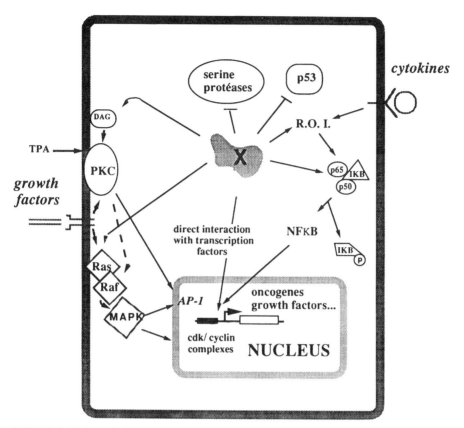

FIGURE 4. Pleiotropic interactions of HBX with cellular factors in the infected hepatocyte, leading to transcription activation of target genes and stimulation of cell proliferation.

mice[56] and that it can transform primary hepatocytes immortalized by the SV40 large T antigen.[57] However, there is no evidence for liver damage in most transgenic mouse lines expressing the X protein in the liver[58] (P. Briand, personal communication), except in the case of a transgenic line established in the outbred CD-1 background.[59] More recent data indicate that X might serve as a cofactor in carcinogen-mediated hepatocarcinogenesis (J. Butel, personal communication) and might accelerate HCC development in transgenic mice expressing the c-*myc* oncogene in the liver (O. Terradillos and M. A. Buendia, unpublished results). Furthermore, as discussed in Section 4, the presence of native or truncated X genes in a majority of integrated viral sequences in HCCs has suggested that deregulated expression of the integrated *trans*-activator might represent one step of the tumorigenic process.[60]

The mode of action of the X protein, its multiple possible interactions with cellular factors, and its contribution to HBV-induced liver tumorigenesis are subjected to intensive investigations but remain controversial issues.

3.3. Endoplasmic Reticulum Localization of HBV Surface Proteins

In productive HBV infections, large amounts of viral envelope proteins are secreted through the ER/Golgi membranes. Accumulation of the viral proteins in this subcellular compartment may occur in chronic hepatitis B, as shown by the appearance of "ground glass cells" in the liver of chronic HBV carriers.[61] The importance of these events has been shown in transgenic mice overexpressing LHBs: the accumulation of toxic amounts of this protein in liver cells leads to chronic liver damage that ultimately evolves to HCC.[62,63] It has been shown that the progressive accumulation of LHBs in the ER is associated with a dramatic expansion of this compartment.[62] It is striking that in other transgenic models, liver lesions and carcinomas induced by liver-specific overexpression of c-*myc* or TGF-α are also associated with marked ER enlargement.[64–66] It has recently been shown that in eukaryotic cells, the accumulation of unfolded proteins in the ER triggers a signaling pathway from the ER to the nucleus, leading to activation of a specific set of transcription factors.[67] It is therefore tempting to speculate that alterations in the normal ER functions of proteins represent a possible pathway of liver tumorigenesis.

In addition, indirect evidence suggests that the viral *trans*-activator MHBs^t (a truncated, nonsecretable form of the pre-S2/S protein produced from integrated HBV sequences) is retained in the ER and might contribute in deregulating the normal cell growth control.[68–71] Taken together, these data provide convincing arguments for a contribution of impaired trafficking of HBV envelope proteins through the ER to HBV-associated carcinogenesis.

4. MUTAGENIC ACTION OF HBV DNA INTEGRATION IN THE HOST GENOME

Integration of HBV DNA into host chromosomes has been observed in about 80% of human HCCs associated with HBV infection. In view of the previously noted similarities between hepadnaviruses and retroviruses, it is tempting to speculate that the integration events might have a direct role in activating nearby cellular protooncogenes, as in many retroviral models.[72] However, it is now established that there is no common genetic locus on human chromosomes representing a preferred target for HBV insertion.[73] Analysis of viral junctions and flanking cellular sequences rather indicate that HBV DNA integration displays some preference for regions containing repeating sequences.[74–76] This feature may reflect a particular chromatin structure that renders these regions preferentially available for integration of HBV DNA, thus determining regional specificity of integration. The finding of extensive DNA rearrangements, including large deletions, chromosomal translocations, or amplifications in the cellular regions flanking the HBV inserts in many HCCs[77–80] has led to the proposal that HBV integration might promote genetic instability in target cells.[81]

Around 20% of HCC arising in HBsAg-positive patients do not exhibit any detectable HBV DNA integration. In some cases, the absence of integrated viral

DNA may be the consequence of genetic rearrangements that accumulate during malignant progression: large deletions following excision of the viral insert may result in the loss of a tumor suppressor gene, thereby conferring growth advantage to the cell. Thus, genetic instability generated by viral integration may be an important step in the pathogenesis of HCC.

4.1. Direct Mechanisms

Surveys of cellular sequences across HBV insertion sites have failed to reveal known protooncogenes. However, in two independent tumors and in an established hepatoma cell line, potential oncogenes were identified at the respective sites of integration.[82–84] In the first described example, HBV DNA was found to be integrated in an exon of a gene related to the family of nuclear receptors for steroid and glucocorticoid hormones.[82,85] This gene was later identified with the gene encoding the receptor β of retinoic acid (RARβ).[86] It seems probable that inappropriate expression of an altered retinoic acid receptor disturbed the normal control of cell growth in this particular tumor. In a second HCC, HBV DNA integration occurred in an intron of the human cyclin A gene, resulting in a strong expression of hybrid HBV/cyclin transcripts.[87,88] Constitutive expression of a chimeric form of cyclin A driven by the strong viral pre-S2/S promoter might lead to uncontrolled DNA synthesis and cell proliferation. Finally, in the PCR/PLF-5 cell line, one of the HBV inserts has disrupted the mevalonate kinase gene, known to play an important role in the isoprenylation of cellular proteins related to growth control, including Ras proteins.[84]

4.2. Indirect Mechanisms

It has recently been suggested that viral integration in human HCC might play a more indirect role by allowing deregulated expression of mutated viral genes. As a consequence of the viral integration process, sequences of the viral X gene are present in a majority (over 70%) of HBV integrated sequences in HCC, and viral/cell junctions near DR1 are associated with the deletion of several residues in the carboxy-terminal part of the X protein. The X transcriptional *trans*-acting function is retained in the truncated gene products.[89,90] Other rearrangements in HBV integrated sequences have been mapped to the S coding region in one-third of human HCCs related to HBV.[71,75] Kekulé *et al.*[91,92] have demonstrated that truncation of the S gene between residues 77 and 221 of the Pre-S2/S product generates a novel transcriptional *trans*-acting capacity. Interestingly, the truncated MHBs protein (MHBs^t) stored in the ER membrane displays a broad spectrum of *trans*-acting activities comparable to that of the viral X protein; it can activate transcription from various cellular promoters, such as those of the c-*myc*, c-*fos*, and epidermal growth factor (EGF) receptor genes, via the PKC pathway.[70] Therefore, the integrated X *trans*-activator and the truncated MHB protein might be implicated in disturbing the control of cellular gene expression in a significant fraction of human HCCs.[71]

5. IMPORTANCE OF *myc* FAMILY GENES IN CARCINOGENESIS INDUCED IN RODENTS BY HEPATITIS-B-LIKE VIRUSES

Eastern woodchucks (*Marmotta monax*) infected in the wild by WHV, or experimentally infected at birth with WHV virions, usually develop HCC within 2 to 4 years.[93] The high oncogenicity of WHV was demonstrated by the development of HCC in 100% of carrier animals maintained in controlled laboratory conditions, in the absence of dietary or environmental cocarcinogens. A significant risk of HCC has also been associated with past WHV infection; in contrast, liver tumors have never been observed in uninfected woodchucks over their 10-year life-span.[94,95] Chronic WHV infection usually provokes only mild to moderate hepatitis and differs mainly from HBV infection by the absence of liver cirrhosis, even in aged carrier animals. The closely related GSHV induces a similar pathology in the natural host (*Spermophilus beecheyi*), but liver tumors develop at a lower rate and after a longer latency period in this model.

Most liver tumors from woodchucks chronically infected with WHV or recovered from previous WHV infection present clonal patterns of viral integration.[94,96,97] Our studies of the viral integration sites have demonstrated that WHV acts mainly as an insertional mutagen of *myc* family genes. Genetic alterations associated with overexpression of c-*myc* were initially observed in three tumors.[98] In two different cases, WHV DNA integration in the immediate vicinity of the c-*myc* gene coding region[99] resulted in enhanced steady-state levels of c-*myc* mRNA. The c-*myc* expression was driven by the normal promoters of the oncogene, indicating that c-*myc* activation in these HCCs was associated with a mechanism of viral enhancer insertion, as previously shown in many examples of retroviral integrations.[100,101] In another tumor, the woodchuck c-*myc* gene was recombined with a cellular sequence termed *hcr*, with no apparent linkage to WHV integration.[102] An identical rearrangement between c-*myc* and *hcr* has recently been described in an independent woodchuck HCC.[103] A survey of 50 woodchuck HCCs for c-*myc* rearrangements has shown only one supplementary case in which WHV DNA was inserted into the c-*myc* locus.[104] It can be estimated now that insertional activation of c-*myc* is involved in about 10% of woodchuck tumors (Fig. 5). The oncogenic efficiency of these integration events has been demonstrated in a transgenic mouse model. Transgenic mice carrying the woodchuck c-*myc* gene and juxtaposed WHV sequences from a woodchuck hepatoma develop HCC at 8 to 12 months of age in virtually all cases.[64]

Further studies have outlined a higher frequency of WHV integrations in the woodchuck N-*myc* genes. Molecular cloning of a viral insertion site has led to the discovery of a second functional N-*myc* gene in the woodchuck genome. This gene, termed N-*myc*2, presents typical features of a processed pseudogene issued from retrotransposition of a spliced N-*myc* transcript.[105] N-*myc*2 represents by far the most frequent target for WHV DNA integration: in more than 40% of tumors, N-*myc*2 was found to be activated by nearby insertion of WHV enhancers, whereas the parental N-*myc* gene was mutated in only one case,[105–107] as illustrated in Fig. 5. Moreover, the gene is overexpressed in a majority of woodchuck HCCs, in the absence of genetic alteration at the N-*myc*2 locus. This observation was recently

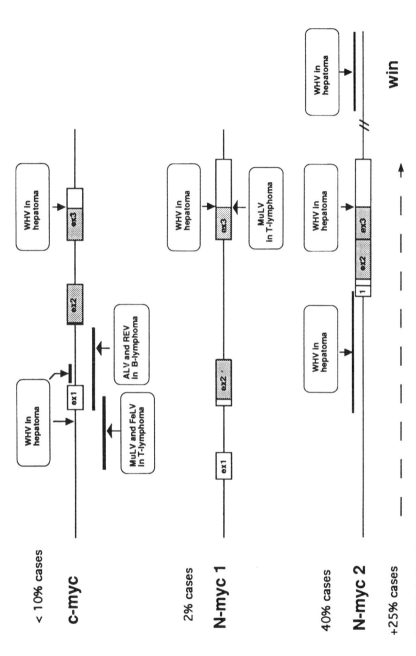

FIGURE 5. Mapping of the WHV integration sites in *myc* genes in 52 woodchuck HCCs, and comparison with frequent retroviral insertion sites in B and T cell lymphomas. The frequency of WHV integration events in each locus is indicated as a percentage above the name of the corresponding *myc* gene, and is shown at the bottom part in the case of *win*, a common WHV insertion locus located about 200 kb downstream of N-*myc*2 on the woodchuck X chromosome.[108]

correlated with frequent WHV integrations in a new locus, located 200 kb away from N-*myc*2 on the woodchuck X chromosome.[108] This locus, termed *win* for WHV insertion site related to N-*myc*2, presents clustered insertions of WHV DNA in about 25% of woodchuck HCCs. Whether the *win* locus contains a growth-related gene or a gene capable of activating N-*myc*2 in *trans*, or whether viral integration in *win* influences the activity of the N-*myc*2 promoter over a long distance, remains to be established.

The reasons for the strong clustering of WHV integration sites into N-*myc*2 and *win* are not yet clear. It seems probable that these loci present a chromatin configuration favorable to viral integration and that the WHV enhancers integrated at either site can efficiently activate expression from the N-*myc*2 promoter.

The N-*myc*2 retroposon is present in a limited number of mammalian species, including squirrels, but not in humans. It is noticeable that insertional mutagenesis of the N-*myc*2 gene has not been observed in ground squirrel tumors induced by GSHV.[109] This might be explained either by reduced ability of GSHV DNA to integrate into the host genome or by a defect of the integrated GSHV sequences to activate the N-*myc*2 gene. The second hypothesis is supported by the finding that GSHV DNA does not select the N-*myc*2 locus for integration in GSHV-induced woodchuck HCCs.[107] However, the ground squirrel/GSHV model also demonstrates that activated expression of *myc* genes plays a predominant role in rodent hepatomas induced by hepadnavirus infections: amplifications of the c-*myc* locus associated with enhanced expression of the oncogene have been detected in about 50% of GSHV-associated squirrel tumors.[109] This process is apparently unlinked to viral integration.[110]

The marked differences in oncogenic efficiencies between the two related rodent hepadnaviruses may be correlated with the ability of WHV, but not of GSHV, to produce mutagenic events at a high frequency. However, these data indicate that activation of *myc* family genes, although achieved through different mechanisms, lies at the meeting point of the tumorigenic process induced by hepadnaviruses in rodent hosts.

6. GENETIC ALTERATIONS AND TUMOR SUPPRESSOR GENES IN LIVER CANCER

A growing number of studies have shown common chromosomal alterations in liver tumors consistent with the loss of tumor suppressor genes. Several regions frequently deleted in HCC have been located on chromosomes 17p, 16p, 16q, 5q, 4q, 1q, 5p, 8q, and 13q.[111–118] This substantial number of allele losses suggests that an equivalent number of tumor suppressor genes are involved in HCC development. Moreover, no other type of tumor has so far been shown to display loss of heterozygosity (LOH) on chromosomes 4 and 16, raising the possibility that such changes are specific for HCC.[119] However, no tumor suppressor gene is known to be systematically inactivated in human HCC, in which only the p53 gene seems to be consistently implicated.[17]

Indeed, the most frequent LOH, occurring in more than 50% of HCCs, has been observed at chromosome 17p, where the p53 gene is located.[113,114] In addition, p53 mutations have been detected in a significant proportion of liver tumors. The p53 protein, "guardian of the genome," allows DNA repair by cell cycle arrest in G_1 phase or, alternatively, triggers programmed cell death.[120] The profiles of p53 mutations exhibit an important heterogeneity, depending on the geographic origin and the etiology of the tumor.[121] Regions of the world where food contamination by AFB1 is endemic (Mozambique, Senegal, or the Qidong area in China, for example) are characterized by a high rate of G-to-T transversions at codon 249, in exon 7 of the p53 gene.[122,123] In Mozambique, this "hotspot" is mutated in about 60% of liver tumors.[121] In other regions of the world mostly free of AFB1 dietary contamination, this p53 mutation is infrequent (around 15%).[124,125] Other pathways of p53 inactivation have been suggested by the finding that the HBX gene product complexes with wild-type p53.[53,54] It should be mentioned that, as in many other neoplastic processes,[126] p53 mutations in liver cancer may be a late event affecting in most cases poorly differentiated HCCs.[127]

Loss of heterozygosity at chromosome 16 occurs in the same proportions on both arms (in around 45% of cases).[113,118] It has been shown that the main affected region is located between the HP (16q22.1) and CTRB loci (16q22.3–q23.2). This LOH was not detected in early HCC, but its frequency increases with tumor size and loss of differentiation; it might therefore represent a late event in tumor progression.[118] The gene encoding E-cadherin, located in 16q, was recently shown to be reduced to hemizygosity in a significant fraction of tumors.[128] Reduced expression of this gene, coding for a cell-adhesion molecule, has been associated with invasive phenotype in human primary tumors of the liver and prostate. Both arms of chromosome 5 are also affected by LOH, in 37% of cases for the short arm and 28% for the long one. Chromosome 5q carries the APC tumor suppressor gene, which is known to be deleted in adenomatous polyposis. However, the deleted locus in HCC does not appear to overlap with the APC locus.[113] The chromosome 4q displays around 35% of LOH, and the most likely location for a tumor suppressor gene is between 4q11–13 and 4q25.[112] Loss of heterozygosity at the Rb locus in 13q has been detected in a substantial proportion of poorly differentiated liver tumors, and only in association with p53 mutations.[127] Conflicting results have been obtained in two other studies reporting less than 10% with rearrangements at the Rb locus.[129,130] Karyotypic abnormalities in 1p have been reported, showing striking correlation with those found in neuroblastomas (1p36) and suggesting the presence of a relevant gene in various types of tumors.[115]

Recently, the discovery that the CDK inhibitor p16/MTS1 is homozygously deleted in a striking proportion of human tumor cell lines has suggested that this gene could be a major contributor to cancer development.[131,132] Although the 9p21 region is not frequently deleted in liver cancer, we have looked for p16/MTS1 rearrangement in hepatoma cell lines and HCC from HBsAg-positive patients. Among six cell lines and 20 HCC samples analyzed by Southern hybridization, none displayed any double deletion or gross rearrangement of the p16/MTS1 gene (P. Pineau and A. Dejean, unpublished observations).

The results of chromosomal change studies indicate that allelic losses are common in HCC and perhaps affect a larger number of loci than in the well-studied colon cancer. The large number of genes potentially involved in multistep liver carcinogenesis may explain the long latency necessary for HCC emergence in humans; it might also reflect a highly redundant growth control of liver cells.

7. CONCLUSIONS

Comparative studies of the hepadnaviruses infecting humans, woodchucks, and squirrels have shown that those viruses, despite closely related structural and biological properties, differ markedly in oncogenicity. The high incidence and rapid onset of hepatocellular carcinoma in woodchucks infected with WHV has been correlated with a direct role of the virus as an insertional mutagen. In addition, the presence of a highly susceptible target site, the woodchuck retrotransposed N-*myc*2 oncogene, probably contributes to accelerate HCC development in WHV-infected woodchucks. The human genome contains no sequence homologous to N-*myc*2, but the absence of viral targeting toward the human c-*myc* and N-*myc* genes in HBV-related HCCs remains a puzzling question because of the high frequency of HBV integrations in human HCCs. Although a direct role of HBV DNA integration in HCC development has been demonstrated only in rare cases, viral integration might have more indirect actions such as long-range mutagenic effects or *trans*-acting effects.

Apart from the large differences in the ability of mammalian hepadnaviruses to provoke insertional activation of cellular oncogenes, it seems probable that common underlying mechanisms are involved in the malignant processes in the three systems. The contribution of viral gene products, such as the envelope proteins and the X *trans*-activator, and that of nonspecific pathways triggered by the necroinflammatory disease induced by persistent HBV infection remains to be determined.

REFERENCES

1. Galibert, F., Mandart, E., Fitoussi, F., Tiollais, P., and Charnay, P., 1979, Nucleotide sequence of the hepatitis B virus genome (subtype ayw) cloned in *E. coli*, *Nature* **281**:646–650.
2. Ticehurst, J. R., Racaniello, V. R., Baroudy, B. M., Baltimore, D., Purcell, R. H., and Feinstone, S. M., 1983, Molecular cloning and characterization of hepatitis A virus cDNA, *Proc. Natl. Acad. Sci. USA* **80**:5885–5889.
3. Choo, Q. L., Kuo, G., Weiner, A. J., Overby, L. R., Bradley, D. W., and Houghton, M., 1989, Isolation of a cDNA clone from a blood-borne non-A, non-B viral hepatitis genome, *Science* **244**:359–362.
4. Wang, K. S., Choo, Q. L., Weiner, A. J., Ou, J. H., Najarian, R. C., Thayer, R. M., Mullenbach, G. T., Denniston, K. J., Gerin, J. L., and Houghton, M., 1986, Structure, sequence and expression of the hepatitis delta (δ) viral genome, *Nature* **323**:508–514.
5. Tsarev, S. A., Emerson, S. U., Reyes, G. R., Tsareva, T. S., Legters, L. J., Malik, I. A., Iqbal, M., and Purcell, R. H., 1992, Characterization of a prototype strain of hepatitis E virus, *Proc. Natl. Acad. Sci. USA* **89**:559–563.
6. Szmuness, W., 1978, Hepatocellular carcinoma and the hepatitis B virus: Evidence for a causal association, *Prog. Med. Virol.* **24**:40–69.

7. Maupas, P., and Melnick, J. L., 1981, Hepatitis B infection and primary liver cancer, *Prog. Med. Virol.* **27**:1–5.
8. Lin, T. M., Tsu, W. T., and Chen, C. J., 1986, Mortality of hepatoma and cirrhosis of liver in Taiwan, *Br. J. Cancer* **54**:969–976.
9. Hsing, A. W., Guo, W., Chen, J., Li, J. Y., Stone, B. J., Blot, W. J., and Fraumeni, J. F. J., 1991, Correlation of liver cancer mortality in China, *Int. J. Epidemiol.* **20**:54–59.
10. Beasley, R. P., Lin, C. C., Hwang, L. Y., and Chien, C. S., 1981, Hepatocellular carcinoma and hepatitis B virus: A prospective study of 22,707 men in Taiwan, *Lancet* **2**:1129–1133.
11. Iijima, T., Saitoh, N., Nobutomo, K., Nambu, M., and Sakuma, K., 1984, A prospective cohort study of hepatitis B surface antigen carriers in a working population, *Gann* **75**:571–573.
12. Sakuma, K., Saitoh, N., Kasai, M., Jitsukawa, H., Yoshino, I., Yamaguchi, M., Nobutomo, K., Yamuni, M., Tsuda, F., Komazawa, T., Nakamura, T., Yoshida, Y., and Okuda, K., 1988, Relative risks of death due to liver disease among Japanese male adults having various statuses for hepatitis B s and e antigen/antibody in serum: A prospective study, *Hepatology* **8**:1642–1646.
13. Bréchot, C., Degos, F., Lugassy, C., Thiers, V., Zafrani, S., Franco, D., Bismuth, H., Trépo, C., Benhamou, J. P., Wands, J., Isselbacher, K., Tiollais, P., and Berthelot, P., 1985, Hepatitis B virus DNA in patients with chronic liver disease and negative tests for hepatitis B surface antigen, *N. Engl. J. Med.* **312**:270–276.
14. Paterlini, P., Gerken, G., Nakajima, E., Terre, S., D'Errico, A., Grigioni, W., Nalpas, B., Franco, D., Wands, J., Kew, M., Pisi, E., Tiollais, P., and Bréchot, C., 1990, Polymerase chain reaction to detect hepatitis B virus DNA and RNA sequences in primary liver cancers from patients negative for hepatitis B surface antigen, *N. Engl. J. Med.* **323**:80–85.
15. Summers, J., Smolec, J. M., and Snyder, R., 1978, A virus similar to human hepatitis B virus associated with hepatitis and hepatoma in woodchucks, *Proc. Natl. Acad. Sci. USA* **75**:4533–4537.
16. Marion, P. L., Oshiro, L. S., Regnery, D. C., Scullard, G. H., and Robinson, W. S., 1980, A virus in beechey ground squirrels that is related to hepatitis B virus of humans, *Proc. Natl. Acad. Sci. USA* **77**:2941–2945.
17. Ozturk, M., 1994, Chromosomal rearrangements and tumor suppressor genes in primary liver cancer, in: *Primary Liver Cancer: Etiological and Progression Factors* (C. Bréchot, ed.), pp. 269–281, CRC Press, Boca Raton.
18. Hollinger, F. B., 1990, Hepatitis B virus, in: *Fields Virology, Vol. 2*, (B. N. Fields, D. M. Knipe, R. M., Chanock, M. S. Hirsch, J. L. Melnick, T. S. Monath, and B. Roizman, eds.), pp. 2171–2236, Raven Press, New York.
19. Israel, J., and London, W. T., 1991, Liver structure, function, and anatomy: Effects of hepatitis B virus, in: *Hepadnaviruses, Molecular Biology and Pathogenesis* (W. S. Mason and C. Seeger, eds.), pp. 1–20, Springer-Verlag, Berlin.
20. Castilla, A., Prieto, J., and Fausto, N., 1991, Transforming growth factor beta 1 and alpha in chronic liver disease, *N. Engl. J. Med.* **324**:933–940.
21. Fourel, G., 1994, Genetic and epigenetic alterations of gene expression in the course of hepatocarcinogenesis, in: *Liver Gene Expression* (L. Tronche and M. Yaniv, eds.), pp. 297–343, R. G. Landes, Austin.
22. Lotze, M. T., Flickinger, J. C., and Carr, B. I., 1993, Hepatobiliary neoplasms, in: *Cancer: Principles and Practice of Oncology* (V. T. De Vita Jr., S. Hellman, and S. A. Rosenberg, eds.), pp. 883–913, J. B. Lippincott, Philadelphia.
23. Popper, H., Shafritz, D. A., and Hoofnagle, J. H., 1987, Relation of the hepatitis B virus carrier state to hepatocellular carcinoma, *Hepatology* **7**:764–772.
24. Edmonson, H. A., 1976, Benign epithelial tumors and tumorlike lesions of the liver, in: *Hepatocellular Carcinoma* (K. Okuda and R. L. Peters, eds.), pp. 309–320, Wiley, New York.
25. Dienes, H. P., Hess, G., Wöorsdörfer, M., Rossol, S., Gallati, H., and Ramadori, G., Büschenfelde, M. Z., 1990, Ultrastructural localization of interferon-producing cells in the liver of patients with chronic hepatitis B, *Hepatology* **13**:321–326.

26. Chisari, F. V., 1991, Multistage hepatocarcinogenesis in hepatitis B virus transgenic mice, in: *Origins of Human Cancer: A Comprehensive Review* (J. Brugge, T. Curran, and F. McCormick, eds.), pp. 727–738, Cold Spring Harbor Laboratory Press, New York.
27. Andus, T., Bauer, J., and Gerok, W., 1991, Effects of cytokines on the liver, *Hepatology* **13**: 364–375.
28. Guidotti, L. G., Guilhot, S., and Chisari, F. V., 1994, Interleukin-2 and alpha/beta interferon down regulate hepatitis B virus gene expression *in vivo* by tumor necrosis factor-dependent and independent pathways, *J. Virol.* **68**:1265–1270.
29. Buendia, M. A., 1992, Hepatitis B viruses and hepatocellular carcinoma, *Adv. Cancer Res.* **59**: 167–226.
30. Ando, K., Moriyama, T., Guidotti, L. C., Wirth, S., Schreiber, R. D., Schlicht, H. J., Huang, S., and Chisari, F. V., 1993, Mechanisms of class I restricted immunopathology. A transgenic mouse model in fulminant hepatitis, *J. Exp. Med.* **178**:1541–1554.
31. Twu, J. S., and Schloemer, R. H., 1989, Transcription of the human beta interferon gene is inhibited by hepatitis B virus, *J. Virol.* **63**:3065–3071.
32. Ames, B. N., and Swirsky Gold, L., 1991, Mitogenesis, mutagenesis and rodent cancer tests, in: *Origins of Human Cancer* (J. Brugge, T. Curran, E. Harlow, and F. McCormick, eds.), pp. 125–135, Cold Spring Harbor Laboratory Press, Cold Spring Harbor, New York.
33. De Flora, S., Izzotti, A., D'Agostini, F., Balansky, R., and Camoirano, A., 1994, Metabolic activation of a cigarette smoke condensate by woodchuck liver, as related to sex, pregnancy, hepatitis virus infection and primary hepatocellular carcinoma, *Mutat. Res.* **324**:153–158.
34. Palmer, C. N. A., Coates, P. J., Davies, S. E., Shephard, E. A., and Phillips, I. R., 1992, Localization of cytochrome p-450 gene expression in normal and diseased liver by *in situ* hybridization of wax embedded archival material, *Hepatology* **16**:682–687.
35. Bartsch, H., and Montesano, R., 1984, Relevance of nitrosamines to human cancer, *Carcinogenesis* **5**:1381–1393.
36. Shimoda, R., Nagashima, M., Sakamoto, M., Yamaguchi, N., Hirohashi, S., Yokata, J., and Kasai, H, 1994, Increased formation of oxidative DNA damage, 8-hydroguanosine in human livers with chronic hepatitis, *Cancer Res.* **54**:3171–3172.
37. Chaudhary, A. K., Nokubo, M., Reddy, G. R., Yeola, S. N., Morrow, J. D., Blair, I. A., and Marnett, L. J., 1994, Detection of endogenous malondialdehyde–deoxyguanosine adducts in human liver, *Science* **265**:1580–1582.
38. Zheng, Y. W., and Yen, B., 1994, Negative regulation of hepatitis B virus gene expression and replication by oxidative stress, *J. Biol. Chem.* **269**:8857–8862.
39. Guidotti, L. G., Ando, K., Hobbs, M. V., Ishikawa, T., Runkel, L., Schreiber, R. D., and Chisari, F. V., 1994, Cytotoxic T lymphocytes inhibit hepatitis B virus gene expression by a noncytolytic mechanism in transgenic mice, *Proc. Natl. Acad. Sci. USA* **91**:3764–3768.
40. Nussler, A., DiSilvio, M., Billiar, T. R., Hoffman, R. A., Geller, D. A., Selby, R., Madariaga, J., and Simmons, R. L., 1992, Stimulation of the nitric oxide synthase pathway in human hepatocytes by cytokines and endotoxin, *J. Exp. Med.* **176**:261–264.
41. Srivatanakul, P., Ohshima, H., Khlat, M., Parkin, M., Sukaryodhin, S., Brouet, I., and Bartsch, H., 1991, *O. viverrini* infestation and endogenous nitrosamines at risk factors for cholangiocarcinoma in Thailand, *Int. J. Cancer* **48**:821–825.
42. Miller, R. H., and Robinson, W. S., 1986, Common evolutionary origin of hepatitis B virus and retroviruses, *Proc. Natl. Acad. Sci. USA* **83**:2531–2535.
43. Chen, H. S., Kanako, S., Girones, R., Anderson, R. W., Hornbuckle, W. E., Tennant, B. C., Cote, P. J., Gerin, J. L., Purcell, R. H., and Miller, R. H., 1993, The woodchuck hepatitis virus X gene is important for establishment of virus infection in woodchucks, *J. Virol.* **67**:1218–1226.
44. Zoulim, F., Saputelli, J., and Seeger, C., 1994, Woodchuck hepatitis virus X protein is required for viral infection *in vivo, J. Virol.* **68**:2026–2030.
45. Blum, H. E., Zhang, Z. S., Galun, E., von Weizsäcker, F., Garner, B., Liang, T. J., and Wands,

J. R., 1992, Hepatitis B virus X protein is not central to the viral life cycle in vitro, *J. Virol.* **66:**1223–1227.
46. Natoli, G., Avantaggiati, M. L., Chirillo, P., Costanzo, A., Artini, M., Balsano, C., and Levrero, M., 1994, Induction of the DNA-binding activity of c-Jun/c-Fos heterodimers by the hepatitis B virus transactivator pX, *Mol. Cell. Biol.* **14:**989–998.
47. Seto, E., Mitchell, P. J., and Yen, T. S. B., 1990, Transactivation by the hepatitis B virus X protein depends on AP-2 and other transcription factors, *Nature* **344:**72–74.
48. Mahé, Y., Mukaida, N., Kuno, K., Akiyama, M., Ikeda, N., Matsushima, K., and Murakami, S., 1991, Hepatitis B virus X protein transactivates human interleukin-8 gene through acting on nuclear factor κB and CCAAT/enhancer-binding protein-like *cis*-elements, *J. Biol. Chem.* **266:** 13759–13763.
49. Maguire, H. F., Hoeffler, J. P., and Siddiqui, A., 1991, HBV X protein alters the DNA binding specificity of CREB and ATF-2 by protein–protein interactions, *Science* **252:**842–844.
50. Siddiqui, A., Gaynor, R., Srinivasan, A., Mapoles, J., and Farr, R. W., 1989, *Trans*-activation of viral enhancers including long terminal repeat of the human immunodeficiency virus by the hepatitis B virus X protein, *Virology* **169:**479–484.
51. Kekulé, A. S., Lauer, U., Weiss, L., Luber, B., and Hofschneider, P. H., 1993, Hepatitis B virus transactivator HBx uses a tumour promoter signalling pathway, *Nature* **361:**742–745.
52. Natoli, G., Avantaggiati, M. L., Chirillo, P., Puri, P. L., Ianni, A., Balsano, C., and Levrero, M., 1994, Ras- and Raf-dependent activation of c-June transcriptional activity by the hepatitis B virus transactivator pX, *Oncogene* **9:**2837–2843.
53. Feitelson, M. A., Zhu, M., Duan, L. X., and London, W. T., 1993, Hepatitis B x antigen and p53 are associated *in vitro* and in liver tissues from patients with primary hepatocellular carcinoma, *Oncogene* **8:**1109–1117.
54. Wang, X. W., Forrester, K., Yeh, H., Feitelson, M. A., Gu, J. R., and Harris, C. C., 1994, Hepatitis B virus X protein inhibits p53 sequence-specific DNA binding, transcriptional activity, and association with transcription factor ERCC3, *Proc. Natl. Acad. Sci. USA* **91:**2230–2234.
55. Takada, S., Kido, H., Fukutomi, A., Mori, T., and Koike, K., 1994, Interaction of hepatitis B virus X protein with a serine protease, tryptase TL2 as an inhibitor, *Oncogene* **9:**341–348.
56. Shirakata, Y., Kawada, M., Fujiki, Y., Sano, H., Oda, M., Yaginuma, K., Kobayashi, M., and Koike, K, 1989, The X gene of hepatitis B virus induced growth stimulation and tumorigenic transformation of mouse NIH3T3 cells, *Jpn. J. Cancer Res.* **80:**617–621.
57. Höhne, M., Schaefer, S., Seifer, M., Feitelson, M. A., Paul, D., and Gerlich, W. H., 1990, Malignant transformation of immortalized transgenic hepatocytes after transfection with hepatitis B virus DNA, *EMBO J.* **9:**1137–1145.
58. Lee, T. H., Finegold, M. J., Shen, R. F., DeMayo, J. L., Woo, S. L. C., and Butel, J. S., 1990, Hepatitis B virus transactivator X protein is not tumorigenic in transgenic mice. *J. Virol.* **64:**5939–5947.
59. Kim, C. M., Koike, K., Saito, I., Miyamura, T., and Jay, G., 1991, HBx gene of hepatitis B virus induces liver cancer in transgenic mice, *Nature* **351:**317–320.
60. Kekulé, A., 1994, Hepatitis B virus transactivator proteins: The "*trans*" hypothesis of liver carcinogenesis, in: *Etiological and Progression Factors of Human Hepatocellular Carcinoma* (C. Bréchot, ed.), pp. 191–210, CRC Press, Boca Raton.
61. Gerber, M., Hadziyannis, S., Vissoulis, C., Schaffner, F., Paronetto, F., and Popper, H., 1974, Electron microscopy and immunoelectron microscopy of cytoplasmic hepatitis B antigen in hepatocytes, *Am. J. Pathol.* **75:**489–502.
62. Chisari, F. V., Fillipi, P., Buras, J., MacLachlan, A., Popper, H., Pinkert, C. A., Palmiter, R. D., and Brinster, R. L., 1987, Structural and pathological effects of synthesis of hepatitis B virus large envelope polypeptide in transgenic mice, *Proc. Natl. Acad. Sci. USA* **84:**6909–6913.
63. Chisari, F. V., Klopchin, K., Moriyama, T., Pasquinelli, C., Dunsford, H. A., Sell, S., Pinkert, C. A., Brinster, R. L., and Palmiter, R. D., 1989, Molecular pathogenesis of hepatocellular carcinoma in hepatitis B virus transgenic mice, *Cell* **59:**1145–1156.

64. Etiemble, J., Degott, C., Renard, C. A., Fourel, G., Shamoon, B., Vitvitski-Trépo, L., Hsu, T. Y., Tiollais, P., Babinet, C., and Buendia, M. A., 1994, Liver-specific expression and high oncogenic efficiency of a c-myc transgene activated by woodchuck hepatitis virus insertion, *Oncogene* **9:**727–737.
65. Murakami, H., Sanderson, N. D., Nagy, P., Marino, P. A., Merlino, G., and Thorgeirsson, S. S., 1993, Transgenic mouse model for synergistic effects of nuclear oncogenes and growth factors in tumorigenesis: Interaction of c-myc and transforming growth factor alpha in hepatic oncogenesis, *Cancer Res.* **53:**1719–1723.
66. Sandgdren, E. P., Palmiter, R. D., Heckel, J. L., Daugherty, C. C., Brinster, R. L., and Degen, J. L., 1991, Complete hepatic regeneration of an albumin–plasminogen activator transgene after somatic deletion, *Cell* **66:**245–256.
67. Mori, K., Ma, W., Gething, M. J., and Sambrook, J., 1993, A transmembrane protein with a $cdc2^+/CDC28$-related kinase activity is required for signaling from the ER to the nucleus, *Cell* **74:**743–756.
68. Huang, Z. M., and Yen, T. S. B., 1993, Dysregulated surface gene expression from disrupted hepatitis B virus genomes, *J. Virol.* **67:**7032–7040.
69. Hildt, E., Urban, S., Lauer, U., Hofschneider, P. H., and Kekulé, A. S., 1993, ER-localization and functional expression of the HBV transactivator MHBs^t, *Oncogene* **8:**3359–3367.
70. Lauer, U., Weiss, L., Hofschneider, P. H., and Kekulé, A. S., 1992, The hepatitis B virus pre-S/S^t transactivator is generated by 3′ truncations within a defined region of the S gene, *J. Virol.* **66:**5284–5289.
71. Schlüter, V., Meyer, M., Hofschneider, P. H., Koshy, R., and Caselman, W. H., 1994, Integrated hepatitis B virus X and truncated preS/S sequences derived from human hepatomas encode functionally active transactivators, *Oncogene* **9:**3335–3344.
72. Tiollais, P., Pourcel, C., and Dejean, A., 1985, The hepatitis B virus, *Nature* **317:**489–495.
73. Tokino, T., and Matsubara, K., 1991, Chromosomal sites for hepatitis B virus integration in human hepatocellular carcinoma, *J. Virol.* **65:**6761–6764.
74. Shaul, Y., Garcia, P. D., Schonberg, S., and Rutter, W. J., 1986, Integration of hepatitis B viruis DNA in chromosome-specific satellite sequences, *J. Virol.* **59:**731–734.
75. Nagaya, T., Nakamura, T., Tokino, T., Tsurimoto, T., Imai, M., Mayumi, T., Kamino, K., Yamamura, K., and Matsubara, K., 1987, The mode of hepatitis B virus DNA integration in chromosomes of human hepatocellular carcinoma, *Genes Dev.* **1:**773–782.
76. Berger, I., and Shaul, Y., 1987, Integration of hepatitis B virus: Analysis of unoccupied sites, *J. Virol.* **61:**1180–1186.
77. Rogler, C. E., Sherman, M., Su, C. Y., Shafritz, D. A., Summers, J., Shows, T. B., Henderson, A., and Kew, M., 1985, Deletion in chromosome 11p associated with a hepatitis B integration site in hepatocellular carcinoma, *Science* **230:**319–322.
78. Tokino, T., Fukushige, S., Nakamura, T., Nagaya, T., Murotsum, T., Shiga, K., Aoki, N., and Matsubara, K., 1987, Chromosomal translocation and inverted duplication associated with integrated hepatitis B virus in hepatocellular carcinomas, *J. Virol.* **61:**3848–3854.
79. Hatada, I., Tokino, T., Ochiya, T., and Matsubara, K., 1988, Co-amplification of integrated hepatitis B virus DNA and transforming gene hst-1 in a hepatocellular carcinoma, *Oncogene* **3:**537–540.
80. Hino, O., Shows, T. B., and Rogler, C. E., 1986, Hepatitis B virus integration site in hepatocellular carcinoma at chromosome 17;18 translocation, *Proc. Natl. Acad. Sci. USA* **83:**8338–8342.
81. Hino, O., Tabata, S., and Hotta, Y., 1991, Evidence for increased *in vitro* recombination with insertion of human hepatitis B virus DNA, *Proc. Natl. Acad. Sci. USA* **88:**9248–9252.
82. Dejean, A., Bougueleret, L., Grzeschik, K. H., and Tiollais, P., 1986, Hepatitis B virus DNA integration in a sequence homologus to v-erbA and steroid receptor genes in a hepatocellular carcinoma, *Nature* **322:**70–72.
83. Wang, J., Chenivesse, X., Henglein, B., and Bréchot, C., 1990, Hepatitis B virus integration in a cyclin A gene in a human hepatocellular carcinoma, *Nature* **343:**555–557.

84. Graef, E., Caselmann, W. H., Wells, J., and Koshy, R., 1994, Insertional activation of mevalonate kinase by hepatitis B virus DNA in a human hepatoma cell line, *Oncogene* **9:**81–87.
85. De Thé, H., Marchio, A., Tiollais, P., and Dejean, A., 1987, A novel steroid thyroid hormone receptor-related gene inappropriately expressed in human hepatocellular carcinoma, *Nature* **330:**667–670.
86. Brand, N., Petkovitch, M., Krust, A., Chambon, P., De Thé, H., Marchio, A., Tiollais, P., and Dejean, A., 1988, Identification of a second human retinoic acid-receptor, *Nature* **332:** 850–853.
87. Wang, Y., Chen, P., Wu, X., Sun, A. L., Wang, H., Zhu, Y. A., and Li, Z. P. 1990, A new enhancer element, ENII, identified in the X gene of hepatitis B virus, *J. Virol.* **64:**3977–3981.
88. Wang, J., Zindy, F., Chenivesse, X., Lamas, E., Henglein, B., and Bréchot, C., 1992, Modification of cyclin A expression by hepatitis B virus DNA integration in a hepatocellular carcinoma, *Oncogene* **7:**1653–1656.
89. Takada, S., and Koike, K., 1990, Trans-activation function of a 3′ truncated X gene–cell fusion product from integrated hepatitis B virus DNA in chronic hepatitis tissues. *Proc. Natl. Acad. Sci. USA* **87:**5628–5632.
90. Wollersheim, M., Debelka, U., and Hofschneider, P. H., 1988, A transactivating function encoded in the hepatitis B virus X gene is conserved in the integrated state, *Oncogene* **3:**545–552.
91. Kekulé, A. S., Lauer, U., Meyer, M., Caselmann, W. H., Hofschneider, P. H., and Koshy, R., 1990, The pre-S2/S region of integrated hepatitis B virus DNA encodes a transcriptional transactivator, *Nature* **343:**457–461.
92. Caselmann, W. H., Meyer, M., Kekulé, A. S., Lauer, U., Hofschneider, P. H., and Koshy, R., 1990, A *trans*-activator function is generated by integration of hepatitis B virus pre-S/S sequences in human hepatocellular carcinoma DNA, *Proc. Natl. Acad. Sci. USA* **87:**2970–2974.
93. Popper, H., Roth, L., Purcell, R. H., Tennant, B. C., and Gerin, J. L., 1987, Hepatocarcinogenicity of the woodchuck hepatitis virus, *Proc. Natl. Acad. Sci. USA* **84:**866–870.
94. Korba, B. E., Wells, F. V., Baldwin, B., Cote, P. J., Tennant, B. C., Popper, H., and Gerin, J. L., 1989, Hepatocellular carcinoma in woodchuck hepatitis virus-infected woodchucks: Presence of viral DNA in tumor tissue from chronic carriers and animals serologically recovered from acute infections, *Hepatology* **9:**461–470.
95. Gerin, J. L., Cote, P. J., Korba, B. E., Miller, R. H., Purcell, R. H., and Tennant, B. C., 1991, Hepatitis B virus and liver cancer: The woodchuck as an experimental model of hepadnavirus-induced liver cancer, in: *Viral Hepatitis and Liver Disease* (F. B. Hollinger, S. M. Lemon, and H. Margolis, eds.), pp. 556–559, Williams & Wilkins, Baltimore.
96. Dejean, A., Vitvitsky, L., Bréhot, C., Trépo, C., Tiollais, P., and Charnay, P., 1982, Presence and state of woodchuck hepatitis virus DNA in liver and serum of woodchucks: Further analogies with human hepatitis B virus, *Virology* **121:**195–199.
97. Hsu, T. Y., Fourel, G., Etiemble, J., Tiollais, P., and Buendia, M. A., 1990, Integration of hepatitis virus DNA near c-*myc* in woodchuck hepatocellular carcinoma, *Gastroenterol. Jpn.* **25:**43–48.
98. Möröy, T., Marchio, A., Etiemble, J., Trépo, C., Tiollais, P., and Buendia, M. A., 1986, Rearrangement and enhanced expression of c-*myc* in hepatocellular carcinoma of hepatitis virus infected woodchucks, *Nature* **324:**276–279.
99. Hsu, T. Y., Möröy, T., Etiemble, J., Louise, A., Trépo, C., Tiollais, P., and Buendia, M. A., 1988, Activation of c-*myc* by woodchuck hepatitis virus insertion in hepatocellular carcinoma, *Cell* **55:**627–635.
100. Payne, G. S., Bishop, J. M., and Varmus, H. E., 1982, Multiple arrangements of viral DNA and an activated host oncogene in bursal lymphomas, *Nature* **295:**209–214.
101. Selten, G., Cuypers, H. T., Zijlstra, M., Melief, C., and Berns, A., 1984, Involvement of c-*myc* in MuLV-induced T cell lymphomas in mice: Frequency and mechanisms of activation, *EMBO J.* **3:**3215–3222.
102. Etiemble, J., Möröy, T., Jacquemin, E., Tiollais, P., and Buendia, M. A., 1989, Fused transcripts of c-*myc* and a new cellular locus, *hcr*, in a primary liver tumor, *Oncogene* **4:**51–57.

103. Hino, O., Kitagawa, T., Nomura, K., Ohtake, K., Yasui, H., Okamoto, N., and Hirayama, Y., 1992, Comparative molecular pathogenesis of hepatocellular carcinomas, in: *Progress in Clinical and Biological Research: Comparative Molecular Carcinogenesis* (A. J. P. Klein-Szanto, M. W. Anderson, J. C., Barrett, and T. J. Slaga, eds.), pp. 173–185, Wiley-Liss, New York.
104. Wei, Y., Ponzetto, A., Tiollais, P., and Buendia, M. A., 1992, Multiple rearrangements and activated expression of c-*myc* induced by woodchuck hepatitis virus integration in a primary liver tumour, *Res. Virol.* **143**:89–96.
105. Fourel, G., Trépo, C., Bougueleret, L., Henglein, B., Ponzetto, A., Tiollais, P., and Buendia, M. A., 1990, Frequent activation of N-*myc* genes by hepadnavirus insertion in woodchuck liver tumours, *Nature* **347**:294–298.
106. Wei, Y., Fourel, G., Ponzetto, A., Silvestro, M., Tiollais, P, and Buendia, M. A., 1992, Hepadnavirus integration: Mechanisms of activation of the N-*myc*2 retrotransposon in woodchuck liver tumors, *J. Virol.* **66**:5265–5276.
107. Hansen, L. J., Tennant, B. C., Seeger, C., and Ganem, D., 1993, Differential activation of *myc* gene family members in hepatic carcinogenesis by closely related hepatitis B virus, *Mol. Cell. Biol.* **13**:659–667.
108. Fourel, G., Couturier, J., Wei, Y., Apiou, F., Tiollais, P., and Buendia, M. A., 1994, Evidence for long-range oncogene activation by hepadnavirus insertion, *EMBO J.* **13**:2526–2534.
109. Transy, C., Fourel, G., Robinson, W. S., Tiollais, P., Marion, P. L., and Buendia, M. A., 1992, Frequent amplification of c-*myc* in ground squirrel liver tumors associated with past or ongoing infection with a hepadnavirus, *Proc. Natl. Acad. Sci. USA* **89**:3874–3878.
110. Transy, C., Renard, C. A., and Buendia, M. A., 1994, Analysis of integrated ground squirrel hepatitis virus and flanking host DNA in two hepatocellular carcinomas, *J. Virol.* **68**:5291–5295.
111. Ding, S. F., Habib, N. A., Dooley, J., Wood, C., Bowles, L., and Delhanty, J. D. A., 1991, Loss of constitutional heterozygosity on chromosome 5q in hepatocellular carcinoma without cirrhosis, *Br. J. Cancer* **64**:1083–1087.
112. Buetow, K. H., Murray, J. R., Israel, J. L., London, W. T., Smith, M., Kew, M., Blanquet, V., Bréchot, C., Redeker, A., and Govindarajah, S., 1989, Loss of heterozygosity suggest tumor suppressor gene responsible for primary hepatocellular carcinoma, *Proc. Natl. Acad. Sci. USA* **86**:8852–8856.
113. Fujimori, M., Tokino, T., Hino, O., Kitagawa, T., Imamura, T., Okamoto, E., Mitsunobu, M., Ishikawa, T., Nakagama, H., Harada, H., Yagura, M., Matsubara, K., and Nakamura, Y., 1991, Allelotype study of primary hepatocellular carcinoma, *Cancer Res.* **51**:89–93.
114. Scorsone, K. A., Zhou, Y. Z., Butel, J. S., and Slagle, B. L., 1992, p53 mutations cluster at codon 249 in hepatitis B virus-positive hepatocellular carcinomas from China, *Cancer Res.* **52**:1635–1638.
115. Simon, D., Knowles, B. B., and Weith, A., 1991, Abnormalities of chromosome 1 and loss of heterozygosity on 1p in primary hepatomas, *Oncogene* **6**:765–770.
116. Slagle, B. L., Zhou, Y. Z., and Butel, J. S., 1991, Hepatitis B virus integration event in human chromosome 17p near the p53 gene identifies the region of the chromosome commonly deleted in virus-positive hepatocellular carcinomas, *Cancer Res.* **51**:49–54.
117. Wang, G. J., Hayward, N. K., Falvey, S., and Cooksley, G. E., 1991, Loss of somatic heterozygosity in hepatocellular carcinoma, *Cancer Res.* **51**:4367–4370.
118. Tsuda, H., Zhang, W., Shimosato, Y., Yokota, J. Terada, M., Sugimura, T., Miyamura, T., and Hirohashi, S., 1990, Allele loss on chromosome 16 associated with progression of human hepatocellular carcinoma, *Proc. Natl. Acad. Sci. USA* **87**:6791–6794.
119. Okuda, K., 1992, Hepatocellular carcinoma: Recent progress, *Hepatology* **15**:948–963.
120. Lane, D. P., 1992, p53, guardian of the genome, *Nature* **358**:15–16.
121. Unsal, H., Yakicier, C., Marçais, C., Kew, M., Volkmann, M., Zentgraf, H., Isselbacher, K. J., and Ozturk, M., 1994, Genetic heterogeneity of hepatocellular carcinoma, *Proc. Natl. Acad. Sci. USA* **91**:822–826.

122. Hsu, I. C., Metcalf, R. A., Sun, T., Welsh, J. A., Wang, N. J., and Harris, C. C., 1991, Mutational hotspot in the p53 gene in human hepatocellular carcinomas, *Nature* **350**:427–428.
123. Bressac, B., Kew, M., Wands, J., and Ozturk, M., 1991, Selective G to T mutations of p53 gene in hepatocellular carcinoma from Southern Africa, *Nature* **350**:429–431.
124. Hosono, S., Chou, M. J., Lee, C. S., and Shih, C., 1993, Infrequent mutation of p53 gene in hepatitis B virus positive primary hepatocellular carcinomas, *Oncogene* **8**:491–496.
125. Buetow, K. H., Sheffield, V. C., Zhu, M., Zhou, T., Shen, F. M., Hino, O., Smith, M., McMahon, B. J., Lanier, A. P., London, W. T., Redeker, A. G., and Govindarajan, S., 1992, Low frequency of p53 mutations observed in a diverse collection of primary hepatocellular carcinomas, *Proc. Natl. Acad. Sci. USA* **89**:9622–9626.
126. Fearon, E. R., and Vogelstein, B., 1990, A genetic model for colorectal tumorigenesis, *Cell* **61**:759–767.
127. Murakami, Y., Hayashi, K., Hirohashi, S., and Sekiya, T., 1991, Aberrations of the tumor suppressor p53 and retinoblastoma genes in human hepatocellular carcinomas, *Cancer Res.* **51**:5520–5525.
128. Slagle, B. L., Zhou, Y. Z., Birchmeier, W., and Scorsone, K. A., 1993, Deletion of the E-cadherin gene in hepatitis B virus positive chinese hepatocellular carcinoma, *Hepatology* **18**:757–762.
129. Nakamura, T., Iwamura, Y., Kaneko, M., Nakagawa, K., Kawai, K., Mitamura, K., Futagawa, T., and Hayashi, H., 1991, Deletions and rearrangements of the retinoblastoma gene in hepatocellular carcinoma, insulinoma and some neurogenic tumors as found in a study of 121 tumors, *Jpn. J. Clin. Oncol.* **21**:325–329.
130. T'ang, A., Varley, J. M., Chakraborty, S., Murphee, A. L., and Fung, Y. K. T., 1988, Structural rearrangement of the retinoblastoma gene in human breast carcinoma, *Science* **242**:263–266.
131. Nobori, T., Miura, K., Wu, D. J., Lois, A., Takabayashi, K., and Carson, D. A., 1994, Deletions of the cyclin-dependent kinase-4 inhibitor gene in multiple human cancers, *Nature* **368**:753–756.
132. Kamb, A., Gruis, N. A., Weaver-Feldhaus, J., Liu, Q., Harshman, K., Tavtigian, S. V., Stockert, E., Day III, R. S., Johnson, B. E., and Skolnick, M. H., 1994, A cell cycle regulator potentially involved in genesis of many tumor types. *Science* **264**:436–440.
133. Raney, A. K., and McLachlan, A., 1991, The biology of hepatitis B virus, in: *Molecular Biology of the Hepatitis B Virus* (A. McLachlan, ed.), pp. 1–38, CRC Press, Boca Raton.
134. Fourel, G., and Tiollais, P., 1994, Molecular biology of hepatitis B virus, in: *Primary Liver Cancer: Etiological and Progression Factors* (C. Bréchot, ed.), pp. 89–124, CRC Press, Boca Raton.

11

Transformation and Tumorigenesis Mediated by the Adenovirus E1A and E1B Oncogenes

ROBERT P. RICCIARDI

1. INTRODUCTION

Adenoviruses and other small DNA tumor viruses, e.g., SV40 virus and papillomaviruses, have evolved efficient strategies for replicating their genomes on entry into their host cells. Because these viruses have limited coding capacities, they cannot produce all of the protein products required to synthesize their new double-stranded DNA genomes and, therefore, must utilize many components of the host-cell machinery. In naturally occurring infections, adenovirus frequently enters noncycling, differentiated epithelial cells. This presents adenovirus with the problem of how to stimulate these quiescent cells to an active growth state in order to create a cellular environment in which viral propagation can take place. The small DNA viruses have devised strategies in which certain of their early viral proteins, i.e., adenovirus E1A and E1B, SV40 T antigen, and papilloma E6 and E7, physically target specific cellular proteins whose function it is to regulate cellular growth.

These same early viral proteins are also responsible for transformation. Adenovirus transformation can result from an abortive infection of nonpermissive cells, where early viral gene products (E1A and E1B) are produced but viral DNA replication fails. Transformation can also occur by transfecting cells with recombinant

---

ROBERT P. RICCIARDI • Department of Microbiology, School of Dental Medicine, and Graduate Program in Microbiology and Virology, University of Pennsylvania, Philadelphia, Pennsylvania 19104.

DNA Tumor Viruses: Oncogenic Mechanisms, edited by Giuseppe Barbanti-Brodano *et al.* Plenum Press, New York, 1995.

plasmids containing early viral genes (e.g., E1A and E1B). Unlike infected cells, the transformed cells do not become lysed but are altered such that they can remain in culture indefinitely. The ability of these early viral proteins to generate continuously proliferating transformed cells is directly related to their ability to perturb cellular growth during lytic infection.

This chapter focuses on the transforming functions encoded by the E1A and E1B proteins of adenovirus. The most salient feature of these transforming viral proteins is that they have multifunctional subdomains that interact with different cellular proteins involved in transcription and cellular growth. Our knowledge of how each of these specific viral–cellular protein interactions contributes to transformation is described.

In addition to contributing to cellular transformation, the E1A gene of a tumorigenic strain of adenovirus (Ad12) provides the transformed cell with the potential to escape immunodetection by causing a reduction in the surface expression of MHC class I antigens. This mechanism, which may be used by Ad12 to persist normally in humans, may likewise contribute to the ability of Ad12-transformed rodent cells to generate tumors in immune-competent hosts. The mechanism by which E1A mediates class I repression is also discussed.

2. ADENOVIRUS TRANSFORMATION

There are over 50 serotypes of human adenoviruses, all of which are capable of transforming nonpermissive rodent cells.[1] Transformation results from the integration of adenovirus E1A and E1B genes, which are encoded in tandem within the first 11 map units (11%) of the 36-kb double-stranded DNA genome[2-5] (Fig. 1). Full transformation by E1A and E1B can be made defective by introducing a mutation into either of these genes.[6] Introduction of the E1A gene alone (0–4.3 mu) is capable of immortalizing primary cells that otherwise would have only a limited lifespan *in vitro*.[4,7-9] However, these E1A-transformed cells fail to grow to high cell density in the absence of E1B and are thus considered to be only partially transformed. Conversely, the E1B gene alone completely fails to transform cells, even when placed under the control of a foreign promoter in order to overcome the normal requirement of E1A *trans*-activation.[10] Interestingly, E1A can cooperate with certain other oncogenes, e.g., polyoma virus middle T antigen or the activated *ras* oncogene, to achieve full transformation, which is manifested as extended proliferation, altered morphology, and growth to high cell density.[11] These findings helped to derive the current notion that transformation is a multistep process involving more than a single oncogene.

3. THE ROLE OF E1A PROTEINS IN TRANSFORMATION

3.1. E1A General Overview

In Ad5, two major E1A proteins of 289R and 243R are synthesized from differentially spliced, overlapping transcripts of 13S and 12S, respectively[12] (Fig.

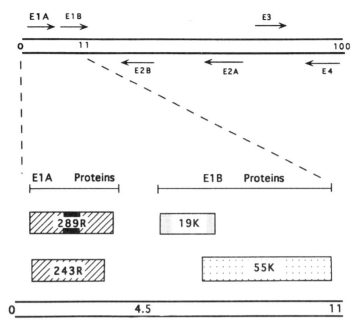

FIGURE 1. The E1A and E1B transforming region of adenovirus. Top: Shown are the genomic location and direction of E1A, E1B, and other early (E) transcription units (arrow, 3' end). E1A and E1B transcripts are synthesized within the first 11 map units of the 36,000-bp linear adenovirus genome. For simplicity, the multiple-spliced overlapping transcripts that are produced from each early region are not indicated. Neither are the late transcripts shown. Bottom: The two major E1A proteins are identical except that the larger 289R protein has a unique internal stretch (black bar) of 46 amino acids. The reading frames of the E1B 19K and 55K proteins are completely different.

2). These E1A proteins are identical except that the larger 289R protein contains a unique internal stretch of 46 amino acids,[13] referred to as conserved region 3 (CR3, residues 140–188). The CR3 domain confers on the 289R protein an ability to stimulate transcription of both viral and cellular promoters.[14–16] Two other evolutionarily conserved regions, CR1 (residues 40–80) and CR2 (residues 121–139), are common to both E1A proteins[17] and also correspond to functional subdomains[18] (Fig. 2). Genetic analysis has shown that CR1[19,20] and CR2[21–24] are required for transformation. This is consistent with the ability of the large and small E1A proteins individually to immortalize cells, although differences in morphology and extent of transformation produced by these viral gene products, separately or together, have been observed.[25–30] Thus, though it may be concluded that CR3 of the larger 289R protein is not essential for transformation, CR3 may in some way enhance the efficiency of transformation.

Coinciding with their requirement for transformation, the CR1 and CR2 domains together with the nonconserved amino terminus of E1A (Fig. 2) are responsible for repression of enhancer-mediated transcription and induction of cellular DNA synthesis.[22–24,31–35] For example, the Ad5 E1A proteins can inhibit the

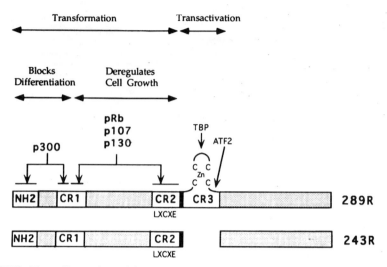

FIGURE 2. The coding regions of the two major E1A proteins of adenovirus 5. The regions of E1A required for transformation (NH2, CR1, and CR2) and *trans*-activation (CR3) are indicated. Both proteins are identical except that the 289R protein has a unique 46-amino-acid region (CR3) produced by differential splicing of the E1A transcripts. The conserved regions (CR1, CR2, CR3) are shared among many adenovirus serotypes. The CR3 *trans*activating region has a Cys_4 zinc finger structure that is required for binding to TBP. Sequences directly C-terminal to the zinc finger are required for binding ATF-2 and certain other DNA recognition transcription factors. The pRb family of proteins (pRb105, p107 and p130) bind to CR1 and CR2. The LXCXE amino acid sequence motif indicated in CR2 is present in other DNA tumor viruses and is required for binding pRb proteins. The p300 protein binds to the amino-terminal sequences (NH_2) of E1A and a portion of CR1 that is not required for binding to pRb proteins. As indicated, E1A is thought to contribute to transformation by (1) deregulating cell growth through the binding of pRb proteins to CR1 and CR2 and (2) blocking differentiation through the binding of p300 to the NH_2 and CR1 regions. The black bar between CR2 and CR3 in the large protein (and following CR2 in the small protein) represents the ~20-amino-acid sequence that is present in adenovirus 12, but not in adenovirus 5, which is required for adenovirus 12 tumorigenesis (see text).

enhancers of the SV40 early promoter[36,37] the immunoglobulin heavy chain,[38] and the E1A promoters.[36,37,39] E1A induction of cellular DNA synthesis[40,41] is abolished when the N-terminal amino acids and CR1 are perturbed.[42,43] As will be described in greater detail, repression of certain types of enhancers and stimulation of DNA synthesis are activities regarded to be intimately associated with the transformation process.

Other features of the E1A proteins are noted. First, the 289R and 243R proteins are each dependent on a C-terminus pentapeptide signal sequence (KRPRP)[44] for efficient delivery into the nucleus.[45] Second, each E1A product exhibits considerable size heterogeneity as a result of posttranslational modifications,[1] which are mainly through phosphorylations of serine residues.[46–49] Although the significance of these phosphorylations has yet to be determined, they do not seem to play a major functional role in E1A transformation or *trans*-activation.[46,49,50]

3.2. The Zinc Finger *Trans*-Activating Domain Binds to TBP and Other Transcription Factors

E1A functions as an immediate early *trans*-activator of adenovirus early genes during infection.[12,51] E1A also possesses the property of a promiscuous *trans*-activator in that it can stimulate transcription from a variety of different viral and cellular promoters.[16,52,53] As alluded to above, the 46-amino-acid CR3 domain present in the larger E1A 289R protein[13] is required for transcriptional activation.[14,15,54] The salient structural feature of CR3 is a single zinc finger formed by the tetrahedral coordination of a zinc ion to four cysteine residues (Fig. 2).[55–57] Mutations that destroy or alter the structure of the zinc finger are defective in *trans*-activation,[56] and extensive genetic analyses of E1A suggested that the zinc finger binds to a cellular transcription factor.[56,58] Indeed, E1A has been shown to bind TBP,[59,60] the TATA box binding protein critical for transcription initiation, and the CR3 subdomain is the principal region of E1A involved in TBP binding.[59] More precisely, the zinc finger of CR3 was shown to be indispensable for TBP binding.[61] Recent studies indicate that E1A also contacts two TBP-associated factors, $dTAF_{110}$ and $dTAF_{250}$.[62] In this sense, the zinc finger of E1A is different from the vast majority of zinc fingers of other proteins[63] in that it does not bind to DNA, at least in a sequence-specific manner,[64,65] but rather serves as a protein-binding motif.

In addition, the carboxyl region of CR3 is thought to interact with the DNA-binding domains of some transcription factors, e.g., ATF-2, which bind to specific recognition sites on E1A-inducible promoters.[66] Surprisingly, an unidentified cellular factor, which is limiting, has also been implicated by genetic analysis to be critical for E1A *trans*-activation.[61] One emerging model to explain how E1A mediates transcriptional *trans*-activation involves a protein–protein bridging interaction between the zinc finger region of CR3 making contact with TBP and TAFs at the initiation site of transcription and the carboxyl region of CR3 making contact with transcription factors that bind upstream of the transcription initiation site. This ability of E1A to act as an adapter that joins cellular transcription complexes helps to explain how the 289R E1A protein functions as a promiscuous *trans*-activator on such a variety of promoters. Detailed resolution of the mechanism of E1A *trans*-activation will require an understanding of the structures and contacts of all of the proteins that participate in this so called "bridging complex."

3.3. The CR1 and CR2 Domains Bind to the pRb Family of Proteins Causing Deregulation of the Cell Cycle

In adenovirus lytically infected cells, the CR1 and CR2 domains (Fig. 2) confer on E1A the ability to stimulate viral and cellular DNA synthesis. In adenovirus-transformed cells, these domains are required for immortalization. These phenotypes depend on the ability of CR1 and CR2 to directly bind to the 105-kDa nuclear retinoblastoma protein pRb.[67–71] The retinoblastoma gene is a tumor suppressor, and loss or alteration of both alleles is associated with a number of human neoplasias.[72] In adenovirus-transformed cells, the binding of E1A to pRb apparently inactivates its function, mimicking the effect of deleting both copies of the retinoblastoma gene.[73]

In addition to E1A,[73] pRb associates with human papillomavirus E7[74] and SV40 T antigen.[75] These otherwise divergent oncoproteins share a common amino acid sequence motif, Leu-X-Cys-X-Glu (where X is any amino acid), which is required for binding to pRb.[74–77] In E1A, both the Leu-X-Cys-X-Glu motif in CR2[78] (Fig. 2) and a sequence in CR1 are needed for the binding of E1A to pRb.[78]

How does the binding of E1A to pRb stimulate cellular DNA synthesis in transformed cells? In normal dividing cells, pRb is phosphorylated in a tightly regulated fashion during different phases of the cell cycle (Fig. 3). During the late G_1 phase, pRb becomes phosphorylated and remains highly phosphorylated throughout S, G_2, and much of M. During G_0 and much of G_1, pRb becomes un- or underphosphorylated.[79–83] Repeated phosphorylations of pRb in mid-late G_1 again removes the cell cycle block associated with unphosphorylated pRb. These phosphorylations appear to be solely regulated by complexes of cyclins and cyclin-dependent kinases.[67,84]

The first clue to reveal that phosphorylation modulates pRb function was the discovery that SV40 T antigen shows a strong preference for binding to the unphosphorylated form of pRb.[85] This finding, along with other evidence,[70] such as the demonstration that senescent human fibroblasts cannot phosphorylate pRb,[86] led to the idea that it is the unphosphorylated form of pRb that blocks progression into late G_1, the step that precedes DNA synthesis and replication. Thus, binding of SV40 T antigen, E1A, and HPV E7 to unphosphorylated pRb would remove this block and permit the phosphorylated form of pRb to prevail and, hence, favor a state of continuous cellular DNA synthesis.

The problem of how the unphosphorylated and phosphorylated forms of pRb differentially modulate the cell cycle started to unravel when it was learned that a transcription factor, E2F, associates with pRb,[87,88] particularly with the unphosphorylated form of pRb.[87] E2F belongs to a family of related transcription factors (E2F-1, E2F-2, E2F-3, DRTF1, and DP1) that have the potential to form heterodimers on E2 sites[89–94] that occur in the promoters of many genes involved in cell-cycle progression and DNA synthesis, e.g., dihydrofolate reductase, DNA polymerase α, thymidine kinase, *cdc2*, *c-myc*, and cyclin A.[95] In the normal cell cycle, complexing of E2Fs with the under- or unphosphorylated forms of pRb[87] that are present during G_0 and early G_1 make the E2F transcription factors unavailable for activating the many genes involved in replication (Fig. 3). Thus, by virtue of their differential abilities to interact with E2Fs, the particular phosphorylated forms of pRb serve as key regulators of the cell cycle.

How does the binding of E1A to pRb perturb the normal cell cycle? E1A protein dissociates the E2F–pRb complex, causing the release of E2F and the formation of an E1A–pRb complex (Fig. 4).[87,88] By liberating E2F transcription factors and preventing their reassociation with pRb, E1A disrupts the normal periodicity of the cell cycle and overcomes the normal growth-restraining mechanisms. In infected cells, this E1A mechanism apparently enables adenovirus to replicate in quiescent cells by stimulating E2 promoter-driven cellular genes that encode replication-dependent protein products (described above) as well as the viral E2 gene that encodes the adenovirus DNA polymerase, elongation factor, and protein primer

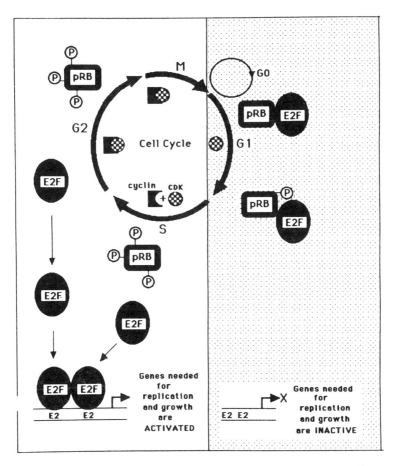

FIGURE 3. Generalized scheme of cell-cycle regulation by the retinoblastoma binding proteins. The stages of the cell cycle are indicated: M (mitosis), G_1, S (DNA synthesis), G_2 and G_0. Also indicated are the interactions of the cyclins and their kinases (CDKs). Shaded side: During G_0 and most of G_1, the E2F transcription factors are bound to the underphosphorylated forms of retinoblastoma proteins, designated here as pRB. These sequestered E2Fs are unable to transcribe genes required for cellular growth and DNA synthesis. Unshaded side: At the end of G_1, S, G_2, and M, pRbs become phosphorylated, liberating the E2Fs, which are then able to stimulate transcription of genes required for cellular growth and DNA synthesis. The associations of cyclins and CDKs at various stages of the cell cycle are also indicated.

used in initiation of viral DNA synthesis.[6] In adenovirus-transformed cells, the disruption of the cell cycle through the liberation of E2F by E1A helps establish a state of continuous proliferation.

Recent studies[96,97] suggest the mechanism by which E1A dissociates the E2F–pRb complex and replaces the liberated E2F with itself, as an E1A–pRb complex.[87,88] In this dissociation–replacement model (Fig. 4), the exchange takes

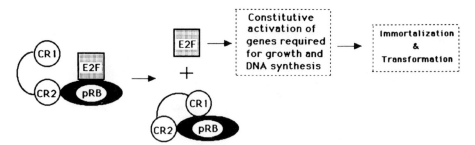

FIGURE 4. Proposed mechanism for how the E1A CR1 and CR2 domains contribute to cellular transformation through the liberation of E2Fs.[96,97] CR2 first binds pRb, and after E2F dissociates, CR1 binds to pRb. E2F cannot reassociate with pRb and is now free to constitutively stimulate transcription of genes involved in cellular growth and replication. This is manifested as uncontrolled cellular proliferation, the hallmark of transformation. This schematic is largely adapted from the study of Ikeda and Nevins.[96]

place in two steps and requires the E1A CR1 and CR2 regions.[96,97] In the first step, CR2 (containing the Leu-X-Cys-X-Glu motif) serves to bring E1A to the E2F–pRb complex, forming a trimeric complex in which CR2 associates with pRb. The second step awaits the normal equilibrium dissociation of E2F from pRb, after which CR1 interacts with a region of pRb that blocks E2F from reassociating with pRb. In this way, E1A drives the formation of free E2F and prevents the reformation of the E2F–pRb complex. Because pRb proteins are now held in association with E1A, the E2F transcription factors are free to constitutively stimulate cell growth, which is the underlying feature of cellular immortalization and transformation (Fig. 4).

It is now realized that E1A may perturb the normal link between transcription and DNA replication by also interacting with other pRb-related proteins. Two of these newly recognized pRb family members are p107[98] and p130.[99–101] The pRb contains an internal 400-amino-acid sequence referred to as the "pocket," which alone is capable of binding E1A, T antigen, and E7 viral proteins.[102–104] The p107 and p130 proteins also contain similar pocket sequences and bind to the same region of E1A.[105,106] Significantly, the pRb, p107, and p130 family of proteins all appear to function as temporal components of cell-cycle progression, and all of these proteins bind to E2F factors.[67,107] Although the details of these temporal associations have yet to be resolved, some studies suggest that pRb is found in association with E2F during G_0/G_1, whereas, p107 complexes with E2F during late G_1 into S phase.[108] It may be relevant that the E2F–p107 complex also contains cyclin A and CDK2[108–111] and that the E1A–p107 "dissociation–exchange" complex also contains cyclin A through its binding to p107.[112] Finally, p130 appears mainly in the G_0 and G_1 phases of the cell cycle in association with cyclins A and E.[99,100] Thus, if it is assumed that the pRb family of proteins have overlapping functions during the cell cycle (Fig. 3), it may be that E1A is required to associate with all of the pRb family proteins in order to effectively maintain the E2F transcription factors in a liberated state.

3.4. The Amino-Terminal and CR1 Domains Bind to p300, Which May Block Differentiation

The amino-terminal region of E1A is also required for transformation[22–24,31,33,34,43] and stimulation of the cell cycle,[32,113–115] but in a manner that is distinct from that brought about by the binding of the pRb family of proteins to CR1 and CR2.[114,116] In addition, the amino-terminal region is responsible for the ability of E1A to mediate transcriptional repression of specific viral and cellular enhancers,[23,33,34,36,37,115,117] the most important of which may be enhancers and regulatory promoter sequences that function in transcribing cellular genes involved in terminal differentiation.[38,117,118] Thus, one explanation to account for the various activities ascribed to the amino-terminal region of E1A may be contained in the general notion that transformation (manifested as uncontrolled cellular proliferation) is the biological converse of terminal differentiation. It is thus possible that, in order for cellular immortalization by E1A to occur, stimulation of cell cycle genes by CR1 and CR2 must be accompanied by repression of cellular differentiation genes by the amino-terminal region (Fig. 2).

Correlated with the E1A amino-terminal region transformation phenotype is the binding of a 300-kDa cellular protein (p300).[78,116,119] E1A amino-terminal residues 1 through 25 as well as a subregion of CR1 [residues 60–80, which are not essential for pRb binding[116]] appear to be singularly important for binding to p300 (Fig. 2), because mutations that alter these sites selectively abolish p300 binding without affecting the binding of other E1A-associated proteins.[116] Studies with the recently cloned p300 cellular gene revealed that the carboxy-terminal region of p300 serves as the binding site for E1A.[120] Most significantly, p300 mutants defective for E1A binding were able to bypass E1A-mediated repression of the SV40 enhancer and restore enhancer activity.[120] One implication of these findings is that p300 functions as a transcriptional adapter that stimulates certain promoter regulatory elements, and recent studies (described below) suggest that these promoters may direct transcription of cellular genes involved in differentiation. Interestingly, p300 contains a bromodomain[120] that, in certain transcriptional adapters, is thought to function in protein–protein interactions.

How might p300 regulate transcription, and what evidence suggests that it activates specific genes involved in terminal differentiation? p300 has been shown to associate with the TATA box binding protein, TBP.[121] In addition, regions of p300 have recently been shown to exhibit strong homology to corresponding regions of CBP,[122] a large bromodomain-containing protein that functions as an adapter between DNA-binding proteins that are activated through signal transduction (i.e., CREB, which is induced by cyclic AMP) and proteins of the transcription initiation complex (i.e., TFIIB).[123,124] It is intriguing that elevated levels of cyclic AMP have been demonstrated to coincide with cell cycle exit and arrest of proliferation and induction of cellular differentiation.[122,125,126] That CBP contains the same region in p300 that is necessary for E1A binding suggests that E1A may also bind CBP and is consistent with the fact that E1A can inhibit cyclic-AMP-mediated gene regulation.[122,127,128]

Taken together, these studies contribute to the notion that transformation by

E1A depends on two separately encoded functions (Fig. 2). The first function is to stimulate the cell cycle and relies on interactions of the CR1 and CR2 domains with the pRb family of proteins. The second function is to prevent transcription of genes involved in terminal differentiation and relies on an interaction of the amino terminus and a portion of CR1 that does not bind pRb, p300, or possibly other related cellular proteins.

4. THE ROLE OF E1B PROTEINS IN TRANSFORMATION

4.1. E1B General Overview

During infection, the E1B promoter is *trans*-activated by the immediate early 289R E1A gene product.[16] The E1B region transcribes two major overlapping RNAs of 13S and 22S, which encode unrelated proteins of 19 and 55 kDa (Fig. 1),[129] both of which are important for efficient viral replication. The 55-kDa protein is necessary for the shutoff of cellular DNA synthesis and, in physical association with another early protein (the 34-kDa, E4 ORF-6 protein), is important for transport of late viral mRNAs from the nucleus to the cytoplasm.[130] The 19-kDa protein appears to protect the newly infected cell from E1A-induced apoptotic effects, which include pronounced degradation of viral and cellular DNAs.[129]

E1B in conjunction with E1A is necessary for adenovirus transformation, and both the 19-kDa and 55-kDa E1B proteins are required for this process to be efficient.[131,132] The prevailing view is that the inability of E1A alone to cause complete transformation results not from a missing component needed to complete the proliferative response but, rather, from an inhibition in cell growth that is imposed by the proliferative response induced by E1A *per se*. As described below, the E1B proteins act in independent ways to circumvent cellular growth inhibition or death (apoptosis) in response to E1A-induced proliferation.[133] Thus, in general terms, for a cell to be rendered completely transformed, E1A is required to produce the proliferative response, while E1B is required to counteract the cell's natural response to this proliferation, which can be negative growth regulation or apoptosis. E1A promotes cell growth while E1B prevents negative growth and cell death.

A central function of the E1B proteins in transformation is to interfere with the normal activity of the cellular p53 protein. The p53 protein is the product of a tumor suppressor gene[134] and is the most prevalent mutated gene found in human cancers.[135] In response to DNA damage, p53 arrests the cell cycle at G_1 to allow for repairs. If damaged DNA is not repaired, then p53 triggers cell death by apoptosis.[136-138] Presumably, without this cell-cycle checkpoint imposed by wild-type p53, damaged DNA would accumulate and lead to genomic instability and transformation (Fig. 5, minus E1B pathway).[139] Indeed, growth of γ-irradiated, p53-deficient cells can be arrested by introduction of wild-type p53.[140]

Wild-type p53 has the ability to function as a transcription factor and most likely regulates expression of growth arrest and DNA-damage-inducible genes (Fig. 5, minus E1B pathway). The presence of a sequence-specific DNA-binding domain and a *trans*-activating domain is consistent with the role of p53 in modulat-

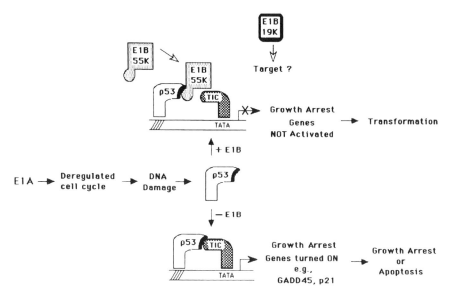

FIGURE 5. Transformation model for how E1B proteins allow E1A-mediated cellular proliferation to be sustained by preventing p53 growth arrest and apoptosis. E1A deregulation of the cell cycle produces DNA damage which may activate DNA damage response proteins (e.g., AT proteins) that, in turn, induce expression of cellular p53. Minus E1B pathway: In the absence of E1B, the p53 protein can function as a transcription factor by binding to a specific recognition site (diagonal stripes) on the promoters of genes that function in growth arrest. The p53 amino-terminal activation domain (black area) may contact the transcription initiation complex (TIC), possibly through the TATA box binding protein, TBP. Apoptosis can occur if DNA damage is not repaired in the p53-induced growth-arrested cell. Plus E1B pathway: In the presence of E1B, the E1B 55K protein binds to p53 and blocks its ability to activate growth-arrest genes. The E1B 19K protein indirectly interferes with p53-induced growth and apoptosis, but its specific target and site of action are not known. Transformation results from a combination of E1A's ability to deregulate the cell cycle and E1B's ability to impair the cell's normal counteractive responses of growth arrest and apoptosis.

ing transcription.[139,141–144] Moreover, p53 exhibits binding to TBP,[145–147] the TATA box-binding protein, which is critical to formation of the RNA polymerase II preinitiation complex. It may be that p53 *trans*-activates only genes that contain a p53-specific DNA binding element (e.g., genes involved in negative growth control such as GADD genes) while repressing a different subset of genes that do not have p53 binding sites (e.g., growth-promoting genes such as *fos* and *jun*).[139] A current model suggests that proteins that are activated in response to DNA damage (referred to as AT proteins because they are defective in patients with ataxia–telangiectasia) induce p53 expression, which *trans*-activates expression of growth arrest and DNA damage-inducible (GADD) genes (Fig. 5, minus E1B pathway).[148] Relevant to this model is the fact that the p53-inducible GADD45 gene has a nearly perfect p53 consensus binding site.[148] Also strongly supportive of the model is the recent finding that p53 can induce a cyclin-dependent protein kinase (CDK) inhibitor protein (p21) that is able to bind to cyclin complexes and inhibit the function of a wide variety of CDKs (Figs. 3 and 5).[149–152]

4.2. E1B 55-kDa Protein Binds to p53 and Blocks Growth Arrest

In transformed cells, the E1B 55-kDa protein stably binds to wild-type cellular p53,[153,154] which blocks the ability of p53 to inhibit cell growth and contributes to the process of transformation (Fig. 5, plus E1B pathway).[132,155] Complexing of p53 with the E1B 55-kDa produces the same effect as mutational inactivation of p53 genes that are found in many human tumors. Other DNA transforming viruses also encode specific products that inhibit p53 function; e.g., p53 binds to SV40 T antigen[156,157] and the human papillomavirus E6 protein.[158] Interestingly, a cellular protein, MDM-2, can also interfere with the ability of wild-type p53 to suppress transformed cell growth.[159–161]

The mechanism by which the E1B 55-kDa protein is thought to inhibit p53 biological activity is by directly interfering with p53's *trans*-activating function (Fig. 5, plus E1B pathway). The E1B 55-kDa protein has been shown to bind to the amino-terminal activation domain of p53,[154] and a study of E1B mutants has revealed a direct correlation between the ability of E1B 55-kDa to block p53 *trans*-activation and to transform cells in cooperation with E1A.[162] Most recently, the E1B 55-kDa protein (55K) was shown to complex with p53 that was bound to its cognate DNA recognition site without disrupting p53 from its DNA site.[163] Moreover, a GAL4–55K fusion protein, tethered to a TK promoter containing GAL4 DNA binding sites, was able to repress transcription.[163] Taken together, these results suggest that the mechanism by which E1B represses *trans*-activation of p53 target promoters involves the binding of E1B–55K to the *trans*-activation domain of p53 while allowing p53 to remain bound to the promoter (Fig. 5, plus E1B pathway). In this configuration, not only would p53 *trans*-activation be neutralized, but transcription from the p53 target promoters would be further repressed by E1B–55K, possibly through inhibition of the activity of the transcription initiation complex.[163]

4.3. E1B 19K Protein Blocks Apoptosis in Response to E1A-Mediated Destruction of DNA

The E1B 19K protein is important for maintaining cell viability during infection because without this protein both viral and cellular DNAs become degraded and exhibit an enhanced cytopathic effect.[164–166] Significantly, it is the expression of E1A that is responsible for inducing DNA degradation.[167,168] In transformed cells, the function of the E1B 19K protein also appears to be geared toward blocking apoptosis, which is caused by E1A-mediated destruction of DNA.[133]

Although the mechanism by which E1B 19K protein inhibits apoptosis is not known, the manner in which it functions is distinct from that of E1B 55K. As mentioned above, the amino acid sequences of these two E1B proteins are completely different, and, unlike E1B 55K, there is no evidence to suggest that E1B 19K binds to p53. However, E1B 19K apparently does block the action of p53 as revealed by its ability to inhibit E1A-induced apoptosis in the presence of the wild-type, but not the mutant form, of a temperature-sensitive p53 protein.[169] Thus, E1B 19K likely prevents cell death by interfering with a critical step in the apoptotic pathway, either before or after the point at which p53 acts (Fig. 5, plus E1B pathway). Interestingly, cells expressing E1A become very susceptible to the tumoricidal cytokine TNF-α, which induces cell death.[170–172] Because the E1B 19K protein can block programmed cell death by both p53 and TNF-α[169,173,174] it is conceivable p53 and TNF-α are part of the same apoptotic pathway.

5. BASIS FOR THE COLLABORATION BETWEEN E1A AND E1B IN TRANSFORMATION

Recent studies in the mouse lens now confirm that loss of pRb function is not sufficient to sustain deregulated cellular proliferation in the presence of intact p53.[175,176] The apoptotic response by p53 is a safeguard mechanism to protect the organism against aberrant cellular proliferation.[177] It is clear to see how adenovirus transformation requires the collaboration of both E1A and E1B products. E1A products deregulate cellular growth and perhaps differentiation by associating the pRb and p300 proteins, while E1B products prevent growth arrest and apoptosis by blocking the normal p53 responses to DNA damage that accompanies this E1A-mediated stimulation of DNA synthesis. To take metaphoric license:

    E1A pushes on the accelerator   &   E1B cuts the brake cables.

6. TUMORIGENESIS: E1A REPRESSION OF MHC CLASS I TRANSCRIPTION AS AN IMMUNE ESCAPE MECHANISM

6.1. Diminished Class I Expression in Ad12-Transformed Cells Provides a Means of Immune Escape from CTLs

A transformed cell is not necessarily capable of causing cancer. In the host, the newly transformed cell must next circumvent the armaments of the immune system in order to proliferate into a tumor. Although all of the more than 50 serotypes of adenovirus are able to transform cells through the action of their E1A and E1B gene products, only a subset of serotypes are capable of actually causing tumors in immunocompetent animals.[6] The reason for this difference in tumorigenic potential had remained a mystery for nearly 30 years following the discovery of adenoviruses until it was learned that rodent cells transformed by the highly tumorigenic Ad12 serotype displayed diminished levels of MHC class I antigens in contrast to cells transformed by the nontumorigenic Ad5 serotype.[178–180]

The class I antigens are cell surface glycoproteins that are expressed on almost all mammalian cells in noncovalent association with $β_2$-microglobulin. Class I antigens play a key role in immune recognition by allowing cytotoxic T lymphocytes

(CTLs) to detect and lyse cells expressing foreign (e.g., viral) proteins. Immune recognition occurs when the class I antigens on target cells (e.g., transformed or virally infected cells) present processed foreign polypeptides to the T-cell receptors on CTLs. If the class I antigens fail to be expressed, the target cells escape detection and destruction by CTLs.

Indeed, Ad12-transformed cells are more resistant to lysis by syngeneic CTLs *in vitro* than are Ad5-transformed cells (Fig. 6).[181] Interferon γ (IFN-γ), which is known to have the general effect of increasing class I expression, has been useful in substantiating these studies. For example, IFN-γ treatment of Ad12-transformed cells overrides the repression, stimulating surface levels of class I antigens[180] and, concomitantly, increasing susceptibility to lysis by syngeneic CTLs *in vitro*.[181] An *in vivo* corollary to these studies is that rodents injected with Ad12-transformed cells have increased survival if the transformed cells are pretreated with IFN-γ.[182,183] In another tumor challenge study, expression of exogenous class I genes in Ad12-transformed cells also had a protective effect.[184] The relationship between low

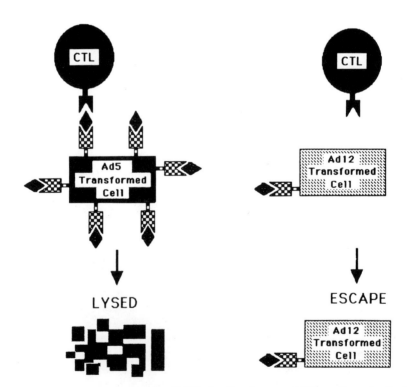

FIGURE 6. A reduction in the levels of MHC class I antigens on Ad12-transformed cells allows them to escape immunosurveillance by CTLs. In Ad5-transformed cells, viral antigen (black diamond) is presented by the class I surface proteins (checked stems) to the T-cell receptor (black stems) on CTLs, resulting in lysis. In Ad12-transformed cells, the diminished surface levels of class I antigens allows them to escape detection by CTLs and contributes to their tumorigenic potential.

class I antigen levels on Ad12-transformed cells and their ability to generate tumors was supported by the demonstration that cells from freshly excised tumors maintained reduced levels of class I antigens.[180]

For Ad12-transformed cells, the necessity of circumventing T-cell immunity as a means of survival in immunocompetent hosts is further underscored by the fact that transformed cells of nononcogenic adenoviruses (e.g., Ad5) form tumors only in athymic (nude) animals, which do not produce T cells,[185] or in animals that have been depleted of their T cells.[186] The adenovirus-specific CTL activity was found to map to the transforming region,[187–189] with E1A providing the target structures for specific CTL recognition.[190] However, it should be noted that the immunodominance of E1A as a CTL target in mice does appear to be haplotype dependent.[191] The carboxyl region of E1A (second exon of Ad5) was required to induce this T-cell response,[192] and an eight-amino-acid synthetic peptide derived from the sequence of the E1A carboxyl region was shown to behave as an immunodominant T-cell determinant to an Ad5-specific CTL (CTL clone 5).[193] Most compelling, in nude mice, tumors that had been induced by the transforming region of Ad5 were eradicated by this E1A-specific CTL clone 5.[193] Presumably, the lowered class I levels in Ad12-transformed tumor cells reduces their capability of presenting foreign peptides to CTLs and provides them with a means of escaping CTL immunosurveillance.

An additional reason, and in some cases an alternative explanation of why Ad12-transformed cells are tumorigenic, is that they are resistant to natural killer (NK) cells.[194–197] But a separate study showed that untransformed control cells, like Ad12-transformed cells, are poorly lysed by NK cells, arguing that Ad12-transformed cells are not more resistant to NK cells but that Ad5-transformed cells are merely more susceptible.[193] This same study[193] demonstrated the involvement of CTLs in eradicating Ad5-induced tumors from nude mice (described above). It is clear that the immune system has available multiple mechanisms for destroying tumor cells, and these mechanisms are meant to be cotriggered to assure survival of the host. However, the elimination of one key mechanism, e.g., CTL recognition, places the host at a considerable disadvantage to the rapidly proliferating transformed cell.

6.2. Ad12 E1A Mediates Down-Regulation of Class I Expression

The Ad12 E1A gene, in the absence of any other adenovirus genes, is capable of class I shutoff.[198] This was demonstrated by introducing into human embryonic kidney cells the Ad12 E1A gene and the transforming gene of a completely different virus (TAg of BKV) for the purpose of achieving complete transformation. Class I HLA-A, -B, and -C antigens and mRNAs were dramatically reduced in cells expressing E1A/TAg but not in the control cells expressing TAg alone.[198] The requirement of Ad12 E1A to mediate class I repression correlates with its requirement in the formation of tumors. By contrast, Ad12 E1B, in the presence of Ad5 E1A, is able to transform cells but is unable to generate tumors in immunocompetent animals.[179,199–201] Mapping studies suggest that AD12 E1A mediated class I repression and tumorigenesis co-map to the first exon of Ad12 E1A and include CR1, CR2 and a 20 amino acid nonconserved domain located between CR2 and CR3 (Fig. 2).[202–206,220]

6.3. In Ad12-Transformed Cells, Class I Transcription Is Down-Regulated by Global Repression of the Class I Enhancer

Ad12 E1A mediates reduced expression of class I antigens and mRNAs in transformed cells by down-regulating transcription from apparently all of the class I promoters,[207,208] i.e., H-2K, D, and L in mouse[180] and HLA-A, -B, and -C in human.[198] Extensive mutational analysis of the class I promoter revealed that the class I enhancer is the target for Ad12 E1A-mediated repression (Fig. 7).[209] This down-regulation appears to affect the transcription and protein-binding activities associated with both of the DNA-binding recognition elements (R1 and R2) that reside within the class I enhancer.[210–215]

The R1 element has an NFκB recognition site that is responsible for most of the positive activity normally associated with the class I enhancer (Fig. 7).[216,217] In Ad12-transformed cells, limited NFκB binding activity to the R1 element[212–214] provides an explanation for its functional inactivity.[212] In a recent study,[212] it was revealed that in Ad12-transformed cells, inactive p50/p50 homodimers predominantly bind to the R1 element, whereas in Ad5-transformed cells, the active p65/p50

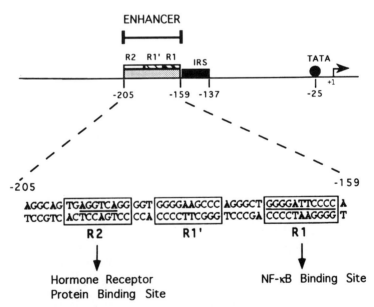

FIGURE 7. The enhancer of the MHC class I promoter is a target for Ad12 E1A-mediated down-regulation of transcription. Indicated are the transcriptional start-site (arrow), TATA box, interferon response element (IRS), and enhancer, which is comprised of three subelements, R1, R1', and R2. The R2 element contains a recognition site for nuclear hormone receptor proteins, and the R1 site has a recognition site for NFκB.

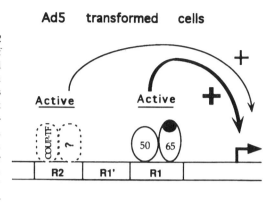

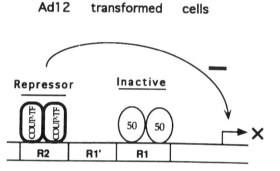

FIGURE 8. Mechanism for Ad12 E1A-mediated down-regulation of class I transcription in transformed cells. The class I promoter in both Ad5- and Ad12-transformed cells is represented by the transcription start site (bent arrow) and enhancer, comprised of the three subelements R1, R1', and R2. In Ad5-transformed cells, active class I transcription is mainly dependent on the R1 element, which binds the positive factor NFκB (p50/p65 heterodimer; p65 activation domain darkened), and to a lesser extent on the R2 element, which exhibits low binding activity to COUP and possibly another nuclear hormone receptor protein. In Ad12-transformed cells, the R1 element binds inactive p50 homodimers, and the R2 element exhibits strong binding activity to homodimers of COUP repressor. The model suggests that Ad12 E1A mediates global down-regulation of the class I enhancer, even though the actual site of Ad12 E1A interaction is unknown.

heterodimer (NFκB) binds R1 (Fig. 8). The Ad12 E1A-mediated block of active NFκB formation must be posttranslational because the cytoplasmic and nuclear levels of the p65 and p50 subunits are similar in both Ad5- and Ad12-transformed cells.[212] Whether p65 is rendered inactive by an inhibitor in Ad12-transformed cells remains to be determined.

The R2 element of the class I enhancer has a recognition sequence that is capable of binding nuclear hormone receptor proteins, e.g., RXRβ and RARβ (Fig. 7).[217,218] The R2 element normally contributes to positive class I enhancer activity, although it seems to be of less importance than the R1 element in this regard.[217] In contrast to Ad5-transformed cells, the R2 element in Ad12-transformed cells appears to regulate transcription negatively[210] and exhibits a much stronger binding activity[209,210] to the nuclear receptor protein COUP,[211] which is known to act as a transcriptional repressor of many promoters (Fig. 8).[219] In Ad12-transformed cells, COUP appears to bind to the R2 element as a homodimer, whereas in Ad5-transformed cells, the dimeric composition of the small amounts of COUP that bind to this element is unknown.[211] Thus, Ad12 E1A mediates repression of the R2 site of the class I enhancer by causing a dramatic increase in the binding activity of the COUP repressor protein.

These studies suggest that in transformed cells, Ad12 E1A causes a global down-regulation of the MHC class I enhancer by affecting individual enhancer sub-elements. At the R1 site there is a block in the binding of the positive-acting transcription factor NFκB, while, concomitantly, at the R2 site there is strong binding of a transcriptional repressor COUP (Fig. 8). There is even a suggestion that the middle R1′ enhancer element, which has a degenerate NFκB binding site, is also functionally down-regulated in Ad12-transformed cells.[212] It is intriguing to consider that global down-regulation of the enhancer is needed to assure transcriptional inactivity of the class I genes under varying physiological conditions. For example, repression by COUP at the R2 site could potentially serve to override temporary increases in NFκB activity at the R1 site resulting from physiological fluctuations of certain cytokines (e.g., TNF-α).[213] Recent studies indicate that the differential COUP and NFκB binding activities are both mediated by sequences within the first exon of Ad12 E1A which include the 20 amino acid nonconserved domain but exclude CR3.[220]

6.4. Viral Persistence as the Biological Basis for Ad12 E1A-Mediated Down-Regulation of Class I Transcription

Why is the class I enhancer down-regulated in Ad12- but not Ad5-transformed cells? The model proposed here and depicted in Fig. 9 attempts to address this enigma. A clue may be provided in considering the basis of persistent infection, an important but poorly understood aspect of adenovirus biology. Once adenoviruses are acquired, usually during early childhood, they have the potential to persist as inapparent infections for many years in certain cells that had, in some way, resisted the usual fate of being lysed during the primary infection. In the case of cells persistently infected by Ad12, the low surface levels of class I proteins because of E1A-mediated down-regulation of class I transcription would favor their escape from immune recognition by CTLs. In the case of cells persistently infected by Ad5, class I surface expression is also postulated to be reduced, but it is thought to occur through an entirely different mechanism that involves the viral 19K glycoprotein encoded by the nontransforming E3 gene[221,222] (Fig. 9). The 19K glycoprotein binds to class I proteins in the endoplasmic reticulum and physically blocks their passage to the cell surface.[223] Here again, reduced class I expression on Ad5 persistently infected cells would make them less apt to be detected by CTLs.[224]

The model presented here (Fig. 9) suggests that viral persistence is the apparent biological basis for Ad12 E1A-mediated repression of class I transcription in Ad12-transformed cells. The difference in tumorigenic potential between Ad5 and Ad12 reflects the difference in the immune escape mechanisms used during persistent infection. Simply, because E1A and E1B (but not E3) are the only viral genes required to transform cells, then, accordingly, only Ad12-transformed cells will exhibit low class I expression and possess high tumorigenic potential.

If this model is valid, what led Ad5 and Ad12 to develop these different strategies for blocking MHC class I expression in persistently infected cells? One appealing speculation is that these different strategies were acquired as Ad5 and Ad12 evolved different tropisms for specific tissues during infection (Fig. 9). It may

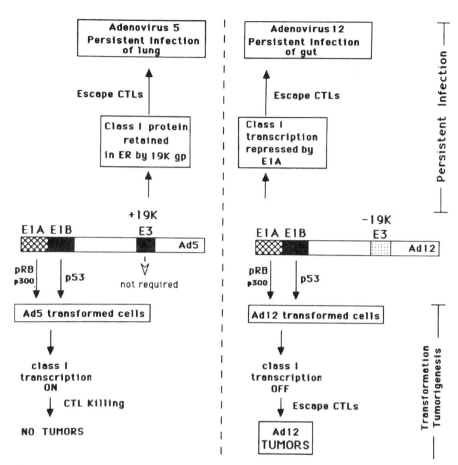

FIGURE 9. Model for how differences in the persistent infection mechanisms suggested for Ad5 and Ad12 account for differences in their tumorigenic potential. Persistent infection: Ad5 (left) and Ad12 (right) have apparent tropisms for lung and enteric cells, respectively. It is proposed that in both persistently infected Ad5 and Ad12 cells, surface levels of class I proteins are reduced, which provides a means of escaping immune detection by CTLs. In Ad5 persistently infected cells, the 19K glycoprotein encoded by the E3 region sequesters class I proteins in the endoplasmic reticulum (ER), whereas in Ad12 persistently infected cells, E1A mediates down-regulation of class I transcription; Ad12 does not appear to encode a functional homologue of the 19K glycoprotein. Transformation/tumorigenesis: the E1A and E1B genes of both viruses are able to transform nonpermissive cells as a result of abortive infection or transfection of the E1A and E1B genes. Transformation occurs as a result of deregulated cell growth by the interaction of E1A and E1B proteins with critical cellular proteins that include pRb, p300, and p53. Ad12-transformed cells are tumorigenic because the Ad12 E1A gene is also able to mediate class I down-regulation, allowing these cells to escape CTL immunosurveillance. By contrast, Ad5-transformed cells are non-tumorigenic because the E3 region, responsible for reduced class I expression in persistent infection, is not required for transformation.

be that adaptation of persistent infection in lung cells caused the respiratory adenoviruses, such as Ad5, to evolve the E3 19K glycoprotein mechanism to interfere with class I expression. By contrast, enteric adenovirus isolates, such as Ad12,[225] evolved the E1A-mediated mechanism of blocking class I expression. Some credence for this hypothesis is based on the finding that transformed cells of two other enteric strains of adenovirus (Ad40 and Ad41) are also reduced in class I expression.[226] Moreover, these enteric adenoviruses do not have apparent homologues corresponding to the Ad5 E3 19K glycoprotein.[227] Also, Ad12 does not appear to encode a functional counterpart of the E3 19K glycoprotein that inhibits the intracellular transport of class I proteins to the cell surface.[228]

7. SUMMARY

The protein products of the E1A and E1B transforming genes of adenovirus interact with cellular proteins that govern cell-cycle regulation, differentiation, growth arrest, and apoptosis. It is the consequences of these protein–protein interactions that lead to immortalization and the transformation phenotype in cells that are not fully permissive for adenovirus infection or in cells in which the E1A and E1B transforming region has been introduced. These viral/cellular protein interactions in the transformed cell appear to reflect similar interactions in the lytically infected cell, where they are responsible for establishing an optimum environment for viral replication and propagation. Through their abilities to perturb normal cellular growth, the E1A and E1B proteins have served as invaluable molecular probes in understanding the regulation of tumor suppressor proteins as well as the direct association between transcription and replication. In addition, the detailed mechanism of how the E1A gene of Ad12 mediates down-regulation MHC class I transcription has provided an explanation of how Ad12-transformed cells can escape immune detection by CTLs and cause tumors. This immune escape mechanism may also be the underlying strategy used by Ad12 to achieve persistent infection. Future studies aimed at further resolving the details by which E1A and E1B usurp cellular processes, predictably, will continue to yield basic insights into how perturbation of key regulatory proteins can tilt cells in the direction of ungoverned proliferation and cancer.

ACKNOWLEDGMENTS. The authors wishes to acknowledge NIH grant CA29797 from the National Cancer Institute and X. Liu and D. Kushner for kind assistance.

REFERENCES

1. Branton, P. E., Bayley, S. T., and Graham, F. L., 1985, Transformation by human adenoviruses, *Biochim. Biophys. Acta* **780**:67–94.
2. Gallimore, P. H., Sharp, P. A., and Sambrook, J., 1974, Viral DNA in transformed cells. II. A

study of the sequences of adenovirus type 2 DNA in nine lines of transformed rat cells using specific fragments of the viral genome, *J. Mol. Biol.* **89:**49–72.
3. Graham, F. L., Van der Eb, A. J., and Heijneker, H. L., 1974, The size and location of the transforming region in human adenovirus DNA, *Nature* **251:**687–691.
4. Graham, F. L., Abrahams, P. J., Mulder, C., Heijneker, H. L., Warnaar, S. O., deVries, F. A. J., Fiers, W., and van der Eb, A. J., 1975, Studies on *in vitro* transformation by DNA and DNA fragments of human adenovirus and SV40, *Cold Spring Harbor Symp. Quant. Biol.* **39:**637–650.
5. Shenk, T., and Flint, J., 1991, Transcriptional and transforming activities of the adenovirus E1A proteins, *Adv. Cancer Res.* **57:**47–85.
6. Williams, J., 1986, Adenovirus genetics, in: *Adenovirus DNA* (W. Doerfler, ed.), Martinus Nijhoff, Boston, pp. 247–309.
7. van der Eb, A. J., van Ormondt, H., Schrier, P. I., Lupker, J. H., Jochemsen, H., van den Elsen, P. J., De Leys, R. J. Maat, J., van Beveren, C. P., Dijkema, R., and De Waard, A., 1979, Structure and function of the transforming genes of human adenovirus and SV40, *Cold Spring Harbor Symp. Quant. Biol.* **44:**383–399.
8. Shiroki, K., Shimojo, H., Sawada, Y., Uemizu, Y., and Fujinaga, K., 1979, Incomplete transformation of rat cells by a small fragment of adenovirus 12 DNA, *Virology* **95:**127–136.
9. Houweling, A., van den Elsen, P. J., and van der Eb, A. J., 1980, Partial transformation of primary rat cells by the left most 4.5% fragment of adenovirus DNA, *Virology* **105:**537–550.
10. van den Elsen, P. H., Houweling, A., and van der Eb, A. J., 1983, Expression of region E1b of human adenoviruses in the absence of region E1A is not sufficient for complete transformation, *Virology* **128:**377–390.
11. Ruley, E., 1983, Adenovirus early region 1A enables viral and cellular transforming genes to transform primary cells in culture, *Nature* **304:**602–606.
12. Berk, A. J., Lee, F., Harrison, T., Williams, J., and Sharp, P. A., 1979, A pre-early adenovirus 5 gene product regulates synthesis of early viral messenger RNAs, *Cell* **17:**935–944.
13. Perricaudet, M., Akusjarvi, G., Virtanen, A., and Pettersson, U., 1979, Structure of two spliced mRNAs from the transforming region of human subgroup C adenoviruses, *Nature* **281:** 694–696.
14. Ricciardi, R. P., Jones, R. L., Cepko, C. L., Sharp, P. A., and Roberts, B. E., 1981, Expression of early adenovirus genes requires a viral encoded acidic polypeptide, *Proc. Natl. Acad. Sci. USA* **78:**6121–6125.
15. Montell, C., Fisher, E. F., Caruthers, M. H., and Berk, A. J., 1982, Resolving the functions of overlapping viral genes by site-specific mutagenesis at a mRNA splice site, *Nature* **295:** 380–384.
16. Berk, A. J., 1986, Functions of adenovirus E1A, *Cancer Surv.* **5:**367–387.
17. Kimelman, D., Miller, J. S., Porter, D., and Roberts, B. E., 1985, E1a regions of the human adenoviruses and of the highly oncogenic simian adenovirus 7 are closely related, *J. Virol.* **53:**399–409.
18. Moran, E., and Mathews, M. B., 1987, Multiple functional subdomains in the adenovirus E1A gene, *Cell* **48:**177–178.
19. Stephens, C., and Harlow, E., 1987, Differential splicing yields novel adenovirus 5 E1A mRNAs that encode 30 kd and 35 kd proteins, *EMBO J.* **6:**2027–2035.
20. Ulfendahl, P. J., Linder, S., Kreivi, J.-P., Nordqvit, K., Sevensson, C., Hultberg, H., and Akusjarvi, G., 1987, A novel adenovirus-2 E1A mRNA encoding a protein with transcription activation properties, *EMBO J.* **6:**2037–2044.
21. Moran, E., Grodzicker, T., Roberts, R. J., Mathews, M. B., and Zerler, B., 1986, Lytic and transforming functions of individual products of the adenovirus E1A gene, *J. Virol* **57:** 765–775.
22. Moran, E., Zerler, B., Harrison, T. M., and Mathew, M. B., 1986, Identification of separate domains in the adenovirus E1A gene for immortalization activity and the activation of virus early genes, *Mol. Cell. Biol.* **6:**3470–3480.

23. Schneider, J. F., Fisher, F., Goding, C. R., and Jones, N. C., 1987, Mutational analysis of the adenovirus E1A gene: The role of transcriptional regulation in transformation, *EMBO J.* **6:**2053–2060.
24. Whyte, P., Ruley, H. E., and Harlow, E., 1988, Two regions of the adenovirus early region are required for transformation, *J. Virol.* **62:**257–265.
25. Ruben, M., Bacchetti, S., and Graham, F. L., 1982, Integration and expression of viral DNA in cells transformed by host-range mutants of adenovirus type 5, *J. Virol.* **41:**674–685.
26. Haley, K. P., Overhauser, J., Babiss, L. F., Ginsberg, H. S., and Jones, N. C., 1984, Transformation properties of type 5 adenovirus mutants that differentially express the E1A gene products, *Proc. Natl. Acad. Sci. USA* **81:**5734–5738.
27. Roberts, B. E., Miller, J. S., Kimelman, D., Cepko, C. L., Lemischka, I. R., and Mulligan R. C., 1985, Individual adenovirus type 5 early region 1A gene products elicit distinct alterations of cellular morphology and gene expression, *J. Virol.* **56:**404–413.
28. Zerler, B., Moran, B., Maruyama, K., Moomaw, J., Grodzicker, T., and Ruley, H. E., 1986, Analysis of adenovirus E1A coding sequences which enable *ras* and *pMT* oncogenes to transform cultures of primary cells, *Mol. Cell. Biol.* **6:**887–889.
29. Montell, C., Courtois, G., Eng, C., and Berk, A. J., 1984, Complete transformation by adenovirus 2 requires both E1A proteins, *Cell* **36:**951–961.
30. Winberg, C., and Shenk, T., 1984, Dissection of overlapping functions within the adenovirus type 5 E1A gene, *EMBO J.* **3:**1907–1912.
31. Lillie, J. W., Green, M., and Green, M. R., 1986, An adenovirus E1A proteins region required for transformation and transcriptional repression, *Cell* **46:**1043–1051.
32. Lillie, J., Loewenstein, P., Green, M. R., and Green, M., 1987, Functional domains of adenovirus type 5 E1A proteins, *Cell* **50:**1091–1100.
33. Subramanian, T., Kuppaswamy, M., Nasser, R. J., and Chinnadurai, G., 1988, An N-terminal region of adenovirus-E1A essential for cell-transformation and induction of an epithelial-cell growth factor, *Oncogene* **2:**105–112.
34. Jelsma, T. N., Howe, J. S., Mymryk, J. S., Evelegh, C. M., Cunniff, N. F. A., and Bayley, S. T., 1989, Sequences in E1A proteins of human adenovirus 5 required for cell transformation, repression of a transcriptional enhancer, and induction of proliferating cell nuclear antigen, *Virology* **170:**120–130.
35. Fahnestock, M. L., and Lewis, J. B., 1989, Genetic dissection of the transactivating domain of the E1a 289R protein of adenovirus type 2, *J. Virol.* **63:**1495–1504.
36. Borrelli, E., Hen, R., and Chambon, P., 1984, Adenovirus-2 E1A products repress enhancer-induced stimulation of transcription, *Nature* **312:**608–612.
37. Velcich, A., and Ziff, E., 1985, Adenovirus E1A proteins repress transcription from the SV40 early promoter, *Cell* **40:**705–716.
38. Hen, R., Borrelli, E., and Chambon, P., 1985, Repression of the immunoglobulin heavy chain enhancer by the adenovirus-2 E1A products, *Science* **230:**1391–1394.
39. Smith, D. H., Kegler, D. M., and Ziff, E. B., 1985, Vector expression of adenovirus type 5 E1A proteins: Evidence for E1A autoregulation, *Mol. Cell. Biol.* **5:**2684–2696.
40. Kaczmarek, L., Ferguson, B., Rosenberg, M., and Baserga, R., 1986, Induction of cellular DNA synthesis by purified adenovirus E1A proteins, *Virology* **152:**1–10.
41. Quinlan, M. P., and Grodzicker, T., 1987, Adenovirus E1A 12S protein induces DNA synthesis and proliferation in primary epithelial cells in both the presence and absence of serum, *J. Virol.* **61:**673–682.
42. Zerler, B., Roberts, R. J., Matthews, M. B., and Moran, E., 1987, Different functional domains of the adenovirus E1A gene are involved in regulation of host cell cycle products, *Mol. Cell. Biol.* **7:**821–829.
43. Smith, D., and Ziff, E., 1988, The amino-terminal region of the adenovirus serotype 5 E1a protein performs two separate functions when expressed in primary baby rat kidney cells, *Mol. Cell. Biol.* **8:**3882–3890.

44. Lyons, R. H., Ferguson, B., and Rosenberg, M., 1987, Pentapeptide nuclear localization signal in adenovirus E1a, *Mol. Cell. Biol.* **7:**2451–2456.
45. Krippl, B., Ferguson, B., Jones, N., Rosenberg, M., and Westphal, H., 1986, Mapping functional domains in adenovirus E1A proteins, *Proc. Natl. Acad. Sci. USA* **82:**7480–7484.
46. Tsukamoto, A. S., Ponticelli, A., Berk, A. J., and Gaynor, R. B., 1986, Genetic mapping of a major site of phosphorylation in adenovirus type 2 E1A proteins, *J. Virol.* **59:**14–22.
47. Tremblay, M. L., McGlade, C. J., Gerber, G. E., and Branton, P. E., 1988, Identification of the phosphorylation sites in early region 1A proteins of adenovirus type 5 by amino acid sequencing of peptide fragments *J. Biol. Chem.* **263:**6375–6383.
48. Smith, C. L., Debouck, C., Rosenberg, M., Culp, J. S., 1989, Phosphorylation of serine residue 89 of human adenovirus E1A proteins is responsible for their characteristic electrophoretic mobility shifts, and its mutation affects biological function, *J. Virol.* **63:**1569–1577.
49. Richter, J. D., Slavicek, J. M., Schneider, J. F., and Jones, N. C., 1988, Heterogeneity of adenovirus 5 E1A proteins: Multiple serine phosphorylations induce slow-migrating electrophoretic variations but do not affect E1A-induced transcriptional activation or transformation. *J. Virol.* **62:**1948–1955.
50. Tremblay, M. L., and Branton, P. E., 1989, Analysis of phosphorylation sites in exon 1 region of E1A proteins of human adenovirus type 5, *Virology* **169:**397–407.
51. Jones, N., and Shenk, T., 1979, An adenovirus type 5 early gene function regulates expression of other early viral genes, *Proc. Natl. Acad. Sci. USA* **76:**3665–3669.
52. Flint, S. J., Sambrook, J., Williams, J., and Sharp, P. A., 1976, Viral nucleic acid sequences in transformed cells *Virology* **72:**456–470.
53. Pei, R., and Berk, A. J., 1989, Multiple transcription factor binding sites mediate adenovirus E1A transactivation, *J. Virol.* **63:**3499–3506.
54. Glenn, G. M., and Ricciardi, R. P., 1985, Adenovirus 5 early region 1A host range mutants hr3, hr4, and hr5 contain point mutations which generate single amino acid substitutions, *J. Virol.* **56:**66–74.
55. Culp, J. S., Webster, L. C., Friedman, D. J., Smith, C. L. Huang, W.-J., Wu, F. Y.-H., Rosenberg, M., and Ricciardi, R. P., 1988, The 289R-amino acid E1A protein of adenovirus binds zinc in a region that is important for *trans*-activation, *Proc. Natl. Acad. Sci. USA* **85:**6450–6454.
56. Webster, L. C., and Ricciardi, R. P., 1991, *Trans*-dominant mutants of E1A provide genetic evidence that the zinc finger of the transactivating domain binds a transcription factor, *Mol. Cell. Biol.* **11:**4287–4296.
57. Webster, L. C., Zhang, K., Chance, B., Ayene, I., Culp, J. S., Huang, W-J,. Wu, F. Y.-H., and Ricciardi, R. P., 1991, Conversion of the E1A Cys_4 zinc finger to a nonfunctional His_2, Cys_2 zinc finger by a single point mutation, *Proc. Natl. Acad. Sci. USA* **88:**9989–9993.
58. Glenn, G. M., and Ricciardi, R. P., 1987, An adenovirus type 5 E1A protein with a single amino acid substitution blocks wild type E1A trans-activation, *Mol. Cell. Biol.* **7:**1004–1011.
59. Lee, W. S., Kao, C. C., Bryant, G. O., Liu, X., and Berk, A. J., 1991, Adenovirus E1A activation domain binds the basic repeat in the TATA box transcription factor, *Cell* **67:**365–376.
60. Horikoshi, N., Maguire, K., Kralli, A., Maldonado, E., Reinberg, D., and Weinmann, R., 1991, Direct interaction between adenovirus E1A protein and the TATA box binding transcription factor IID, *Proc. Natl. Acad. Sci. USA* **88:**5124–5128.
61. Geisberg, J. V., Lee, W. S., Berk, A. J., and Ricciardi, R. P., 1994, The zinc finger region of the adenovirus E1A transactivating domain complexes with the TATA box binding protein, *Proc. Natl. Acad. Sci. USA* **91:**2488–2492.
62. Geisberg, J. V., Chen, J. L., and Ricciardi, R. P., 1995, Subregions of the adenovirus E1A transactivation domain target multiple components of the TFIID complex, *Mol. Cell. Biol.* **15:**6283–6290.
63. Klug, A., and Rhodes, D., 1987, "Zinc finger": A novel motif for nucleic acid recognition, *Trends Biochem. Sci.* **12:**464–469.
64. Chatterjee, P. K., Bruner, M., Flint, S. J., and Harter, M. L., 1988, The DNA binding properties of an adenovirus 289R E1A protein, *EMBO J.* **7:**835–841.

65. Ferguson, B., Krippl, B., Andrisani, O., Jones, N., Westphal, H., and Rosenberg, M., 1985, E1A 13S and 12S mRNA products in *Escherichia coli* both function as nucleus localized transcription activators but do not directly bind DNA, *Mol. Cell. Biol.* **5:**2653–2661.
66. Liu, F., and Green, M. R., 1994, Promoter targeting by adenovirus E1a through interaction with different cellular DNA-binding domains, *Nature* **368:**520–525.
67. Hinds, P. W., and Weinberg, R. A., 1994, Tumor suppressor genes, *Curr. Opin. Genet. Dev.* **4:**135–141.
68. Nevins, J. R., 1994, Cell cycle targets of the DNA tumor viruses, *Curr. Opin. Genet. Dev.* **4:**130–134.
69. Moran, E., 1993, Interaction of adenoviral E1A proteins with pRB and p53, *FASEB J.* **7:**880–885.
70. Winman, K. G., 1993, The retinoblastoma gene: Role in cell cycle control and cell differentiation, *FASEB J.* **7:**841–845.
71. Livingston, D. M., Kaelin, W., Chittenden, T., and Qin, X., 1993, Structural and functional contributions to the G_1 blocking acting of the retinoblastoma protein, *Br. J. Cancer* **68:**264–268.
72. Klein, G., 1993, Genes that can antagonize tumor development, *FASEB J.* **7:**821–825.
73. Whyte, P., Buchkovich, K. J., Horowitz, J. M., Friend, S. H., Raybuk, M., Weinberg, R. A., and Harlow, E., 1988, Association between an oncogene and an anti-oncogene; the adenovirus E1A proteins bind to the retinoblastoma gene product, *Nature* **334:**124–129.
74. Dyson, N., Howley, P. M., Munger, K., and Harlow, E., 1989, The human papilloma virus-16 E7 oncoprotein is able to bind to the retinoblastoma gene product, *Science* **243:**934–937.
75. DeCaprio, J. A., Ludlow, J. W., Figge, J., Shew, J.-Y., Huang, C.-M., Lee, W.-H., Marsilio, E., Paucha, E., and Livingston, D. M., 1988, SV40 large tumor antigen forms a complex with the product of the retinoblastoma susceptibility gene, *Cell* **54:**275–283.
76. Stabel, S., Argos, P., and Philipson, L., 1985, The release of growth arrest by microinjection of adenovirus E1A DNA, *EMBO J.* **4:**2329–2336.
77. Moran, E., 1988, The region of SV40 large T antigen can substitute for a transforming domain of the adenovirus E1A products, *Nature* **334:**167–170.
78. Whyte, P., Williamson, N. M., and Harlow, E., 1989, Cellular targets for transformation by the adenovirus E1A proteins, *Cell* **56:**67–75.
79. Buchkovich, K., Duffy, L. A., and Harlow, E., 1989, The retinoblastoma protein is phosphorylated during specific phases of the cell cycle, *Cell* **58:**1097–1105.
80. Chen, P.-L., Scully, P., Shew, J.-Y., Wang, J., and Lee, W.-H., 1989, Phosphorylation of the retinoblastoma gene product is modulated during the cell cycle and cellular differentiation, *Cell* **58:**1193–1198.
81. DeCaprio, J. A., Ludlow, J. W., Lynch, D., Furukawa, Y., Griffin, J., Piwnica-Worms, H., Huang, C.-M., and Livingston, D. M., 1989, The product of the retinoblastoma susceptibility gene has properties of a cell cycle regulatory element, *Cell* **58:**1085–1095.
82. Mihara, K., Cao, X., Yen, A., Chandler, S., Driscoll, B., Murphree, A. L., Tang, A., and Fung, Y., 1989, Cell cycle-dependent regulation of phosphorylation of the human retinoblastoma gene product, *Science* **246:**1300–1303.
83. Ludlow, J. W., Shon, J., Pipas, J. M., Livingston, D. M., and DeCaprio, J. A., 1990, The retinoblastoma susceptibility gene product undergoes cell cycle-dependent dephosphorylation and binding to and release from SV40 large T, *Cell* **60:**387–396.
84. La Thangue, N. B., 1994, DP and E2F proteins: Components of a heterodimeric transcription factor implicated in cell cycle control, *Curr. Opin. Cell Biol.* **6:**443–450.
85. Ludlow, J. W., DeCaprio, J. A., Huang, C.-M., Lee, W.-H., Paucha, E., and Livingston, D. M., 1989, SV40 large T antigen binds preferentially to the underphosphorylated member of the retinoblastoma susceptibility gene product family, *Cell* **56:**57–65.
86. Stein, G. H., Beeson, M., and Gordon, L., 1980, Failure to phosphorylate the retinoblastoma gene product in senescent human fibroblasts, *Science* **249:**666–669.

87. Chellappan, S., Hiebert, S., Mudryj, M., Horowitz, J., and Nevins, J., 1991, The E2F transcription factor is a cellular target for the RB protein, *Cell* **65**:1053–1061.
88. Bandura, L. R., and La Thangue, N. B., 1991, Adenovirus E1A prevents the retinoblastoma gene product from complexing with a cellular transcription factor, *Nature* **351**:494–497.
89. Helin, K., Lees, J. A., Vidal, M., Dyson, N., Harlow, E., and Fattaey, A., 1992, A cDNA encoding a pRb-binding protein with properties of the transcription factor E2F, *Cell* **70**:337–350
90. Kaelin, W. G., Krek, W., Sellers, W. R., DeCaprio, J. A., Ajchenbaum, F., Fuchs, C. S., Chittenden, T., Li, Y., Farnham, P. J., Blanar, M. A., Livingston, D. M., and Flemington, E. K., 1992, Expression cloning of a cDNA encoding a retinoblastoma-binding protein with E2F-like properties, *Cell* **70**:351–364.
91. Shan, B., Zhu, X., Chen, P.-L., Durfee, T., Yang, Y., Sharp, D., and Lee, W-H., 1992, Molecular cloning of a gene with properties of the transcription factor E2F, *Mol. Cell. Biol.* **12**:5620–5631.
92. Ivey-Hoyle, M., Conroy, R., Huber, H., Goodhart, P., Oliff, A., and Haimbrook, D. C., 1993, Cloning and characterization of E2F-2, a novel protein with the biochemical properties of transcription factor E2F, *Mol. Cell. Biol.* **13**:7802–7812.
93. Lees, J. A., Saito, M., Vidal, M., Valentine, M., Look, T., Harlow, E., Dyson, N., and Helin, K., 1993, The retinoblastoma protein binds to a family of E2F transcription factors, *Mol. Cell. Biol.* **13**:7813–7825.
94. Girling, R., Partridge, J. F., Bandura, L. R., Burden, N., Totty, N. F., Hsuan, J. J., and LaThangue, N. B., 1993, A new component of the transcription factor DRTF1/E2F, *Nature* **362**:83–87.
95. Nevins, J. R., 1992, E2F: A link between the Rb tumor suppressor protein and viral oncoproteins, *Science* **258**:424–429.
96. Ikeda, M. A., and Nevins, J. R., 1993, Identification of distinct roles for separate E1A domains in disruption of E2F complexes, *Mol. Cell. Biol.* **13**:7029–7035.
97. Fattaey, A. R., Harlow, E., and Helin, K., 1993, Independent regions of adenovirus E1A are required for binding to and dissociation of E2F-protein complexes, *Mol. Cell. Biol.* **13**:7267–7277.
98. Ewen, M. E., Xing, Y., Lawrence, J. B., and Livingston, D. M., 1991, Molecular cloning, chromosomal mapping and expression of the cDNA for p107, a retinoblastoma gene product-related protein, *Cell* **66**:1155–1164.
99. Li, Y., Graham, C., Lacy, S., Duncan, A. M. V., and Whyte, P., 1993, The adenovirus E1A-associated 130 kDa protein is encoded by a member of the retinoblastoma gene family and physically interacts with cyclins, *Genes Dev.* **7**:2366–2377.
100. Cobrinik, D., Whyte, P., Peeper, D. S., Jacks, T., and Weinberg, R. A., 1993, Cell cycle-specific association of E2F with the p130 E1A-binding protein, *Genes Dev.* **7**:2392–2404.
101. Hannon, G. J., Demetrick, D., and Beach, D., 1993, Isolation of the Rb-Related p130 through its interaction with CDK2 and cyclins, *Genes Dev.* **7**:2378–2391.
102. Hu, Q., Dyson, N., and Harlow, E., 1990, The regions of the retinoblastoma protein needed for binding to adenovirus E1A or SV40 large T antigen are common sites for mutations, *EMBO J.* **9**:1147–1155.
103. Huang, S., Wang, N., Tseng, B. Y., Lee, W., and Lee, E. H., 1980, Two distinct and frequently mutated regions of retinoblastoma protein are required for binding to SV40 T antigen, *EMBO J.* **9**:1815–1822.
104. Kaelin, W. G. Jr., Ewen, M. E., and Livingston, D. M., 1990, Definition of the minimal simian virus 40 large T antigen- and adenovirus E1A-binding domain in the retinoblastoma gene product, *Mol. Cell. Biol.* **10**:3761–3769.
105. Giordano, A., McCall, C., Whyte, P., and Franza, B. R., 1991, Human cyclin A and the retinoblastoma protein interact with similar but distinguishable sequences in the adenovirus E1A product, *Oncogene* **6**:481–486.
106. Dyson, N., and Harlow, E., 1992, Adenovirus E1A targets key regulators of cell proliferation, *Cancer Surv.* **2**:161–195.

107. La Thangue, N. B., 1994, DRTF1/E2F: An expanding family of heterodimeric transcription factors implicated in cell-cycle control, *Trends Biochem. Sci.* **19:**108–114.
108. Shirodkar, S., Ewen, M. E., DeCaprio, J. A., Morgan, J., Livingston, D. M., and Chittenden, T., 1992, The transcription factor E2F interacts with the retinoblastoma product and a p107–cyclin A complex in a cell cycle-regulated manner, *Cell* **68:**157–166.
109. Hinds, P. W., Mittnacht, S., Dulic, V., Arnold, A., Reed, S. I., and Weinberg, R., 1992, Regulation of retinoblastoma protein functions by ectopic expression of human cyclins, *Cell* **70:**993–1006.
110. Devoto, S. H., Mudryj, M., Pines, P., Hunter, T., and Nevins, J. R., 1992, A cyclin A-specific protein kinase complex possesses sequence-specific DNA binding activity: p33^{cdk2} is a component of the E2F–cyclin A complex, *Cell* **68:**167–176.
111. Cao, L., Faha, B., Dembski, M., Tsai, L.-H., Harlow, E., and Dyson, N., 1992, Independent binding of the retinoblastoma protein and p107 to the transcription factor E2F, *Nature* **355:**176–179.
112. Faha, B., Harlow, E., and Lees, E., 1993, The adenovirus E1a-associated kinases consist of cyclin E-p33^{cdk2} and cyclin A-p33^{cdk2}, *J. Virol.* **67:**2456–2465.
113. Moran, E., and Zerler, B., 1988, Interactions between cell growth-regulating domains in the products of the adenovirus E1A oncogene, *Mol. Cell. Biol.* **8:**1756–1764.
114. Howe, J. A., Mymryk, J. S., Eagan, C., Branton, P. E., and Bayley, S. T., 1990, Retinoblastoma growth suppressor and a 300-kDa protein appear to regulate cellular DNA synthesis, *Proc. Natl. Acad. Sci. USA* **87:**5883–5887.
115. Stein, R. W., Corrigan, M., Yaciuk, P., Whelan, J., and Moran, E., 1990, Analysis of E1A-mediated growth regulation functions: Binding of the 300-kilodalton cellular product correlates with E1A enhancer repression function and DNA synthesis-inducing activity, *J. Virol.* **64:**4421–4427.
116. Wang, H.-G., Rikitake, Y., Carter, M. C., Yaciuk, P., Abraham, S. E., Zerler, B., and Moran, E., 1993, Identification of specific adenovirus E1A N-terminal residues critical to the binding of cellular proteins and the control of cell growth, *J. Virol.* **67:**476–488.
117. Stein, R. W., and Ziff, E. B., 1987, Repression of insulin gene expression of adenovirus type 5 E1A proteins, *Mol. Cell. Biol.* **7:**1164–1170.
118. Webster, K. A., Muscat, G. E. O., and Kedes, L., 1988, Adenovirus E1A products suppress myogenic differentiation and inhibit transcription from muscle-specific promoters, *Nature* **332:**553–557.
119. Egan, C., Jelsma, T. N., Howe, J. A., Bayley, S. T., Ferguson, B., and Branton, P. E., 1988, Cellular mapping of protein-binding sites on the products of early-region 1A of human adenovirus type 5, *Mol. Cell. Biol.* **8:**3955–3959.
120. Eckner, R., Ewen, M. E., Newsome, D., Gerdes, M., DeCaprio, J. A., Lawrence, J. B., and Livingston, D. M., 1994, Molecular cloning and functional analysis of the adenovirus E1A-associated 300-kD protein (p300) reveals a protein with properties of a transcriptional adaptor, *Genes Dev.* **8:**869–884.
121. Abraham, S. E., Lobo, S., Yaciuk, P., Wang, H.-G., and Moran, E:, 1993, p300, and p300-associated proteins, are components of TATA-binding protein (TBP) complexes, *Oncogene* **8:**1639–1647.
122. Arany, Z., Sellers, W. R., Livingston, D. M., and Eckner, R., 1994, E1A-associated p300 and CREB-associated CBP belong to a conserved family of coactivators, *Cell* **77:**799–800.
123. Kwok, R. P. S., Lundbald, J. R., Chrivia, J. C., Richards, J. P., Bachinger, H. P., Brennan, R. G., Roberts, S. G. E., Green, M. R., and Goodman, R. H., 1994, Nuclear protein CBP is a coactivator for the transcription factor CREB, *Nature* **370:**223–226.
124. Arias, J., Alberts, A. S., Brindle, P., Claret, F. X., Smeal, T., Karin, M., Feramisco, J., and Montminy, M., 1994, Activation of cAMP and mitogen responsive genes relies on a common nuclear factor, *Nature* **370:**226–229.
125. Burgering, B. M. T., Pronk, G. J., van Weeren, P. C., Chardin, P., and Bos, J. L., 1993, cAMP antagonizes p21 *ras*-directed activation of extracellular signal-regulated kinase 2 and phosphorylation of mSos nucleotide exchange factor, *EMBO J.* **12:**4211–4220.

126. Cho-Chung, Y. S., Clair, T., Tortora, G., and Yokazaki, H., 1991, Role of site-selective cAMP analogs in the control and reversal of malignancy, *Pharmacol. Ther.* **50:**1–33.
127. Janaswami, P. M., Kalvakolanu, D. V. R., Zhang, Y., and Sen, G. C., 1992, Transcriptional repression of interleukin-6 gene by adenoviral E1A proteins, *J. Biol. Chem.* **267:**24886–24891.
128. Kalvakolanu, D. V. R., Liu, J., Hanson, R. W., Harter, M. L., and Sen, G. C., 1992, Adenovirus E1A represses the cyclic AMP-induced transcription of the gene for phosphoenolpyruvate carboxykinase (GTP) in hepatoma cells, *J. Biol. Chem.* **267:**2530–2536.
129. Stillman, B., 1986, Functions of the adenovirus E1B tumor antigens, *Cancer Surv.* **5:**389–404.
130. Boulanger, P. A., and Blair, G. E., 1991, Expression and interactions of human adenovirus oncoproteins, *Biochem. J.* **275:**281–299.
131. Senear, A. W., and Lewis, J. B., 1986, Morphological transformation of established rodent cell lines by high-level expression of the adenovirus type 2 E1a gene, *Mol. Cell. Biol.* **6:**1253–1260.
132. Barker, D. B., and Berk, A. J., 1987, Adenovirus proteins from both E1B reading frames are required for transformation of rodent cells by viral infection and DNA transfection, *Virology* **156:**107–121.
133. White, E., 1993, Regulation of apoptosis by the transforming genes of the DNA tumor virus adenovirus, *Proc. Soc. Exp. Biol. Med.* **204:**30–39.
134. Levine, A. J., Momand, J., and Finlay, C. A., 1991, The p53 tumor suppressor gene, *Nature* **351:**453–456.
135. Vogelstein, B., and Kinzler, K. W., 1992, p53 function and dysfunction, *Cell* **70:**523–526.
136. Kastan, M. B., Onyckwere, O., Sidransky, D., Vogelstein, B., and Craig, R. W., 1991, Participation of p53 protein in the cellular response to DNA damage, *Cancer Res.* **51:**6304–6311.
137. Hartwell, L., 1992, Defects in a cell cycle checkpoint may be responsible for the genomic instability of cancer cells, *Cell* **71:**543–546.
138. Lin, D., Shields, M. T., Ullrich, S. J., Appella, E., and Mercer, W. E., 1992, Growth arrest induced by wild-type p53 protein blocks cells prior to or near the restriction point in late G1 phase, *Proc. Natl. Acad. Sci. USA* **89:**9210–9214.
139. Zambetti, G. P., and Levine, A. J., 1993, A comparison of the biological activities of wild-type and mutant p53, *FASEB J.* **7:**855–865.
140. Kuerbitz, S. J., Plunkett, B. S., Walsh, W. V., and Kastan, M. B., 1992, Wild-type p53 is a cell cycle checkpoint determinant following irradiation, *Proc. Natl. Acad. Sci. USA* **89:**7491–7495.
141. Farmer, G., Bargonetti, J., Zhu, H., Friedman, P., Prywes, R., and Prives, C., 1992, Wild-type p53 activates transcription *in vitro*, *Nature* **358:**83–86.
142. Kern, S. E., Pietenpol, J. A., Thiagalingam, S., Seymour, A., Kinzler, K. W., and Vogelstein, B., 1992, Oncogenic forms of p53 inhibit p53-regulated gene expression, *Science* **256:**827–830.
143. Funk, W. D., Pak, D. T., Karas, R. H., Wright, W. E., and Shay, J. W., 1992, A transcriptionally active DNA-binding site for human p53 protein complexes, *Mol. Cell. Biol.* **12:**2866–2871.
144. Zambetti, G. P., Bargonetti, J., Walker, K., Prives, C., and Levine, A. J., 1992, Wild-type p53 mediates positive regulation of gene expression through a specific DNA sequence element, *Genes Dev.* **6:**1143–1152.
145. Seto, E., Usheva, A., Zambetti, G. P., Momand, J., Horikoshi, N., Weinmann, R., Levine, A. J., and Shenk, T., 1992, Wild-type p53 binds to the TATA-binding protein and represses transcription, *Proc. Natl. Acad. Sci. USA* **89:**12028–12032.
146. Liu, X., Miller, C. W., Koeffler, P. H., and Berk, A. J. 1993, The p53 activation domain binds the TATA box-binding polypeptide in holo-TFIID and a neighboring p53 domain inhibits transcription, *Mol. Cell. Biol.* **13:**3291–3300.
147. Truant, R., Xiao, H., Ingles, J., Greenblatt, J., 1993, Direct interaction between the transcriptional activation domain of human p53 and the TATA box-binding protein, *J. Biol. Chem.* **268:**2284–2287.
148. Kastan, M. B., Zhan, Q., El-Deiry, W. S., Carrier, F., Jacks, T., Walsh, W. V., Plunkett, B. S., Vogelstein, B., and Fornace, A. J., Jr., 1992, A mammalian cell cycle checkpoint pathway utilizing p53 and GADD45 is defective in ataxia–telangiectasia, *Cell* **71:**587–597.
149. Harper, J. W., Adami, G. R., Wei, N., Keyomarsi, K., and Elledge, S. J., 1993, The p21 cdk-interacting protein Cip1 is a potent inhibitor of G_1 cyclin-dependent kinases, *Cell* **75:**805–816.

150. El-Deiry, W. S., Tokino, T., Velculescu, V. E., Levy, D. B., Parsons, R., Trent, J. M., Lin, D., Mercer, W. E., Kinzler, K. W., and Vogelstein, B., 1993, WAF1, a potent mediator of p53 tumor suppression, *Cell* **75**:817–825.
151. Xiong, Y., Hannon, G. J., Zhang, H., Casso, D., Kobayashl, R., and Beach, D., 1993, p21 is a universal inhibitor of cyclin kinases, *Nature* **366**:701–704.
152. Gu, Y., Turck, W., and Morgan, D. O., 1993, Inhibition of CDK2 activity *in vivo* by an associated 20K regulatory subunit, *Nature* **366**:707–710.
153. Sarnow, P., Ho, Y. S., Williams, J., and Levine, A. J., 1982, Adenovirus E1B-58Kd tumor antigen and SV40 large tumor antigen are physically associated with the same 54Kd cellular protein in transformed cells, *Cell* **28**:387–394.
154. Kao, C. C., Yew, P. R., and Berk, A. J., 1990, Domains required for *in vitro* association between the cellular p53 and the adenovirus 2 E1B 55K proteins, *Virology* **179**:806–814.
155. White, E., and Cipriani, R., 1990, Role of adenovirus E1B proteins in transformation: Altered organization of intermediate filaments in transformed cells that express the 19-kilodalton protein, *Mol. Cell. Biol.* **10**:120–130.
156. Tan, T.-H., Wallis, J., and Levine, A. J., 1986, Identification of the p53 protein domain involved in the formation of SV40 large T antigen p53 protein complex, *J. Virol.* **59**:574–583.
157. Schmeig, F. I., and Simmons, D. T., 1988, Characterization of the *in vitro* interaction between SV40 T antigen and p53: Mapping the p53 binding site, *Virology* **164**:132–140.
158. Werness, B. A., Levine, A. J., and Howley, P. M., 1990, The E6 proteins encoded by human papillomavirus types 16 and 18 can complex p53 *in vitro, Science* **248**:76–79.
159. Momand, J., Zambetti, G. P., Olson, D. C., George, D. L., and Levine, A. J., 1992, The *mdm-2* oncogene product forms a complex with the p53 protein and inhibits p53 transactivation, *Cell* **69**:1237–1245.
160. Fakharzadeh, S. S., Trusko, S. P., and George, D. L., Tumorigenic potential associated with enhanced expression of a gene that is amplified in a mouse tumor cell line, *EMBO J.* **10**:1565–1569.
161. Finlay, C. A., 1993, The *mdm-2* oncogene can overcome wild-type p53 suppression of transformed cell growth, *Mol. Cell. Biol.* **13**:301–306.
162. Yew, P. R., and Berk, A. J., 1992, Inhibition of p53 transactivation required for transformation by adenovirus early 1B protein, *Nature* **357**:82–85.
163. Yew, P. R., Liu, X., and Berk, A. J., 1994, Adenovirus E1B oncoprotein tethers a transcriptional repression domain to p53, *Genes Dev.* **8**:190–202.
164. White, E., Grodzicker, T., and Stillman, B. W., 1984, Mutations in the gene encoding the adenovirus E1B 19K tumor antigen cause degradation of chromosomal DNA, *J. Virol.* **52**:410–419.
165. Pidler, S., Logan, J., and Shenk, T., 1984, Deletion of the gene encoding the adenovirus 5 early region 1B—21,000-molecular weight polypeptide leads to degradation of viral and cellular DNA, *J. Virol.* **52**:664–671.
166. Takemori, N., Cladaras, C., Bhat, B., Conley, A. J., and Wold, W. S., 1984, Cyt gene of adenovirus 2 and 5 is an oncogene for transforming function in early region E1B and encodes the E1B 19,000 molecular-weight polypeptide, *J. Virol.* **52**:793–805.
167. White, E., and Stillman, B., 1987, Expression of the adenovirus E1B mutant phenotype is dependent on the host cell and on synthesis of E1A proteins, *J. Virol.* **61**:426–435.
168. White, E., Cipriani, R., Sabbatini, P., and Denton, A., 1991, The adenovirus, E1B 19-kilodalton protein overcomes the cytotoxicity of E1A proteins, *J. Virol.* **65**:2968–2978.
169. Debbas, M., and White, E., 1993, Wild-type p53 mediates apoptosis by E1A, which is inhibited by E1B, *Genes Dev.* **7**:546–554.
170. Beutler, B., and Cerami, A., 1986, Cachectin and tumor necrosis factor as two sides of the same biological coin, *Nature* **320**:584–588.
171. Chen, M.-J., Holskin, B., Strickler, J., Gorniak, J., Clark, M. A., Johnson, P. J., Mitcho, M., and Shalloway, D., 1987, Induction by E1A oncogene expression of cellular susceptibility to lysis by TNF, *Nature* **330**:581–583.

172. Wold, W. S., 1993, Adenovirus genes that modulate the sensitivity of virus infected cells to lysis by TNF, *J. Cell. Biochem.* **53**:329–335.
173. Hashimoto, S., Ishii, A., and Yonehara, S., 1991, The E1B oncogene of adenovirus confers cellular resistance to cytotoxicity of tumor necrosis factor and monoclonal antibodies anti-Fas antibody, *Int. Immunol.* **3**:343–351.
174. White, E., Sabbatini, P., Debbas, M., Wold, W. S., Kusher, D. I., and Gooding, L., 1992, The 19-kilodalton adenovirus E1B transforming protein inhibits programmed cell death and prevents cytolysis by tumor necrosis factor α, *Mol. Cell. Biol.* **12**:2570–2580.
175. Morgenbesser, S. D., Williams, B. O., Jacks, T., and DePinho, R. A., 1994, p53-dependent apoptosis produced by Rb-deficiency in the developing mouse lens, *Nature* **371**:72–74.
176. Pan, H., and Griep, A. E., 1994, Altered cell cycle regulation in the lens of HPV-16 E6 or E7 transgenic mice: Implications for tumor suppressor gene function in development, *Genes Dev.* **8**:1285–1299.
177. White, E., 1994, p53 guardian of Rb, *Nature* **371**:21–22.
178. Schrier, P. I., Bernards, R., Vaessen, R. T. M. J., Houweling, A., and van der Eb, A. J., 1983, Expression of class I major histocompatibility antigens switched off by highly oncogenic adenovirus 12 in transformed rat cells, *Nature* **305**:771–775.
179. Bernards, R., Schrier, P. I., Houweling, A., Bos, J. L., and van der Eb, A. J., 1983, Tumorigenicity of cells transformed by adenovirus type 12 by evasion of T-cell immunity, *Nature* **305**:776–779.
180. Eager, K. B., Williams, J., Breiding, D., Pan, S., Knowles, B., Appella, E., and Ricciardi, R. P., 1985, Expression of histocompatibility antigens H-2K, -D, and -L is reduced in adenovirus-12-transformed mouse cells and is restored by interferon-γ, *Proc. Natl. Acad. Sci. USA* **82**:5525–5529.
181. Yewdell, J. W., Bennink, J. R., Eager, K. B., and Ricciardi, R. P., 1988, CTL recognition of adenovirus transformed cells infected with influenza virus: Lysis by anti-influenza CTL parallels adenovirus-12-induced suppression of class I MHC molecules, *Virology* **162**:236–238.
182. Tanaka, K., Isselbacher, K. J., Khoury, G., and Jay, G., 1985, Reversal of oncogenesis by the expression of a major histocompatibility complex class I gene, *Science* **228**:26–30.
183. Hayashi, H., Tanaka, K., Jay, F., Khoury, G., and Jay, G., 1985, Modulation of the tumorigenicity of human adenovirus-12-transformed cells by interferon, *Cell* **43**:263–267.
184. Tanaka, K., Hayashi, H., Hamada, C., Khoury, G., and Jay, G., 1986, Expression of major histocompatibility complex class I antigens as a strategy for the potentiation of immune recognition of tumor cells, *Proc. Natl. Acad. Sci. USA* **83**:8723–8727.
185. Bernards, R., and van der Eb, A., 1984, Adenovirus transformation and oncogenicity, *Biochim. Biophys. Acta* **783**:187–204.
186. Harwood, L., and Gallimore, P., 1975, A study of the oncogenicity of adenovirus type 2 transformed rat embryo cells, *Int. J. Cancer* **16**:498–508.
187. Gallimore, P. H., and Paraskeva, C., 1979, A study to determine the reasons for differences in the tumorigenicity of rat cell lines transformed by adenovirus 2 and adenovirus 12, *Cold Spring Harbor Symp. Quant. Biol.* **44**:703–713.
188. Shiroki, K., Shimojo, H., Maeta, Y., and Hamada, C., 1979, Tumor-specific transplantation and surface antigen in cells transformed by the adenovirus 12 DNA fragments, *Virology* **99**:188–191.
189. Gallimore, P. H., and Williams, J., 1982, An examination of adenovirus type 5 mutants for their ability to induce group C adenovirus tumor-specific transplantation antigenicity in rats, *Virology* **120**:146–156.
190. Bellgrau, D., Walker, T. A., and Cook, J. L., 1988, Recognition of adenovirus E1A gene products on immortalized cell surfaces by cytotoxic T lymphocytes, *J. Virol.* **62**:1513–1519.
191. Rawle, F. C., Knowles, B. B., Ricciardi, R. P., Brahmacheri, V., Duerksen-Hughes, P., Wold, W. S., and Gooding, L. R., 1991, Specificity of the mouse cytotoxic T lymphocyte response to adenovirus 5, *J. Immunol.* **146**:3977–3984.
192. Urbanelli, D., Sawada, K., Raskova, J., Jones, N. C., Shenk, T., and Raska, K., 1989, C-terminal domain of the adenovirus E1A oncogene product is required for induction of cytotoxic T lymphocytes and tumor-specific transplantation immunity, *Virology* **173**:607–614.

193. Kast, W. M., Offringa, R., Peters, P. J., Voordouw, A. C., Meloen, R. H., van der Eb, A. J., and Melief, C. J. M., 1989, Eradication of adenovirus E1-induced tumors by E1A-specific cytotoxic T lymphocytes, *Cell* **59**:603–614.
194. Sawada, Y., Fohring, B., Shenk, T. E., and Raska, K., 1985, Tumorigenicity of adenovirus-transformed cells: Region E1A of adenovirus 12 confers resistance to natural killer cells, *Virology* **147**:413–421.
195. Cook, J. L., May, D., Lewis, A. M., and Walker, T. A., 1987, Adenovirus E1A gene induction of susceptibility to lysis by natural killer cells and activated macrophages in infected rodent cells, *J. Virol.* **61**:3510–3520.
196. Cook, J. L., and Lewis, A. M., 1984, Differential NK cell and macrophage killing of hamster cells infected with nononcogenic adenovirus, *Science* **224**:612–615.
197. Kenyon, D. J., Doughtery, J., and Raska, K., 1991, Tumorigenicity of adenvorius-transformed cells and their sensitivity to tumor necrosis factor alpha and NK/LAK cell cytolysis, *Virology* **180**:818–821.
198. Vasavada, R., Eager, K. B., Barbanti-Brodano, G., Caputo, A., and Ricciardi, R. P., 1986, Adenovirus type 12 early region 1A proteins repress class I HLA expression in transformed human cells, *Proc. Natl. Acad. Sci. USA* **83**:5257–5261.
199. Bernards, R., Vaessen, M. J., van der Eb, A. J., and Sussenbach, J. S., 1983, Construction and characterization of an adenovirus type 5/adenovirus type 12 recombinant virus, *Virology* **131**:30–38.
200. Sawada, Y., Raska, K., and Shenk, T., 1988, Adenovirus type 5 and adenovirus type 12 recombinant viruses containing heterologous E1 genes are viable, transform cells, but are not tumorigenic in rats, *Virology* **166**:281–284.
201. Jochemsen, A. G., de Wit, C. M., Bos, J. L., and van der Eb, A. J., 1986, Transforming properties of a 15-kDa truncated Ad12 E1A gene product, *Virology* **152**:375–383.
202. Jochemsen, A. G., Bos, J. L., and van der Eb, A. J., 1984, The first exon of region E1A genes of adenovirus 5 and 12 encodes a separate functional protein domain, *EMBO J.* **3**:2923–2927.
203. Pereira, D. S., Jelinek, T., and Graham, F. L., 1994, The adenovirus E1A-associated p300 protein is differentially phosphorylated in Ad12 E1A-compared to AD5 E1A-transformed rat cells, *Int. J. Oncol.* **5**:1197–1205.
204. Jelinek, T., Pereira, D. S., and Graham, F. L., 1994, Tumorigenicity of adenovirus-transformed cells is influenced by at least two regions of adenovirus type 12 early region 1A, *J. Virol.* **68**:888–896.
205. Telling, G. C., and Williams, J., 1994, Constructing chimeric type12/type5 adenovirus E1A genes and using them to identify an oncogenic determinant of adenovirus type 12, *J. Virol.* **68**:877–887.
206. Vasavada, R., and Ricciardi, R. P., 1994, A nonconserved region of adenovirus type 12 E1A is required for down-regulation of MHC class I transcription in transformed cells, unpublished.
207. Friedman, D. J., and Ricciardi, R. P., 1988, Adenovirus type 12 E1A gene represses accumulation of MHC class I mRNA at the level of transcription, *Virology* **165**:303–305.
208. Ackrill, A. M., and Blair, G. E., 1988, Regulation of major histocompatibility class I gene expression at the level of transcription in highly oncogenic adenovirus transformed rat cells, *Oncogene* **3**:483–487.
209. Ge, R., Kralli, A., Weinmann, R., and Ricciardi, R. P., 1992, Down-regulation of the major histocompatibility complex class I enhancer in adenovirus type 12-transformed cells is accompanied by an increase in factor binding, *J. Virol.* **66**:6969–6978.
210. Kralli, A., Ge, R., Graeven, U., Ricciardi, R. P., and Weinmann, R., 1992, Negative regulation of the major histocompatibility complex class I enhancer in adenovirus type 12-transformed cells via a retinoic acid response element, *J. Virol.* **66**:6979–6988.
211. Liu, X., Ge, R., Westmoreland, S., Cooney, A. J., Tsai, S. Y., Tsai, M.-J., and Ricciardi, R. P., 1994, Negative regulation by the R2 element of the MHC class I enhancer in adenovirus-12 transformed cells correlates with high levels of COUP-TF binding, *Oncogene* **9**:2183–2190.

212. Liu, X., Ge, R., and Ricciardi, R. P., 1996, Evidence for the involvement of a nuclear NF-κB inhibitor in global down-regulation of the MHC class I enhancer in Ad12-transformed cells, *Mol. Cell. Biol.*, in press.
213. Nielsch U., Zimmer, S. G., and Babiss, L. E., 1991, Changes in NK-κB and ISGF3 DNA binding activities are responsible for differences in MHC and β-IFN gene expression in Ad5-versus Ad12-transformed cells, *EMBO J.* **10:**4169–4175.
214. Meijer, I., Boot, A. J. M., Mahabir, G., Zantema, A., and van der Eb, A. J., 1992, Reduced binding activity of transcription factor NF-kappa-B accounts for MHC class I repression in adenvorius type 12 E1-transformed cells, *Cell. Immunol.* **145:**56–65.
215. Ackrill, A. M., and Blair, G. E., 1989, Nuclear proteins binding to an enhancer of the major histocompatibility class I promoter: Differences between highly oncogenic and nononcogenic adenovirus transformed rat cells, *Virology* **172:**643–646.
216. Kimura, A., Israel, A., Bail, O. L., and Kourilsky, P., 1986, Detailed analysis of the mouse H-2K^b promoter: Enhancer-like sequences and their role in the regulation of class I gene expression, *Cell* **44:**261–272.
217. Hamada, K., Gleason, S. L., Levi, B., Hirschfeld, S., Appella, E., and Ozato, K., 1989, H-2RIIBP, a member of the nuclear hormone receptor superfamily that binds to both the regulatory element of major histocompatibility class I genes and the estrogen response element, *Proc. Natl. Acad. Sci. USA* **86:**8289–8293.
218. Segars, J. H., Nagata, T., Bours, V., Medin, J. A., Franzoso, G., Blanco, J. C. G., Drew, P. D., Becker, K. G., An, J., Tang, T., Stephany, D. A., Neel, B., Siebenlist, U., and Ozato, K., 1993, Retinoic acid induction of major histocompatibility complex class I genes in N tera-2 embryonal carcinoma cells involves induction of NK-kappa B (P50-P65) and retinoic acid receptor beta–retinoid × receptor beta heterodimers, *Mol. Cell. Biol.* **13:**6157–5169.
219. Wang, L.-H., Ing, N. H., Tsai, S. Y., O'Malley, B. W., and Tsai, M.-J., 1991, The COUP-TFs compose a family of functionally related transcription factors, *Gene Expr.* **1:**207–216.
220. Kushner, D. B., Pereira, D. S., Liu, X., Graham, F. L., and Ricciardi, R. P., 1996, The first exon of Ad12 E1A excluding the transactivation domain mediates differential binding of COUP-TF and NF-κB to the MHC class I enhancer in transformed cells, *Oncogene*, in press.
221. Andersson, M., Paabo, S., Nilsson, T., and Peterson, P. A., 1985, Impaired intracellular transport of class I MHC antigens as a possible means for adenoviruses to evade immune surveillance, *Cell* **43:**215–222.
222. Burgert, H., and Kvist, S., 1985, An adenovirus type 2 glycoprotein blocks cell surface expression of human histocompatibility class I antigens, *Cell* **41:**987–997.
223. Cox, J. H., Bennink, J. R., and Yewdell, J., 1991, Retention of adenovirus E19K glycoprotein in the endoplasmic reticulum is essential to its ability to block antigen presentation, *J. Exp. Med.* **174:**1629–1637.
224. Burgert, H., Maryanski, J. L., and Kvist, S., 1987, "E3/19K" glycoprotein of adenovirus type 2 glycoprotein inhibits lysis of cytotoxic T lymphocytes by blocking cell surface expression of histocompatibility class I antigens, *Proc. Natl. Acad. Sci. USA* **84:**1356–1360.
225. Horwitz, M. S., 1990, Adenoviridae, and their replication, in: *Virology*, Vol. 2 (B. Fields and D. Knipe, eds.), Raven Press, New York, pp. 1679–1721.
226. Cousin, C., Winter, N., Gomes, S. A., and D'Halluin, J. C., 1991, Cellular transformation by E1 genes of enteric adenoviruses, *Virology* **181:**277–287.
227. Davidson, A. J., Telford, E. A. R., Watson, M. S., McBride, K., and Mautner, V., 1993, The DNA sequence of adenovirus type 40, *J. Mol. Biol.* **234:**1308–1316.
228. Paabo, S., Nilsson, T., and Peterson, P. A., 1986, Adenoviruses of subgenera B, C, D, and E modulate cell-surface expression of major histocompatibility complex I antigens, *Proc. Natl. Acad. Sci. USA* **83:**9665–9669.

12

Current Developments in the Molecular Biology of Marek's Disease Virus

MEIHAN NONOYAMA† and AKIKO TANAKA

1. INTRODUCTION

Marek's disease virus (MDV) is an avian herpesvirus that induces lymphoproliferative disease in chicken, its natural host.[1,2] The MDV once was a threat to the poultry industry because of the great economic damage caused by destruction of chicken flocks by infection. However, in the 1970s, vaccination of chickens by use of the nononcogenic MDV serotype II or serotype III (herpesvirus of turkey, HVT) or by attenuated serotype I strain of oncogenic MDV effectively prevented the spread of Marek's disease (MD) in chicken flocks. Although vaccination solved much of the problem caused by MDV-induced disease, vaccine breaks still occur, and economic loss to the poultry industry currently amounts to over $100 million annually.[3]

From the immunologic point of view, vaccination to prevent MDV infection is a very interesting and important model.[4] Vaccination is effective both by enhanced elicitation of cytotoxic T cells against tumor cells as well as by humoral immunity against MDV. Much important research has taken place in this field. In contrast, the molecular biology study of MDV, i.e., the characterization of the virus and understanding the mechanism of tumor induction, has been somewhat slow. Recently, there have been a number of reports on the viral genes responsible for productive infection as well as on the potential tumor-inducing genes of MDV. Comparison of the oncogenic mechanism of MDV with that of Epstein–Barr virus (EBV) is also

†Deceased.

MEIHAN NONOYAMA and AKIKO TANAKA • Tampa Bay Research Institute, St. Petersburg, Florida 33716.

DNA Tumor Viruses: Oncogenic Mechanisms, edited by Giuseppe Barbanti-Brodano *et al.* Plenum Press, New York, 1995.

important because both viruses cause lymphoproliferative disease by infection. Thus, in this chapter, recent developments on the molecular aspects of the MDV genome and gene expression are emphasized and discussed.

2. GENOMIC STRUCTURE OF MDV DNA SEROTYPE I

Marek's disease virus is an avian herpesvirus that induces T-cell lymphoma. The establishment of a lymphoblastoid cell line containing the MDV genome is possible from tumor tissue, but *in vitro* immortalization has not been successful. Because MDV is lymphotropic, it has previously been classified as a γ-herpesvirus.[5] However, with the aid of electron microscopy[6] and by mapping with restriction enzymes,[7] MDV has been classified more properly as an α-herpesvirus; i.e., it contains the short and long terminal and inverted repeats as well as unique regions TR_L-U_L-IR_L-IR_S-U_S-TR_S.

The restriction mapping of HVT DNA has also been established.[8] The structures of HVT and MDV DNA resemble those of typical α-herpesviruses such as herpes simplex virus (HSV) types 1 and 2. Reciprocal hybridization analysis of MDV and HVT showed collinearity within the unique regions only and not within the inverted repeat regions.[8,9] A detailed analysis of the junction area contained in IR_L-IR_S revealed a specific sequence (GGGTTA) in both MDV and HVT DNA[10] that is also the terminal sequence of the human chromosome telomere.[11,12] This telomeric sequence was also detected in human herpesvirus 6 (HHV-6) DNA,[13] which belongs to the β-herpesvirus subfamily. The junction sequence in IR_L-IR_S consists of an "a"-like sequence as found in HSV DNA. In the case of HSV, this "a" repeat sequence is found in the junction of IR_L and IR_S, and the "a" sequence of the junction region of HSV DNA plays a role in the cleavage of concatameric genomes and encapsidation of virion DNA.[14] In the case of MDV, the "a"-like sequence consists of $DR1$-U_b-$(DR2)_{26}$ $(DR4)_2$-U_c-$DR1$. The junction region contains ten units of "a"-like sequence, and the terminal region contains five units.[10] The DR2 sequence contains the GGGTTA telomeric motif. It is hypothesized that this sequence may be functionally homologous to the DR2 sequence of HSV, which is a repeated component of the "a"-like sequence of the junction region, and therefore may play a role in the cleavage of concatameric genomes.

A unique feature of MDV DNA is a promoter region for the H gene family, which is located in the IR_L region of BamHI-H. Analysis of the DNA structure in the promoter region revealed that it resembled the structure of ori-lyt of HSV DNA.[15] A DNA transfection experiment demonstrated that on transfection of this DNA fragment, replication of DNA was inhibited.[16] It was also found that the MDV DNA nuclear antigen (MDNA) binding region is located in subfragment EcoRI-c of BamHI-A in the U_S region.[17] This region may coincide with the replication origin (ori-p) of the MDV genome, which exists in plasmid form in latently infected cells.[18]

3. INTERACTION WITH RETROVIRUS

A number of reports have described a potential relationship between MDV and avian leukosis virus (ALV).[19–21] Normally, enhanced expression of ALV antigen and ALV-associated RNA were observed in infected birds. Some reports have suggested that enhanced complete manifestation of MD requires infection by both MDV and ALV. However, Calneck and Payne[22] reported a lack of correlation between MD tumor induction and expression of endogenous ALV genome. A recent report concerning enhanced incidence of ALV-associated lymphoid leukosis (LL) following vaccination of chickens with a bivalent MDV preparation, particularly with MDV serotype II, drew attention to such MDV–ALV interaction.[23] In conjunction with the enhancement of LL in chicken by serotype II MDV vaccination by a bivalent vaccine, there is a report that the Rous sarcoma virus (RSV) long terminal repeat (LTR) is significantly enhanced by MDV infection,[24] and enhancement is at a much higher rate with serotype II than with serotype I or III. These *in vivo* observations, together with the *in vitro* demonstration of enhanced transactivation of RSV LTR by MDV infection, clearly established the fact that MDV infection can enhance ALV expression. Further, it was also demonstrated that a 20- to 28-bp region of the RSV LTR containing a pentanucleotide repeat element was the target sequence for the MDV *trans*-activation factor.[25] This MDV factor has not been identified thus far. Such interactions of herpesviruses and retroviruses may occur in human pathogenesis, and this system serves as a good model system for the understanding of the interaction between these two virus families. Such interaction in a human environment has been demonstrated between HSV or cytomegalovirus (CMV) and human immunodeficiency virus and human T-cell leukemia virus, i.e., *trans*-activation of these retrovirus promoters by herpesviruses.[26,27]

The interaction of MDV with ALV is not limited to *trans*-activation. Indeed, recently Jones *et al.*[28] demonstrated the integration of retrovirus DNA into MDV DNA by Southern blot hybridization and DNA sequence analysis. Specifically, they provided evidence for the insertion of reticuloendotheliosis virus (REV) DNA into the MDV genome. A sequence homologous to REV LTR was found in the MDV genome. Sequence analysis revealed that REV LTR was located in the junction between U_S and IR_S or TR_S of the MDV genome. Coinfection studies of MDV and REV indicated that REV LTR became integrated into MDV DNA, and the specific area of integration was again in the junction region between U_S and TR_S.[29] Such integration was observed only for the LTR region of REV and not for other regions. They also found no evidence for a similar relationship with other avian retroviruses. These observations may raise the possibility that integration of REV LTR into MDV DNA may enhance an MDV gene for tumorigenicity. It is interesting to note that REV LTR is already integrated into MDV DNA in nature.

4. MAREK'S DISEASE VIRUS GENE EXPRESSION IN LYTICALLY INFECTED CELLS

It is known that there are two antigens expressed, A and B, in MDV-infected cell cultures. When pathogenic MDV is serially passaged in chick embryo fibroblast (CEF) cultures, the virus is attenuated and is no longer able to induce tumors in chicken. The highly passaged attenuated virus does not express A antigen.[30] The A antigen is a glycoprotein of 56–65 kDa[31,32] and has been mapped to the BamHI-B fragment of the U_L region.[33] The amino acid sequence as deduced by open reading frames (ORFs) has a significant homology at the C-terminal regions with

HSV-1 gC[34] and varicella–zoster virus (VZV) gpV.[35] The virus-neutralizing B antigen was located in the BamHI-I_3-K_3 region of the MDV genome and was a leftward transcript.[36,37] The B antigen is a homologue of HSV gB.[38] Fowlpox virus expressing the B antigen of MDV elicited neutralizing antibodies against MDV and protected chickens against challenge with oncogenic MDV.[39]

BamHI-D in U_L region contains a novel spliced ORF that also reacts with B-antigen-specific monoclonal antibody.[40] A phosphoprotein, pp38, originally detected by Silva and Lee[41] and Ikuta et al.,[42] has been mapped to BamHI-H from IR_L to U_L as a leftward transcript.[43] This transcript and proteins were detected in both productively and latently infected cells. The function of this protein is not known. A phosphorylated pp24 protein sharing the same N-terminal polypeptide as pp38 has been found in BamHI-D from TR_L to U_L region.[44,45]

As previously mentioned, when the pathogenic virus was serially passaged in CEF culture, the virus lost its oncogenicity and became attenuated. Studies have indicated that the pattern of restriction enzyme fragments of MDV DNA of oncogenic and attenuated virus strains differed in BamHI-D and -H. Specifically, in BamHI-D and -H within the TR_L and IR_L regions, there was a 132-bp repeat unit existing as two units in pathogenic MDV DNA, whereas the attenuated DNA showed 10–100 units of the 132-bp repeat.[46–49] There is a strong correlation between oncogenicity and the number of 132-bp repeat units. If this repeat unit is amplified, the virus will not induce lymphoma in chicken. Studies were focused on the relationship of gene expression within BamHI-H region to amplification of the 132-bp repeat.

First, the expression of pathogenic virus DNA was examined by Northern blot analysis. There are three major transcripts thus far identified: 1.8-, 3.0-, and 3.8-kb RNA.[50] A cDNA library was constructed with CEF infected with pathogenic MDV. The library was screened, and the isolated cDNAs were classified by size, location in BamHI-H and -I_2 fragments of MDV genomic DNA, and sequence analysis. Sequence analysis indicated that the isolated cDNAs were derived from rightward transcripts of the BamHI-H gene family. Mapping of the H gene family transcripts by isolated cDNAs is presented in Fig. 1. Two species of cDNA were isolated that belong to 1.8-kb RNA: (1) 1.69 kb, which has no splicing within the 132-bp repeat unit,[51] and (2) 1.5 kb, which has the same promoter and termination sequence but has a splice site within the 132-bp repeat unit and its termination site extended to the adjacent BamHI-I_2 fragment. Isolated 1.92-kb cDNA shares the same termination site as the 1.5-kb cDNA, but this cDNA clone is only a fragment of the 3.0-kb transcript. Isolated 2.2-kb cDNA corresponded to the largest size unspliced transcript, 3.8-kb RNA. Again, this cDNA shares a 3′-termination site with the previous cDNAs. Surrounding the 132-bp repeat unit in BamHI-H were four types of transcripts that share the same promoter and termination site, except for the unspliced 1.8-kb transcript termination site.

This family of transcripts is unique to pathogenic strains of MDV because (1) in the case of attenuated virus, which has amplification of the 132-bp repeat, this H gene family transcription (1.8-, 3.0-, and 3.8-kb mRNAs) was abolished, and instead only 0.4-kb mRNA was detected[15]; and (2) mapping of the 0.4-kb mRNA transcript revealed that this transcript had the same promoter and initiation site and a new termination site about 117 bp downstream of the initiation site.[15] Amplification

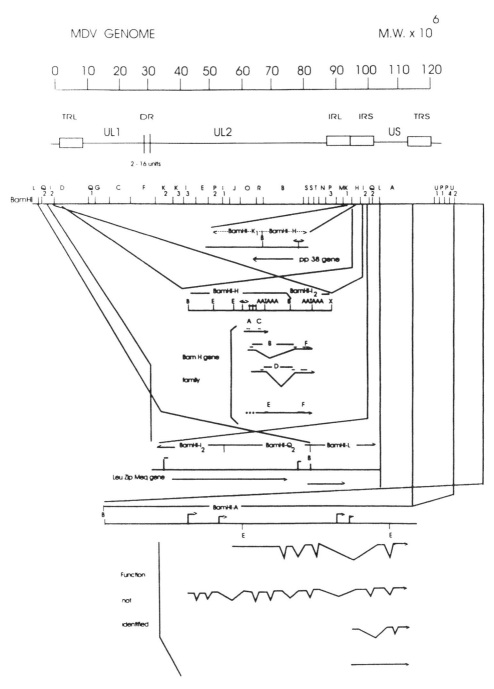

FIGURE 1. Summary of latency-associated viral transcripts. Bent arrow, potential promoter; open rectangles, 132-bp repeat.

of the 132-bp repeat caused premature termination of H gene family transcription and resulted in formation of 0.4-kb mRNA. The region surrounding the 3' end of 0.40 kb RNA contains a sequence that is capable of forming stem–loop structures. A poly(A) signal and GT-rich sequence present in the same region may have created a new termination site for those H gene transcripts when the 132-bp repeat is amplified.

Premature termination, which generates truncated transcripts, has been known to be a regulatory mechanism for genes in both prokaryotic and eukaryotic systems. In the case of c-*myc* and SV40 genes, the leader sequence can exist in two alternative configurations.[52,53] These configurations consist of either a single stem–loop structure, which allows readthrough, or a pair of stem–loop structures, which participate in transcription termination.[54] The termination of transcription in these genes is found to occur on the 3' side of the second stem–loop structure at a stretch of oligo(U). It may be possible that DNA amplification in this region induces a change in the DNA structure, leading to (or directly responsible for) truncation of the transcript. Amplification of the 132-bp direct repeat may induce stress within the DNA molecule, which is relieved by formation of a stem–loop structure. On the other hand, amplification may relieve stress in the DNA molecule and prevent formation of the stem–loop structure, thereby exposing a poly(A) signal and a GT-rich region, which are required for efficient formation of mRNA 3'-termini. The precise mechanism of premature termination is not known; however, it is interesting to observe that this type of gene regulation strongly correlates with virus pathogenicity.

Thus far six small ORFs have been identified within this H gene family of transcripts by sequence analysis.[55] However, which ORFs are expressed in pathogenic virus-infected CEF is not known. ORF A of the H gene family (BHa) has been studied to a certain extent concerning growth promotion and expression in infected cells. A recent study indicated that BHa gene is expressed in both infected CEF and immortalized lymphoblastoid cells as a 7-kDa protein, as predicted from the ORF.[56] It is worth mentioning that BHa protein is immunoprecipitated with a 37-kDa cellular protein in productively infected cells, whereas in lymphoblastoid cells such 37-kDa immunoprecipitation was not observed. This suggests that BHa protein plays a different role in productively infected cells and latently infected cells. When BHa gene is placed under CMV promoter, it confers growth promotion to transfected CEF cells, although CEF cells could not be immortalized, suggesting that BHa is capable of promoting growth in transfected cells.[50] Transformation of NIH3T3 cells with BHa gene was not successful.

Virus gene expression in infected cells can be classified as immediate early (IE), early (E), and late (L). Generally, the cycle of virus infection is completed at the time the progeny virus is secreted into the culture medium. However, mature virus (cell-associated virus) remains in the cells in the MDV system. Because of this phenomenon, it is difficult to classify the above three types of viral transcripts in infected cells. Nevertheless, some studies have attempted to classify the expression of transcripts by using cell-associated virus-infected cells.[57,58] Immediate early transcripts were found in BamHI-A (IR_S and TR_S), -H, and -D (IR_L and TR_L) as well as -F (U_L). A homologue to ICP4 of HSV, an immediate early regulatory protein of HSV-1 and -2, has also been reported in the IR_S region of BamHI-A.[59]

5. VIRAL GENE EXPRESSION IN LATENTLY INFECTED CELLS

Lymphoblastoid cell lines can be established from MDV-induced lymphoma tumor tissue. A number of established lymphoblastoid cell lines are currently available. Some of the cell lines do not produce virus and exist as nonproductive cell lines. In these lymphoblastoid cell lines, virus DNA exists in a latent form,[18] and the viral genes are expressed only in a restricted manner.[60] An early study indicated that viral DNA in latently infected cell lines exists as a circular plasmid form and is not integrated into the host cell DNA.[18] Existence of virus DNA as a plasmid form was rather novel when an EBV-transformed B-lymphoblastoid cell line was first examined,[61] and it was somewhat surprising because the genome of most oncogenic DNA viruses, as exemplified by SV40, is typically integrated into the host cell DNA. An experiment using synchronized lymphoblastoid cells revealed that, although the virus DNA exists as a plasmid form, these viruses replicate under the control of cellular DNA replication machinery at the early S phase.[62] A recent study reported that MDV DNA is also present in integrated form in some latently infected cell lines.[63]

In the EBV system, EBV nuclear antigen 1 (EBNA-1) plays a role in stabilizing plasmid viral DNA in latently infected cells, as it binds to the ori-p region of EBV.[64] We have identified a nuclear antigen for MDV in the T-lymphoblastoid cell lines MKT and MSB, with a mass of 28 kDa.[17] The MDNA binds a subfragment of BamHI-A, subfragment EcoRI-c, which is located in the U_S region. Whether this EcoRI-c binding region is the ori-p binding region is not known, and the function of MDNA is not presently understood.

Lymphoblastoid cells latently infected by MDV express virus genes at a restricted level. For example, in the case of the MKT cell line, DNA–RNA reassociation kinetics showed that MDV genes are expressed at a level of 10–15% of the virus genome.[65] In a recent study by Sugaya et al.,[66] the transcriptional activity of MDV in MKT cells was restricted to TR_L and IR_L and TR_S and IR_S. The H gene family transcripts, which are unique to the pathogenic strain of MDV, were expressed in latently infected cells, but not to the extent detected in productively infected cells. The most abundant transcripts among the expressed regions were BamHI-I_2 in IR_L and TR_L and BamHI-A in IR_S and TR_S, which are expressed to a certain degree. Based on size determinations, some of the transcripts in BamHI-I_2 were present in lytically infected cells; however, several viral transcripts were uniquely found in latently infected cells. On the other hand, viral expression in IR_S and TR_S were commonly found in both lytic and latently infected cells.

The highlight of virus transcripts in latently infected cells could be summarized as follows. The most actively transcribed region is BamHI-I_2, which is located in IR_L and TR_L. In the right-half region of BamHI-I_2 fragment, transcripts in both directions were observed to be abundant by Northern blot hybridization. Also, other transcripts were made in both the leftward and rightward directions from the left-half region of BamHI-I$_2$, in addition to the previously mentioned transcript of BamHI-H gene. At the right-end half of BamHI-I$_2$ region, the rightward transcript encodes a protein with a leucine zipper and a DNA-binding motif as identified by means of its ORF. The mass of this protein is 40 kDa, and a recent study by Jones et al.[67] indicates that this gene encodes a protein of 362 amino acids that has a leucine zipper sequence and an upstream domain rich in basic amino acids; both such motifs are characteristic of the *fos/jun* family of transcriptional activators. These authors have developed an antiserum against the synthetic peptide deduced from the DNA sequence and were able to demonstrate the presence of the 40-kDa protein in latently infected cells.

cDNA cloning has led to the isolation of cDNAs hybridizing to 2.5-, 0.8-, and 0.6-kb transcripts,[68] spanning from BamHI-Q$_2$ to -L within IR$_L$ and TR$_L$, near to the junction between IR$_S$ and IR$_L$. The 0.6-kb transcript contains an ORF of 107 amino acids, the function of which is not known. cDNA cloning in the IR$_S$ region gave a complicated pattern, although a 1.5-kb cDNA with an ORF of 94 amino acids has been reported,[69] and this ORF has been shown to be expressed in MDV-infected cells. So far there has been evidence for four different types of cDNA, varying in size from 1.5 to 2.3 kb, and they exhibit extensive splicing and a common terminal region close to the junction of IR$_S$ and U$_S$. This shows that most of the viral transcripts in latently infected cells in the IR$_S$ region are rightward and stop at the same terminal region, which is close to the junction of IR$_S$ and U$_S$.[70] In the IR$_S$ region, the HSV ICP4 homologue to MDV is known to exist as an ORF and is expressed in productively infected cells.[59] However, thus far, this gene has not been shown to be expressed in latently infected cells. The function of these transcripts remains to be shown in a future study.

6. CONCLUSION

In summary, MDV is a unique herpesvirus that induces lymphoma in chickens within 6–8 weeks after infection. An MDV infection causes immortalization of lymphocytes, leading to tumor induction. In the case of EBV, tumor induction and immortalization are clearly separate processes, and EBV gene expression in latently infected cells contributes only to the immortalization. Tumor induction is probably caused by translocation or mutation of c-*myc* or by another indirect mechanism.[71] In contrast, for the induction of MDV-related lymphoma, infection of chicken with MDV always results in lymphoma within 6–8 weeks, suggesting that MDV itself possesses a tumor-inducing gene. For that matter, expression of the MDV H gene family is unique to the pathogenic virus and is not found in the attenuated virus. Such gene expression may be expected to contribute to tumor induction. It should be noted that this gene family is expressed more strongly in productively infected cells than in latently infected cells, and recent experiments suggest that interaction of the H gene family products with cellular proteins may be different in the two systems and may trigger different effects in the cells. Other proteins such as a leucine zipper protein, which is encoded by BamHI-I$_2$, may play an important regulatory role in latently infected cells for tumor induction. Last, it is important to recall that MDV not only is tumorigenic but also induces atherosclerosis in chicken.[72] This characteristic is another important aspect of MDV. Recently, herpesvirus infection has been implicated in human atherosclerosis.[73] As MDV is indeed capable of inducing atherosclerosis, this may be an important indication for the relationship of the herpesvirus to atherosclerosis in humans. Thus, MDV may serve as a model system for the study of atherosclerosis.

REFERENCES

1. Churchill, A. E., and Biggs, P. M., 1967, Agent of Marek's disease in tissue culture, *Nature* **215**: 528–530.
2. Nazerian, K., and Burmester, B. R., 1968, Electron microscopy of a herpes-virus associated with the agent of Marek's disease in cell culture, *Cancer Res.* **28**:2454–2462.
3. Purchase, H. G., 1985, Clinical disease and its economic impact, in: *Marek's Disease Virus* (L. N. Payne, ed.), Martinus Nihjoff, Ithaca, NY, pp. 17–42.
4. Okazaki, W., Purchase, H. G., and Burmester, B. R., 1970, Protection against Marek's disease by vaccination with a herpesvirus of turkeys (HVT), *Avian Dis.* **14**:413–429.
5. Roizman, B., Carmichael, S., de Thé, G., Masic, M., Nahmias, A., Plowright, W., Rapp, F., Sheldrick, P., Takashi, M., Terni, M., and Wolfe, K., 1978, Provisional classification of herpesviruses, in: *Oncogenesis of Herpesviruses III, Part II* (G. de Thé, W. Henle, and F. Rapp, eds.), International Agency for Research on Cancer, Lyon, pp. 1079–1082.
6. Cebrian, J., Kaschka-Dietrich, C., Berthelot, N., and Sheldrick, P., 1982, Inverted repeat nucleotide sequences in the genomes of Marek disease virus and the herpesvirus of the turkey, *Proc. Natl. Acad. Sci. USA* **79**:555–558.
7. Fukuchi, K., Sudo, M., Lee, Y.-S., Tanaka, A., and Nonoyama, M., 1984, Structure of Marek's disease virus DNA: Detailed restriction enzyme map, *J. Virol.* **51**:102–109.
8. Igarashi, T., Takahashi, M., Donovan, J., Jessip, J., Smith, M., Kirai, K., Tanaka, A., Nonoyama, M., 1987, Restriction enzyme map of herpesvirus of turkey DNA and its collinear relationship with Marek's disease virus DNA, *Virology* **157**:351–358.
9. Sakaguchi, M., Urakawa, T., Hirayama, Y., Miki, N., Yamamoto, M., and Hirai, K., 1992, Sequence determination and genetic content of an 8.9-kb restriction fragment in the short unique region and the internal inverted repeat of Merek's disease virus type 1 DNA, *Virus Genes* **6**:365–378.
10. Kishi, M., Bradley, G., Jessip, J., Tanaka, A., and Nonoyama, M., 1991, Inverted repeat regions of Marek's disease virus DNA possess a structure similar to that of the *a* sequence of herpes simplex virus DNA and contain host cell telomere sequences, *J. Virol.* **65**:2791–2797.
11. Roberts, L., 1988, Chromosomes: The ends in view, *Science* **240**:982–983.
12. Moyzis, R. K., Buckingham, J. M., Cram, L. C., Dani, M., Deaven, L. L., Jones, M. D., Meyne, J., Ratliff, R. L., and Wu, J.-R., 1988, A highly conserved repetitive DNA sequence, $(TTAGGG)_n$, present at the telomeres of human chromosomes, *Proc. Natl. Acad. Sci. USA* **85**:6622–6626.
13. Kishi, M., Harada, H., Takahashi, M., Tanaka, A., Hayashi, M., Nonoyama, M., Josephs, S. F., Buchbinder, A., Schachter, F., Ablashi, D. V., Wong-Staal, F., Salahuddin, S. Z., and Gallo, R. C., 1988, a repeat sequence, GGGTTA, is shared by DNA of human herpesvirus 6 and Marek's disease virus, *J. Virol.* **62**:4824–4827.
14. Mocarski, E. S., and Roizman, B., 1982, Structure and role of the herpes simplex virus DNA termini in inversion, circularization and generation of virion DNA, *Cell* **31**:89–97.
15. Bradley, G., Hayashi, M., Lancz, G., Tanaka, A., and Nonoyama, M., 1989, Structure of the Marek's disease virus BamHI-H gene family: Genes of putative importance for tumor induction, *J. Virol.* **63**:2534–2542.

16. Morgan, R. W., Cantello, J. L., Claessens, J. A. J., and Sondermeyer, P., 1991, Inhibition of Marek's disease virus DNA transfection by a sequence containing an alphaherpesvirus origin of replication and flanking transcriptional regulatory elements, *Avian Dis.* **35:**70–81.
17. Wen, L. T., Tanaka, A., and Nonoyama, M., 1988, Identification of Marek's disease virus nuclear antigen in latently infected lymphoblastoid cells, *J. Virol.* **62:**3764–3771.
18. Tanaka, A., Silver, S., and Nonoyama, M., 1978, Biochemical evidence of the nonintegrated status of Marek's disease virus DNA in virus-transformed lymphoblastoid cells of chicken, *Virology* **88:**19–24.
19. Peters, W. P., Kufe, D., Schlow, J., Frankel, J. W., Prickett, C. O., Groupe, V., and Spiegelman, S., 1973, Biological and biochemical evidence for an interaction between Marek's disease herpesvirus and avian leukosis virus *in vivo, Proc. Natl. Acad. Sci. USA* **70:**3175–3178.
20. Campbell, W. F., Kufe, D. W., Peters, W. P., Spiegelman, S., and Frankel, J. W., 1978, Contrasting characteristics of Marek's disease herpesvirus isolated from chickens with and without avian leukosis virus infection, *Intervirology* **10:**11–23.
21. Campbell, W. F., and Frankel, J. W., 1979, Enhanced oncornavirus expression in Marek's disease tumors from specific-pathogen-free chickens, *J. Natl. Cancer Inst.* **62:**323–328.
22. Calneck, B. W., and Payne, L. N., 1976, Lack of correlation between Marek's disease tumor induction and expression of endogenous avian RNA tumor virus genome, *Int. J. Cancer* **17:**235–244.
23. Bacon, L. D., Witter, R. L., and Fadly, A. M., 1989, Augmentation of retrovirus induced lymphoid leukosis by Marek's disease herpesvirus in white leghorn chickens, *J. Virol.* **63:**504–512.
24. Tieber, V. L., Zalinskis, L. L., Silva, R. F., Finkelstein, A., and Coussens, P. M., 1990, Transactivation of the Rous sarcoma virus long terminal repeat promoter by Marek's disease virus, *Virology* **179:**719–727.
25. Banders, U. T., and Coussens, P. M., 1992, Interactions between Marek's disease virus and avian leukosis, in: *Proceedings of the Worlds Poultry Congress, Amsterdam*, Poonsen and Loogerinen, Amsterdam, pp. 62–66.
26. Ho, W.-Z., Harouse, J. M., Rando, R. F., Gönczöl, E., Srinivasan, A., and Plotkin, S. A., 1990, Reciprocal enhancement of gene expression and viral replication between human cytomegalovirus and human immunodeficiency virus type 1, *J. Gen. Virol.* **71:**97–103.
27. Rice, S. A., and Knipe, D. M., 1988, Gene-specific transactivation by herpes simplex virus type 1 alpha protein ICP27, *J. Virol.* **62:**3814–3823.
28. Jones, D., Lee, L., Liu, J.-L., Kung, H.-J., and Tillotson, J. K., 1992, Marek disease virus encodes a basic-leucine zipper gene resembling the *fos/jun* oncogenes that is highly expressed in lymphoblastoid tumors, *Proc. Natl. Acad. Sci. USA* **89:**4042–4046.
29. Kost, R., Jones, D., Isfort, R., Witter, R., and Kung, H.-J., 1993, Retrovirus insertion into herpesvirus: Characterization of a Marek's disease virus harboring a solo LTR, *Virology* **192:**161–169.
30. Churchill, A. E., Chubb, R. C., and Baxendale, W., 1969, The attenuation, with loss of antigenicity, of the herpes-type virus of Marek's disease, *J. Gen. Virol.* **4:**557–564.
31. Davidson, I., Malkinson, M., and Becker, Y., 1988, Marek's disease virus serotype-1 antigens A and B and their unglycosylated precursors detected by Western blot analysis of infected cells, *Virus Genes* **2:**5–18.
32. Isfort, R. J., Kung, H.-J., and Velicer, L. F., 1987, Identification of the gene encoding Marek's disease herpesvirus A antigen, *J. Virol.* **61:**2614–2620.
33. Binns, M. M., and Ross, N. L. J., 1989, Nucleotide sequence of the Marek's disease virus (MDV) RB-1B A antigen gene and the identification of the MDV A antigen as the herpes simplex virus-1 glycoprotein C homologue, *Virus Res.* **12:**371–382.
34. Frink, R. J., Eisenberg, R., Cohen, G., and Wagner, E. K., 1983, Detailed analysis of the portion of herpes simplex virus type 1 genome encoding glycoprotein, C., *J. Virol.* **45:**634–647.
35. Kinchinton, P. R., Remenick, J., Ostrove, J. M., Straus, S. E., Ruyecan, W. T., and Hay, J., 1986, Putative glycoprotein gene of varicella–zoster virus with variable copy numbers of a 42-base-pair repeat sequence has homology to herpes simplex virus glycoprotein C, *J. Virol.* **59:**660–668.

36. Niikura, M., Matsuura, Y., Endoh, D., Onuma, M., and Mikami, T., 1992, Expression of the Marek's disease virus (MDV) homolog of glycoprotein B of herpes simplex virus by a recombinant baculovirus and its identification as the B antigen (gp100, gp60, gp49) of MDV, *J. Virol.* **66:** 2631–2638.
37. Ross, L. J. N., Sanderson, M., Scott, S.D., Binns, M. M., Doel, T., and Milne, B., 1989, Nucleotide sequence and characterization of the Marek's disease virus homologue of glycoprotein B of herpes simplex virus. *J. Gen. Virol.* **70:**1789–1804.
38. Pellett, P. E., Kouzoulas, K. G., Pereira, L., and Roizman, B., 1985, The anatomy of the herpes simplex virus 1 strain F glycoprotein B gene: Primary sequence and predicted protein structure of the wild type and of monoclonal antibody-resistant mutants, *J. Virol.* **53:**243–253.
39. Nazerian, K., Lee, L. F., Yanagida, N., and Ogawa, R., 1992, Protection against Marek's disease by a fowlpox virus recombinant expressing the glycoprotein B of Marek's disease virus, *J. Virol.* **66:**1409–1413.
40. Becker, Y., Asher, Y., Tabor, E., Davidson, I., and Malkinson, M., 1994, Open reading frames in a 4556 nucleotide sequence within MDV-1 BamHI-D DNA fragment: Evidence for splicing of mRNA from a new viral glycoprotein gene, *Virus Genes* **8:**55–69.
41. Silva, R. F., and Lee, L. F., 1984, Monoclonal antibody-mediated immunoprecipitation of proteins from cells infected with Marek's disease virus or turkey herpesvirus, *Virology* **136:** 307–320.
42. Ikuta, K., Nakajima, K., Naito, M., Ann, S. H., Ueda, S., Kato, S., and Hirai, K., 1985, Identification of Marek's disease virus-specific antigens in Marek's disease lymphoblastoid cell lines using monoclonal antibody against virus-specific phosphorylated polypeptides, *Int. J. Cancer* **35:**257–264.
43. Chen, X., Sondermeijer, P. J. A., and Velicer, L. F., 1992, Identification of a unique Marek's disease virus gene which encodes a 38-kilodalton phosphoprotein and is expressed in both lytically infected cells and laterally infected lymphoblastoid tumor cells, *J. Virol.* **66:**85–94.
44. Makimura, K., Peng, F.-Y., Tsuji, M., Hasegawa, S., Kawai, Y., Nonoyama, M., and Tanaka, A., 1994, Mapping of Marek's disease virus genome: Identification of junction sequences between unique and inverted repeat regions, *Virus Genes* **8:**15–24.
45. Zhu, G.-S., Iwata, A., Gong, M., Ueda, S., and Hirai, K., 1994, Marek's disease virus type 1-specific phosphorylated proteins pp38 and pp24 with common amino acid termini are encoded from the opposite junction regions between the long unique and inverted repeat sequences of viral genome, *Virology* **200:**816–820.
46. Fukuchi, K., Tanaka, A., Schierman, L. W., Witter, R. L., and Nonoyama, M., 1985, The structure of Marek's disease virus DNA: Presence of unique expansion in nonpathogenic viral DNA, *Proc. Natl. Acad. Sci. USA* **82:**751–754.
47. Silva, R. F., and Witter, R. L., 1985, Genomic expansion of Marek's disease virus DNA is associated with serial *in vitro* passage, *J. Virol.* **54:**690–696.
48. Maotani, K., Kanamori, A., Ikuta, K., Ueda, S., Kato, S., and Hirai, K., 1986, Amplification of a tandem direct repeat within inverted repeats of Marek's disease virus DNA during serial *in vitro* passage, *J. Virol.* **58:**657–660.
49. Ross, N., Binns, M. M., Sanderson, M., and Schat, K. A., 1993, Alterations in DNA sequence and RNA transcription of the BamHI-H fragment accompany attenuation of oncogenic Marek's disease herpesvirus, *Virus Genes* **7:**33–51.
50. Peng, F.-Y., Donovan, J., Specter, S., Tanaka, A., and Nonoyama, M., 1993, Prolonged proliferation of primary chicken embryo fibroblasts transfected with cDNAs from the BamHI-H gene family of Marek's disease virus, *Int. J. Oncol.* **3:**587–591.
51. Iwata, A., Ueda, S., Ishihama, A., and Hirai, K., 1992, Sequence determination of cDNA clones of transcripts from the tumor-associated region of the Marek's disease virus genome, *Virology* **187:**805–808.
52. Bentley, D. L., and Groudine, M., 1988, Sequence requirements from premature termination of transcription in the human c-*myc* gene, *Cell* **53:**245–256.
53. Hay, N., Hagit, S.-D., and Aloni, Y., 1982, Attenuation in the control of SV40 gene expression, *Cell* **29:**183–193.

54. Eick, D., and Bornkamm, G., 1986, Transcriptional arrest within the first exon is a fast control mechanism in c-*myc* gene expression, *Nucleic Acids Res.* **14:**8331–8346.
55. Peng, F.-Y., Bradley, G., Tanaka, A., Lancz, G., and Nonoyama, M., 1992, Isolation and characterization of DNAs from BamHI-H gene family RNAs associated with the tumorigenicity of Marek's disease virus, *J. Virol.* **66:**7389–7396.
56. Peng, F.-Y., Specter, S., Tanaka, A., and Nonoyama, M., 1994, A 7 kDa protein encoded by the BamHI-H gene family of Marek's disease virus is produced in lytically and latently infected cells, *Int. J. Oncol.* **4:**799–802.
57. Maray, T., Malkinson, M., and Becker, Y., 1988, RNA transcripts of Marek's disease virus (MDV) serotype-1 in infected and transformed cells, *Virus Genes* **2:**49–68.
58. Schat, K. A., Buckmaster, A., and Ross, L. J. N., 1989, Partial transcription map of Marek's disease herpesvirus in lytically infected cells and lymphoblastoid cell lines, *Int. J. Cancer* **44:**101–109.
59. Anderson, A. S., Francesconi, A., and Morgan, R. W., 1992, Complete nucleotide sequence of the Marek's disease virus ICP4 gene, *Virology* **189:**657–667.
60. Silver, S., Tanaka, A., and Nonoyama, M., 1979, Transcription of the Marek's disease virus genome in a nonproductive chicken lymphoblastoid cell line, *Virology* **93:**127–133.
61. Nonoyama, M., and Pagano, J. S., 1972, Separation of Epstein–Barr viral DNA from large chromosomal DNA in non-virus producing cells, *Nature [New Biol.]* **238:**169.
62. Lau, R. Y., and Nonoyama, M., 1979, Replication of the resident Marek's disease virus genome in synchronized non-producer MKT-1 cells, *J. Virol.* **33:**912–914.
63. Delecluse, H.-J., and Hammerschmidt, W., 1993, Status of Marek's disease virus in established lymphoma cell lines: Herpesvirus integration is common, *J. Virol.* **67:**82–92.
64. Yates, J., Warren, N., Reisman, D., and Sugden, B., 1984, A *cis*-acting element from the Epstein–Barr virus genome that permits stable replication of recombinant plasmids in latently infected cells. *Proc. Natl. Acad. Sci. USA* **81:**3806–3810.
65. Silver, S., Smith, M., and Nonoyama, M., 1979, Transcription of the Marek's disease virus genome in virus-induced tumors, *J. Virol.* **30:**84–89.
66. Sugaya, K., Bradley, G., Nonoyama, M., and Tanaka, A., 1990, Latent transcripts of Marek's disease virus are clustered in the short and long repeat regions, *J. Virol.* **64:**5773–5782.
67. Jones, D., Isfort, R., Witter, R., Kost, R., and Kung, H.-J., 1993, Retroviral insertions into a herpesvirus are clustered at the junctions of the short repeat and short unique sequences, *Proc. Natl. Acad. Sci. USA* **90:**3855–3859.
68. Ohashi, K., Zhou, W., O'Connell, P. H., and Schat, K. A., 1994, Characterization of a Marek's disease virus BamHI-L-specific cDNA clone obtained from a Marek's disease lymphoblastoid cell line, *J. Virol.* **68:**1191–1195.
69. Ohashi, K., O'Connell, P. H., and Schat, K. A., 1994, Characterization of Marek's disease virus BamHI-A-specific cDNA clones obtained from a Marek's disease lymphoblastoid cell line, *Virology* **199:**275–283.
70. McKie, E., Ubukata, E., Hasegawa, S., Zhang, S., Nonoyama, M., and Tanaka, A., 1995, The transcripts from the sequences flanking the short component of Marek's Disease virus during latent infection from a unique family of 3′-coterminal RNAs, *J. Virol.* **69:**1310–1314.
71. Taub, R., Mulding, C., Battey, J., Murphy, W., Vasicek, T., Lenoir, G. M., and Leder, P., 1984, Activation and somatic mutation of the translocated c-*myc* gene in Burkitt lymphoma cells, *Cell* **36:**339–348.
72. Fabricant, C. G., Fabricant, J., Litrenta, M. M., and Minick, C. R., 1978, Virus-induced atherosclerosis, *J. Exp. Med.* **148:**335–340.
73. Melnick, J. L., Dreesman, G. R., McCollum, C. H., Petrie, B. L., Burek, J., and DeBakey, M. E., 1983, Cytomegalovirus antigen within human arterial smooth muscle cells, *Lancet* **2:**644–647.

13

Oncogenic Transformation of T Cells by *Herpesvirus saimiri*

PETER G. MEDVECZKY

1. INFECTION OF T CELLS *IN VIVO* AND *IN VITRO* BY *HERPESVIRUS SAIMIRI*

Herpesvirus saimiri (HVS) is a ubiquitous agent of squirrel monkeys (*Saimiri sciureus*); this virus can be reproducibly isolated from the peripheral blood of apparently healthy animals.[1] The most intriguing feature of HVS is its ability to induce acute T-cell lymphomas or lymphoid leukemias. These experimental malignancies develop within a few months after inoculation of the virus into New World monkeys and New Zealand white rabbits.[2-4] Most T lymphocytes transformed *in vivo* or *in vitro* by HVS are CD8 positive,[5-9] although some CD4 cell lines have been reported,[9] suggesting that HVS encodes genes responsible for the preferential and efficient immortalization of CD8 T cells.

Although HVS-associated malignancies are restricted to T cells, a much wider range of cell types can be persistently infected by HVS,[10,11] suggesting that T-cell specificity of tumors is not determined by a T-cell-specific cellular receptor. Similarly, HVS can cause experimental lymphomas in certain monkey and rabbit species, but cells infectable with HVS are found in a wider range of species including humans.

Tumor-derived cells isolated from HVS-infected animals and T cells infected and immortalized *in vitro* display characteristics of malignant lymphocytes. Unlike uninfected lymphocytes HVS-immortalized or -transformed T cells can be cultured in media without interleukin 2 (IL-2), and cell growth is sustained practically indefinitely. These cultures are also morphologically distinct from normal T cells in the peripheral blood, and immortalized cells are typically enlarged, resembling lymphoblasts and usually grow in large clumps. Immortalized T cells not only grow

---

PETER G. MEDVECZKY • Department of Medical Microbiology and Immunology, University of South Florida, Tampa, Florida 33612-4799.

DNA Tumor Viruses: Oncogenic Mechanisms, edited by Giuseppe Barbanti-Brodano *et al.* Plenum Press, New York, 1995.

much longer in tissue culture than uninfected T cells but contain circular episomal viral genomes, suggesting an active role of the virus and its genes in the process, and transformed cells are oncogenic in syngeneic animals.[8,9,12–16] Another easily measurable consequence of immortalization of T cells by HVS is abrogation of the requirement for IL-2 for growth. Sudden removal of IL-2 followed by incubation in IL-2-free medium induces apoptosis of normal T-cell cultures[17]; however, such treatment has no effect on the viability of the HVS-infected cultures.[8]

Strains of HVS from different subgroups are markedly different in their immortalizing potential, as first revealed by studies using common marmoset T cells. Viruses of groups A and C (virus strains are grouped according to DNA sequence homology of a region required for transformation and are discussed later) can efficiently and reproducibly immortalize common marmoset peripheral blood lymphocytes (PBL) *in vitro*.[13,15] Group B strains can immortalize PBL only at a low frequency, and group-B-immortalized cells require IL-2 for optimal growth.[15]

Immortalization of human cells by group C strains 484-77 and 487-77 has also been reported; group A or B strains were negative in these experiments.[8,9] Human T lymphocytes infected by HVS can be maintained in tissue culture without stimulation by lectins or antigen for a prolonged period of time.[8,9] However, the long-term maintenance of HVS-immortalized human T cells in tissue culture is a much more difficult task than that of immortalized marmoset cells, and special conditions and media are required. Fickenscher and Fleckenstein developed a special growth medium for the transformation of human T cells that contains 45% RPMI 1640, 45% GC medium, 10% fetal calf serum, glutamine, and 40 U/ml IL-2 (Boehringer Mannheim) (B. Fleckenstein, personal communication). GC medium is available from Vitromex GmbH (Dr. F. Zimmermann, Adilgestrasse 33, D-94474 Vilshofen, Germany). AIM-V medium (readily available from Gibco-BRL Inc.) supplemented with 10% fetal calf serum, 500 U/ml recombinant IL-2, and antibiotics was also found suitable for the immortalization and long-term maintenance of HVS-infected human T cells (M. Medveczky and P. Medveczky, unpublished observation).

As mentioned earlier, HVS-immortalized T cells are usually $CD8^+$ and $CD4^-$, which is unique in comparison with the phenotype of other virally transformed lymphocytes. It is consistent with their phenotype that HVS-transformed T cells are strongly cytotoxic against various tumor cell targets[8,9,18] and thus resemble lymphokine-activated killer (LAK) cells, which have been under intense investigation. However, unlike uninfected LAK cells, which require IL-2 for induction of cytotoxic activity, HVS-immortalized T cells are strongly cytotoxic with or without IL-2. Therefore, to reflect IL-2 independence and cytotoxicity of these cells, the term HAK for herpesvirus-activated killer cell was proposed.[8]

2. THE STRUCTURE OF THE VIRAL DNA: ARRANGEMENT AND ORIGIN OF VIRAL GENES

As shown in Fig. 1, the genome of HVS consists of about 113 kb of unique sequences (called L-DNA, about 35% G + C content) flanked by tandem repeats (H-DNA, about 71% G + C).[19] At least 75 open reading frames (ORFs), which are

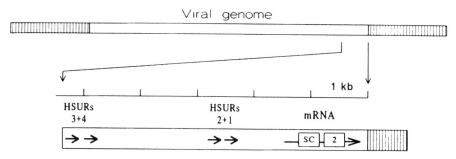

FIGURE 1. Schematic representation of the viral genome. The unique L-DNA region of *H. saimiri*, depicted as a long open box, is flanked by repetitive H-DNA. The right-end region is enlarged, and corresponding transcripts mapped in transformed rabbit T cells are shown as arrows. SC, SCOL open reading frame; 2, ORF2, the IL-11-like open reading frame.

likely to be expressed as proteins, are encoded by the L-DNA of HVS.[20] The genome of HVS encodes various genes commonly described in all herpesviruses, including several enzymes involved in viral DNA metabolism, structural proteins such as capsid components, and various glycoproteins.[20] These common herpesvirus genes presumably function during the lytic replication cycle.

Among the human herpesviruses, Epstein–Barr virus (EBV) is the closest relative of HVS. Many of the HVS ORFs are collinear with those of EBV as revealed by DNA sequencing.[20] However, in the genome of HVS, these blocks of genes with homology to EBV are interrupted with clusters of HVS-specific sequences, which often display significant homology with host sequences, and these cellular homologues are discussed later in detail. On the other hand, HVS lacks most genes thought to be associated with immortalization of B cells by EBV; no homologues of EBV genes EBNA-1, EBNA-2, EBNA-3, LP, or LMP (genes expressed in EBV-immortalized B cells) have been identified in the genome of HVS.[20] Therefore the transformation-associated genes of EBV and HVS appear to be unrelated, and one can hypothesize that these two agents diverged and evolved as B-cell-specific and T-cell-specific viruses from a common ancestor.

DNA sequencing and sequence comparison revealed that most genes from the rightmost 7 kb L-DNA sequence of the HVS genome (orientation is identical with EBV) and several ORFs encoded by other regions of the genome show significant homology with cellular genes[20–24] (Table I). Some of these viral homologues, such as the seven U-type small RNAs (HSURs), SCOL/STP, ECLF2, and ORF2, are probably involved in oncogenicity and/or T-cell activation by the virus. These putative transforming HVS genes are discussed later in greater detail. The second group of genes includes ECRF3, ORF15, and CCHP, which are probably important in survival of virus-infected cells against host immune attack. A third group of genes of cellular origin are DHFR and TS, which may function in nucleotide metabolism.

These data leave little doubt that HVS is a transducing virus similar to acutely transforming retroviruses. The transduced genes of HVS have no introns and thus have most likely been transduced from mRNA transcribed to cDNA by an unknown

TABLE I
Herpesvirus saimiri Genes with Significant Homology to Cellular Genes

Name of viral gene	Strain/ group	Cellular homologue	Putative function	Reference
HSUR1-7	A, B, C	U-type small nuclear RNAs	IL-2, IL-4 mRNA stabilization?	21, 23, 25
SCOL, ORF1, STP-C	C	Collagen	Transforming membrane protein	22, 25
ECLF2 homologue	A	Cyclin D	Transformation?	20, 26
ECRF3	A	IL-8 receptor B	Immunoregulation?	20, 27
ORF2	C	IL-11? MDM-2?	IL-2 independence IL-11? p53 binding?	22; T. Lund, P. Geck, M. Medveczky, and P. Medeczky (unpublished data)
ORF13	A	CTL-8	Unknown	20, 28
ORF15	A	CD59 (membrane inhibitor of reactive lysis)	Unknown	20
CCPH	A	CD21, CD46, CD35, CD55, C4b-binding protein	Inhibition of complement function	20
Dihydrofolate reductase	A, C	Dihydrofolate reductase	DNA metabolism	24
Thymidylate synthase	A	Thymidylate synthase	DNA metabolism	29

reverse transcriptase. The virus encodes no known reverse transcriptase, and one can only speculate that perhaps either an endogenous or exogenous squirrel monkey retrovirus reverse transcriptase is responsible for the generation and insertion of these cellular sequences into the HVS genome.

3. CIRCULARIZATION, DELETIONS, AND METHYLATION OF THE VIRAL GENOME IN TRANSFORMED T CELLS

In latent/persistent infection, which occurs in T lymphocytes, the viral genome is a circular episome. Tumor tissues and cell lines established from tumors or by *in vitro* immortalization carry multiple copies of the viral DNA in covalently closed circular form.[8,9,15,30-32] No evidence is available that HVS would integrate into the host genome.

Although viral episomes in transformed T cells are very stable, large spontaneous deletions in the middle of the L-DNA are commonly found in tumor cell lines maintained in tissue culture.[14,30,33] In contrast, the leftmost and the rightmost approximately 15 kb L-DNA sequences do not suffer of such deletions, suggesting that the middle of the genome is not essential for the maintenance of the transformed state, and perhaps, sequences relevant to immortalization and maintenance of the circular episome are located in the left and right H–L DNA junctions.

Most of the viral genes encoded by HVS are inactive in immortalized T cells, and only a limited number of gene products can be detected. Extensive methylation of the episomal HVS DNA at C–G residues in immortalized T cells has been described, which is thought to correlate with the lack of gene expression in mammalian cells.[34–36] On the other hand, a few unmethylated sites of the right end region have been also reported in tumor cells,[36] suggesting that some selected genes at the right end of L-DNA are expressed.

4. DNA VARIABILITY AND MAPPING OF A REGION OF THE VIRAL GENOME INVOLVED IN ONCOGENIC TRANSFORMATION

Similarly to other herpesviruses, HVS strains can be readily distinguished by restriction enzyme cleavage site polymorphism.[37] However, the right-end 7-kb "oncogenic" region is much more variable than the rest of the genome.[38] On the basis of DNA hybridization experiments, the various isolates of HVS have been classified into DNA homology groups A, B, and C[38] [group C was formerly called group non-A, non-B[38]]. Comparison of DNA sequences confirmed this finding, and no homology was found between group A (strain 11) and group C strains (484-77 and 488-77) within the rightmost 2.5 kb of L-DNA.[22,25] To explain the lack of homology between group A and C strains in the right-end 1.8-kb L-DNA region, we favor the hypothesis that different cellular sequences have been acquired by these strains, because there is strong evidence that HVS contains sequences highly homologous to cellular genes in its genome.

As discussed earlier, DNA variability correlates with the transforming ability of strains: group C strains appear to be the most potent oncogenic/transforming agents, group A strains rank second, and B strains rank last in this respect. Tumor formation in rabbits also correlates with DNA grouping. New Zealand white rabbit experiments showed that the group C strain 484-77 is highly oncogenic in New Zealand white rabbits, whereas group A or B viruses are not oncogenic in these rabbits.[4]

To localize the region of the genome conferring the highly oncogenic phenotype to strain 484-77 (group C), strain B–C recombinants have been constructed. Two recombinants consisting of strain B virus DNA, in which the right-end 9 kb of unique DNA is replaced by group C virus DNA, were oncogenic in rabbits.[4] These experiments showed that the right-end 9-kb DNA of the group C strain contains gene(s) relevant to the transforming and oncogenic potential.

Several studies with deletion mutants also revealed that the right-end genomic segment is important for oncogenic transformation.[8,13,39,40,42] Large deletions of the right-end sequences of the L-DNA result in loss of immortalization of T cells and oncogenicity by HVS, and such transformation-deficient deletion mutants of all three virus subgroups have been described. A common feature of these mutants is that they are entirely competent for lytic replication. All data with deletion mutants strongly suggest that protein products corresponding to the ORFs are involved in the process of immortalization of T cells. Deletion of an ORF called STP (for saimiri transforming protein) in strain 11 of group A correlated with loss of oncogenic and transforming potential of the virus.[13,39,40,42] Deletion mapping of a group C strain showed the importance of two ORFs, ORF1 and ORF2.[8] ORF1 encodes a collagen-like oncoprotein product SCOL, and ORF2 an IL-11-like protein; these genes and their role in transformation are further discussed below.

5. EXPRESSION OF AN mRNA AND ITS COLLAGEN-LIKE ONCOPROTEIN PRODUCT SCOL AND AN IL-11-LIKE PROTEIN IN TRANSFORMED T CELLS

Recently, significant progress has been made to reveal gene expression of the oncogenic region of the group C strain 484-77. A 1.2-kb polyadenylated virus-specific RNA is transcribed from the rightmost 2 kb of L-DNA in lymphocytes transformed by the highly oncogenic group C strain 484-77.[22] The 1.2-kb transcript codes for the two ORFs, ORF1 and ORF2.[22]

ORF1 is termed either SCOL[22] (for saimiri collagen) for strain 484-77 or STP-C[25] (saimiri transforming protein C) for strain 488-77. The SCOL protein is composed of three domains: an acidic amino terminus, a central collagen-like region, and a hydrophobic carboxy end. The SCOL protein is expressed in all C-virus-induced tumor cells and *in vitro* immortalized T cells tested.[43,44] The subcellular target of SCOL is not entirely clear; although the highest amount of this protein was found in the cytoplasm[43,44] in association with the Golgi apparatus,[43] it is perhaps also expressed on the outer surface of tumor cells.[44]

Several studies using different approaches proved that SCOL is a viral oncogene. SCOL is expressed in tumor-bearing animals as revealed by antibody responses.[44] Jung *et al.*[45] demonstrated that collagen-like sequences of strain 488-77 overexpressed by retrovirus can transform rat kidney cells, and these transformed cells were oncogenic in nude mice. In this study the corresponding ORF from group A strain 11 also induced tumors in nude mice. Experiments with transgenic mice also provided strong evidence for the role of SCOL in transformation; a strain of mice carrying the SCOL transgene developed various tumors.[46] Paradoxically, no T-cell tumors were observed in these transgenic mice.[46]

It is unknown whether any of the three domains of SCOL, including the collagen-like region, are dispensable for oncogenic transformation. One can hypothesize a potential interaction of the collagen domain with surface adhesion molecules or with intracellular proteins that may trigger signal transduction.

The origin of the collagen-like domain in the SCOL sequence is also uncertain. The repeat is perhaps the result of a recent amplification of an 18-bp sequence by the virus; this recent amplification of the collagen 18-bp unit is supported by the fact that the 18-bp sequence is perfectly repeated nine times.[22] Alternatively, the repeats may be amplified by the host cell and then transduced by the virus. Interestingly, some collagen exons are 54-nucleotides in length as in the chicken collagen gene, where most of the 50 exons contain 54 nucleotides. In HVS, the 162-nucleotide sequence appears to be the result of exactly three reiterations of a 54-nucleotide unit.[22] The absence of splicing of the viral collagen-like domain may refer to a reverse transcription step prior to integration into the viral genome.

A second protein (ORF2) coded by the bicistronic 1.2-kb mRNA was detected by *in vitro* assays.[44] Moreover, recent studies proved that ORF2 is expressed in rabbit tumor cells and is associated with cell membranes.[47] Interestingly, the predicted ORF2 protein sequence shows homology to interleukin 11 (IL-11), which was recently cloned[48,49] and shown to be functionally similar to IL-6.[50] Interleukin 11 stimulates colony formation of hematopoietic blast cells, augments IL-3-mediated proliferation of blast cells, and shortens the G_o period for stem cells.[51,54] This lymphokine also synergizes with IL-4 to stimulate stem cells.[52] It is interesting to note that HVS-transformed T cells secrete IL-4[55]; if the viral sequence turns out to be functionally similar to IL-11, it could synergize with IL-4.

It is very likely that not only SCOL but other genes of HVS including ORF2 also contribute to the transforming potential of the virus. The SCOL coding sequence and the IL-11-like ORF2 protein are both required for IL-2-independent growth as tested by deletion mutants in human T cells.[8] Moreover, recent studies using deletion mutants revealed that not only SCOL and ORF2 but two small RNAs are also required for oncogenicity in rabbits (M. Medveczky, P. Geck, and P. Medveczky, unpublished results).

6. VIRAL SMALL RNAs IN TRANSFORMED CELLS AND IDENTIFICATION OF CELLULAR PROTEINS THAT BIND BOTH VIRAL AUUUA REPEATS AND THE 3′ END OF UNSTABLE mRNAs OF LYMPHOKINES

Four viral nonpolyadenylated HSURs are expressed in tumor cells transformed by strain 484-77.[23] Similar transcripts (seven small RNAs) have been described in marmoset cell lines transformed by strain 11[20,21,56] (group A). Although the functions of these HSURs have not been elucidated yet, some data suggest involvement of these transcripts in mRNA stabilization. Sequence comparison between group A and C strains showed that the small RNAs 1 and 2 encode conserved AUUUA repeats.[21,23,25,57] Interestingly, the same AUUUA repeats occur at the 3′ noncoding regions of growth factor, lymphokine, and protooncogene mRNAs; these sequence motifs are involved in rapid mRNA degradation, and several *trans*-acting factors have been shown to complex with this 3′ noncoding mRNA sequence. Removal of the AU-rich region confers greater stability to mRNA produced from transfected constructs.[58] Similarly, addition of a short DNA segment containing AUUUA to stable mRNA destabilized the mRNA.[59]

As shown by *in vitro* binding assays, the HSUR AU-rich repeats form specific complexes with a 32-kDa and a 70-kDa cellular RNA-binding protein. The 32-kDa factor has been implicated in mRNA destabilization and binds HSURs of both group A and group C strains.[57,60,61] It has been proposed that HSURs and lymphokine and oncogene mRNA sequences compete for the 32-kDa AUUUA-specific binding protein,[57,60] and as a consequence of such competition, the 32-kDa destabilizing factors could be titrated out by HSURs, which would allow mRNA to stabilize.

A lectin-inducible AUUUA specific novel 70-kDa binding factor has also been identified (termed AUBF70) that binds both the 3' noncoding region of IL-4 mRNA as well as the AUUUA repeats of a viral small RNA.[61,62] The HSURs and the IL-4 sequence compete for the 70-kDa AUUUA-specific binding protein.[61] However, it is more likely that AUBF70 is not a destabilizing factor but a stabilizing positive regulator, and HSURs facilitate its transport from the cytoplasm to the nucleus[62] (Fig. 2). According to this hypothesis, AUBF70 forms a complex with HSURs in the cytoplasm, and the complex is then transported into the nucleus. In the nucleus, AUBF70 then binds oncogene and/or lymphokine mRNAs, which have higher affinity for 70 kDa; this binding would result in mRNA stabilization. This hypothesis is supported by the following data: (1) U-RNAs are known to be exported to the cytoplasm, where they complex with proteins before returning to the nucleus[63,64]; (2) HSUR1 and HSUR2 contain AUUUA repeats and are detectable in both the nucleus and cytoplasm[23,62]; (3) AUBF70 is inducible by mitogens in T cells[62]; (4) the IL-4 mRNA 3' end exhibits tenfold higher affinity for binding than HSUR, indicating that the IL-4 sequence could easily take up proteins from the HSUR complex.[61] However, further experiments are required to unravel the role of HSURs in mRNA stabilization.

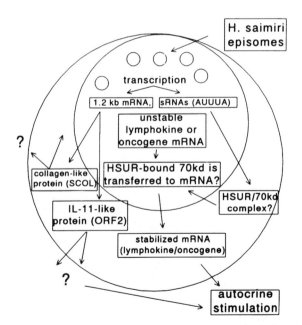

FIGURE 2. Transformation of T cells by HVS: a hypothetical model.

7. SECRETION OF LYMPHOKINES AND EXPRESSION OF THEIR RECEPTORS

The relatively autonomous growth of malignant cells has been known for many years; it has been suggested that cells could become malignant by the endogenous production of polypeptide growth factors acting on their producer cells via external receptors. This mechanism is termed "autocrine secretion."[65,66] Long-term growth of activated uninfected T cells can be supported by several lymphokines, including IL-2 and IL-4.[67,68] Therefore, constitutive up-regulation of these growth factors can result in uncontrolled growth of T cells. For example, the long terminal repeat of gibbon leukemia virus is integrated into the IL-2 gene of MLA 144 cells; MLA 144 cells secrete high levels of IL-2.[69]

Lymphocytes transformed by HVS cannot survive at low cell density, and a T-cell growth-promoting activity was found in conditioned media, suggesting that autocrine secretion of growth factor(s) could be involved.[70] Attempts to demonstrate constitutive high-level expression of IL-2 by HVS-transformed cells have failed, although small amounts of IL-2 mRNA can be detected.[55,70,71] Studies have also demonstrated that high-affinity receptors for IL-2 are present in both *in vitro* immortalized cells and tumor cells.[70] The growth of these cells is strongly inhibited by a monoclonal antibody to the receptor for IL-2, suggesting that IL-2 may be involved in autocrine secretion and proliferation of HVS-transformed cells.[55,70,71]

Some of the growth-promoting activity found in conditioned media is probably not IL-2 but is IL-4 because (1) IL-4 mRNA can be detected,[55,71] (2) supernatants were tested positive in a highly specific IL-4 biological assay,[55] and (3) HVS-transformed cells can be stimulated by IL-4.[55] Cell lines transformed by the highly oncogenic strain 484-77 (group C) secrete the highest amount of IL-4 activity, whereas no growth-promoting factor was secreted by group B transformed cells.[55] As discussed above, the predicted amino acid sequence ORF2 is homologous to IL-11. Autocrine stimulation by synergizing IL-4 and IL-11 is an exciting possibility. Taken together, a large set of data suggests that lymphokines and their receptors (IL-2, IL-4, IL-11?) are involved in the maintenance of the transformed state of the cell, but more experiments are required to prove a direct link between viral genes and lymphokine secretion.

Recent experiments suggest a possible role for SCOL and ORF2 in the induction of lymphokines. The IL-2-independent growth of HVS-infected cells requires both the SCOL and ORF2 sequences, as revealed by analysis of deletion-mutant-infected T cells.[8] To examine the effect of HVS deletions on lymphokine gene expression, IL-2 and IL-4 mRNA levels were compared in T cells infected with a series of deletion mutants by polymerase-chain-reaction-based assays.[55] Expression of IL-4 mRNA was readily observed in wild-type-infected samples; however, SCOL and ORF2 mutant-infected cells did not express detectable IL-4.[55] Similarly, SCOL and ORF2 were both required for IL-2 gene expression.[55]

The role of IL-2 and IL-4 in the process of oncogenic transformation in the HVS model is unclear. One plausible explanation is that these lymphokines, especially IL-2, may act as autocrine growth factors; IL-4, however, not only is a growth factor but selectively stimulates development of Th2 T-cell subsets.[72] It is proposed that IL-4 secretion in tumor-bearing animals could shift the balance of immune responses from Th1 subsets representing cellular T-cell responses toward Th2 humoral immunity in HVS-infected animals.[55] Some of the cell lines that secrete IL-4 (1670 and 70N2) have been isolated from monkey tumors. If IL-4 is secreted by virus-transformed tumor cells *in vivo*, it would cause a shift in T-cell subsets that could help tumor cells to escape from elimination by Th1-mediated cellular immune responses.

8. CONCLUSIONS: POSSIBLE MECHANISMS OF TRANSFORMATION

Oncogenic transformation by HVS is a distinctly unique phenomenon. The T-cell-specific human retroviruses HTLV-I and HTLV-II integrate in the host genome; however, there is no evidence for integration of the HVS DNA into the cellular genome. Because the disease caused by HVS is extremely acute, and tumors are polyclonal, it is also unlikely that integration of viral enhancer sequences adjacent to a cellular protooncogene would be the mechanism of transformation by this virus. Although EBV and HVS are probably evolved from a common ancestor, their target cell specificity is quite different. In addition, the various genes involved in transformation by EBV and HVS are entirely unrelated, which further supports the idea that HVS is a unique transforming agent.

Studies on the SCOL gene established that it is a viral oncogene; however, it is very likely that this collagen-like protein is not the only factor responsible for growth transformation. Because deletion of SCOL cannot completely eliminate IL-2 independence,[8] other sequences in the viral genome must also be involved. Removal of the IL-11-like ORF2 resulted in a phenotype similar to the SCOL mutant.[8] These data suggest that the two proteins somehow cooperate in T-cell stimulation. Small RNAs are also likely candidates as cofactors in the process of transformation through their AUUUA motifs. In addition, preliminary experiments with deletion mutants also suggest that the two small RNAs with AUUUA repeats are required for oncogenicity in rabbits (M. Medveczky, P. Geck, and P. Medveczky, unpublished data). Detailed studies regarding the vast majority of the viral genome have not been published yet, and it is possible that several genes outside the well-studied right end of HVS genome are also involved in transformation of T cells.

It is also worth emphasizing another unique feature, i.e., that oncogenes acquired by HVS from the cellular genome are highly variable. The viral collagen-like SCOL and the IL-11-like protein are found only in subgroup C of HVS, and both are absent in group A and B strains. It appears, therefore, that the rightmost area, where these genes are located, is a "hot spot" to acquire host DNA sequences. It is also clear, however, that even if these genes are of cellular origin, they are significantly diverged from their cellular ancestor. Discovery of viral oncogenes in retroviruses and identification of their cellular counterparts (protooncogenes) represented a historic breakthrough in cancer research. It is possible that cellular homologues of these HVS genes are also involved in regulation of growth of normal cells or could be activated in human cancers. Future studies are required to identify putative cellular homologues of SCOL and HSUR genes and several other HVS genes that may have evolved from cellular counterparts.

Application of HVS as an immortalizing agent is also a useful research tool for various immunologic studies, and among many possibilities, cloning CD8 lymphocytes is one of the important tasks that can be achieved by this method. Infection of human T cells with HVS also provides a model that could facilitate studies on growth regulation of human cytotoxic T cells; CD8 cells are important effector cells in immune responses against infectious diseases and cancer.

It still remains a hypothesis that aberrant regulation of lymphokines and/or their receptors is involved in immortalization and tumorigenicity by HVS. The details of molecular mechanisms by which viral gene products of HVS activate T cells are not yet fully understood. Information learned in this model should provide us with insight regarding T-cell activation and lymphomagenesis that should help us to understand how human lymphomas develop. This can contribute to improved diagnostic and therapeutic approaches.

REFERENCES

1. Melendez, L., Daniel, M. D., Hunt, R. D., and Garcia, F. G., 1968, An apparently new herpesvirus from primary kidney cultures of the squirrel monkey (*Saimiri sciureus*), *Lab. Anim. Care* **18**:374–381.
2. Melendez, L., Hunt, R. D., Daniel, M. D., Garcia, F., and Fraser, C. E. O., 1969, *Herpesvirus saimiri* II. Experimentally induced malignant lymphoma in primates, *Lab. Anim. Care* **19**:378–386.
3. Daniel, M. D., Melendez, L. V., Hunt, R. D., King, N. W., Anver, M., Fraser, C. E. O., Barahona, H. H., and Baggs, R. B., 1974, *Herpesvirus saimiri*. VII. Induction of malignant lymphoma in New Zealand white rabbits, *J. Natl. Cancer Inst.* **53**:1803–1807.
4. Medveczky, M. M., Szomolany, E., Hesselton, R., DeGrand, D., Geck, P., and Medveczky, P. G., 1989, *Herpesvirus saimiri* strains from three DNA subgroups have different oncogenic potentials in New Zealand white rabbits, *J. Virol.* **63**:3601–3611.
5. Kiyotaki, M., Desrosiers, R. C., and Letvin, N. L., 1986, *Herpesvirus saimiri* strain 11 immortalizes a restricted marmoset T8 lymphocyte subpopulation *in vitro*, *J. Exp. Med.* **164**:926–931.
6. Neubauer, R. H., Briggs, C. J., Noer, K. B., and Rabin, H., 1983, Identification of normal and transformed lymphocyte subsets of nonhuman primates with monoclonal antibodies to human lymphocytes, *J. Immunol.* **130**:1323–1329.
7. Rabin, H., Wallen, W. C., Neubauer, R. H., and Epstein, M. A., 1975, Comparisons of surface markers on *Herpesvirus saimiri* associated lymphoid cells of nonhuman primates and established human lymphoid cell lines. *Bibl. Haematol.* **40**:367–374.
8. Medveczky, M. M., Geck, P., Sullivan, J. L., Serbousek, D., Djeu, J. Y., and Medveczky, P. G., 1993, IL-2 independent growth and cytotoxicity of *Herpesvirus saimiri* infected human CD8 cells and involvement of two open reading frames of the virus, *Virology* **196**:402–412.
9. Biesinger, B., Müller-Fleckenstein, I., Simmer, B., Lang, G., Wittmann, S., Platzer, E., Desrosiers, R. C., and Fleckenstein, B., 1992, Stable growth transformation of human T lymphocytes by *Herpesvirus saimiri*, *Proc. Natl. Acad. Sci. USA* **89**:3116–3119.
10. Ablashi, D. V., Armstrong, G. R., Heine, U., and Manaker, R. A., 1971, Propagation of *Herpesvirus saimiri* in human cells, *J. Natl. Cancer Inst.* **47**:241–244.
11. Simmer, B., Alt, M., Buckreus, I., Berthold, S., Fleckenstein, B., Platzer, E., and Grassmann, R., 1991, Persistence of selectable *Herpesvirus saimiri* in various human haematopoetic and epithelial cell lines, *J. Gen. Virol.* **72**:1953–1958.

12. Fleckenstein, B., Muller, I., and Werner, J., 1977, The presence of *Herpesvirus saimiri* genomes in virus-transformed cell, *Int. J. Cancer* **19**:546–554.
13. Desrosiers, R. C., Silva, D. P., Waldron, L. M., and Letvin, N. L., 1986, Nononcogenic deletion mutants of *Herpesvirus saimiri* are defective for *in vitro* immortalization, *J. Virol.* **57**:701–705.
14. Kaschka-Dierich, C., Werner, F. J., Bauer, I., and Fleckenstein, B., 1982, Structure of nonintegrated, circular *Herpesvirus saimiri* and *Herpesvirus ateles* genomes in tumor cell lines, and *in vitro*-transformed cells, *J. Virol.* **44**:295–310.
15. Szomolanyi, E., Medveczky, P., and Mulder, C., 1987, *In vitro* immortalization of marmoset cells with three subgroups of *Herpesvirus saimiri*, *J. Virol.* **61**:3485–3490.
16. Ablashi, D. V., Schirm, S., Fleckenstein, B., Faggioni, A., Dahlberg, J., Rabin, H., Loeb, W., Armstrong, G., Peng, J. W., and Aulakh, G., 1985, *Herpesvirus saimiri*-induced lymphoblastoid cell line: Growth characteristics, virus persistence, and oncogenic properties, *J. Virol.* **55**: 623–633.
17. Duke, R. C., and Cohen, J. J., 1986, IL-2 addiction: Withdrawal of growth factor activates a suicide program in dependent T cells, *Lymphokine Res.* **5**:289–299.
18. Johnson, D. R., and Jondal, M., 1981, Herpesvirus-transformed cytotoxic T-cell lines, *Nature* **291**: 81–83.
19. Bornkamm, G. W., Delius, H., Fleckenstein, B., Werner, F. J., and Mulder, C., 1976, Structure of *Herpesvirus saimiri* genomes: Arrangement of heavy and light sequences in the M-genome, *J. Virol.* **19**:154–161.
20. Albrecht, J.-C., Nicholas, J., Biller, D., Cameron, K. R., Biesinger, B., Newman, C., Wittmann, S., Craxton, M. A., Coleman, H., Fleckenstein, B., and Honess, R. W., 1992, Primary structure of the *Herpesvirus saimiri* genome, *J. Virol.* **66**:5047–5058.
21. Lee, S. I., Murthy, S. C. S., Trimble, J. J., Desrosiers, R. C., and Steitz, J. A., 1988, Four novel U RNAs are encoded by a herpesvirus, *Cell* **54**:599–607.
22. Geck, P., Whitaker, S. A., Medveczky, M. M., and Medveczky, P. G., 1990, Expression of collagenlike sequences by a tumorvirus, *Herpesvirus saimiri*, *J. Virol.* **64**:3509–3515.
23. Geck, P., Whitaker, S. A., Medveczky, M. M., Last, T. J., and Medveczky, P. G., 1993, Small RNA expression from the oncogenic region of a highly oncogenic strain of *Herpesvirus saimiri*, *Virus Genes* **8**:25–34.
24. Trimble, J. J., Murthy, S. C. S., Bakker, A., Grassmann, R., and Desrosiers, R. C., 1988, A gene for dehydrofolate reductase in a herpesvirus, *Science* **239**:1145–1147.
25. Biesinger, B., Trimble, J., Desrosiers, R. C., and Fleckenstein, B., 1990, The divergence between two oncogenic *Herpesvirus saimiri* strains in a genomic region related to the transforming phenotype, *Virology* **176**:505–514.
26. Nicholas, J., Cameron, K. R., and Honess, R. W., 1992, *Herpesvirus saimiri* encodes homologues of G protein-coupled receptors and cyclins. *Nature* **355**:362–365.
27. Murphy, P. M., and Tiffany, H. L., 1991, Cloning of complementary DNA encoding a functional human interleukin-8 receptor, *Science* **253**:1280–1283.
28. Rouvier, E., Luciani, M.-F., Mattei, M.-G., Denizot, F., and Golstein, P., 1993, CTLA-8, cloned from an activated T cell, bearing AU-rich messenger RNA instability sequences, and homologous to a *Herpesvirus saimiri* gene, *J. Immunol.* **150**:5445–5456.
29. Honess, R. W., Bodemer, W., Cameron, K. R., Miller, H.-H., Fleckenstein, B., and Randall, R. E., 1986, The A + T-rich genome of *Herpesvirus saimiri* contains a highly conserved gene for thymidylate synthase, *Proc. Natl. Acad. Sci. USA* **83**:3604–3608.
30. Fleckenstein, B., and Mulder, C., 1980, Molecular aspects of *Herpesvirus saimiri* and *Herpesvirus ateles*, in: *Viral Oncology 1980* (G. Klein, ed.), Raven Press, New York, pp. 799–812.
31. Gardella, T., Medveczky, P., Sairenji, T., and Mulder, C., 1984, Detection of circular and linear herpesvirus DNA molecules in mammalian cells by gel electrophoresis, *J. Virol.* **50**:248–254.
32. Schirm, S., Muller, I., Desrosiers, R. C., and Fleckenstein, B., 1984, *Herpesvirus saimiri* DNA in a lymphoid cell line established by *in vitro* transformation, *J. Virol.* **49**:938–946.

33. Derosiers, R. C., 1981, *Herpesvirus saimiri* DNA in tumor cells—deleted sequences and sequence rearrangements, *J. Virol.* **39**:497–509.
34. Desrosiers, R. C., Mulder, C., and Fleckenstein, B., 1979, Methylation of *Herpesvirus saimiri* DNA in lymphoid tumor cell lines, *Proc. Natl. Acad. Sci. USA* **76**:3839–3843.
35. Youssoufian, H., and Mulder, C., 1981, Detection of methylated sequences in eukaryotic DNA with the restriction endonucleases SmaI and XmaI, *J. Mol. Biol.* **150**:133–136.
36. Desrosiers, R. C., 1982, Specifically unmethylated cytidylic–guanylate sites in *Herpesvirus saimiri* DNA in tumor cells, *J. Virol.* **43**:427–435.
37. Desrosiers, R. C., and Falk, L. A., 1982, *Herpesvirus saimiri* strain variability, *J. Virol.* **43**:352–356.
38. Medveczky, P., Szomolanyi, E., Desrosiers, R. C., and Mulder, C., 1984, Classification of *Herpesvirus saimiri* into three groups based on extreme variation in a DNA region required for oncogenicity, *J. Virol.* **52**:938–944.
39. Daniel, M. D., Silva, D., Koomey, J. M., Mulder, C., Fleckenstein, B., Tamulevich, R., King, N. W., Hunt, R. D., Sehgal, P., and Falk, L. A., 1979, *Herpesvirus saimiri*: Strain SMHI modification of oncogenicity, in: *Advances in Comparative Leukemia Research (I)*, (D. S. Yohn, B. A. Lapin, and I. Blakeslee, eds.), Elsevier/North-Holland, Amsterdam, pp. 395–396.
40. Koomey, J. M., Mulder, C., Burghoff, R. L., Fleckenstein, B., and Desrosiers, R. C., 1984, Deletion of DNA sequences in a nononcogenic variant of *Herpesvirus saimiri*, *J. Virol.* **50**:662–665.
41. Rabin, H., Hopkins, R. F., III, Desrosiers, R. C., Ortaldo, J. R., Djeu, J. Y., and Neubauer, R. H., 1984, Transformation of owl monkey T cells *in vitro* with *Herpesvirus saimiri*, *Proc. Natl. Acad. Sci. USA* **81**:4563–4567.
42. Murthy, S. C. S., Trimble, J. J., and Desrosiers, R. C., 1989, Deletion mutants of *Herpesvirus saimiri* define an open reading frame necessary for transformation, *J. Virol.* **63**:3307–3314.
43. Jung, J. U., and Desrosiers, R. C., 1991, Identification and characterization of the *Herpesvirus saimiri* oncoprotein STP-C488, *J. Virol.* **65**:6953–6960.
44. Medveczky, M. M., Geck, P., and Medveczky, P., 1993, Expression of the collagen-like putative oncoprotein of *Herpesvirus saimiri* in transformed T cells, *Virus Genes* **7**:349–365.
45. Jung, J. U., Trimble, J. J., King, N. W., Biesinger, B., Fleckenstein, B., and Desrosiers, R. C., 1991, Identification of transforming genes of subgroup A and C strains of *Herpesvirus saimiri*, *Proc. Natl. Acad. Sci. USA* **88**:7051–7055.
46. Murphy, C., Kretschmer, C., Biesinger, B., Beckers, J., Jung, J., Desrosiers, R. C., Muller-Hermelink, H. K., Fleckenstein, B. W., and Ruther, U., 1994, Epithelial tumors induced by a *Herpesvirus* oncogene in transgenic mice, *Oncogene* **9**:221–226.
47. Lund, T., Medveczky, M. M., Geck, P., and Medveczky, P. G., 1995, A Herpesvirus saimiri protein required for IL-2 independence is associated with membranes of transformed T cells, *J. Virol.* **69**:4495–4499.
48. McKinley, D., Wu, Q., Yang-Feng, T., and Yang, Y. C., 1992, Genomic sequence and chromosomal location of human interleukin-11 gene (IL-11), *Genomics* **13**:814–819.
49. Paul, S. R., Bennett, F., Calvetti, J. A., Kelleher, K., Wood, C. R., O'Hara, R. M., Jr., Leary, A. C., Sibley, B., Clark, S. C., Williams, D. A., and Yang, Y.-C., 1990, Molecular cloning of a cDNA encoding interleukin-11, a stromal cell-derived lymphopoietic and hematopoietic cytokine, *Proc. Natl. Acad. Sci. USA* **87**:7512.
50. Baumann, H., and Schendel, P., 1991, Interleukin-11 regulates the hepatic expression of the same plasma protein genes as interleukin-6, *J. Biol. Chem.* **266**:20424–20427.
51. Mushashi, M. T., Yang, Y. C., Paul, S. R., Clark, S. C., Sudo, T., and Ogawa, M., 1991, Direct and synergistic effects of interleukin-11 on murine hematopoiesis in culture, *Proc. Natl. Acad. Sci. USA* **88**:765–769.
52. Mushashi, M., Clark, S. C., Sudo, T., Urdal, D. L., and Ogawa, M., 1991, Synergistic interactions between interleukin-11 and interleukin-4 in support of proliferation of primitive hematopoietic progenitors of mice, *Blood* **78**:1448–1451.

53. Schibler, K. R., Yang, Y. C., and Christensen, R. D., 1992, Effect of interleukin-11 on cycling status and clonogenic maturation of fetal and adult hematopoietic progenitors, *Blood* **80:** 900–903.
54. Yin, T., Myazawa, K., and Yang, Y. C., 1992, Characterization of interleukin-11 receptor and protein kinase phosphorylation induced by interleukin-11 in mouse 3T3-L1 cells, *J. Biol. Chem.* **267:**8347–8351.
55. Chou, C.-S., Medveczky, M. M., Geck, P., DeGrand, D., Vercelli, D., and Medveczky, P. G., 1994, Expression of IL-2 and IL-4 in T lymphocytes transformed by *Herpesvirus saimiri*, *Virology* **208:**418–426.
56. Murthy, S., Kamine, J., and Desrosiers, R. C., 1986, Viral-encoded small RNAs in *Herpesvirus saimiri* induced tumors, *EMBO J.* **5:**1625–1632.
57. Geck, P., Malter, J., Medveczky, M. M., Last, T. J., Whitaker, S., and Medveczky, P. G., 1990, Viral U-like RNAs in transformed T cells may alter mRNA stability in: *Abstracts VIIIth International Congress on Virology* P42-007, p. 360.
58. Raj, N. B. K., and Pitha, P. M., 1983, Two levels of regulation of beta-interferon gene expression in human cells, *Proc. Natl. Acad. Sci. USA* **80:**3923–3927.
59. Shaw, G., and Kamen, R., 1986, A conserved AU sequence from the 3' untranslated region of GM-CSF mRNA mediates selective mRNA degradation, *Cell* **46:**659–667.
60. Myer, V. E., Lee, S. I., and Steitz, J. A., 1992, Viral small nuclear ribonucleoproteins bind a protein implicated in messenger RNA destabilization, *Proc. Natl. Acad. Sci. USA* **89:**1296–1300.
61. Geck, P., Medveczky, M. M., Chou, C.-S., Brown, A., and Medveczky, P. G., 1994, *Herpesvirus saimiri* small RNA and interleukin-4 mRNA AUUUA repeats compete for sequence-specific factors including a novel 70kd protein, *J. Gen. Virol.* **75:**2293–2301.
62. Chou, C.-S., Geck, P., Medveczky, M. M., Hernandez, O., and Medveczky, P. G., 1994, *Herpesvirus saimiri* small RNA complex with an inducible 70kd AUUUA-binding factor, *Arch Virol.* **140:** 415–435.
63. Madore, S. J., Wieben, E. D., and Pederson, T., 1984, Intracellular site of U1 small nuclear RNA processing and ribonucleoprotein assembly, *J. Cell. Biol.* **98:**188–192.
64. Zeller, R., Nyffenegger, T., and De Robertis, E. M., 1983, Nucleocytoplasmic distribution of snRNPs and stockpiled snRNA-binding proteins during oogenesis and early development in *Xenopus laevis*, *Cell* **32:**425–434.
65. Heldin, C.-H., and Westermark, B., 1984, Growth factors: Mechanism of action and relation to oncogens, *Cell* **37:**9–20.
66. Sporn, M. B., and Todaro, G. J., 1980, Autocrine secretion and malignant cells, *N. Engl. J. Med.* **303:**878–880.
67. Smith, K. A., 1984, Interleukin 2, *Annu. Rev. Immunol.* **2:**319–333.
68. Spits, H., Yssel, H., Takebe, Y., Arai, N., Yokota, T., Lee, F., Arai, K.-I., Banchereau, J., and de Vries, J. E., 1987, Recombinant interleukin 4 promotes the growth of human T cells, *J. Immunol.* **139:**1142–1147.
69. Chen, S. J., Holbrook, N. J., Mitchell, K. F., Vallone, C. A., Greengard J. S., Crabtree, G. R., and Lin, Y., 1985, A viral long terminal repeat in the interleukin 2 gene of a cell line that constitutively produces interleukin 2, *Proc. Natl. Acad. Sci. USA* **82:**7284–7288.
70. Medveczky, P., and Medveczky, M., 1989, Expression of IL-2 receptor in T cells transformed by strains of *Herpesvirus saimiri* representing three DNA subgroups, *Intervirology* **30:**213–226.
71. De Carli, M., Berthold, S., Fickenscher, H., Fleckenstein, I. M., D'Elios, M. M., Gao, Q., Biagiotti, R., Giudizi, M. G., Kalden, J. R., Fleckenstein, B., Romagnani, S., and Prete, G. D., 1993, Immortalization with *Herpesvirus saimiri* modulates the cytokine secretion profile of established Th1 and Th2 human T cell clones, *J. Immunol.* **151:**5022–5030.
72. Swain, S. L., Weinberg, A. D., English, M., and Huston, G., 1990, IL-4 directs the development of Th2-like helper effectors, *J. Immunol.* **145:**3796–3806.

14

Transformation and Mutagenic Effects Induced by Herpes Simplex Virus Types 1 and 2

LAURE AURELIAN

1. INTRODUCTION

Compelling evidence indicates that cancer is a multistep process resulting from the accumulation of many genetic defects in the tumor progenitor cell. Malignant transformation, also mediated by DNA and RNA viruses, is an *in vitro* model of carcinogenesis in animals. The events triggering transformation often result from the transcription of genes, called oncogenes. These are part of a virus or are altered cellular genes, called protooncogenes, that transmit growth signals from the extracellular environment to the cell nucleus. The oncogenes of animal DNA tumor viruses are integral parts of the viral genome. Those of the retroviruses are normal or slightly modified cellular genes that were appropriated from the cell or were activated in the host cell by virus infection. Here we briefly review the available evidence on transformation by the human herpes simplex viruses (HSV), with particular emphasis on the various functions with which it is associated.

2. TRANSFORMATION BY INACTIVATED HSV

Original transformation studies with HSV-1 and HSV-2 were done by Duff and Rapp[1,2] with inactivated virus. These and subsequent studies with inactivated virus

---

LAURE AURELIAN • Virology/Immunology Laboratories, Department of Pharmacology and Experimental Therapeutics, University of Maryland School of Medicine, Baltimore, Maryland 21201-1192; and Departments of Biochemistry and Comparative Medicine, The Johns Hopkins Medical Institutions, Baltimore, Maryland 21205.

DNA Tumor Viruses: Oncogenic Mechanisms, edited by Giuseppe Barbanti-Brodano *et al.* Plenum Press, New York, 1995.

used primary rodent (rat or hamster) cells and/or the mouse aneuploid NIH3T3 cell line. The transformants were oncogenic, and they maintained and expressed viral DNA.[1–11] Skinner[12] was the first to propose the hit-and-run hypothesis in order to explain his failure to detect HSV DNA in transformed cells. Subsequently, Minson et al.[13] showed that this resulted from the successive loss of HSV DNA sequences on passage of the transformed cells. A possible interpretation for these findings is provided by the observations of Manak et al.,[14] who demonstrated that the different cellular alterations characteristic of transformed cells are induced by distinct HSV functions. These studies were done with HSV-2, the DNA of which was substituted with bromodeoxyuridine so that it could be fragmented (by exposure to near ultraviolet light) at various times after infection. The HSV-2 functions expressed at 2–4 hr postinfection (p.i.) were found to cause focus formation in 1% serum while other functions, maximally expressed at 4–6 hr p.i., induced anchorage-independent/neoplastic lines. Only 2/45 (4%) of the foci grew into anchorage-independent lines, and they did not retain HSV DNA. However, neoplastic lines induced by HSV functions expressed at 4–6 hr p.i. were HSV DNA positive. In detailed analyses, focus formation in 1% serum was evidenced by 11 neoplastic lines and some of their clonal derivatives.[14,15] Further studies of the clonal derivatives indicated that anchorage-independent growth and tumorigenicity are evidenced only by those clones that retain and express HSV DNA as indicated by the presence of viral DNA, mRNA, and the large subunit of HSV-2 ribonucleotide reductase (RR1)[5,6,14,15] designated ICP10 (Table I). These findings indicate that virion proteins and/or early (0–4 hr p.i.) HSV functions cause focus formation in

TABLE I
Properties of Primary Hamster Cells Transformed by HSV-2

Characteristic	HT4B cl1	HT4B cl2	HT4B cl3	HT8A cl4	HT8A cl1
Predominant morphology	Epithelioid	Epithelioid	Fibroblast	Fibroblast	Fibroblast
Saturation density ($\times 10^6$ cells/cm^2)	0.6	3.4	4.2	4.7	1.0
Growth in 1% serum[a] (% CE)	15	20	25	37	33
Anchorage independence[b] (% CE)	1	10	21	28	0
Tumorigenicity[c]	0	8/11 (72)	3/6 (50)	4/7 (57)	0
Latent period	–	13 weeks	10 weeks	9 weeks	–
Viral DNA[d]	–	+	+	+	–
Viral mRNA[e]	0	13 ± 5	21 ± 5	31 ± 6	0
ICP10[f]	0	+	+	+	0

[a]100 cells were seeded/35-mm dish. Foci (>40 cells) were counted on day 10. Results are cloning efficiency (CE) expressed as number of colonies/number cells seeded × 100%.
[b]10^4 cells were seeded/35-mm dish, and colonies (>20 cells) were counted at 2 to 3 weeks. Results are cloning efficiency (CE) expressed as number of colonies × 100/10^4 (%).
[c]Newborn hamsters were injected subcutaneously with 2 × 10^6 cells. Parentheses represent percentage.
[d]Spot blot hybridization with ICP10 PK oncogene.
[e]In situ hybridization with HSV-2 DNA. Grains were counted for 100 cells. Results are number of grains/cell ± SEM.
[f]Western blot with anti-LA-1 antibody (recognizes ICP10 PK residues 13–26).

1% serum that does not depend on the maintenance of HSV DNA. This in itself constitutes an increased risk for progression to oncogenicity mediated by cellular functions. Focus formation may result from mutagenesis,[16–23] gene amplification,[24–30] increased DNA synthesis,[31–33] and/or induction of cellular genes[34–40] including endogenous retroviruses,[41–44] all of which are evidenced by both HSV-1 and HSV-2. However, HSV-2 functions expressed somewhat later (4–6 hr p.i.) induce anchorage-independent growth and tumorigenicity. DNA sequences that encode at least one of these functions (ICP10) are retained and expressed in the transformed cells.[5,6,14,15]

3. TRANSFORMING HSV GENES

Studies of the mechanism of HSV-induced transformation sought to identify the transforming HSV genes. The data are summarized in Tables II and III. Temperature-sensitive (*ts*) HSV-1 mutants defective in various genes were studied in primary or secondary rat embryo cells.[8,9,45,46] These included *ts*K and *ts*D, both of which are DNA-negative mutants with defects in the immediate early (IE) gene IE175. At the nonpermissive temperature they produce four functional IE proteins [IE110 (ICP4), IE68 (ICP22), IE63 (ICP27), and IE12 (ICP47)] and a functional RR1.[47–49] However, the high level of cytotoxicity seen with these mutants,[50] coupled with low efficiency of transformation, made accurate analysis impossible. More recently, Bauer *et al.*[21] used a focus formation in agar assay in order to evaluate the transforming potential of a HSV-1 *ts* mutant in the U9 gene, which codes for a DNA-binding protein required for virus DNA synthesis.[51] Consistent with previous conclusions for HSV-2,[5,6,14,15] these investigators found that early (IE and/or delayed early, DE) HSV-1 gene expression is sufficient to induce focus formation that does not depend on the maintenance and expression of viral DNA. On the other hand, recent studies of HSV-2 *ts* mutants in ICP10 (*ts*5-152 and 859/152) support the conclusion that maintenance and expression of HSV-2 DNA sequences that code for IPC10 are required for the establishment of anchorage-independent/tumorigenic lines.[52]

Three regions of the viral genome were identified as having transforming potential using HSV DNA fragments. One of these, known as the morphological transformation region (*mtr*) I, was identified in HSV-1, and it maps in the *Xba*I *f* [map units (m.u.) 0.29–0.45][53] or *Bgl*II *i* (m.u. 0.311–0.415)[54] DNA restriction fragment. The *mtr*I caused focus formation in primary hamster and NIH3T3 cells, and at least a portion thereof was retained and expressed in the transformed cells.[53,54] However, late cell passages were not studied for the presence of viral DNA. Genes within *mtr*I (UL23–UL29) code for thymidine kinase (TK), the major DNA-binding protein (mDBP), glycoprotein gB, a protease, and two virion structural proteins.[51] The mDBP is a likely candidate for the *mtr*I function because it binds DNA,[55] promotes organization of nuclear structures involved in viral and cellular DNA replication,[56] destabilizes duplex DNA during origin unwinding and replication fork movement,[57] and has recombinase function.[58] However, recent studies suggest that the focus formation potential of *mtr*I results from the mutagenic function of a virion protein.[59]

TABLE II
Transformation by HSV-2

Strain	Map location (m.u.)	Gene function/features	References
333	0.58–0.62	RR2	54
HG52	0.58–0.62	RR2	62
G	0.58–0.62	RR2	75
HG52	0.58–0.62; 0583–0.596	RR2	69, 70
333	0.58–0.62	RR2	68
S-1	0.419–0.58	UL31–UL36 and RR1	78
333	0.419–0.525	UL31–UL36	64
333	0.533–0.58	RR1	64, 79
333	0.554–584	RR1	77
333	0.53–0.58	RR1	81; L. Kucera (personal communication)
333	0.569–0.576	ICP10 PK oncogene	52, 82
333	0.572–0.5766	Amino acid residues 70–501 of ICP10 PK	80
333	737 bp left of 0.6	IS-like structure	67, 138
333	0.567–0.570 (486TF)	IS structure; DNA repeats; promoter elements	133, 135
HG52	Infectious virus	Mutagenesis at HGPRT gene	22, 23
G, 333	Infectious virus	Amplication of DNA (SV40 or HPV18)	29
G	Infectious virus	Activation of DNA synthesis	31
333	0.58–0.62	Activation of DNA synthesis	33
HG52 186, 333	Infectious virus or dsDNA	Induction of cellular genes in transformed and immortalized cells	34, 36–40
333	0.5–0.75	Induction of serum amyloid-A-related mRNA	35
Savage	Total DNA	Induction of MuX	41
333	0.05, 0.42–0.58, and 0.8–1.0	Homology to cellular DNA sequences	63, 79
pCsla1	0.5719–0.5725	Cloned from human cells	76

The other two transforming regions in the HSV genome are in HSV-2. They are mtrII, mapping[54,60–62] in the BglII N DNA restriction fragment (m.u. 0.58–0.62) and a region confusingly referred to as mtrIII[63] that maps in the BglII C restriction fragment (m.u. 0.54–0.58)[64] of HSV-2 DNA (Fig. 1). The BglII C and BglII N fragments respectively encode the large (RR1, ICP10) and small (RR2) subunits of HSV-2 ribonucleotide reductase, and the colinear 3′ ends of the two transcripts map within the BglII N fragment.[65] Accordingly, Huszar and Bacchetti[66] speculated that RR is involved in HSV-2-induced transformation. However, because both RR subunits are needed for enzymatic activity, and they are not encompassed within either of the two transforming DNA fragments, RR activity is not likely to be involved in transformation.

TABLE III
Transformation by HSV-1

Strain	Map location (m.u.)	Gene function/features of DNA	References
F	0.29–0.45	UL23-UL31, gB, TK, DBP, DNA polymerase	53
STH and MP	0.311–0.415	UL23-UL29	54
McIntyre	Infectious virus	Mutagenesis	16
17 and *ts* mutants	0.39–0.42	Mutagenesis; DNA polymerase	26, 27, 172
17		UL30 mutagenesis; UL29, UL5, UL8, UL42, and UL52	28
MP	0.38–0.392	Reversion of bacterial mutation; function unknown	17
17; synt McIntyre	Infectious virus	Amplication of integrated SV40 DNA and flanking cellular sequences in rodent cells	24–27
KOS		Amplification of HPV-18 DNA sequences in cervical cell lines	30
17		Increased expression of cellular genes; function unknown	174
Patton	0.29–0.32 0.46–0.49 0.92–0.97	Induction of MuX	41
MP	Mainly S segment and inverted repeats of S	Homology to mouse and human DNA	63
Paton	Around 0.83	Homology to mouse DNA	151, 173
KOS	Left of 0.80	Homology to human DNA	153–156

4. RR2 DOES NOT CAUSE NEOPLASTIC TRANSFORMATION

*Bgl*II *N* restriction fragment causes focus formation in NIH3T3 or rat embryo cells.[54,62,67,68] When present in lines established from these foci, HSV DNA sequences were detected in very low copy number, and they were lost on passage in culture and/or in tumor derivatives.[61,62,67,68] Saavedra and Kessous-Elbaz[69] suggested that the complete *Bgl*II *N* fragment is "poisonous" for cells, as its presence actually depressed selectable transformation with the neomycin resistance (neo^R) gene. They cotransfected immortalized NIH3T3 cells with a plasmid that encodes only RR2 (m.u. 0.583–0.596) and the neo^R gene and analyzed the isolated clones for retained RR2 sequences, their expression, and a tumorigenic phenotype. They found that neither the retention of the RR2 gene nor its expression correlates with tumorigenic conversion of these immortalized cells.[70]

Galloway *et al.*[67] found that focus formation in NIH3T3 and rat embryo cells is mediated by a 737-bp fragment within *Bgl*II *N* (BC24) that does not lie within coding sequences. They proposed that BC24 is a small stem–loop structure bounded by direct repeats that contains a region resembling an insertion sequence (IS). They suggested (hit-and-run hypothesis) that the IS sequence acts as a mutagen, and this mutagenesis produces the first of a series of events (activation of a cellular oncogene, enhancement of cellular functions) that lead to the expression of the transformed phenotype.[67,71] The stem–loop structure proposed for BC24 resembles the P elements of *Drosophila*.[72] However, it is not large enough to encode a transposition function,[73] it does not conform to the classical description of an insertion sequence, and its structure is unstable because of base mismatches. Indeed, using a computer program that recognizes IS sequences and other transposable elements, Shillitoe et al.[17] found that IS-like structures are common in many regions of HSV DNA, thereby raising the question of why the IS structure in BC24 is able to cause focus formation if other such structures do not. Furthermore, *Bgl*II N, BC24, and the RR2 gene did not have mutagenic activity,[20,23,74] arguing against the hit-and-run hypothesis.

How do we explain the original observation that the *Bgl*II N restriction fragment causes focus formation? An interesting interpretation is provided by the findings of Becjek and Conley,[75] who detected extrachromosomal DNA after transfection of *Bgl*II N sequences with the selectable neo^R gene and rescued a plasmid that contained both *Bgl*II N sequences and cell sequences with homology to *Bgl*II C. The rescued plasmids could transform cells 1000 times more efficiently than the *Bgl*II N fragment alone, suggesting that *Bgl*II N causes the rearrangement of a protooncogene (homologous to sequences within *Bgl*II C), thereby inducing its activation and causing the resulting neoplastic lines. The recent cloning of sequences homologous to the HSV-2 RR1 gene (ICP10) from human tissues[76] supports this interpretation.

5. NEOPLASTIC TRANSFORMATION IS A MULTISTEP PROCESS

Several studies indicate that escape from senescence (immortalization) is an early event in neoplastic transformation *in vitro*, which is mediated by viral genes other than those required for the full expression of a tumorigenic phenotype. In accord with these observations, the cloned *Eco*RI/*Hind*III AE restriction fragment of HSV-2 DNA (m.u. 0.419–0.525) that represents 64% of *Bgl*II C (Fig. 1) caused the immortalization of primary hamster cells, but the cells remained nontumorigenic.[64,77] Lines displaying anchorage-independent growth and tumorigenicity were established by transfection of the immortalized cells with a 4.4-kb segment within the *Bam*HI E restriction fragment of HSV-2 DNA (m.u. 0.554–0.584) that overlaps 16% of the right hand of *Bgl*II N (Fig. 1) and encodes ICP10.[52,64,77–81] The transforming 4.4-kb segment was retained and expressed in the transformed cells.[52,64,77–82] Immortalized cells transfected with the 2.0-kb (m.u. 0.533–0.546) and 1.0-kb (m.u. 0.546–0.533) subfragments of *Bam*HI E did not acquire anchorage-independent growth and tumorigenic potential.[77] However, the transforming potential did not require the entire ICP10 coding gene, but, rather, it localized in fragments *Sal*I/*Hpa*I (m.u. 0.569–0.576)[52,82] or *Pst*I C (m.u. 0.572–0.577)[80] that code for the first 411–500 amino acid residues of ICP10[83] (Fig. 1). In rat cells transformed with *Bam*HI E, extrachromosomal DNA was found to contain the left-hand 70% of HSV-2 *Bam*HI E fused to cellular DNA.[79] This further supports the interpretation[75,76] that normal cells contain DNA sequences (a putative proto-

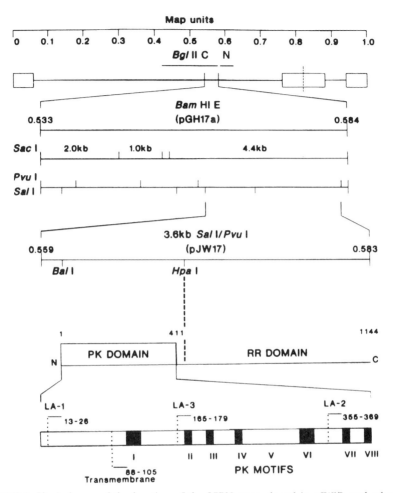

FIGURE 1. Physical map of the location of the ICP10 gene cloned in pJW17, and schematic representation of ICP10 as a multifunctional protein. Two domains are identified: an amino-terminal PK domain (residues 1 to 411) and a carboxy-terminal RR domain (residues 411 to 1144). The unique HpaI site within ICP10 represents the 3' end of the minimal transforming region of HSV-2 and cuts the gene after the codon for residue 417. The expanded PK domain highlights regions of homology with catalytic PK motifs (I–VIII). The TM (transmembrane) and synthetic peptides LA-1, LA-3, and LA-2 used in protein studies are identified.

oncogene) that are homologous to the transforming HSV-2 gene ICP10, which can be targeted for genetic alterations associated with neoplastic transformation.

6. HSV-2 GENES THAT CAUSE CELLULAR IMMORTALIZATION

The HSV-2 immortalizing functions were localized at position 0.419–0.525 on the DNA map.[64,77,78] Genes localized at this site are unique long region (UL) 31–

UL36.[51] The function of UL32 is unknown. The product of UL33 is required for the assembly of full virus capsids.[84] UL35 encodes the 12-kDa capsid protein,[85] and UL36 encodes the large tegument protein ICP1/2.[86] It is therefore unlikely that these genes are involved in immortalization. On the other hand, UL31 codes for a 34-kDa phosphoprotein that partitions with the nuclear matrix,[87] and inasmuch as the nuclear matrix may be involved in the regulation of messenger RNA maturation and transport and in gene expression,[88] UL31 may be involved in immortalization. Protein products of immortalizing genes that localize to the nuclear matrix were described for adenovirus (E1A)[89] and simian virus 40 (SV40) (large T antigen).[90] UL34 could also be associated with immortalization. Its gene product is a 30-kDa phosphoprotein that complexes with cellular 25- to 35-kDa phosphoproteins. It has been suggested that the unphosphorylated UL34 protein functions as a regulatory protein by activating a cellular kinase or inactivating a phosphatase.[91] In the context of immortalization such a cellular protein could be cyclin A, which complexes with a 33-kDa protein kinase (PK, cdk2) to play a role in G_1/S phase transition. Indeed, cyclin A is associated with the transcription factor E2F during S phase, in a complex that has a lower transcriptional activity than free E2F. By complexing with cyclin A,[92] the adenovirus immortalizing protein E1A disrupts the complex and releases E2F, thereby influencing the regulation of genes important for immortalization. The masses of the UL31 and UL34 phosphoproteins are consistent with that (27 kDa) of the immortalization-associated HSV-2 protein.[93] However, further studies are required in order to elucidate the mechanism of HSV-2-induced immortalization.

7. THE ICP10 PK ONCOGENE

The HSV-2 RR1 protein ICP10 differs from its counterparts in eukaryotic and prokaryotic cells and in other viruses in that it possesses a unique amino-terminal region that causes a 50% increase in molecular weight and is encoded by the *Sal*I/*Hpa*I transforming DNA fragment (Fig. 1). Computer-assisted analysis of the predicted amino acid sequence of ICP10[83,94] revealed the presence of PK catalytic motifs (I through VIII) clustered within the unique amino terminus, at residues 1 to 411. Motif I, at residues 105 to 110, is a near-consensus Gly-X-Gly-X-X-Gly found in many nucleotide-binding proteins; it is followed at residue 112 by Val, consistent with other PKs.[95] Residues 174 to 176 (motif II) include the invariant Lys that appears to be involved in phosphotransfer reactions.[96] In other Pks, all substitutions at this site have resulted in loss of PK activity.[95] Other motifs include the triplet Ala-Pro-Glu (motif VIII) that contains the invariant residue Glu, considered a key PK catalytic domain indicator,[98] and the invariant or nearly invariant residues in motifs VI and VII that have also been implicated in ATP binding.[97] Motif VI contains residues that are specifically conserved in either the serine/threonine (Ser/Thr) or the tyrosine (Tyr) PKs. The ICP10 motif is similar to the consensus Asp-Leu-Lys-Glu-Asn, considered a strong indicator of Ser/Thr specificity,[95] and phosphoamino acid analysis of *in vitro* phosphorylated ICP10 confirmed that it is a Ser/Thr PK.[83] Computer-assisted analysis of the ICP10 promoter also identified cis-response elements associated with the regulation of IE gene expression and functional AP-1 elements. Both of these are associated with the regulation of the ICP10 PK function.[100,101]

By using constructs (Fig. 2) that express the entire ICP10 protein (pJW17), its amino-terminal domain (m.u. 0.569–0.576; pJW32), its carboxy-terminal domain (m.u. 0.576–584; pJW31), and an expression-negative frameshift mutant (pJW21) and antibodies to synthetic peptides that represent amino acid residues at various sites within ICP10, we showed that the 57- to 60-kDa amino-terminal domain of ICP10 is a functional Ser/Thr-specific PK.[83] The following evidence supports the interpretation that ICP10 has intrinsic kinase activity that does not reside in a contaminant protein. First, both auto- and transphosphorylating Ser/Thr activities were detected in cells transfected with the expression vectors pJW17 or pJW32 but not pJW31 or pJW21, as determined by immunocomplex kinase assays[83,94] and in immunoprecipitates of metabolically labeled transfected cells.[102] It is unlikely that kinase activity was caused by a contaminating protein because it was not seen in similar assays of cells transfected with pJHL9 (Fig. 2), which is deleted in all eight catalytic motifs and expresses a 95-kDa protein.[102] Second, a PK (pp29^{lal}) that is antigenically and structurally identical to the first 280 amino acid residues of the ICP10 PK (Fig. 2) was expressed in *Escherichia coli*.[103] Because most bacterial PKs have a structure totally unrelated to eukaryotic PKs,[104] it is highly unlikely that the putative contaminant is structurally identical in both eukaryotic and prokaryotic cells. Finally, PK activity was retained by ICP10 separated by SDS-PAGE and transferred to a membrane filter, and it specifically bound the ^{14}C-labeled ATP analogue p-fluorosulfonylbenzoyl 5′-adenosine (FSBA), a binding abolished by another analogue (AMP-PNP).[102] Significantly, the ICP10 expression vector pJW17, pJW32, which expresses the ICP10 PK domain (amino acid residues 1–411), or a vector (*Pst*I C) that expresses amino acid residues 70–501 converted immortalized hamster, rat, or human cells to an anchorage-independent/neoplastic phenotype. The vectors pJW31 and pJHL9, which express the ICP10 RR domain, and the frameshift (expression negative) mutant pJW21 did not impart neoplastic potential. The viral DNA (ICP10 PK oncogene) was retained and expressed in the transformed cells.[52,80,82]

8. ICP10 PK IS A NOVEL KINASE

ICP10 PK is a novel Ser/Thr kinase that functions in phosphotransfer reactions with regions of the molecule distinct from those previously described.[95] Thus, catalytic motifs I/II are not absolutely essential for the kinase activity of ICP10.[102] Site-directed mutants pJHL2 (motif I: Gly^{106}) and pJHL4 (motif II: Lys^{176}) and a mutant deleted in both motifs (pJHL17) (Fig. 2) retained auto- and transphosphorylating activities *in vitro* and *in vivo*, albeit at a significantly reduced level. We do not believe that *in vivo* phosphorylation of p140^I, p140^{II}, or p136$^{I/II}$ (respectively expressed by pJHL2, pJHL4, and pJHL17) is catalyzed by an unrelated coprecipitated protein because activity was decreased and phosphorylated proteins were not seen in precipitates from cells transfected with the PK-deleted mutant pJHL9

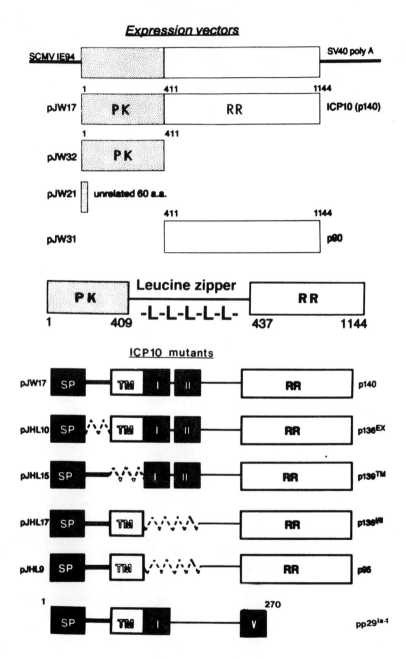

studied in parallel.(102) Supporting the conclusion that the invariant Lys (Lys^{176}) is not absolutely required for ICP10 PK activity is the observation that p140^{II} still binds [^{14}C]FSBA,(102) and kinase activity is evidenced by p140^{II} expressed in *E. coli*, although both are reduced.(103)

However, catalytic motifs I/II contribute toward optimization of kinase activity because (1) K_m values (calculated from immunocomplex kinase assays) were increased by mutation of Lys^{176} or by deletion of motifs I and II and (2) *in vivo* kinase activity was lower in cells transfected with the mutants in motifs I/II than with wild-type ICP10. The contribution of catalytic motifs I/II presumably involves the ability to respond to Mn^{2+} ions.(102) The alternate ATP binding site is Lys^{259} and Pk activity is only lost by mutation of both Lys residues. Gly^{209} is the Mn^{2+} chelation site (Nelson *et al.*, in preparation). Based on the finding that the ICP10 protein expressed in bacteria (pp29^{lal}) retains PK activity although it lacks catalytic motifs VI–VIII, we conclude that ICP10 kinase activity requires only catalytic motifs III to V.(103)

The HSV-1 RR1 (ICP6) is also a chimera protein that consists of a PK domain at the amino terminus and the RR domain at the carboxyl terminus.(83,94,105,106) However, ICP10 has both auto- and transphosphorylating activities,(52,83,94,102,103) whereas the ICP6 PK activity is only autophosphorylating.(105) Factors responsible for the failure of ICP6 to evidence transphosphorylating activity are unclear: PK catalytic motifs I and II are not conserved in ICP6,(83) and its kinase activity has different ion requirements than that of ICP10.(83,105,106) However, it is unlikely that their loss is responsible for the failure of ICP6 to evidence transphosphorylation because the motifs are not required for the ICP10 auto- or transphosphorylating activities.(102) A more likely interpretation is that ICP6 does not bind certain substrates involved in the transmission of the growth signal to the nucleus and therefore does not transphosphorylate them. Consistent with this interpretation, the ICP6 ATP binding site is outside of the PK catalytic loop(175) and computer-assisted analysis of the ICP10 PK minigene-predicted amino acid sequence revealed a consensus SH3 binding motif(108) that may be involved in protein–protein interactions(109,110) at residues 150–159 [x-P-xx-PPPΨxP (where Ψ denotes a hydrophobic amino acid)].(81) This motif is missing in ICP6.(82) Presumably the failure of ICP6 to evidence binding/transphosphorylating activity is responsible for its failure to cause neoplastic transformation.

←—————————————————————————————————————

FIGURE 2. Schematic representation of the ICP10 PK expression vectors and their mutants. ICP10 (p140) is a chimera consisting of the PK (residues 1–411) and RR (residues 411–1144) domains. It is expressed by vector pJW17. Vectors pJW32 and pJW31, respectively, express the 60- and 90-kDa PK and RR domains. pJW21 is an expression-negative frameshift mutant. The junction of the PK and RR domains consists of a leucine zipper at residues 409–437 that functions to stabilize the RR activity. Additional features include a signal peptide (SP), an extracellular domain (residues 1–85), a transmembrane domain (TM) (residues 85–105), and PK catalytic motifs I–VIII. pJHL10 expresses p136^{Ex} deleted in the extracellular domain; pJHL17 expresses p136$^{I/II}$ deleted in catalytic motifs I/II; pJHL15 expresses p139^{TM} deleted in the TM; and pJHL9 expresses p95 deleted in all PK catalytic motifs. The minimal size of the ICP10 PK is pp29^{lal}, which encompasses the SP, TM, and PK catalytic motifs I–V (residues 1–280).

9. ICP10 PK IS A GROWTH FACTOR RECEPTOR

Additional features revealed by computer-assisted analysis of the predicted ICP10 amino acid sequence are a potential transmembrane helical segment at residues 85 to 105, four potential N-glycosylation sites (one in the extracellular domain), and features of a signal peptide, including a positively charged residue (Arg^4), followed by a short (nine residues) hydrophobic core ending with the Ala-X-Ala motif (Ala^{11}-Gly^{12}-Ala^{13}),[94] the most common signal peptidase recognition site found in membrane-associated proteins.[99] These features suggest that ICP10 may behave as growth factor receptor. Also, like the growth factor receptor kinases that can be activated to transforming potential,[111–114] the ICP10 PK domain (1) is myrystylated[94]; (2) its single transmembrane (TM) segment (at amino acid residues 85–105) is followed by a basic residue (Arg^{107})[83,94] that is thought to anchor the membrane-spanning helix[113,114]; (3) amino acid residues within its extracellular domain (upstream of the TM) have a modulatory effect on the PK activity[94]; and (4) its kinase activity is greatly enhanced by polylysine and Mn^{2+} ions.[94] The extracellular domain is not required for PK activity as evidenced with a mutant (pJHL10) deleted in residues 14–85 (Fig. 2). However, the signal peptide is essential for both expression and kinase activity.[102,103]

Consistent with the conclusion that it functions as a growth factor receptor kinase, ICP10 is located in the plasma membranes of transformed human cells. Studies of the polarity of the ICP10 association with the plasma membranes indicate that amino acid residues 1–84 (upstream of the TM) are extracellular and that residues 106–1144 downstream of the TM are intracellular.[82,102] Indeed, cells transformed by ICP10 stain in membrane immunofluorescence with antibody to ICP10 residues 13–26 [(LA-1) also called ICP10 PK oncoprotein (Fig. 1)] upstream of the TM but not with antibody to ICP10 residues 165–179 [LA-3 (Fig. 1)] downstream of the TM (Fig. 3).

The growth factor receptor function of ICP10 is essential for transforming potential. Thus, TM deletion abrogated the ability of ICP10 to localize to the plasma membranes as well as its PK activity.[82,102] The TM-deleted protein (p139^{TM}) (Fig. 2) was PK negative, although it was expressed as well as ICP10. Because the TM is not the ICP10 autophosphorylation site,[102] loss of PK activity by TM deletion may reflect the loss of membrane localization. Alternative interpretations for the loss of PK activity by TM deletion are the potential involvement of the TM in the genera-

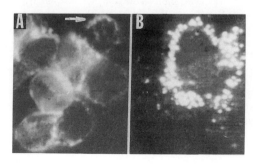

FIGURE 3. Membrane (A) ad fixed-cell (B) immunofluorescence of ICP10-transformed human cells with anti-*LA-1* antibody reveals patchy aggregates (arrow) characteristic of growth factor receptors.

tion of a functional conformation with the PK catalytic motifs or the possibility that deletion of the TM segment produces an allosteric effect so that the tertiary structure of the PK catalytic motifs is detrimentally altered. Ongoing studies with ICP10 mutants containing distinct nuclear[115] or cytoskeleton association signals[116] that alter subcellular localization should help differentiate between these two interpretations. In any case, it is noteworthy that the TM-deleted mutant (pJHL15) that expresses p139^{TM} (Fig. 2) is transformation negative.[82,102] Human cells that constitutively express p139^{TM} or the ICP10 RR domain (p95) (Fig. 2) did not evidence anchorage-independent growth, indicating that the PK activity is required for transformation.[82,102] Involvement of the TM in PK activity and transforming potential was also reported for other growth factor receptor oncoproteins.[117] Computer-assisted analysis using the PROTDIST algorithm indicates that ICP10 PK[118] is a novel member of the family of TM-containing Ser/Thr kinases that include the activin receptor[119] and the receptor for transforming growth factor β.[120]

10. SIGNALING PATHWAYS IN ICP10 PK MINIGENE TRANSFORMED CELLS

Recent studies indicate that *ras* activation, involving specific protein–protein interactions, is required for receptor-mediated mitogenic and oncogenic processes.[121,122] On the positive side, Sos1 functions as a guanine nucleotide-releasing factor that promotes the conversion of the inactive *ras*-GDP to the active GTP-bound state. Sos1 recruitment and binding to growth factor receptors (in a complex with a 25-kDa adaptor protein designated Grb2) facilitates *ras* activation.[123–125] On the negative side, *ras*–GAP [GTPase-activating protein; influences the catalytic activity of the *ras* protooncoprotein by promoting its return from the active (GTP-bound) to inactive (GDP-bound) state] stimulates the otherwise weak intrinsic GTPase activity of *ras*, thereby promoting its return to an inactive GDP-bound state.[126,127] Possibly, effective signal transmission requires both Sos1 binding and the inactivation of *ras*–GAP by phosphorylation or complexing with a phosphorylated p190 species designated *ras*-GAP-p190.[128,129]

Consistent with the conclusion that the ability of the ICP10 PK oncogene to transform cells results from its growth factor receptor kinase function, immortalized human cells transfected with pJHL15 (which expresses the transmembrane deleted PK-negative p139^{TM}) and JHL15 cells that constitutively express p139^{TM} and are also PK negative did not evidence anchorage-independent growth[82,102] We conclude that PK activity is required for transformation because anchorage-independent growth was also not evidenced by JHL9 cells that express p95, which is deleted in the PK catalytic domain but retains the TM segment. Electron microscopy of immunogold-stained cells (with antibody to an epitope within the PK domain) confirmed the conclusion that ICP10, but not its TM-deleted mutant (p139^{TM}), is located on the cell surface and is rapidly internalized in multivesicular endosomes characteristic of ligand-activated growth factor receptors. However, the putative ICP10 ligand has yet to be identified.[118]

The mechanism of transformation by the ICP10 PK oncogene is similar to that described for growth factor receptor oncogenes, and it involves signaling pathways that include *ras* activation. Thus, using immunoprecipitation/immunoblotting experiments such as those routinely used to identify proteins in signaling complexes,[124,130,131] we found that Sos1 binds ICP10 and is coprecipitated with it from transformed cells. p139^{TM} does not bind Sos1. Possibly Sos1 binding requires membrane localization and/or available SH3 binding sites that are obscured by conformational changes resulting from TM deletion. Immunoprecipitation/immunoblotting assays also indicated that *ras*–GAP binds to ICP10 and serves as its phosphorylation substrate; p139^{TM} binds *ras*–GAP equally well, but it fails to phosphorylate it. Furthermore, a 190-kDa phosphorylated protein that may be *ras*–GAP-p190 also coprecipitated with *ras*–GAP only from JHLa1 cells. Consistent with the conclusion that the phosphorylated (and/or p190-complexed) *ras*–GAP has a decreased negative regulatory potential, the levels of GTP hydrolysis were significantly lower in JHLa1 than in JHL15 cells. The outcome of these interactions was a major increase in the levels of activated (GTP-bound) *ras* in ICP10-transformed JHLa1 cells as compared to JHL15 cells that express the PK-negative p139^{TM} and are not transformed.

11. THE ICP10 PK ONCOGENE IS AN HSV-APPROPRIATED CELLULAR GENE

Sequences homologous to a portion of the ICP10 PK oncogene were cloned from human tissue,[76] suggesting that the PK minigene may have evolved from a cellular gene that was inserted into the polypeptide-coding region of an ancestral HSV RR1, such as that reported for the human somatomammotropin gene.[132] Presumably the upstream (5′) recombination site is within the promoter of the HSV-2 RR1 gene, thereby explaining the presence in this promoter of *cis*-response elements such as AP-1[100,101] that are characteristic of cellular but not HSV genes. The C-terminus (3′) recombination site at the junction of the PK and RR presumably occurred within the promoter region of the ancestral HSV-2 RR1, thereby giving rise to an open reading frame (ORF) that contains enhancer core and functional promoter elements.[101,133] Indeed, when cloned into expression vectors, the ORF region had functional promoter elements and bound cellular factors.[133,134] Additionally, a 95-bp fragment derived from the ICP10 PK minigene was shown to function as a silencer when placed adjacent to a heterologous promoter.[135]

Implicit in the interpretation that the PK minigene originated from a homologous cellular gene is the assumption that by participating in the viral life cycle, the cellular gene provided a functional advantage that justified its conservation. A critical question, therefore, is the role of the ICP10 kinase activity in the virus life cycle. A leucine zipper motif at the junction of the ICP10 PK and RR domains (Fig. 2) was presumably generated through the incorporation of the PK minigene, and it is critical for RR complex formation and enzymatic activity.[107] However, the functional significance of the PK domain does not depend on the structural integrity of the fusion protein, as there is evidence that the PK and RR activities are separated by proteolysis of the RR1 protein during infection.[83,136]

12. ICP10 PK EXPRESSION IN HSV-2-INFECTED CELLS

The finding that ICP10 is a chimera protein that has both membrane-associated growth factor receptor PK activity and nucleus-associated RR activity raises the question whether the PK is functional in HSV-2-infected cells. Computer-assisted analysis of the ICP10 amino acid sequence according to the PEST SCORE algorithm revealed the presence of motifs (at residues 177–212) that are associated with rapid intracellular protein degradation (score = 23.15) and a short half-life.[137] Similar sequences were also identified in the HSV-1 protein at residues 212 to 243 (PEST SCORE 6.65) but not in the HSV RR2 nor in the mammalian or the Epstein–Barr virus or varicella–Zoster virus RR1 proteins.[83] These findings suggest that in HSV-infected cells, the RR and PK functions in ICP10 might become dissociated. Indeed, we found that at 0 hr p.i., ICP10 represented 25.9% of the total labeled proteins in HSV-2-infected cells. At 5 hr it was no longer detectable. On the other hand, the 55-kDa PK domain of ICP10 increased from 14.4% at 0 hr to 35.4% at 5 hr. This suggests that the PK and RR domains of at least some of the ICP10 protein synthesized in HSV-2-infected cells are rapidly dissociated by proteolytic cleavage. It is unclear whether the dissociated RR domain can still complex with RR2 because an intact leucine zipper at the junction of the PK and RR domains (Fig. 2) is required for complexation and RR activity.[107] However, both the entire ICP10 and its PK domain were detected in purified membranes from HSV-2-infected cells.[83] ICP6 also undergoes proteolytic degradation in HSV-1-infected cells.[136]

13. MUTAGENESIS AND GENE AMPLIFICATION

Inactivated HSV-1 is mutagenic for the hypoxanthine–guanine phosphoribosyltransferase (HGPRT) gene of the permissive human rhabdosarcoma cells RD-176, and HSV-2 increases the mutation frequency of the HGPRT gene in the nonpermissive XC cell line.[22,23] It was suggested that mutagenesis is not an HSV function, as introduction of dsDNA *per se* mutagenizes the HGPRT gene.[74,138,139] However, by analogy to the Ames test for mutagenic activity,[140] Shillitoe et al.[17] introduced the *Bam*HI g fragment of HSV-1 into the expression vectors pUC7, pUC8, and pUC9 in both orientations and identified a higher rate (up to 39-fold) of reversion of a frameshift mutation in *E. coli* caused by only one of these constructs. More recently, these investigators demonstrated that mutagenesis depends on HSV-1 binding to the cell surface and disassembly of the virus particles, but expression of virus genes is not necessary. They identified a virion protein encoded within *mtr*I that may have the mutagenic property.[59] Similar conclusions were also reached by Clarke and Clements,[20] who found that mutagenic potential localizes to a component of the incoming HSV-1 or HSV-2 virion or is an effect exerted by the viral DNA itself.

Spontaneous mutants differed from those induced by HSV-1 in several properties. Thus, point mutations accounted for 63% of spontaneous as compared to 44% of HSV-1-induced mutations, and deletions of DNA were seen at a 44% frequency in HSV-1-induced mutations as compared to 29% in spontaneous mutants. A size increase (from 118 to 4500 bp) was seen only for HSV-l-induced mutants, and it was created by duplication of plasmid DNA or (in half of the mutants) by insertion of sequences derived from cellular DNA. The inserted DNA contained multiple repeats such as α-satellite Alu or Kpn repeats. There was no preferred site for recombination or similarities between the inserted sequences.[18,19] This finding indicates that HSV-1 stimulates nonhomologous recombination between cellular DNA sequences.[18,19]

Gene amplification is thought to be an essential step in the alteration of a normal to an oncogenic cell.[141] In a manner similar to chemical carcinogens,[142] inactivated HSV-1 was shown to cause amplification of the viral genes in SV40-transformed cells[25] and amplification of integrated human papillomovirus 18 DNA sequences in HeLa, C4-1, and C4-11 cells.[30] Amplification, monitored by *in situ* hybridization and Southern blot analysis, was more pronounced than that induced by the carcinogen benzo(a)pyrene and similar to that induced by 4-NQO. Original studies indicated that the expression of the HSV-1 polymerase gene is required for gene amplification. Drugs acting on the HSV polymerase abolished the amplification potential, and DNA polymerase-negative mutants did not induce gene amplification.[26] However, more recent studies suggest that six HSV-1 genes acting together are necessary and sufficient for DNA amplification. These are UL30 (DNA polymerase), UL29 (mDBP), UL5, UL8, UL42, and UL52.[28] Because the last four genes are not present in *mtr*I, it is unlikely that gene amplification is the mechanism for *mtr*I-induced focus formation. However, the possibility is not excluded that HSV-1-induced focus formation depends on its ability to cause gene amplification.

14. ACTIVATION OF DNA SYNTHESIS AND INDUCTION OF CELLULAR GENES

The term "firone" has been introduced by Varshavsky[143] to describe substances that activate cellular replicons outside the normal course of their replication during the cell cycle. Agents (such as chemical carcinogens) that damage DNA or inhibit its replication act as firones and induce gene amplification following their removal from the treated cells.[144] Indeed, chemical carcinogens that arrest the cell cycle early in the S phase cause amplification of resident SV40 genomes and cellular sequences such as c-Ha-*ras* and *dhfr*.[142] It has been suggested that the perturbation of DNA replication causes accumulation of the DNA-synthetic machinery, the misfiring of replicons, and consequent aberrant replication of specific DNA sequences.[143,144] Danovich and Frenkel[29] considered the possibility that HSV, which causes arrest of host DNA replication, acts as a firone in cells in which the infection aborts. In these cells, aberrantly amplified host DNA sequences could undergo further rearrangements,[144] and the cells containing amplified or translocated activated oncogenes would be further selected for advantageous growth. Consistent with such an interpretation, HSV can activate DNA synthesis and induce unscheduled cellular DNA synthesis[31,32] (apparently caused by both mDBP and cellular proteins),[32] thereby potentially initiating an event that facilitates later stages in the process of acquiring an oncogenic potential.[145] Both HSV-1 and HSV-2 were shown to induce replication of plasmids containing the SV40 DNA replication origin and the large T gene in nonpermissive cells[29] and the amplification of DNA sequences devoid of a replication origin.[25,28–30]

Filion et al.[34] and Gervais and Suh[35] identified several clones in cDNA libraries representing mRNAs transcribed at increased levels in HSV-2 transformation, and two heat shock proteins were overexpressed in immortalized rat cells.[38,39] However, their exact role in transformation, if any, is unclear. Potentially significant is the ability of HSV-2 to induce c-*fos*,[40] which is a component of the AP-1 transcription factor associated with transformation.[146] HSV also increases c-*jun* expression and AP-1 activity.[147] The nuclear oncoprotein c-*myc* was also overexpressed in cells transformed by the ICP10 PK oncogene,[77] and inactivated HSV-2 as well as the *Bam*HI *E* restriction fragment of HSV-2 DNA induced phospholipase and cyclooxygenase functions in transformed rat cells.[36] Finally, by analogy to cells transformed by chemical carcinogens,[148] DNA from cells transformed by HSV-1 or by *Bgl*II *N* is hypomethylated, a finding associated with some viral DNA synthesis.[149] We find that intracisternal A particles (IAP)[150] are undermethylated in cells immortalized by the HSV-2 *Eco*RI/*Hind*III *AE* restriction fragment and in cells neoplastically transformed by the ICP10 PK oncogene. In the latter, the IAP sequences were also transcriptionally active, consistent with the ability of the ICP10 PK to initiate and/or cause transcriptional activation of cellular genes.

15. ACTIVATION OF ENDOGENOUS VIRUSES

An HSV infection activated an endogenous C-type virus (MuX) from a feline cell line (F81) transformed by the Moloney strain of murine sarcoma virus,[41,44] and HSV-1 sequences that map within *mtr*I were shown to induce MuX in a line of BALB/c mouse cells.[42] Regions of the HSV-1 genome, located within 0.46 to 0.49 and 0.92 to 0.97 m.u., had equal reactivation potential. The induction of the MuX function could be the result of mutagenesis by *mtr*I or other HSV DNA sequences and/or the activation of gene expression by other as yet unidentified mechanisms.

16. HOMOLOGY OF HSV DNA WITH CELL DNA SEQUENCES

Homologous DNA sequences provide a mechanism by which the virus may integrate into the host cell by homologous recombination. Such recombination may be involved in the enhancement of cellular gene activity or in the deregulation of genes encoding proteins important in cellular control. In the case of HSV-1, considerable homology has been detected in the long (L) and short (S) inverted and terminal repeat regions of the genome to both human and mouse cell DNA.[63,151–153] This homology is not caused by G + C-rich sequences.[154] Jones et al.[155] and Parks et al.[156] have shown homology between human 28S RNA and HSV DNA from the inverted repeat region of the L segment. Homology between mouse and human cell DNAs and the L and S inverted repeat regions and the center of the L region of HSV-2 DNA has also been reported.[63,79] However, HSV DNA integration does not seem to be a likely mechanism for HSV-mediated transformation, as it is a relatively rare event. On the other hand, the ICP10 PK oncogene (located in the center of the L region of HSV-2 DNA) presumably originated from a cellular gene,[76] suggesting that the HSV transforming potential results from an appropriated cellular gene.

17. ANIMAL MODELS OF HSV CARCINOGENESIS

Treatment with tetradecanoyl phorbol acetate and inactivated HSV-1 was shown to cause papilloma lesions on the mouse lip to progress to carcinomas,[157] suggesting that HSV-1 may have a cocarcinogenic potential for orofacial tissues. Consistent with this conclusion, both HSV-1 and HSV-2 were shown to increase the carcinogenic potential of dimethylbenzanthracene (DMBA) in the hamster buccal pouch epithelium, and immunization with HSV reduced this cocarcinogenic effect.[158,159] The cocarcinogenic effect appeared to involve c-*myc* activation because normal cells and a tumor line induced by DMBA alone expressed a single *myc* transcript, and tumors established by DMBA together with HSV-1 or HSV-2 expressed both transcripts.[159] The HSV cocarcinogenic effect may result from promotion of the chemical carcinogen-induced activation of protooncogenes and the inactivation of the p53 tumor suppressor gene.[160]

Wentz *et al.*[161] showed that ultraviolet-light-inactivated HSV-1 or HSV-2 applied to the mouse cervix could induce cervical cancer and premalignant (dysplasia) lesions. Prior immunization of the mice with HSV-1 prevented both dysplasia and carcinoma.[162] HSV-2 DNA or *Bgl*II-cleaved HSV-2 DNA also induced dysplasia and cervical carcinoma.[161] HEp-2 or calf thymus DNA given in a similar regimen had no effect. Similar results were obtained by Chen *et al.*,[164] who also demonstrated that HSV-2 vaccination protects mice against the development of cervical carcinoma, reducing the tumor incidence from 50% to 19%. Although Meignier *et al.*[165] did not find that HSV induces cervical tumors in mice, this may reflect differences in the pathological criteria used for tumor diagnosis in these animals.

18. CONCLUSIONS

The ability of the herpes simplex viruses to cause neoplastic transformation is not in dispute. Their oncogenic potential is significant both from a clinical standpoint and within the context of the development of vaccines and antiviral chemotherapy. However, the mechanism of HSV-induced neoplastic transformation has been more difficult to elucidate. The data discussed in this review provide persuasive evidence that the HSVs contribute to cancer causation by virtue of their multiple tumorigenic functions including their ability to cause chromosome breakage, mutagenize cells, amplify cellular DNA sequences, and switch on or modify the expression of host cell proteins that are directly or indirectly involved in the regulation of cell proliferation. The exact identity of the HSV functions that cause these cellular alterations is still unclear and may include IE and DE functions as well as virion proteins. Additionally, and most significantly, HSV-2 carries its own oncogene (ICP10 PK) that is located in the middle of the L segment of the DNA. This oncogene, which we have termed *LA-1*,[166] is appropriated from the cell and functions as a growth factor receptor kinase that signals through *ras*, as do other growth factor receptor oncogenes. The *LA-1* oncogene (ICP10 PK oncogene) converts immortalized human and rodent cells to a neoplastic phenotype, and its PK activity is critical for transformation. The finding of amplified and rearranged cellular DNA sequences homologous to the *LA-1* oncogene in cells transfected with mutagenic HSV-2 DNA suggests that activation of a putative *LA-1* protooncogene is another mechanism whereby HSV-2 can cause neoplastic transformation. Cloning of the *LA-1* protooncogene and its further study should provide much needed information on the role of HSV-2 in human cancer.[46,166]

The carcinogenic and cocarcinogenic potential of HSV in animal models and its association with human cancer,[46,166,167] support the conclusion that HSV is a significant risk factor for tumor development. In this context it is particularly relevant that the *LA-1* oncogene is expressed in human cervical cancer tissues,[168–171] and its expression is associated with the carcinogenic process.[168–170] Better understanding of the role of *LA-1* oncogene signaling pathways in cancer development will offer opportunities for improved diagnosis/prognosis and therapeutic intervention. For instance, antisense therapy with modified oligodeoxynucleotides that target LA-1 can be used as specific and effective means of inhibiting tumor cells whose growth depends on *LA-1* oncogene expression. Another strategy involves administration of monoclonal antibodies that induce *LA-1* down-regulation. Tumors that overexpress the *LA-1* oncoprotein might also be targeted with radioisotopes or toxins linked to monoclonal antibodies against *LA-1*. Such a strategy would have the advantage of being both tumoricidal and tumor specific.

ACKNOWLEDGMENTS. The studies done in this laboratory were supported by Public Health Service Grant CA39691 from the National Cancer Institute.

REFERENCES

1. Duff, R., and Rapp, F., 1973, Oncogenic transformation of hamster cells after exposure to herpes simplex virus type 2, *Nature New Biol.* **233:**48–50.
2. Duff, R., and Rapp, F., 1973, Oncogenic transformation of hamster embryo cells after exposure to inactivated herpes simplex virus type 1, *J. Virol.* **12:**209–217.
3. Darai, G., and Munk, K., 1976, Neoplastic transformation of rat embryo cells with herpes simplex virus, *Int. J. Cancer* **18:**469–481.
4. Kutinova, L., Vonka, V., and Broucek, J., 1973, Increased oncogenicity and synthesis of herpesvirus antigens in hamster cells exposed to herpes simplex type-2 virus, *J. Natl. Cancer Inst.* **50:**759–766.
5. Kimura, S., Flannery, V. L., Levy, B., and Schaffer, P. A., 1975, Oncogenic transformation of primary hamster cells by herpes simplex virus type 2 (HSV-2) and an HSV-2 temperature-sensitive mutant, *Int. J. Cancer* **15:**786–798.

6. Flannery, V. L., Courtney, R. J., and Schaffer, P. A., 1977, Expression of an early, nonstructural antigen of herpes simplex virus in cells transformed *in vitro* by herpes simplex virus, *J. Virol.* **21:**284–291.
7. Collard, W., Thornton, H., and Green, M., 1973, Cells transformed by human herpesvirus type 2 transcribe virus specific RNA sequences shared by herpesvirus types 1 and 2, *Nature New Biol.* **243:**264–266.
8. Macnab, J. C. M., 1974, Transformation of rat embryo cells by temperature-sensitive mutants of herpes simplex virus, *J. Gen. Virol.* **24:**143–153.
9. Macnab, J. C. M., 1979, Tumor production by HSV-2 transformed lines in rats and the varying response to immunosuppression, *J. Gen. Virol.* **43:**39–56.
10. Macnab, J. C. M., Visser, L., Jamieson, A. T., and Hay, J., 1980. Specific viral antigens in rat cells transformed by herpes simplex virus type 2 and in rat tumours induced by inoculation of transformed cells, *Cancer Res.* **40:**2074–2079.
11. Kessous, A., Bibar-Hardy, V., Suh, M., and Simard, D., 1979, Analysis of chromosome nucleic acids and polypeptides in hamster cells transformed by herpes simplex type 2, *Cancer Res.* **39:**3225–3234.
12. Skinner, G. B. R., 1976, Transformation of primary hamster embryo fibroblasts by type 2 herpes simplex virus: Evidence for a hit and run mechanism, *Br. J. Exp. Pathol.* **57:**361–376.
13. Minson, A. C., Thouless, M. E., Eglin, R. P., and Darby, G., 1976, The detection of virus DNA sequences in a herpes type 2 transformed hamster cell line (333-8-9), *Int. J. Cancer* **17:**493–500.
14. Manak, M. M., Aurelian, L., and Ts'o, P. O. P., 1981, Focus formation and neoplastic transformation by herpes simplex virus type 2 inactivated intracellularly by 5-bromo-2'-deoxyuridine and near UV light, *J. Virol.* **40:**289–300.
15. Aurelian, L., Manak, M. M., McKinlay, M., Smith, C. C., Klacsmann, K. T., and Gupta, P. K., 1981, "The herpesvirus hypothesis"—Are Koch's postulates satisfied? *Gynecol. Oncol.* **12:**S56–S87.
16. Schlehofer, J. R., and zu Hausen, H., 1982, Induction of mutations within the host genome by partially inactivated herpes simplex virus type 1, *Virology* **122:**471–475.
17. Shillitoe, E. J., Matney, T. S., and Conley, A. J., 1986, Induction of mutations in bacteria by a fragment of DNA from herpes simplex virus type 1, *Virus Res.* **6:**181–191.
18. Huang, C. B., and Shillitoe, E. J., 1990, DNA sequence of mutations induced in cells by herpes simplex virus type-1, *Virology* **178:**180–188.
19. Huang, C. B., and Shilitoe, E. J., 1991, Analysis of complex mutations induced in cells by herpes simples virus type-1, *Virology* **181:**620–629.
20. Clarke, P., and Clements, J. B., 1991, Mutagenesis occurring following infection with herpes simplex virus does not require virus replication, *Virology* **182:**597–606.
21. Bauer, G., Kahl, S., Sawhney, I. S., Hofler, P., Gerspach, R., and Matz, B., 1992, Transformation of rodent fibroblasts by herpes simplex virus: presence of morphological transforming region I (*mtr*I) is not required for the maintenance of the transformed state, *Int. J. Cancer* **51:**754–760.
22. Pilon, L., Langelier, Y., and Royal, A., 1986, Herpes simplex virus type 2 mutagenesis: Characterisation of mutants induced at the hprt locus of non-permissive XC cells, *Mol. Cell. Biol.* **6:**2977–2983.
23. Pilon, L., Kessous-Elbaz, A., Langelier, Y., and Royal, A., 1989, Transformation of NIH 3T3 cells by herpes simplex type 2 *Bgl*II N fragment and subfragments is independent from induction of mutation at the hprt locus, *Biochem. Biophys. Res. Commun.* **159:**1249–1255.
24. Gerspach, R., and Matz, B., 1988, Herpes simplex virus-directed overreplication of chromosomal DNA physically linked to the simian virus 40 integration site of a transformed hamster cell line, *Virology* **165:**282–285.
25. Schlehofer, J. R., Gissmann, L., Matz, B., and zur Hausen, H., 1983, Herpes simplex virus induced amplification of SV40 sequences in transformed Chinese hamster embryo cells, *Int. J. Cancer* **32:**99–103.
26. Matz, B., Schlehofer, J. R., and zur Hausen, H., 1984, Identification of a gene function of herpes simplex virus type I essential for amplification of simian virus 40 DNA sequences in transformed hamster cells, *Virology* **134:**328–337.

27. Matz, B., Schlehofer, J. R., zur Hausen, H., Huber, B., and Fanning, E., 1985, HSV and chemical carcinogen-induced amplification of SV40 DNA sequences in transformed cells is cell line dependent, *Int. J. Cancer* **35:**521–525.
28. Heilbronn, R., and zur Hausen, H., 1989, A subset of herpes simplex virus replication genes induces DNA amplification within the host cell genome, *J. Virol.* **63:**3683–3692.
29. Danovich, R. M., and Frenkel, N., 1988, Herpes simplex virus induces the replication of foreign DNA, *Mol. Cell. Biol.* **8:**3272–3281.
30. Brandt, C. R., McDougall, J. K., and Galloway, D. A., 1987, Synergistic interactions between human papilloma virus type-18 sequences, herpes simplex virus infection and chemical carcinogen treatment, in: *Papillomaviruses, Cancer Cells*, Vol. 5 (B. M. Steinberg, J. L. Brandsma, and L. B. Taichman, eds.), Cold Spring Harbor Laboratory, New York, pp. 179–186.
31. Marcon, M. J., and Kucera, L. S., 1979, Stimulation of human cell DNA synthesis by defective herpes simplex virus type 2, *Virology* **98:**364–372.
32. Kulomaa, P., Paavonen, J., and Lehtinen, M., 1992, Herpes simplex virus induces unscheduled DNA synthesis in virus-infected cervical cancer cell lines, *Res. Virol.* **143:**351–359.
33. Lee, P.-G., Chang, J.-Y., Yen, M.-S., Cheng, Y. C., and Nutter, L. M., 1988, Enhancement of herpes simplex virus type 2 (HSV-2) DNA synthesis in infected cells that constitutively express the *Bgl*II-*N* region of the HSV-2 genome, *Virus Genes* **2:**269–281.
34. Filion, M., Skup, D., and Suh, M., 1988, Specific induction of cellular gene transcription in herpes simplex virus type 2-transformed cells, *J. Gen. Virol.* **69:**2011–2019.
35. Gervais, C., and Suh, M., 1990, Serum amyloid A protein-related mRNA expression in herpes simplex virus type 2-transformed hamster cells, *Mol. Cell. Biol.* **10:**4412–4414.
36. Roddick, V. L., Krebs, C. R., Kucera, L. S., Daniel, L. W., and Waite, M., 1988, Phospholipid-sensitive, Ca^{2+}-dependent protein kinase activity in rat embryo fibroblasts transformed by herpes simplex virus type 2, *Oncology* **45:**197–201.
37. Patel, R., Chan, W. L., Kemp, L. M., La Thangue, N. B., and Latchman, D. S., 1986, Isolation of cDNA clones derived from a cellular gene transcriptionally induced by herpes simplex virus, *Nucleic Acids Res.* **14:**5629.
38. Macnab, J. C. M., Orr, A., and La Thangue, N. B., 1985, Cellular proteins expressed in herpes simplex virus transformed cells also accumulate on herpes simplex virus infection, *EMBO J.* **4:**3223–3228.
39. La Thangue, N. B., and Latchman, D. S., 1988, A cellular protein related to heat-shock protein 90 accumulates during herpes simplex virus infection and is overexpressed in transformed cells, *Exp. Cell Res.* **178:**169–179.
40. Goswami, B. B., 1987, Transcriptional induction of proto-oncogene *fos* by HSV-2, *Biochem. Biophys. Res. Commun.* **143:**1055–1062.
41. Boyd, A. L., Derge, J. G., and Hampar, B., 1978, Activation of endogenous type C virus in BALB/c mouse cells by herpesvirus DNA, *Proc. Natl. Acad. Sci. USA* **75:**4558–4562.
42. Boyd, A. L., Enquist, L., Vande Woude, G. F., and Hampar, B., 1980, Activation of mouse retrovirus by herpes simplex virus type 1 cloned DNA fragments, *Virology* **103:**228–231.
43. Hampar, B., 1981, Transformation induced by herpes simplex virus: A potentially novel type of virus-cell interaction, *Adv. Cancer Res.* **35:**27–47.
44. Hampar, B., Aaronson, S. A., Derge, J. G., Chakrabarty, M., Showalter, S. D., and Dunn, C. Y., 1976, Activation of an endogenous mouse type C virus by ultraviolet-irradiated herpes simplex virus types 1 and 2, *Proc. Natl. Acad. Sci. USA* **73:**646–650.
45. Macnab, J. C. M., 1975, Transformed cell lines produced by temperature sensitive mutants of herpes simplex types 1 and 2, in: *Oncogenesis and Herpesviruses II* (G. de The, M. A. Epstein, and H. zur Hausen, eds.), IARC, Lyon, pp. 227–236.
46. Macnab, J. C. M., 1987, Herpes simplex virus and human cytomegalovirus: Their role in morphological transformation and genital cancers, *J. Gen. Virol.* **68:**2525–2550.
47. Marsden, H. S., Crombie, I. K., and Subak-Sharpe, J. H., 1976, Control of protein synthesis in herpesvirus-infected cells: Analysis of the polypeptides induced by wild type and sixteen temperature-sensitive mutants of HSV strain 17, *J. Gen. Virol.* **31:**347–372.

48. Preston, C. M., 1979, Control of herpes simplex virus type 1 mRNA synthesis in cells infected with wild-type virus or the temperature sensitive mutant *ts*K, *J. Virol.* **29:**275–284.
49. Preston, V. G., 1981, Fine-structure mapping of herpes simplex virus type 1 temperature-sensitive mutations within the short repeat regions of the genome, *J. Virol.* **39:**150–161.
50. Schek, N., and Bachenheimer, S. L., 1985, Degradation of cellular mRNAs induced by a virion-associated factor during herpes simplex virus infection of Vero cells, *J. Virol.* **55:**601–610.
51. McGeoch, D. J., Dalrymple, M. A., Davison, A. J., Dolan, A., Frame, M. C., McNab, D., Perry, L. J., Scott, J. E., and Taylor, P., 1988, The complete DNA sequence of the long unique region in the genome of herpes simplex virus type 1, *J. Gen. Virol.* **69:**1531–1574.
52. Smith, C. C., Kulka, M., Wymer, J. P., Chung, T. D., and Aurelian, L., 1992, Expression of the large subunit of herpes simplex virus type 2 ribonucleotide reductase (ICP10) is required for virus growth and neoplastic transformation, *J. Gen. Virol.* **73:**1417–1428.
53. Camacho, A., and Spear, P. G., 1978, Transformation of hamster embryo fibroblasts by a specific fragment of the herpes simplex virus genome, *Cell* **15:**993–1002.
54. Reyes, G. R., La Femina, R., Hayward, S. D., and Hayward, G. S., 1979, Morphological transformation by DNA fragments of human herpesviruses: Evidence for two distinct transforming regions in HSV-1 and HSV-2 and lack of correlation with biochemical transfer of the thymidine kinase gene, *Cold Spring Harbor Symp. Quant. Biol.* **44:**629–641.
55. O'Donnell, M. E., Elias, P., Funnell, B. E., and Lehman, I. R., 1987, Interaction between the DNA polymerase and single-stranded DNA-binding protein (infected cell protein 8) of herpes simplex virus 1, *J. Biol. Chem.* **262:**4260–4266.
56. de Bruyn Kops, A., and Knipe, D. M., 1988, Formation of DNA replication structures in herpes virus-infected cells required a viral DNA binding protein, *Cell* **55:**857–868.
57. Boehmer, P. E., and Lehman, I. R., 1993, Herpes simplex virus type 1 ICP8: Helix-destabilizing properties, *J. Virol.* **67:**711–715.
58. Dutch, R. E., and Lehman, I. R., 1993, Renaturation of complementary DNA strands by herpes simplex virus type 1 ICP8, *J. Virol.* **67:**6945–6949.
59. Shillitoe, E. J., Zhang, S., Wang, G., and Hung, C. B., 1993, Functions and proteins of herpes simplex virus type-1 that are involved in raising the mutation frequency of infected cells, *Virus Res.* **27:**239–251.
60. Macnab, J. C. M., and McDougall, J. K., 1980, Transformation by herpesviruses, in: *The Human Herpesviruses* (A. J. Nahmias, W. R. Dowdle, and R. F. Schinazi, eds.), Elsevier/North-Holland, New York, p. 634.
61. Galloway, D. A., and McDougall, J. K., 1981, Transformation of rodent cells by a cloned DNA fragment of herpes simplex virus type 2, *J. Virol.* **38:**749–760.
62. Cameron, I. R., Park, M., Dutia, B. M., Orr, A., and MacNab, J. C. M., 1985, Herpes simplex virus sequences involved in the initiation of oncogenic morphological transformation of rat cells are not required for maintenance of the transformed state, *J. Gen. Virol.* **66:**517–527.
63. Peden, K., Mounts, P., and Hayward, G. S., 1982, Homology between mammalian cell DNA sequences and human herpesvirus genomes detected by a hybridisation procedure with high complexity probe, *Cell* **31:**71–80.
64. Jariwalla, R. J., Aurelian, L., and Ts'o, P. O. P., 1983, Immortalisation and neoplastic transformation of normal diploid cells by defined cloned DNA fragments of herpes simplex virus type 2, *Proc. Natl. Acad. Sci. USA* **80:**5902–5906.
65. McLauchlan, J., and Clements, J. B., 1983, DNA sequence homology between two co-linear loci on the HSV genome which have different transforming abilities, *EMBO J.* **2:**1953–1961.
66. Huszar, D., and Bacchetti, S., 1983, Is ribonucleotide reductase the transforming function of herpes simplex virus 2? *Nature* **302:**76–79.
67. Galloway, D. A., Nelson, J. A., and McDougall, J. K., 1984, Small fragments of herpesvirus DNA with transforming activity contain insertion sequence-like structures, *Proc. Natl. Acad. Sci. USA* **81:**4736–4740.
68. van den Berg, F. M., van Amstel, P. J., and Walboomers, J. M. M., 1985, Construction of rat cell lines that contain potential morphologically transforming regions of the herpes simplex virus type 2 genome, *Intervirology* **24:**199–210.

69. Saavedra, C., and Kessous-Elbaz, A., 1985, Retention of herpes simplex virus type II sequences in BglII n transformed cells after cotransfection with a selectable marker, *EMBO J.* **4:**3419–3426.
70. Kessous-Elbaz, A., Pelletier, M., Cohen, E. A., and Langelier, Y., 1989, Retention and expression of the left end subfragment of the herpes simplex virus type 2 BglII N DNA fragment do not correlate with tumorigenic conversion of NIH 3T3 cells, *J. Gen. Virol.* **70:**2171–2177.
71. Galloway, D. A., and McDougall, J. K., 1983, The oncogenic potential of herpes simplex viruses: Evidence for a "hit and run" mechanism, *Nature* **302:**21–24.
72. Rubin, G., 1983, Dispersed repetitive DNAs in drosophila, in: *Mobile Genetic Elements* (J. A. Shapiro, ed.), Academic Press, New York, pp. 329–361.
73. Lewin, B., 1983, *Genes*, John Wiley & Sons, New York, pp. 603–604.
74. Brandt, C. R., Buonagura, F. M., McDougall, J. K., and Galloway, D. A., 1987, Plasmid mediated mutagenesis of a cellular gene in transfected eukaryotic cells, *Nucleic Acids Res.* **15:**561–573.
75. Becjek, B., and Conley, A. J., 1986, A transforming plasmid from HSV-2 transformed cells contains rat DNA homologous to the HSV-1 and HSV-2 genomes, *Virology* **154:**41–55.
76. Smith, C. C., Wymer, J. P., Luo, J. H., and Aurelian, L., 1991, Genomic sequences homologous to the protein kinase region of the bifunctional herpes simplex virus type 2 protein ICP10, *Virus Genes* **5:**215–226.
77. Hayashi, Y., Iwasaka, T., Smith, C. C., Aurelian, L., Lewis, G. K., and Ts'o, P. O. P., 1985, Multistep transformation by defined fragments of herpes simplex virus type 2 DNA: Oncogenic region and its gene product, *Proc. Natl. Acad. Sci. USA* **82:**8493–8497.
78. Jariwalla, R. J., Aurelian, L., and Ts'o, P. O. P., 1980, Tumorigenic transformation induced by a specific fragment of DNA from herpes simplex virus type 2, *Proc. Natl. Acad. Sci. USA* **77:**2279–2283.
79. Jariwalla, R. J., Taczos, B., Jones, C., Ortiz, J., and Salimi-Lopez, S., 1986, DNA amplification and neoplastic transformation mediated by a herpes simplex DNA fragment containing cell related sequences, *Proc. Natl. Acad. Sci. USA* **83:**1738–1742.
80. Ali, M. A., McWeeney, D., Milosavljevic, A., Jurka, J., and Jariwalla, R. J., 1991, Enhanced malignant transformation induced by expression of a distinct protein domain of ribonucleotide reductase large subunit from herpes simplex virus type 2, *Proc. Natl. Acad. Sci. USA* **88:**8257–8261.
81. Krebs, C. R., Waite, M., Jariwalla, R., and Kucera, L. S., 1987, Induction of cellular functions in spontaneously immortalised rat 2 cells transfected with cloned herpes simplex virus type 2 (HSV-2) DNA, *Carcinogenesis* **8:**183–185.
82. Smith, C. C., Luo, J. H., Hunter, J. C. R., Ordonez, J. V., and Aurelian, L., 1994, The transmembrane domain of the large subunit of HSV-2 ribonucleotide reductase (ICP10) is required for protein kinase activity and transformation-related signaling pathways that result in *ras* activation. *Virology* **200:**598–612.
83. Chung, T. D., Wymer, J. P., Smith, C. C., Kulka, M., and Aurelian, L., 1989, Protein kinase activity associated with the large subunit of herpes simplex virus type 2 ribonucleotide reductase (ICP10), *J. Virol.* **63:**3389–3398.
84. Al-Kobaisi, M. F., Rixon, F. J., McDougall, I., and Preston, V. G., 1991, The herpes simplex virus UL33 gene product is required for the assembly of full capsids, *Virology* **60:**1018–1026.
85. McNabb, D. S., and Courtney, R. J., 1992, Identification and characterization of the herpes simplex virus type 1 virion protein encoded by the UL35 open reading frame, *J. Virol.* **66:**2653–2663.
86. McNabb, D. S., and Courtney, R. J., 1992, Analysis of the UL36 open reading frame encoding the large tegument protein (ICP1/2) of herpes simples virus type 1, *J. Virol.* **66:**7581–7584.
87. Chang, J. E., and Roizman, B., 1993, The product of the U_L31 gene of herpes simplex virus 1 is a nuclear phosphoprotein which partitions with the nuclear matrix, *J. Virol.* **67:**6348–6356.
88. Berezney, R., 1991, The nuclear matrix: A heuristic model for investigating genomic organization and function in the cell nucleus, *J. Cell Biochem.* **47:**109–123.
89. Chatterjee, P. K., and Flint, S. J., 1986, Partition of E1A proteins between soluble and structural fractions of adenovirus-infected and -transformed cells, *J. Virol.* **60:**1018–1026.

90. Schirmbeck, R., and Deppert, W., 1989, Nuclear subcompartmentalization of simian virus 40 large T antigen: Evidence for *in vivo* regulation of biochemical activities, *J. Virol.* **63**:2308–2316.
91. Purves, F. C., Spector, D., and Roizman, B., 1992, U_L34, the target of the herpes simplex virus U_S3 protein kinase, is a membrane protein which in its unphosphorylated state associates with novel phosphoproteins, *J. Virol.* **66**:4295–4303.
92. Tsai, L.-H., Harlow, E., and Meyerson, M., 1991, Isolation of the human cdk2 gene that encodes the cyclin A and adenovirus E1A-associated p33 kinase, *Nature* **353**:174–177.
93. Iwasaka, T., Smith, C., Aurelian, L., and Ts'o, P. O. P., 1985, The cervical tumor-associated antigen (ICP-10/AG-4) is encoded by the transforming region of the genome of herpes simplex virus type 2, *Jpn. J. Cancer Res.* **76**:946–958.
94. Chung, T. D., Wymer, J. P., Kulka, M., Smith, C. C., and Aurelian, L., 1990, Myristylation and polylysine-mediated activation of the protein kinase domain of the large subunit of herpes simplex virus type 2 ribonucleotide reductase (ICP10), *Virology* **179**:168–178.
95. Hanks, S. K., Quinn, A. M., and Hunter, T., 1988, The protein kinase family: Conserved features and deduced phylogeny of the catalytic domains, *Science* **241**:42–51.
96. Kamps, M. P., and Sefton, B. M., 1986, Neither arginine nor histidine can carry out the function of lysine-295 in the ATP-binding site of $p60^{src}$, *Mol. Cell. Biol.* **6**:751–757.
97. Brenner, S., 1987, Phosphotransferase sequence homology, *Nature* **329**:21.
98. Hunter, T., and Cooper, J. A., 1986, Viral oncogenes and tyrosine phosphorylation, in: *The Enzymes: Control by Phosphorylation*, 3rd ed., vol. 17, part A, (P. D. Boyer and E. G. Krebs, eds.), Academic Press, Orlando, pp. 191–246.
99. Perlman, D., and Halverson, H. O., 1983, A putative signal peptide recognition site and sequence in eukaryotic and prokaryotic signal peptides, *J. Mol. Biol.* **167**:391–409.
100. Wymer, J. P., Chung, T. C., Chang, Y. N., Hayward, G. S., and Aurelian, L., 1989, Identification of immediate-early-type *cis*-response elements in the promoter for the ribonucleotide reductase large subunit from herpes simplex virus type 2, *J. Virol.* **63**:2773–2784.
101. Wymer, J. P., Aprhys, C. M. J., Chung, T. D., Feng, C.-P., Kulka, M., and Aurelian, L., 1992, Immediate early and functional AP-1 *cis*-response elements are involved in the transcriptional regulation of the large subunit of herpes simplex virus type 2 ribonucleotide reductase (ICP10), *Virus Res.* **23**:253–270.
102. Luo, J. H., and Aurelian, L., 1992, The transmembrane helical segment but not the invariant lysine is required for the kinase activity of the large subunit of herpes simplex virus type 2 ribonucleotide reductase, *J. Biol. Chem.* **267**:9645–9653.
103. Luo, J. H., Smith, C. C., Kulka, M., and Aurelian, L., 1991, A truncated protein kinase domain of the large subunit of the herpes simplex type 2 ribonucleotide reductase (ICP10) expressed in *Escherichia coli*, *J. Biol. Chem.* **266**:20976–20983.
104. Stock, J. B., Ninfa, A. J., and Stock, A. M., 1989, Protein phosphorylation and regulation of adaptive responses in bacteria, *Microbiol. Rev.* **53**:450–490.
105. Conner, J., Cooper, J., Furlong, J., and Clements, J. B., 1992, An autophosphorylating but not transphosphorylating activity is associated with the unique N terminus of the herpes simplex virus type 1 ribonucleotide reductase large subunit, *J. Virol.* **66**:7611–7516.
106. Paradis, H., Gaudreau, P., Massie, B., Lamarche, N., Guilbault, C., Gravel, S., and Langeier, Y., 1991, Affinity purification of active subunit 1 of herpes simplex virus type 1 ribonucleotide reductase exhibiting a protein kinase function, *J. Biol. Chem.* **266**:9647–9651.
107. Chung, T. C., Luo, J. H., Wymer, J. P., Smith, C. C., and Aurelian, L., 1991, Leucine repeats in the large subunit of herpes simplex virus type 2 ribonucleotide reductase (RR; ICP10) are involved in RR activity and subunit complex formation, *J. Gen. Virol.* **72**:1139–1144.
108. Ren, R. B., Mayer, B. J., Cicchetti, P., and Baltimore, D., 1993, Identification of a ten-amino acid proline-rich SH3 binding site, *Science* **259**:1157–1161.
109. Koch, C. A., Anderson, D., Moran, M. F., Ellis, C., and Pawson, T., 1991, SH2 and SH3 domains: Elements that control interactions of cytoplasmic signaling proteins, *Science* **252**:668–674.

110. Mayer, B. J., and Baltimore, D., 1993, Signaling through SH2 and SH3 domains, *Trends Cell Biol.* **3**:8–13.
111. Wang, L. H., Lin, B., Jong, S. M. J., Dixon, D., Ellis, L., Roth, R. A., and Rutter, W. J., 1987, Activation of transforming potential of the human insulin receptor gene, *Proc. Natl. Acad. Sci. USA* **84**:5725–5729.
112. Gherzi, R., Sesti, G., Andraghetti, G., DePirro, R., Lauro, R., Adezati, L., and Cordera, R., 1989, An extracellular domain of the insulin receptor β-subunit with regulatory function on protein-tyrosine kinase, *J. Biol. Chem.* **264**:8627–8635.
113. Yarden, Y., and Ullrich, A., 1988, Growth factor receptor tyrosine kinases, *Annu. Rev. Biochem.* **57**:442–478.
114. Yarden, Y., 1990, Receptor-like oncogenes: Functional analysis through novel experimental approaches, *Mol. Immunol.* **27**:1319–1324.
115. Kalderon, D., Roberts, B. L., Richardson, W. D., and Smith, A. E., 1984, A short amino acid sequence able to specify nuclear location, *Cell* **39**:499–509.
116. Fukui, Y., O'Brien, M. C., and Hanafusa, H., 1991, Deletions in the SH2 domain of p60$^{v\text{-}src}$ prevent association with the detergent insoluble cellular matrix, *Mol. Cell. Biol.* **11**:1207–1213.
117. Rettenmier, C. W., Roussel, M. F., Quinn, C. O., Kitchingman, G. R., Look, A. T., and Sherr, C. J., 1985, Transmembrane orientation of glycoproteins encoded by the v-*fms* oncogene, *Cell* **40**:971–981.
118. Hunter, J. C. R., Smith, C. C., Bose, D., Kulke, M., Broderick, R., and Aurelian, L., 1995, Intracellular internalization and signaling pathways triggered by the large subunit of HSV-2 ribonucleotide reductase (ICP10), *Virology* **210**:345–360.
119. Matthews, L. S., and Vale, W. W., 1991, Expression cloning of an activin receptor, a predicted transmembrane serine kinase, *Cell* **65**:973–982.
120. Wrana, J. L., Attisano, L., Caracamo, J., Zentella, A., Doody, J., Laiho, M., Wang, X.-F., and Massague, J., 1992, TGFβ signals through a heteromeric protein kinase receptor complex, *Cell* **71**:1003–1014.
121. Satoh, T., Endo, M., Nakafuku, M., Akiyama, T., Yamamoto, T., and Kaziro, Y., 1990, Accumulation of p21^{ras} GTP in response to stimulation with epidermal growth factor and oncogene products with tyrosine kinase activity, *Proc. Natl. Acad. Sci. USA* **87**:7926–7929.
122. Satoh, T., Endo, M., Nakafuku, M., Nakamura, S., and Kaziro, Y., 1990, Platelet-derived growth factor stimulates formation of active p21^{ras} GTP complex in Swiss mouse 3T3 cells, *Proc. Natl. Acad. Sci. USA* **87**:5993–5997.
123. Lowenstein, E. J., Daly, R. J., Batzer, A. G., Li, W., Margolis, B., Lammers, R., Ullrich, A., Scholnick, E., Bar-Sagi, D., and Schlesinger, J., 1992, The SH2 and SH3 domain containing protein Grb2 links receptor tyrosine kinases to *ras* signaling, *Cell* **70**:431–442.
124. Li, N., Batzer, A., Daly, R., Yajnik, V., Skolnik, E., Chardin, P., Bar-Sagi, D., Margolis, B., and Schlessinger, J., 1993, Guanine-nucleotide-releasing factor hSos1 binds to Grb2 and links receptor tyrosine kinases to Ras signalling, *Nature* **363**:85–88.
125. Chardin, P., Carnonis, J. H., Gale, N. W., VanAelst, L., Schlesinger, J., Willer, M. H., and Bar-Sagi, D., 1993, Human Sos1—a guanine nucleotide exchange factor for *ras* that binds to Grb2, *Science* **260**:1338–1343.
126. McCormick, F., 1989, Ras GTPase activating protein: Signal transmitter and signal terminator, *Cell* **56**:5–8.
127. Trahey, M., and McCormick, F., 1987, A cytoplasmic protein stimulates normal N-*ras* p21 GTPase, but does not affect oncogenic mutants, *Science* **238**:542–545.
128. Ellis, C., Moran, M., McCormick, F., and Pawson, T., 1990, Phosphorylation of GAP and GAP-associated proteins by transforming and mitogenic tyrosine kinases, *Nature* **343**:377–381.
129. Moran, M. F., Polakis, P., McCormick, F., Pawson, T., and Ellis, C., 1991, Protein-tyrosine kinases regulate the phosphorylation, protein interactions, subcellular distribution, and activity of p21^{ras} GTPase-activating protein, *Mol. Cell. Biol.* **11**:1804–1812.

130. Kaplan, D. R., Morrison, D. K., Wong, G., McCormick, F., and Williams, L. T., 1990, PDGF β-receptor stimulates tyrosine phosphorylation of GAP and association of GAP with a signaling complex, *Cell* **61**:125–133.
131. Buday, L., and Downward, J., 1993, Epidermal growth factor regulates p21^{ras} through the formation of a complex of receptor, Grb2 adapter protein, and Sos nucleotide exchange factor, *Cell* **73**:611–620.
132. Selby, M. J., Barta, A., Baxter, J. D., Bell, G. L., and Eberhardt, N. L., 1984, Analysis of a major human chorionic somatomammotropin gene. Evidence for two functional promoter elements, *J. Biol. Chem.* **259**:13131–13138.
133. Jones, C., Ortiz, J., and Jariwalla, R. J., 1986, Localization and comparative nucleotide sequence analysis of the transforming domain in herpes simplex virus DNA containing repetitive genetic elements, *Proc. Natl. Acad. Sci. USA* **83**:7855–7859.
134. Jones, C., Zhu, F., and Dhanwada, K. R., 1993, Analysis of a herpes simplex virus 2 fragment from the open reading frame of the large subunit of ribonucleotide reductase with transcriptional regulatory activity, *DNA Cell Biol.* **12**:127–137.
135. Jones, C., 1989, The minimal transforming fragment of HSV-2 mtrIII can function as a complex promoter element, *Virology* **169**:346–353.
136. Lankinen, H., Telford, E., MacDonald, D., and Marsden, H., 1989, The unique N-terminal domain of the large subunit of herpes simplex virus ribonucleotide reductase is preferentially sensitive to proteolysis, *J. Gen. Virol.* **70**:3159–3169.
137. Rogers, S., Wells, R., and Rechsteiner, M., 1986, Amino acid sequences common to rapidly degraded proteins: The PEST hypothesis, *Science* **234**:364–368.
138. Galloway, D. A., Buonaguro, F. M., Brandt, C. R., and McDougall, J. K., 1985, Herpes simplex virus and cytomegalovirus: Unconventional DNA tumor viruses, in: *DNA Tumor Viruses, Cancer Cell*, Vol. 4 (M. Botcham, T. Grodzicker, and P. A. Sharp, eds.), Cold Spring Harbor Laboratory, New York, pp. 355–361.
139. McDougall, J. K., Beckmann, A. M., and Galloway, D. A., 1985, The enigma of viral nucleic acids in genital neoplasia, in: *Viral Etiology of Cervical Cancer* (R. Peto and H. zur Hausen, eds.), Cold Spring Harbor Laboratory, New York, pp. 199–209.
140. Ames, B. N., McCann, J., and Yamasaki, E., 1975, Methods for detecting carcinogens and mutagens with the salmonella/mammalian microsome mutagenicity test, *Mutat. Res.* **31**:347–364.
141. Schimke, R. T. (ed.), 1982, *Gene Amplification*, Cold Spring Harbor Laboratory, New York.
142. Lavi, S., 1981, Carcinogen-mediated amplification of viral DNA sequences in simian virus 40-transformed Chinese hamster embryo cells, *Proc. Natl. Acad. Sci. USA* **78**:6144–6148.
143. Varshavsky, A., 1981, On the possibility of metabolic control of replicon "misfiring": Relationship to emergence of malignant phenotypes in mammalian cell lineages, *Proc. Natl. Acad. Sci. USA* **78**:742–746.
144. Schimke, R. T., Sherwood, S. W., Hill, A. B., and Johnston, R. N., 1986, Overreplication and recombination of DNA in higher eukaryotes: Potential consequences and biological implications, *Proc. Natl. Acad. Sci. USA* **83**:2157–2161.
145. Walker, A. I., Hunt, T., Jackson, R. J., and Anderson, C. W., 1985, Double stranded DNA induces the phosphorylation of several proteins including the 90,000 mol. wt. heat shock protein in animal cell extracts, *EMBO J.* **4**:139–145.
146. Schuermann, M., Neuberg, M., Hunter, J. B., Jenuwein, T., Ryseck, R. P., and Muller, R., 1989, The leucine repeat motif in Fos protein mediates complex formation with Jun/AP-1 and is required for transformation, *Cell* **56**:507–516.
147. Jang, K. L., Pulverer, B., Woodgett, J. R., and Latchman, D. S., 1991, Activation of the cellular transcription factor AP-1 in herpes simplex virus infected cells is dependent on the viral immediate-early protein ICP0, *Nucleic Acids Res.* **19**:4879–4883.
148. Hsiao, W.-L., W., Galtoni-Celli, S., and Weinstein, I. B., 1985, Effect of 5-azacytidine on the progressive nature of cell transformation, *Mol. Cell. Biol.* **5**:1800–1803.

149. Macnab, J. C. M., Adams, R. L. P., Rinaldi, A., Orr, A., and Clark, L., 1988, Hypomethylation of host cell DNA synthesized after infection or transformation of cells by herpes simplex virus, *Mol. Cell. Biol.* **8:**1443–1448.
150. Wilson, S. M., and Kuff, E. L., 1972, A novel DNA polymerase activity found in association with intracisternal A-type particles, *Proc. Natl. Acad. Sci. USA* **69:**1531–1536.
151. Gomez-Marquez, J., Puga, A., and Notkins, A. L., 1985, Regions of the terminal repetitions of the herpes simplex virus type 1 genome, *J. Biol. Chem.* **260:**3490–3495.
152. Jones, T. R., and Hyman, R. W., 1986, Sequences in the proximal IR_L of herpes simplex virus DNA hybridise to human DNA, *Virus Res.* **4:**369–375.
153. Spector, D. J., Jones, T. R., Parks, C. L., Deckhut, A. M., and Hyman, R. W., 1987, Hybridisation between a repeated region of herpes simplex virus type 1 DNA containing the sequence $[GGC]_n$ and heterodisperse cellular DNA and RNA, *Virus Res.* **7:**69–82.
154. Jones, T. R., and Hyman, R. W., 1983, Specious hybridisation between herpes simplex virus DNA and human cellular DNA, *Virology* **131:**555–560.
155. Jones, T. R., Parks, C. L., Spector, D. J., and Human, R. W., 1985, Hybridisation of herpes simplex virus DNA and human ribosomal DNA and RNA, *Virology* **144:**384–197.
156. Parks, C. L., Jones, T. R., Gonzalez, I. L., Schmickel, R. D., Hyman, R. W., and Spector, D. J., 1986, A simple repetitive sequence common to herpes simplex virus type 1 and human ribosomal DNAs, *Virology* **154:**381–388.
157. Burns, J. C., and Murray, B. K., 1981, Conversion of herpetic lesions to malignancy by ultraviolet exposure and promoter application, *J. Gen. Virol.* **55:**305–313.
158. Min, B. M., Kim, K., Cherrick, H. M., and Park, N. H., 1991, Three cell lines from hamster buccal pouch tumors induced by topical 7,12-dimethylbenz(a)anthracene, alone or in conjunction with herpes simplex virus inoculation, *In Vitro* **27A:**128–136.
159. Park, K., Cherrick, H. M., Min, B. M., and Park, N. H., 1990, Active HSV-1 immunization prevents the cocarcinogenic activity of HSV-1 in the oral cavity of hamsters, *Oral Surg. Oral Med. Oral Pathol.* **70:**186–191.
160. Park, N. H., Li, S. L., Xie, J. F., and Cherrick, H. M., 1992, *In vitro* and animal studies of the role or viruses in oral carcinogenesis, *Oral Oncol. Eur. J. Cancer* **28B:**145–152.
161. Wentz, W. B., Reagan, J. W., Heggie, A. D., Fu, Y.-S., and Anthony, D. D., 1981, Induction of uterine cancer with inactivated herpes simplex virus types 1 and 2, *Cancer* **48:**1783–1790.
162. Wentz, W. B., Heggie, A. D., Anthony, D. D., and Reagan, J. W., 1983, Effect of prior immunisation on induction of cervical cancer in mice by herpes simplex virus type 2, *Science* **222:**1128–1129.
163. Anthony, D. D., Wentz, W. B., Reagan, J. W., and Meggie, A. D., 1989, Induction of cervical neoplasia in the mouse by herpes simplex virus type 2 DNA, *Proc. Natl. Acad. Sci. USA* **86:**4520–4524.
164. Chen, M., Dong, C., Liu, Z., Skinner, G. B. R., and Hartley, C. E., 1986, Efficacy of vaccination with Skinner vaccine towards the prevention of herpes simplex virus induced cervical carcinomas in an experimental mouse model, *Vaccine* **4:**249–252.
165. Meignier, B., Norrild, B., Thuning, C., Warren, J., Frenkel, N., Nahmias, A. J., Rapp, F., and Roizman, B., 1986, Failure to induce cervical cancer in mice by long term frequency vaginal exposure to live or inactivated herpes simplex viruses, *Int. J. Cancer* **38:**387–394.
166. Aurelian, L., 1993, Genital cancer and Langerhans cells, *In Vivo* **7:**297–304.
167. Hildesheim, A., Mann, V., Brinton, L., Szklo, M., Reeves, W. C., and Rawls, W. E., 1991, Herpes simplex virus type 2: A possible interaction with human papillomavirus types 16/18 in the development of invasive cervical cancer, *Int. J. Cancer* **49:**335–340.
168. Flanders, R. T., Kucera, L. S., Raben, M., and Ricardo, M. J., Jr., 1985, Immunologic characterization of herpes simplex virus type 2 antigens ICP10 and ICSP11/12, *Virus Res.* **2:**245–260.
169. Te Velde E. R., and Aurelian, L., 1987, Antibodies to the herpes simplex virus type 2 induced tumor-associated antigen AG-4 as markers of recurrence in cervical cancer, *Tumour Biol.* **8:**26–33.

170. Sainz de la Cuesta, R., Reed, T. P., Brothman, J. C., and Dubin, N. H., 1983, LA-1 oncogene: A possible new prognostic index for evaluating cervical squamous intraepithelial lesions, *J. Reprod. Med.* **38:**173–178.
171. Aurelian, L., 1986, Seroepidemiologic association of HSV-2 with cervical cancer: Transforming viral genes, in: *Herpes and Papillomaviruses* (G. De Palo, F. Rilke, and M. Zur Heusen, eds.), New York, Raven Press, pp. 63–82.
172. Heilbronn, R., Schlehofer, J. R., Yalkinoglu, A. O., and zur Hausen, H., 1985, Selective DNA amplification induced by carcinogens (initiators): Evidence for a role of proteases and DNA polymerase alpha, *Int. J. Cancer* **36:**85–91.
173. Puga, A., Cantin, E. M., and Notkins, A. L., Homology between murine and human cellular DNA sequences and the terminal repetition of the S component of herpes simplex virus type I DNA, *Cell* **31:**81–87.
174. Macnab, Y. C. M., 1987, Herpes simplex virus and human cytomegalovirus: Their role in morphological transformation and genital cancers, *J. Gen. Virol.* **68:**2525–2550.
175. Cooper, J., Conner, J. and Clemants, J. B., 1995, Characterization of the morel protein kinase present in the R1 subunit of herpes simplex virus ribonucleotide reductase, *J. Virol.* **69:**4979–4985.

15

Herpes Simplex Virus as a Cooperating Agent in Human Genital Carcinogenesis

DARIO DI LUCA, E. CASELLI, and E. CASSAI

1. INTRODUCTION

Initial suggestions that herpes simplex virus type 2 (HSV-2) could be involved in the development of cervical cancer derived from seroepidemiologic studies that found greater amounts of antibodies against HSV-2 in cervical cancer patients than in controls.[1] Although findings from a prospective study failed to confirm this association,[2] other studies described a two- to fourfold excess risk of cervical cancer among HSV-seropositive women.[3,4] In addition to epidemiologic evidence, laboratory studies support a carcinogenic role of HSV-2: the virus contains DNA sequences able to transform cells *in vitro* [reviewed by Macnab[5]]. However, a consensus concerning a direct role for HSV in genital carcinogenesis has not been attained. The inability to identify a unique fragment of HSV DNA that morphologically transforms cells and the evidence that *in vitro* transformed cells remain morphologically transformed in the absence of detectable HSV DNA sequences have contributed to the hypothesis that viral DNA is not essential to maintaining the transformed state.[6,7] In addition, viral DNA-transforming sequences have been detected in only 10–30% of human genital neoplasias.[8–13]

In the meantime, there has been a growing interest in the association between human papillomavirus (HPV) and genital tumors. A possible role for HPV in the etiology of genital cancer was proposed initially on the basis of virological considerations.[14] Human papillomavirus DNA sequences (most notably of the types 16 and

---

DARIO DI LUCA, E. CASELLI, and E. CASSAI • Institute of Microbiology, University of Ferrara, I-44100 Ferrara, Italy.

DNA Tumor Viruses: Oncogenic Mechanisms, edited by Giuseppe Barbanti-Brodano *et al.* Plenum Press, New York, 1995.

18) are present in the majority of premalignant and malignant genital tissues.[15] The study of papillomavirus malignancies in animal models[16] and in human beings,[15] together with the observations on the pattern of HPV-induced transformation of *in vitro* cultured cells,[17,18] have raised the hypothesis that additional endogenous or exogenous modifications of the host cell are required for malignant progression of infected cells. Because of its characteristics, HSV could be considered one of the mutagenic cofactors postulated to play a role in HPV neoplastic promotion. Recent reports seem to confirm the hypothesis of a cooperation between these two viruses in the progression of a malignant phenotype.

Because it is difficult to assess the role of HSV in human cancer without, at the same time, briefly examining our present knowledge of the ability of HSV to transform cultured cells, the first part of this chapter describes studies on cell transformation and the implications of recent results in this field. A more detailed analysis of these aspects can be found in Chapter 14 by L. Aurelian. Second, we review the reports concerning HSV involvement in genital neoplasia, with special regard to the seroepidemiologic data and the detection of HSV-specific macromolecules in such tumors.

2. CELL TRANSFORMATION

2.1. Transforming Potential

Since the initial reports that HSV-1 and HSV-2 could morphologically transform rodent cells,[20,21] there have been several reports confirming and extending those observations using inactivated virus, temperature-sensitive mutants, sub- or supraoptimal infection temperatures, sheared viral DNA, or subgenomic fragments [reviewed by Minson[22]]. Cells transformed by HSV exhibit a wide range of properties: in some cases viral DNA, RNA, and proteins were found, but in other studies no viral sequences or proteins could be detected.[23,24] These results suggested that HSV can transform cells with a hit-and-run mechanism; viral sequences would initiate the transformation process, which would subsequently progress without the need to maintain viral sequences.[9]

The experiments designed to detect the transforming sequences of HSV genome have defined essentially three distinct transforming regions: HSV-1 DNA fragment *Bgl*II I [map units (m.u.) 0.31–0.42], designated *mtr*I (morphological transformation region I), coding for the major DNA-binding protein[6]; HSV-2 DNA fragment *Bgl*II N (m.u. 0.58–0.63), designated *mtr*II[7,25]; and HSV-2 DNA fragment *Bgl*II C (m.u. 0.42–0.58), designated *mtr*III, containing both an immortalizing and a transforming function (the *Bam*HI E subfragment) and coding for the viral protein designated infected cell protein (ICP)10.[26]

Recent reports have shown that ICP10 possesses a protein kinase activity similar to many retroviral oncogene products such as v-*src* or v-*abl*[27] and that the N-terminal domain of ICP10 is responsible for enhancement of transformation.[28] However, *mtr*III DNA is capable of transformation in the absence of protein-coding functions through the activity of sequence motifs functioning as transcriptional promoter elements, possibly altering the expression of cellular genes.[29]

Several laboratories have obtained cell transformation with HSV-2 *mtr*II, in the absence of retention of viral sequences or expression of viral proteins.[6,7,30] Interestingly, this transforming fragment contains stem–loop structures flanked by direct repeats, resembling insertion-like elements,[31] suggesting that HSV transformation may be mediated by modulation of the expression of cellular genes or, alternatively, by promoting recombination. An insertion-like structure was also found in the minimal transforming region of *Bam*HI *E* fragment of *mtr*III[32]; this subfragment contains repeated sequences that resemble DNA elements known to play a role in genetic recombination and control of gene activity in mammalian cells.

2.2. Interactions with Cell DNA and Other Viruses

Different mechanisms by which HSV may interact with cells and alter their phenotype have been suggested on the basis of experimental evidence. These include induction of mutations within the host cell genome, amplification of DNA sequences, and activation of cellular genes or endogenous retroviruses. The mutagenic potential of HSV is shown by its ability to induce genetic mutations,[33,34] chromosomal aberrations,[33] and stimulation of host cell DNA synthesis and repair.[35] As a consequence of DNA damage and repair, DNA amplification is frequently observed. In fact, HSV-2 infection of SV40-transformed hamster or human cells induces amplification of the integrated SV40 DNA sequences[36] similar to chemical carcinogens.[37] Recent results show that HSV infection of human cervical carcinoma cell lines containing integrated HPV-18 sequences induces a two- to eightfold amplification of HPV-18 DNA.[38]

Concerning the ability of HSV to alter cellular gene expression, HSV genes encode for *trans*-acting factors (TIF, ICP0, ICP4, ICP27) that activate the immediate early genes of HSV[39] as well as cellular promoters.[40,41] In addition, previous observations had shown that HSV can activate endogenous retrovirus genes in murine and hamster cells,[42] and it was suggested that the HSV-1 sequences responsible for this induction act as mutagens.[43] More recently, it was also shown that HSV activates the expression of the human immunodeficiency virus type 1 and human T-cell lymphotropic virus types 1 and 2 long terminal repeats.[44,45] Finally, several studies have shown a possible interaction between HSV and HPV. The *mtr*II DNA fragment of HSV-2 is capable of converting immortal genital epithelial cells containing integrated HPV-16 sequences into tumorigenic cells but is ineffective on normal cells.[46] The HSV-2 *mtr*III is able to convert hamster cells morphologically transformed by HPV-16 or -18 DNA to a fully tumorigenic phenotype[47]; it also enhances the transformation of human fibroblasts and keratinocytes immortalized by HPV-16 or -18 DNA.[48,49] Herpes simplex virus is able to alter the expression of the HPV genome; in fact, HSV-1-encoded *trans*-activators have proven capable of activating expression from the promoter located in the long control region of HPV-16 and -18,[50,51] and the infection of HeLa cells with HSV-1 modifies HPV-18 gene expression by reducing the amount of specific mRNAs.[52]

In summary, these data indicate that papilloma and herpes viruses interact at the molecular level and suggest that HSV could play a role in oncogenesis and neoplastic progression.

3. HERPES SIMPLEX VIRUS AND GENITAL NEOPLASMS

3.1. Seroepidemiologic Evidence

The epidemiology of genital neoplasia has been extensively studied and appears to be related to a number of risk factors most of which are linked to sexual activity, thus suggesting a sexually transmitted agent as a putative etiological agent for these tumors. These risk factors include multiple sexual partners, early age at first sexual intercourse, high parity, early age at the first pregnancy, low socioeconomic status, use of oral contraceptives, and smoking [reviewed by Brinton[53]]. In the 1980s, the hypothesis that certain types of HPV were related to human genital cancers gained rapid support. Based on experimental and clinical evidence, high-risk types of HPV were identified and recognized as causal factors in cervical cancer,[19] and their identification has allowed adequate consideration of the role of the other putative possible risk factors.

Since it was first suggested, on the basis of cytological observations, that genital herpes infection might be related to cervical cancer,[54] many seroepidemiologic surveys have been performed, most of which were case-control groups in which the prevalence of HSV-2 antibodies and/or antigens in women with cervical neoplasia were compared to that in women free of such lesions. Different problems have at times characterized these surveys. In fact, the cases were selected from all socioeconomic classes, and a wide range of ages were represented; in some works invasive cancers and all grades of intraepithelial neoplasia were analyzed separately, and in other works they were not; the cases were not always matched to controls in terms of age, race, and socioeconomic status. Some reports failed to show an epidemiologic association between HSV infection and genital neoplasia,[2,55] but several others showed a higher prevalence and titer of antibodies against HSV-2 in patients with cervical carcinoma or with premalignant lesions than in controls.[1,56,57] The relative frequency of HSV-2 seropositivity among patients and controls varies with the population studied, and some surveys report a certain number of cancer patients with no signs of previous HSV-2 infection.[58] In addition, it has recently been shown that sera from patients with cervical cancer contain antibodies to a 40-kDa polypeptide that is inducible by HSV-2 infection but is a cell-coded tumor-specific polypeptide.[59]

Well-controlled epidemiologic studies, conducted in different geographic areas, suggested a relationship between HSV seropositivity and cervical cancer. Kjaer et al.[60] observed that the higher incidence of cervical cancer in Greenland compared to Denmark correlates with a higher HSV-2 seroprevalence. The results of work performed in Australia showed a positive relationship between exposure to HSV and *in situ* cervical cancer development, with a relative risk (RR) of 2.3.[61] An association between seroreactivity to HSV and cancer patients was also seen in case-control studies performed in Uganda[62] and in China, where women with HSV-2 antibodies had a 30% increased risk of cervical cancer.[63] It is debated whether the higher antibody titers in cancer patients merely reflect the common sexual transmission for HSV and the factors inducing genital neoplasia. However, a recent study in Latin America detected a significant interactive effect between HSV-2 and

HPV-16/18 positivity for the risk of cervical cancer.[64] Compared with women testing negative for both viruses, those positive for HSV-2 alone had a RR of 1.2, those positive for HPV-16/18 DNA alone had a RR of 4.3, and those positive for both had a RR of 8.8.[64] These data support the hypothesis that HSV-2 might be one of a number of agents capable of participating in the etiology of cervical cancer.

3.2. Herpes Simplex Virus-Specific Macromolecules in Genital Neoplasms

3.2.1. Antigens

Several studies examined cervical cancers for the presence of viral antigens. Two antigens (AG-4 and ICP11/12) are of particular interest because they are also found in HSV-2-transformed cells.

AG-4 is an HSV-2-specific antigen immunologically identical to the virus-coded protein ICP10 already mentioned. This protein, encoded by the *mtr*III fragment of the HSV-2 genome, is expressed in cells transformed by *mtr*III,[26] and it was identified in cervical tissue from 65–85% of neoplastic patients but not in normal cells.[65] Furthermore, it was shown that monoclonal antibodies directed against ICP10 stained atypical cells in 55% of patients with mild dysplasia and in 100% of those with more severe cervical lesions, while normal squamous, metaplastic inflammatory, or koilocytic cells did not stain with these antibodies.[66] Our results also showed the presence of ICP10 in 48.5% of cervical intraepithelial neoplasia (CIN) and 52.9% of cervical carcinoma tissues,[67] and in 33.3% and 66.6% of vulvar intraepithelial neoplasia (VIN) and vulvar carcinomas, respectively.[68]

The other antigen frequently detected in cervical tumor cells is ICP 11/12, the HSV-2 major DNA-binding protein. Its presence was detected in cells transformed by inactivated HSV-2,[24] in 31–50% of patients with cervical tumors,[69] and in 33.3% and 75% of patients with VIN or vulvar carcinomas, respectively.[68]

3.2.2. Herpes Simplex Virus DNA Sequences

Because transformed cells contain less than one copy of HSV homologous sequences per diploid genome, one approach used for demonstrating the presence of HSV genetic sequences in tumor tissue was research on specific RNA. By *in situ* hybridization, using the whole HSV-2 DNA as a probe, HSV RNA was detected in neoplastic cervical cells with frequencies ranging from 35%[70] to 72%.[71] Subsequently, by the use of cloned DNA fragments as probes, it was possible to show that these viral transcripts were preferentially associated with specific regions of the HSV genome,[69] located at m.u. 0.1–0.4, 0.58–0.63, and 0.82–0.85. It is interesting to note that two of these regions correspond to the *in vitro* transforming regions *mtr*I and *mtr*II, thus establishing a link between transformation studies and the analysis of tumor tissues.

The first report showing the presence of HSV-2 DNA in an invasive cervical carcinoma[8] stimulated further research in this field. Subsequent reports described negative findings,[72,73] possibly because of dilution of the tumor cells in the connective matrix and poor sensitivity of the probes (the entire HSV-2 genome was utilized). However, the use of specific cloned fragments of HSV-2 DNA made it possible to obtain high-specific-activity probes able to detect, by Southern blot hybridization, small homologous DNA fragments present at less than one copy per diploid genome. The presence of HSV-2 DNA was shown in one of eight cervical tumors,[10] where the sequences corresponded to the HSV-2 *Bgl*II *N* fragment, and in 5 of 16 cervical tumors,[74] two of which were positive for *Bgl*II *N*, one for *Bgl*II *J* (m.u. 0.31–0.39), and two for both probes. Healthy genital tissues were consistently negative. Antigens and DNA sequences of HSV-2 were also identified in vulvar tissue from patients with invasive or premalignant lesions.[11,75]

To study the possible cooperation between HSV and HPV in tumor development, we analyzed several specimens for the simultaneous presence of DNA of both viruses in a large number of normal, preneoplastic, and neoplastic tissues derived from cervix and vulva.[11,13,67,68,76,77] The samples were examined for homology to HSV and HPV DNA by Southern blot or dot–blot hybridization in high-stringency conditions. Besides HPV-16 and -18 DNAs, the following HSV DNA fragments were employed as probes: (1) HSV-2 *Bgl*II *N* (m.u. 0.58–0.63); (2) HSV-2 *Bgl*II *O* (m.u. 0.38–0.42), a fragment overlapping a portion of the HSV-1 *Bgl*II *I* region with transforming activity[6]; (3) HSV-2 *Bgl*II *C* (m.u. 0.42–0.58); (4) HSV-2 *Bam*HI *E* (m.u. 0.53–0.58); (5) HSV-2 *Bgl*II *G* (m.u. 0.20–0.31), selected to represent viral DNA sequences with no known transforming activity; (6) HSV-1 *Bgl*II *I* (m.u. 0.31–0.42), selected because in our geographic area up to 40% of genital HSV infections are caused by HSV-1.[78] The results obtained are summarized in Table I.

Cervical intraepithelial neoplasia contained sequences homologous to *Bgl*II *N* and *Bam*HI *E*, respectively, in 14% and 4% of the samples. Invasive cervical carcinomas contained sequences positive to *Bgl*II *N* in 13% of the samples, to *Bam*HI *E* in 17%, and to *Bgl*II *O* in 7% of the cases. DNAs of HPV-16 or -18 were present in 67% of CIN and 49% of invasive carcinomas, thus confirming the strong association of specific HPV types with cervical neoplasia.[19]

When possible, the tissues were enriched for tumor cells by cleaning the neoplastic samples of contaminating stromal tissue.[13] In fact, one of the initial experiments had shown that homology to HSV-2 could be detected with high frequency in those lesions from which the stroma had been removed (31% compared to 4%), thus suggesting that careful cleaning of the neoplastic sample was crucial for detecting the presence of HSV in preinvasive lesions. The removal of contaminating stroma from biopsies also appears to be important for the detection of HPV DNA. In fact, positive samples increase from 47% to 88% (Table I).

The results further show that 84% of VIN and 13% of invasive carcinomas also contained sequences homologous to HPV-16 DNA. The strong association of HPV-16 with VIN suggests that HPV plays a causative role in this type of lesion, but the presence of HPV in the host cell appears insufficient to induce cancer: additional damaging endogenous or exogenous events are required, and the presence of the viral genome may not be necessary to maintain the malignant phenotype and the progression after transformation.[19]

When premalignant and malignant vulvar lesions were analyzed for the presence of HSV,[11,68,76,77] we observed that 25% and 17% of VIN were positive for *Bgl*II *N* and *Bgl*II *O*, respectively, and 14% and 23% of invasive carcinomas contained

TABLE I
Presence of Viral DNA in Cervical and Vulvar Lesions[11,13,67,68,77]

	HPV-16/18	HSV-2		
		BglII N	BamHI E	BglII O
Cervical intraepithelial neoplasia	24/36 (67%) 15/17 (88%)[a] 9/19 (47%)[b]	5/37 (14%) 4/13 (31%)[a] 1/24 (4%)[b]	1/27 (4%)	0/8
Cervical invasive carcinoma	45/91 (49%)	8/60 (13%)	4/24 (17%)	1/14 (7%)
Vulvar intraepithelial neoplasia	16/19 (84%)	4/16 (25%)	0/4	1/6 (17%)
Vulvar invasive carcinoma	4/32 (13%)	2/14 (14%)	0/8	3/13 (23%)

[a]Without stroma.
[b]With stroma.

sequences homologous to those HSV-2 probes. Interestingly, about 75% of cervical biopsies positive for the presence of HSV-2 DNA also contained sequences hybridizing to HPV-16 or -18, suggesting that the two viral agents could indeed interact in cells undergoing neoplastic transformation.

Healthy tissues from tumor-bearing patients, used as controls, did not hybridize to any of the viral probes, showing that the hybridization detected did not reflect nonspecific cross-reactions between human and viral DNA; this was also shown by a study in which labial or cerebral tumors were hybridized to BglII N without giving any positive reaction.[12]

The presence of HSV-2 in cervical specimens has been confirmed using the polymerase chain reaction (PCR) technique in which the amplification of a region of BglII N fragment attains high sensitivity levels.[79] DNA of HSV-2 was found in 30% of abnormal cervical tissues, suggesting that the previous failure to demonstrate HSV DNA probably reflected the poor sensitivity of the techniques then available.

Finally, a highly intriguing result recently reported shows that HSV and HPV integration sites are nonrandomly distributed in cervical carcinoma but are located near fragile sites, protooncogenes, and specific cancer chromosome breakpoints.[80] A statistically significant association appears to exist between breakpoints observed in cervical carcinoma and viral breakage sites, thus suggesting that both viruses play an important role in carcinogenesis and/or development of karyotypic abnormalities in cervical cancers.

4. CONCLUSIONS

The data available indicate that HSV can induce cell transformation; however, viral DNA sequences cannot always be detected in the transformed cells, suggesting that retention of HSV DNA fragments is not required to maintain the transformed state. Unlike other oncogenic viruses, a single viral DNA sequence or gene product responsible for neoplastic changes has not been detected, thus rendering the research in this field particularly complex and confusing. Besides the trivial, but not unimportant, explanation related to the poor sensitivity of the techniques used until a few years ago, one more attractive explanation for these results could be that HSV is involved in the initiation of neoplastic development.

In fact, several lines of evidence now support a multistep progression of a normal cell to a fully malignant phenotype. This process would include different phases, namely, initiation, promotion, and progression. During the first stage, cellular DNA is irreversibly damaged in the form of mutations, insertions, deletions, or rearrangements without giving phenotypic manifestations. In the second stage, modifying factors act on the initiated cell, leading to a neoplastic phenotype. During the last phase, the neoplastic cell evolves and gives rise to a tumor lineage. Most of the data reported to date point to the fact that HSV may act as an initiator of the neoplastic process: it can induce chromosomal aberrations,[33] DNA amplification,[38] and alterations of cellular and viral gene expression.[39–45,50–52] Interestingly, insertion-like elements were detected within the HSV-2 transforming fragments *Bgl*II *N* and *Bam*HI *E*.[31,32] In addition, it has recently been shown that HSV integration sites are often located near critical points in the host cell genome, such as protooncogenes, fragile sites, and specific chromosome breakpoints.[80]

In light of these findings, a possible mechanism of transformation could involve the insertion in the cell genome of short HSV sequences (not always detectable by poorly sensitive techniques) that could activate the expression of cellular oncogenes or cause irreversible chromosomal alterations.

Recent studies have revealed a strong association between cervical cancer and HPV infection. However, several observations suggested that HPV infection alone is not sufficient for the induction and progression of malignancies: malignant growth seems to require the synergistic action of carcinogens, such as chemical substances, irradiation, or other oncogenic viruses.[15,16] The data available on the HPV mode of action in transforming *in vitro* cultured cells seems to confirm this hypothesis. In fact, long-time *in vitro* cultivation of HPV immortalized cells may lead to the development of malignant clones.[19] These clones are characterized by chromosomal instability and aneuploidy, suggesting that additional endogenous events may have acted as cofactors in malignant progression. Alternately, the transformation of the infected cells toward a fully tumorigenic phenotype can be mediated by exogenous factors such as the presence of transforming regions of the HSV genome.[46–49] Moreover, analysis of the DNA extracted from genital tumors shows the presence of HPV malignant types in about 90% of samples tested; however, the presence of HPV-16 and -18 infection, even in cytologically normal women, as detected by PCR technique,[79] points out that the precise mode of action of these viruses in the oncogenic process still needs to be elucidated.

Therefore, HPV infection could act as a promoter in genital cancer under the influence of initiating events such as HSV infection. This hypothesis has now also found some epidemiologic support, because the risk of developing cervical cancer is consistently higher for women testing positive for both viruses, compared to those positive for HSV or HPV alone.[64] Further epidemiologic and molecular studies are, therefore, required to address this hypothesis, by using the accurate and sensitive methods now available for virus detection.

ACKNOWLEDGMENTS. The studies performed in our laboratory were supported by grants from A.I.R.C. (Associazione Italiana per la Ricerca sul Cancro) and from C.N.R. (Consiglio Nazionale delle Ricerche), "Special Project A.C.R.O. (Applicazioni Cliniche della Ricerca Oncologica)."

REFERENCES

1. Kaufman, R. H., and Adam, E., 1986, Herpes simplex virus and human papilloma virus in the development of cervical cancer, *Clin. Obstet. Gynecol.* **29:**678:692.
2. Vonka, V., Kanka, J., Hirsch, I., Zavadova, H., Krcmar, M., Suchankova, A., Rezacova, D., Broncek, J., Press, M., Domorazkova, E., Svoboda, B., Havrankova, A., and Jelinek, J., 1984, Prospective study on the relationship between cervical neoplasia and herpes simplex type-2 virus. II. Herpes simplex type-2 antibody presence in sera taken at enrollment, *Int. J. Cancer* **33:**61–66.
3. Dale, G. E., Coleman, R. M., Best, J. M., Benetato, B. B. B., Drew, N. C., Chinn, S., Papacosta, A. O., and Nahmias, A. J., 1988, Class-specific herpes simplex virus antibodies in sera and cervical secretions from patients with cervical neoplasia: A multi-group comparison, *Epidemiol. Infect.* **100:**455–465.
4. Slattery, M. L., Overall, J. C., Abbott, T. M., French, T. K., Robinson, L. M., and Gardner, J., 1989, Sexual activity, contraception, genital infections and cervical cancer: Support for a sexually transmitted disease hypothesis, *Am. J. Epidemiol.* **130:**248–258.
5. Macnab, J. C. M., 1987, Herpes simplex virus and human cytomegalovirus: Their role in morphological transformation and genital cancers, *J. Gen. Virol.* **68:**2525–2550.
6. Reyes, G. R., La Femina, S. D., and Hayward, G. S., 1980, Morphological transformation by DNA fragments of human herpesviruses: Evidence for two distinct transforming regions in HSV-1 and HSV-2 and lack of correlation with biochemical transfer of thymidine kinase gene, *Cold Spring Harbor Symp. Quant. Biol.* **44:**629–641.
7. Galloway, D. A., and McDougall, J. K., 1981, Transformation of rodent cells by a cloned DNA fragment of herpes simplex virus type 2, *J. Virol.* **38:**749–760.
8. Frenkel, N., Roizman, B., Cassai, E., and Nahmias, A., 1972, A DNA fragment of herpes simplex 2 and its transcription in human cervical tissue, *Proc. Natl. Acad. Sci. USA* **69:**3784–3789.
9. Galloway, D. A., and McDougall, J. K., 1983, The oncogenic potential of herpes simplex viruses: Evidence for a *hit-and-run* mechanism, *Nature* **302:**21–24.
10. Park, M., Kitchener, H. C., and Macnab, J. C. M., 1983, Detection of herpes simplex virus type 2 DNA restriction fragments in human cervical carcinoma tissue, *EMBO J.* **2:**1029–1034.
11. Manservigi, R., Cassai, E., Deiss, L. P., Di Luca, D., Segala, V., and Frenkel, N., 1986, Sequences homologous to two separate transforming regions of herpes simplex virus DNA are linked in two human genital tumors, *Virology* **155:**192–201.
12. Rotola, A., Di Luca, D., Monini, P., Manservigi, R., Tognon, M., Virgili, A. R., Segala, V., Trapella, G., and Cassai, E., 1986, Search for HSV DNA in genital, cerebral and labial tumors, *Eur. J. Cancer Clin. Oncol.* **10:**1256–1265.
13. Di Luca, D., Rotola, A., Pilotti, S., Monini, P., Caselli, E., Rilke, F., and Cassai, E., 1987, Simultaneous presence of herpes simplex and human papilloma virus sequences in human genital tumors, *Int. J. Cancer* **40:**763–768.
14. zur Hausen, H., 1976, Condyloma acuminata and human genital cancer, *Cancer Res.* **36:**794.
15. Crum, C. P., Mitao, M., Levine, R. V., and Silverstein, S., 1985, Cervical papillomaviruses segregate within morphologically distinct precancerous lesions, *J. Virol.* **5:**675–681.
16. Jarrett, W. F. H., McNeil, P. E., Grimshaw, W. T. R., Selman, I. E., and McIntyre, W. I. M., 1978, High incidence area of cattle cancer with a possible interaction between an environmental carcinogen and a papillomavirus, *Nature* **274:**215–217.
17. Munger, K., Phelps, W. C., Bubb, V., Howley, P. M., and Schlegel, R., 1989, The E6 and E7 genes

of human papillomavirus type 16 together are necessary and sufficient for transformation of primary human keratinocytes, *J. Virol.* **63:**4417–4421.
18. Pecoraro, G., Lee, M., Morgan, D., and Defendi, V., 1991, Evolution of *in vitro* transformation and tumorigenesis of HPV16 and HPV18 immortalized primary cervical epithelial cells, *Am. J. Pathol.* **138:**1–8.
19. zur Hausen, H., 1991, Human papillomaviruses in pathogenesis of anogenital cancer, *Virology* **184:**9–13.
20. Duff, R., and Rapp, F., 1971, Oncogenic transformation of hamster cells after exposure to herpes simplex virus type 2, *Nature [New Biol.]* **233:**48–50.
21. Duff, R., and Rapp, F., 1973, Oncogenic transformation of hamster embryo cells after exposure to inactivated herpes simplex virus type 1, *J. Virol.* **12:**209–217.
22. Minson, A. C., 1984, Cell transformation and oncogenesis by herpes simplex virus and human cytomegalovirus, *Cancer Surv.* **3:**91–111.
23. Galloway, D. A., and McDougall, J. K., 1990, Alterations in the cellular phenotype induced by herpes simplex virus, *J. Med. Virol.* **31:**36–42.
24. Lewis, J. G., Kucera, L. S., Eberle, R., and Courtney, R. J., 1982, Detection of herpes simplex virus type 2 glycoproteins expressed in virus-transformed rat cells, *J. Virol.* **42:**275–282.
25. Cameron, I. R., and Macnab, J. C. M., 1980, Transformation studies using defined fragments of herpes simplex virus type 2, in: *The Human Herpesviruses. An Interdisciplinary Perspective* (A. J. Nahmias, W. R. Dowdle, and F. R. Schinazi, eds.), Elsevier, New York, pp. 634–649.
26. Hayashi, Y., Iwasaka, T., Smith, C. C., Aurelian, L., Lewis, G. K., and Ts'o, P. O. P., 1985, Multistep transformation by defined fragments of herpes simplex virus type 2 DNA: Oncogenic region and its gene product, *Proc. Natl. Acad. Sci. USA* **82:**8493–8497.
27. Chung, T. D., Wymer, J. P., Smith, C. C., Kulka, M., and Aurelian, L., 1989, Protein kinase activity associated with the large subunit of herpes simplex virus type 2 ribonucleotide reductase (ICP10), *J. Virol.* **63:**3389–3398.
28. Ali, M. D., McWeeney, D., Milosavljevic, A., Jurka, J., and Jariwalla, R. J., 1991, Enhanced malignant transformation induced by expression of a distinct protein domain of ribonucleotide reductase large subunit from herpes simplex virus type 2, *Proc. Natl. Acad. Sci. USA* **88:**8257–8261.
29. Jones, C., 1989, The minimal transforming fragment of HSV-2 *mtr*III can function as a complex promoter element, *Virology* **169:**346–353.
30. Cameron, I. R., Park, M., Dutia, B. M., Orr, A., and Macnab, J. C. M., 1985, Herpes simplex virus sequences involved in the initiation of oncogenic morphological transformation of rat cells are not required for maintenance of the transformed state. *J. Gen. Virol.* **66:**517–527.
31. Galloway, D. A., Nelson, J. A., and McDougall, J. K., 1984, Small fragments of herpesvirus DNA with transforming activity contain insertion sequence-like structures, *Proc. Natl. Acad. Sci. USA* **81:**4736–4740.
32. Jones, C., Ortiz, J., and Jariwalla, R. J., 1986, Localization and comparative nucleotide sequence analysis of the transforming domain in herpes simplex virus DNA containing repetitive genetic elements, *Proc. Natl. Acad. Sci. USA* **83:**7855–7859.
33. Hampar, B., and Ellison, S. A., 1961, Chromosomal aberrations induced by an animal virus, *Nature* **192:**145–147.
34. Brandt, C. R., Buonaguro, F. M., McDougall, J. K., and Galloway, D. A., 1987, Plasmid mediated mutagenesis of a cellular gene transfected into eukaryotic cells, *Nucleic Acids Res.* **15:**561–573.
35. Nishiyama, Y., and Rapp, F., 1981, Repair replication of viral and cellular DNA in herpes simplex virus type 2 infected human embryonic and xeroderma pigmentosum cells, *Virology* **110:**466–475.
36. Schlehofer, J. R., Gissman, L., Metz, B., and zur Hausen, H., 1983, Herpes simplex virus-induced amplification of SV-40 sequences in transformed hamster embryo cells, *Int. J. Cancer* **32:**99–103.
37. Lavi, S., 1981, Carcinogen mediated amplification of viral DNA sequences in simian virus 40 transformed Chinese hamster embryo cells, *Proc. Natl. Acad. Sci. USA* **78:**6144–6148.
38. Brandt, C. R., McDougall, J. K., and Galloway, D. A., 1987, Synergistic interaction between

human papillomavirus type 18 sequences, herpes simplex virus infection and chemical carcinogen treatment, in: *Papillomaviruses, Cancer Cells,* Vol. 5 (B. Steinberg J. Brandsma, and L. Taichman, eds.), Cold Spring Harbor Laboratory Press, New York, pp. 179–186.
39. Batterson, W., and Roizman, B., 1983, Characterization of the herpes simplex virion associated factor responsible for the induction of α genes, *J. Virol.* **46:**371–377.
40. O'Hare, P., and Hayward, G. S., 1985, Evidence for a direct role for both the 175,000- and 110,000-molecular-weight immediate-early proteins of herpes simplex virus in the transactivation of delayed early promoters, *J. Virol.* **53:**751–760.
41. Latchmann, D. S., Estridge, J. K., and Kemp, L. M., 1987, Transcriptional induction of the ubiquitin gene during herpes simplex virus infection is dependent upon the viral immediate early protein ICP4, *Nucleic Acids Res.* **15:**7283–7293.
42. Hampar, B., Aaronson, S. A., Derge, J. G., Chakrabarty, M., Showalter, S. D., and Dunn, C. Y., 1976, Activation of an endogenous mouse type C virus by ultraviolet irradiated herpes simplex virus type 1 and 2, *Proc. Natl. Acad. Sci. USA* **73:**646–650.
43. Boyd, A. L., Enquist, L., Vande Wonde, G. F., and Hampar, B., 1980, Activation of mouse retrovirus by herpes simplex virus type 1 cloned DNA fragments, *Virology* **103:**228–231.
44. Chen, I. S. Y., Cann, A. J., Shah, N. P., and Gaynor, R. B., 1985, Functional relation between HTLV-IIx and adenovirus E1A proteins in transcriptional activation, *Science* **230:**570–573.
45. Ostrove, J. M., Leonard, J., Weck, K. E., Rabson, A. B., and Gendelman, H. E., 1987, Activation of the human immunodeficiency virus by herpes simplex virus type 1, *J. Virol.* **61:**3726–3732.
46. Di Paolo, J. A., Woodworth, C. D., Popescu, N. C., Koval, D. L., Lopez, J. V., and Doniger, J., 1990, HSV-2 induced tumorigenicity in HPV-16 immortalized human genital keratinocytes, *Virology* **177:**777–779.
47. Iwasaka, T., Yokoyama, M., Hayashi, Y., and Sugimori, H., 1988, Combined herpes simplex virus type 2 and human papillomavirus type 16 or 18 deoxyribonucleic acid leads to oncogenic transformation, *Am. J. Obstet. Gynecol.* **159:**1251–1255.
48. Dhanwada, K. R., Veerisetty, V., Zhu, F., Razzaque, A., Thompson, K. D., and Jones. C., 1992, Characterization of primary human fibroblasts transformed by human papilloma virus type 16 and herpes simplex virus type 2 DNA sequences, *J. Gen. Virol.* **73:**791–799.
49. Dhanwada, K. R., Garrett, L., Smith, P., Thompson, K. D., Doster, A., and Jones, C., 1993, Characterization of human keratinocytes transformed by high risk human papillomavirus type 16 or 18 and herpes simplex virus type 2, *J. Gen. Virol.* **74:**955–963.
50. McCusker, D. T., and Bacchetti, S., 1988, The responsiveness of human papillomavirus upstream regulatory region to herpes simplex virus immediate early proteins, *Virus Res.* **11:**199–207.
51. Gius, D., and Laimins, L. A., 1989, Activation of human papillomavirus type 18 gene expression by herpes simplex virus type 1 viral transactivators and a phorbol ester, *J. Virol.* **63:**555–563.
52. Karlen, S., Offord, E. A., and Beard, P., 1993, Herpes simplex virions interfere with the expression of human papillomavirus type 18 genes, *J. Gen. Virol.* **74:**965–973.
53. Brinton, L. A., 1992, Epidemiology of cervical cancer—an overview, in: *The Epidemiology of Cervical Cancer and Human Papillomavirus* (N. Munoz, F. X. Bosch, K. Shah, and A. Meheus, eds.), IARC Scientific Publication 119, International Agency for Research on Cancer, Lyon, pp. 3–22.
54. Naib, Z. M., Nahmias, A. J., and Josey, W. E., 1966, Cytology and histopathology of cervical herpes simplex infection, *Cancer* **19:**1026–1031.
55. Krcmar, M., Suchankova, A., Kanka, J., and Vonka, V., 1986, Prospective study on the relationship between cervical neoplasia and herpes simplex type-2 virus. III. Presence of herpes simplex type-2 antibody in sera of subjects who developed cervical neoplasia later in the study, *Int. J. Cancer* **38:**161–165.
56. Rotola, A., Gerna, G., Di Luca, D., Virgili, A. R., Manservigi, R., and Cassai, E., 1983, Herpes simplex virus and human cancer. III. Search for relationship of herpes simplex antibodies and cervical dysplasia and labial neoplasia, *Tumori* **69:**83–89.
57. Rawls, W. E., Bacchetti, S., and Graham, F. C., 1977, Relation of herpes simplex viruses to human malignancies, *Curr. Top. Microbiol. Immunol.* **77:**72–95.

58. Rawls, W. E., Clarke, A., Smith, K. O., Docherty, J. J., Gilman, S. C., and Graham, S., 1980, Specific antibodies to herpes simplex virus type 2 among women with cervical cancer, *Cold Spring Harbor Conf. Cell Prolif.* **7**:117–133.
59. Macnab, J. C. M., Nelson, J. S., Daw, S., Hewitt, R. E. P., Lucasson, J. F., and Shirodaria, P. V., 1992, Patients with cervical cancer produce an antibody response to an HSV-inducible tumor-specific cell polypeptide, *Int. J. Cancer* **50**:578–584.
60. Kjaer, S. K., De Villiers, E. M., Hangaard, B. J., Christensen, R. B., Teisen, C., Moller, K. A., Poll, P., Jensen, H., Vestergaard, B. F., Lynge, E., and Jensen, O. M., 1988, Human papillomavirus, herpes simplex virus and cervical cancer incidence in Greenland and Denmark. A population-based cross-sectional study, *Int. J. Cancer* **41**:518–524.
61. Brock, K. E., MacLennan, R., Brinton, L. A., Melnick, J. L., Adam, E., Mock, P. A., and Berry, G., 1989, Smoking and infectious agents and risk of *in situ* cervical cancer in Sydney, Australia, *Cancer Res.* **49**:4925–4928.
62. Schmauz, R., Okong, P., De Villiers, E. M., Dennin, R., Brade, C., Lwanga, S. K., and Owor, R., 1989, Multiple infections in cases of cervical cancer in a high-incidence area in tropical Africa, *Int. J. Cancer* **43**:805–809.
63. Peng, H., Liu, S., Mann, V., Rohan, T., and Rawls, W., 1991, Human papillomavirus type 16 and 33, herpes simplex virus type 2 and other risk factors for cervical cancer in Sichuan province, China, *Int. J. Cancer* **47**:711–716.
64. Hildesheim, A., Mann, V., Brinton, L. A., Szklo, M., Reeves, W. C., and Rawls, W. E., 1991, Herpes simplex virus type 2: A possible interaction with human papillomavirus types 16/18 in the development of invasive cervical cancer, *Int. J. Cancer* **49**:335–340.
65. Costa, S., Smith, C. C., Taylor, S., Aurelian, L., and Orlandi, C., 1986, Intracellular localization and serological identification of a HSV-2 protein in cervical cancer, *Eur. J. Gyn. Oncol.* **7**:1–12.
66. Costa, S., D'Errico, A., Grigioni, W. F., Orlandi, C., Smith, C. C., Mancini, A. M., and Aurelian, L., 1987, Monoclonal antibody to HSV-2 protein as an immunodiagnostic marker in cervical cancer, *Cancer Detect. Prevent. [Suppl.]* **1**:189–205.
67. Di Luca, D., Costa, S., Monini, P., Rotola, A., Terzano, P., Savioli, A., Grigioni, W., and Cassai, E., 1989, Search for human papillomavirus, herpes simplex virus and c-*myc* oncogene in human genital tumors, *Int. J. Cancer* **43**:570–577.
68. Costa, S., Rotola, A., Terzano, P., Poggi, M. G., Di Luca, D., Aurelian, L., Cassai, E., and Orlandi, C., 1990, Search for herpes simplex virus 2 and human papillomavirus genetic expression in vulvar neoplasia, *J. Reprod. Med.* **35**:1108–1112.
69. MacDougall, J. K., Crum, C. P., Fenoglio, C. M., Goldstein, L. C., and Galloway, D. A., 1982, Herpesvirus specific RNA and protein in carcinoma of the uterine cervix, *Proc. Natl. Acad. Sci. USA* **79**:3853–3857.
70. McDougall, J. K., Galloway, D. A., and Fenoglio, C. M., 1980, Cervical carcinoma: Detection of herpes simplex virus RNA in cells undergoing neoplastic change, *Int. J. Cancer* **25**:1–8.
71. Eglin, R. P., Sharp, F., MacLean, A. B., Macnab, J. C. M., Clements, J. B., and Wilkie, N. M., 1981, Detection of RNA complementary to herpes simplex virus DNA in human cervical squamous cell neoplasms, *Cancer Res.* **41**:3597–3603.
72. Pagano, J. S., 1975, Disease and mechanisms of persistent DNA virus infection: Latency and cellular transformation, *J. Infect. Dis.* **132**:209–223.
73. Cassai, E., Rotola, A., Meneguzzi, G., Milanesi, G., Garsia, S., Remotti, G., and Rizzi, G., 1981, Herpes simplex virus and human cancer. I. Relationship between human cervical tumors and herpes simplex type 2, *Eur. J. Cancer* **17**:685–693.
74. McDougall, J. K., Smith, P., Tamimi, H. K., Tolentino, E., and Galloway, A., 1984, Molecular biology of the relationship between herpes simplex virus 2 and cervical cancer, *The Role of Viruses in Human Cancer*, Vol. II (G. Giraldo and E. Beth, eds.), Elsevier, New York, pp. 59–71.
75. Cabral, G. A., Marciano-Cabral, F., Fry, D., Lumpkin, C. K., Mercer, L., and Gloperud, D., 1982, Expression of herpes simplex type 2 antigens in premalignant and malignant human vulvar cells, *Am. J. Obstet. Gynecol.* **143**:611–619.

76. Di Luca, D., Rotola, A., Monini, P., and Cassai, E., 1988, Sequences homologous to HSV and HPV nucleic acid in genital neoplasia, in: *Herpes and Papilloma Viruses*, Vol. 2 (G. De Palo, F. Rilke, and H. zur Hausen, eds.), Raven Press, New York, pp. 129–140.
77. Pilotti, S., Rotola, A., D'Amato, L., Di Luca, D., Shah, K. V., Cassai, E., and Rilke, F., 1990, Vulvar carcinomas: Search for sequences homologous to human papillomavirus and herpes simplex virus DNA, *Mod. Pathol.* **3**:442–447.
78. Taparelli, F., Squadrini, F., Cassai, E., Tognon, M., and Fornaciari, A., 1985, Comparaison entre les methodes de laboratoires serologique et moleculaires pour la caracterisation des virus de l'herpes simplex, *Med. Mal. Infect.* **9**:509–513.
79. Bevan, I. S., Blomfield, P. I., Johnson, M. A., Woodman, C. B. J., and Young, L. S., 1989, Oncogenic viruses and cervical cancer, *Lancet* **2**:907–908.
80. De Braekeleer, M., Sreekantaiah, C., and Haas, O., 1991, Herpes simplex virus and human papillomavirus sites correlate with chromosomal breakpoints in human cervical carcinoma, *Cancer Genet. Cytogenet.* **59**:135–137.

16

Human Cytomegalovirus
Aspects of Viral Morphogenesis and of Processing and Transport of Viral Glycoproteins

K. RADSAK, H. KERN, B. REIS, M. RESCHKE, T. MOCKENHAUPT, and M. EICKMANN

1. INTRODUCTION

Human cytomegalovirus (HCMV) is a ubiquitous agent that rarely induces symptoms of disease in the immunocompetent host in spite of its persistence. However, it frequently causes severe disease in immunocompromised individuals.[1,2] The HCMV is classified as a β-herpesvirus and as such exhibits typical properties in cultured cells, i.e., species as well as cell type specificity, induction of cell enlargement, and a comparatively long infectious cycle,[3] i.e., delayed release of progeny virus and cell death occurring only many days after infection. Another feature that distinguishes HCMV in particular from other herpesviruses is its pronounced genomic complexity, with 240 kbp versus about 150 kbp in e.g., herpes simplex virus (HSV).[4] Like HSV, and in contrast to cytomegaloviruses of other species, HCMV has an isomerizing genome representative of the E group of herpesviridae genomes with complexity of structure increasing from group A to E.[5]

The complete nucleotide sequence of laboratory strain AD169 has been determined, showing 208 potential open reading frames.[6] Because of the inefficient growth of HCMV in cultured cells, molecular dissection of the details of the virus–host cell interaction has been lagging behind that of other herpesviruses, though

---

K. RADSAK, B. REIS, M. RESCHKE, T. MOCKENHAUPT, and M. EICKMANN • Institut für Virologie, Philipps-Universität, D-35037 Marburg, Germany.   H. KERN • Institut für Zellbiologie, Philipps-Universität, D-35037 Marburg, Germany.

DNA Tumor Viruses: Oncogenic Mechanisms, edited by Giuseppe Barbanti-Brodano *et al.* Plenum Press, New York, 1995.

considerable progress has been made in recent years. Virus entry occurs by pH-independent fusion of the viral envelope with the surface membrane of the host cell.[7] Little is known about the mechanisms of uncoating and delivery of the viral nucleic acid to the nucleus. It has been shown, however, that tegument components and the nucleic acid of the infecting virus are transported within less than 1 hr into the nucleus, where one of the proteins, pp71, appears to function in *trans*-activation of specific viral genes,[8] an observation that again reminds one of the HSV system.[9] The general course of the HCMV reproductive cycle, including cascade-like regulation of immediate-early, early, and late gene expression, resembles that of the prototype herpesvirus HSV, with distinct differences that appear to be responsible for the unique phenotype including the extended length of the infectious cycle. For details the reader is referred to comprehensive textbooks.[4]

Regarding the complex events of morphogenetic maturation of HCMV, assembly of nucleocapsids through various stages[10] in the infected cell nucleus is followed by their egress across the nuclear envelope (Fig. 1); this process requires nucleocapsid budding at the inner nuclear membrane, resulting in delivery of enveloped virus particles with immature envelope proteins into the nuclear cisterna (perinuclear space)[11] as originally described in early studies on HSV.[12,13] Controversial views have been reported in regard to the next phases. (1) It was suggested that enveloped virions maintain their luminal localization and leave the nuclear envelope, which is contiguous with the endoplasmic reticulum, to follow

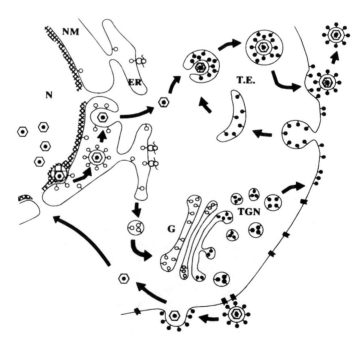

FIGURE 1. Scheme of the various host-cell compartments participating in the infectious cycle of HCMV. Nucleus (N), nuclear membranes (NM), endoplasmic reticulum (ER), Golgi apparatus (G), *trans*-Golgi network (TGN), early tubular endosome (TE).

the constitutive cellular secretory pathway through the Golgi complex for release by transport vesicles at the infected cell surface membrane.[14] According to this notion, processing of viral envelope glycoproteins was thought to take place on the virus particles during passage through the Golgi. It has to be pointed out that ultrastructural evidence for the presence of enveloped viral particles in the Golgi compartments of infected cells is lacking. Accumulation of naked nucleocapsids in the cytoplasm of infected cells was considered a result of aborted maturation or viral reentry.[15] (2) The alternative view postulates a process of deenvelopment of cisternal virus at the outer nuclear membrane, which is part of the endoplasmic reticulum, for delivery of naked nucleocapsids into the cytoplasm, where a second envelopment takes place at cytoplasmic cisternae, and subsequent vesicular transport and release at the infected cell surface. This concept implies that the first, transient envelopment serves the transit through the nuclear envelope (transport budding)[11]; naked cytoplasmic nucleocapsids thus represent fully functional structures that are bypassing the Golgi complex to acquire their final envelope during the second envelopmental process at the cytoplasmic cisternae (maturational budding; Fig. 1). Late after infection, these cisternal membranes are decorated with mature viral membrane glycoproteins that have been processed separately during synthesis in the endoplasmic reticulum and during their obligatory passage through the Golgi compartments.

The available experimental evidence obtained in several systems including α- and β-herpesviruses favors the latter, more complex model of two consecutive envelopmental events during herpesvirus maturation.[16,17] Furthermore, it would be difficult to reconcile the present knowledge of cellular transport and processing of membrane glycoproteins with the notion of particle-bound maturation of viral envelope glycoproteins. The objective of this chapter is to summarize experimental data on the cellular transport and processing of dominant envelope glycoproteins in the context of the envelopmenting events during HCMV morphogenesis.

2. EGRESS OF HCMV NUCLEOCAPSIDS THROUGH THE NUCLEAR ENVELOPE: TRANSPORT BUDDING

2.1. HCMV-Induced Alteration of the Nuclear Envelope

It is generally accepted that for nuclear egress, late postinfection (p.i.) cytomegalovirus (CMV) nucleocapsids like those of other herpesviruses traverse the nuclear envelope, which consists of the nuclear lamina and the inner and outer nuclear membranes.[18] Exit through the nuclear pores has been suggested for murine CMV but appears to be the exception.[19] In the case of HCMV, the ultrastructural alteration of the nuclear envelope typical for this event[20] is comparable to that observed for HSV[21]: the nuclear lamina underlying the inner leaflet of the nuclear membranes appears to be thickened in focal areas that are independent of juxtaposition of nucleocapsids and possibly represent lesions for viral exit. Interestingly, during this phase of the infectious cycle, the state of phosphorylation of the major lamina components, lamins A and C, changes late p.i. Biosynthetic labeling of infected interphase cells allowed net hyperphosphorylation to be observed,[22] as was dephosphorylation of specific epitopes in the N-terminal portion of lamins A and C.[20] In uninfected growing cells, the nuclear envelope, including the lamina, displays a dynamic behavior over the cell cycle, undergoing disassembly and reassembly, processes that appear to be controlled by the state of phosphorylation of the lamins.[23] Human-CMV-induced changes in interphase may possibly result from the action of viral kinases/phosphatases, which produce focal relaxation of the lamina for nucleocapsid interaction with the modified inner nuclear membrane. However, additional experimental data are needed to address this point.

Transport budding of herpesvirus nucleocapsids requires translocation of viral membrane glycoproteins to the inner nuclear membrane of infected cells.[24–26] The pathway by which this is achieved is still a matter of discussion: a likely, although not proven, mechanism would be transport by lateral diffusion via the nuclear pores.[27] Translocation of the major HCMV membrane glycoproteins, gB and gH, which are conserved in the herpesvirus family, into the inner nuclear membrane has been demonstrated with various techniques.[28,29] In the case of HCMV gB, immunoelectron microscopy following preembedding staining of permeabilized infected cells was used to identify this viral product in the nuclear compartment.[28] The molecular mechanisms involved in the process of transport budding are still unclear. Several questions are pertinent for a better understanding of this event. For instance, are all viral glycoproteins displayed at the inner nuclear membrane, or only a specific subset? If so, are there endogenous nuclear targeting or retention signals on viral membrane glycoproteins? In addition, do glycoproteins on extracellular virions differ from those on viral particles in the perinuclear space? Which are the viral (and/or cellular) products that interact during the budding event? Are specific tegument components that may be acquired subsequent to nucleocapsid assembly needed for interaction with the nucleoplasmic glycoprotein tail? At the ultrastructural level, most of the nuclear nucleocapsids show a clear lining in contrast to those in the cytoplasm, which exhibit a "fuzzy" contour indicative of adherent material (tegument proteins).[11] Precise biochemical comparison of the respective purified viral structures should help us to eventually approach a better understanding of the molecular processes involved in nuclear budding. An experimental model to study the inherent mechanisms might be digitonin-treated infected cells,[30] in which nuclear functions, including transport budding into the perinuclear space, is unimpaired in the absence of a functional cytoplasm (unpublished observation).

2.2. Human CMV Maturation in the Presence of Inhibitors of Glycoprotein Processing and Transport

Human-CMV DNA replication, in contrast to that of other herpesviruses, exhibits a peculiar sensitivity to inhibitors of glycosylation[31,32] and of glycoprotein maturation, such as monensin[33] and brefeldin A (BFA).[11] This may indicate that regulation of HCMV DNA synthesis involves vectorial nuclear transport of a glycosylated viral or cellular product.[11] Regarding this effect, the use of inhibitors to study the role of glycoproteins for viral morphogenetic events is limited. It has been shown that BFA, which leads to morphological and functional disintegration of the Golgi apparatus,[34] does not affect HCMV transport budding,[11] an observation that is in line with data obtained for other herpesviruses.[16,17] This is not unexpected, as this drug does not hinder functions of the endoplasmic reticulum (RER), i.e., biosynthesis and cellular transport of immature glycoproteins into the nuclear membranes, which are contiguous with the RER. On the other hand, BFA treatment drastically interfered with cytoplasmic envelopmental processes, producing accumulations of nucleocapsids and dense bodies[11] in the cytoplasm of HCMV-infected cells, and eventually abolished release of virions.[11] Under the experimental conditions used, viral glycoproteins must have been present in cellular smooth membranes before drug treatment was initiated.[11] This suggests that cytoplasmic nucleocapsid envelopment possibly requires adaptor-like proteins that bind the surface of the nucleocapsid or dense body to the modified membranes; BFA could remove such adaptors, as shown recently for the Golgi structure in general.[35]

3. IDENTIFICATION OF CYTOPLASMIC CISTERNAE ENGAGED IN HCMV MATURATIONAL BUDDING

3.1. Visualization of Early Endosomal Cisternae in Human Fibroblasts

Ultrastructural studies have consistently demonstrated that HCMV maturational envelopment takes place at cytoplasmic cisternae, often localized in the vicinity of the Golgi stacks.[11,36] On the basis of this observation, the conclusion was drawn that the cellular compartment most likely involved in this event was the *trans*-Golgi network (TGN), a compartment involved in the final step of protein sorting in the cellular secretory pathway. Recently it was shown for several established animal cell lines that an endocytotic cellular compartment, the early tubular endosome (TE), which is found in close vicinity to the microtubule-organizing center (MTOC) and to the TGN, can be clearly distinguished from the exocytotic pathway by the use of a fluid-phase marker such as horseradish peroxidase (HRP).[37,38] This finding stimulated experiments to redefine the cellular compartment of HCMV maturation either as cisternae of the exocytotic TGN or possibly as tubules of the neighboring early endosome. The presence of TE in primary human fibroblasts, which are highly permissive for virus propagation, was demonstrated by incubation with HRP and subsequent visualization at the ultrastructural level of the marker in tubular cellular structures that were morphologically distinct from Golgi vesicles, the latter being devoid of the marker.[39] Further evidence that these endosomal cisternae represented a different cellular compartment was obtained in BFA-treated fibroblasts: in the presence of BFA, the TE was structurally extended and remained largely intact, whereas the Golgi cisternae were entirely disintegrated into small vesicles.[39]

3.2. Herpesvirus Envelopment at Cisternae of the TE

Ultrastructural examination of HRP-incubated HCMV-infected fibroblasts late p.i. showed that wrapping of viral particles—nucleocapsids as well as dense bodies—occurred exclusively at cisternae filled with the tracer (Fig. 2).[39] The extent of accumulation of HRP-labeled endosomal structures was significantly enhanced over that seen in uninfected cells, particularly in the pericentriolar region. In addition, virus envelopment continued for at least 1–2 hr in the presence of BFA in spite of disintegration of the Golgi (Fig. 3), suggesting that this process was in fact independent of a functional Golgi machinery for a considerable time interval. Subsequent cessation may indicate depletion of the endosomal compartment of viral components essential for viral maturation, in particular envelope glycoproteins, which are not correctly processed and transported in the presence of BFA.[11] By this experimental approach it could be shown that the identical structures in infected human fibroblasts were also involved in maturational envelopment of the α-herpesviruses, herpes simplex virus and varicella–zoster virus (Fig. 4).[40] Extension of this work is needed also to define the site of herpesvirus maturational envelopment for cells other than human fibroblasts.

4. CELLULAR TRANSPORT AND FUNCTIONAL DOMAINS OF HCMV GLYCOPROTEIN B

4.1. Transport and Processing of HCMV Glycoprotein B

Characterization of HCMV glycoproteins has lagged behind that of other herpesviruses, particularly that of HSV, because of special properties of this infectious agent, primarily its slow and inefficient growth and its strict species and cell type specificity.[3] Most of the available information concerns the major HCMV envelope glycoprotein B (gB), a structural and possibly functional homologue of HSV gB. (At the Fourth International HCMV Congress in Paris, 1993, it was suggested that the nomenclature of viral products be related to the respective coding reading frame; accordingly, gB would be synonymous with gpUL55. Here the original designation will be used to facilitate the relation to other herpesviruses.) Human CMV gB is the main target of neutralizing antibodies elicited during natural infection[41] and is thought to play an essential role for the initial virus–host cell interaction and viral cell-to-cell spread.[42] With regard to this dominant functional role, it is considered a prime candidate for a subunit vaccine.[43] This fact stimulated research on the molecular biology and immunology of HCMV gB. The amino acid (aa) sequence of the gB molecule (906 aa; HCMV strain AD169)[44] was deduced from the nucleotide (nt) sequence of the coding gene and exhibits typical features of a membrane-anchored glycoprotein: an N-terminal hydrophobic leader segment (aa 1–19), two potential membrane anchor domains (hd1, aa 714–747; hd2, aa 751–771), and consensus signals for 17 N-glycosylation sites, which are, with the exception of one, located in the N-terminal portion. In HCMV-infected cell cultures, gB biosynthesis begins at about 24–36 hr p.i.; the 100-kDa polypeptide backbone is cotranslationally N-glycosylated and subject to formation of intra- and intermolecular disulfide linkages in the rough endoplasmic reticulum (RER), which increases the molecular mass of the molecule under reducing conditions to about 160 kDa and that of the nonreduced complexed

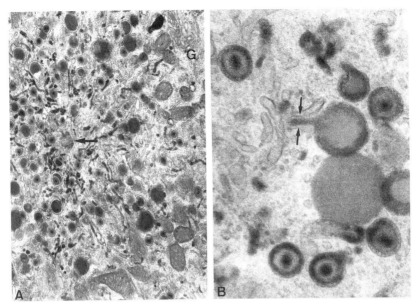

FIGURE 2. Envelopment of cytoplasmic nucleocapsids by tubules of the early endosome (TE). The electron micrograph in A shows the TE near the centriole (arrow) contrasted with the fluid-phase marker HRP in an HCMV-infected fibroblast 72 hr p.i. in an overview. Nucleocapsids and dense bodies are observed in the vicinity but outside the Golgi. B shows nucleocapsids or dense bodies enveloped by, or in the process of envelopment (arrows) by, tubular structures of the TE that are filled with HRP. Magnification: A, 8000×; B, 68,750×.

homodimer (oligomer) to more than 300 kDa.[45] Trimming of the carbohydrate side chains and addition of complex sugars occur during Golgi passage, and, unlike gB of HSV, HCMV gB is processed by specific proteolytic cleavage at aa 459 in the TGN.[44,46] This event produces an N-terminal product of about 100 kDa and a C-terminal product of about 55 kDa, which are disulfide-linked.[47] Mainly the mature molecules and little of the uncleaved precursor are exposed at the surface of infected cells.[45]

4.2. Function of C-Terminal Hydrophobic Domains of HCMV gB

Eukaryotic expression of the HCMV gB gene in various systems has revealed that all the information for processing is contained in the gB molecule, and thus this approach has been used increasingly to study functional domains of gB in more detail. Regarding membrane anchorage, it has been shown by constitutive expression of mutagenized HCMV gB genes in human astrocytoma cells that deletion of the C-terminal portion including hd1 and hd2 yields a secretory gB-specific product.[46] This observation also supported the notion that the C-terminal portion of 135 aa is located in the cytoplasm. More precise analysis demonstrated that

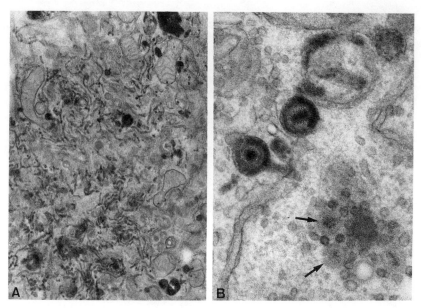

FIGURE 3. Effect of brefeldin A (BFA) on HCMV maturational envelopment. Incubation of HCMV-infected fibroblasts (96 hr p.i.) in HRP for 3 hr in the presence of 1 μg BFA/ml induces an expansion of the TE (A) and permits continuation of nucleocapsid envelopment by TE cisternae (B). The Golgi complex is vesiculated, and some nonenveloped viral particles are observed in its vicinity (arrows in B). Magnification: A, 8000×; B, 82,500×.

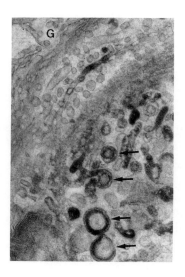

FIGURE 4. Maturational envelopment of α-herpesvirus nucleocapsids by tubular structures of the TE. The electron micrograph shows budding of HSV-2 nucleocapsids (12 hr p.i.) into tubules contrasted with fluid-phase marker (arrows) in the vicinity of the *trans*-Golgi network. Magnification, 70,000×.

deletion of hd2 alone, but not of hd1, led to secretion of fully processed and oligomerized HCMV gB (Fig. 5); this observation suggested that, in analogy to the recent report on HSV gB,[48] hd2 is essential as well as sufficient for HCMV membrane anchoring.[49] As to the functional role of hd1 of the HCMV gB molecule, alignment of its sequence with those of fusogenic peptides of various viral glycoproteins revealed a striking similarity of the positioning of glycine residues (Fig. 6), indicating a possible functional analogy.[42,49] There is presently no other attractive model for the pH-independent fusion between viral envelope and host-cell surface membrane mediated by herpesvirus membrane glycoproteins.[50]

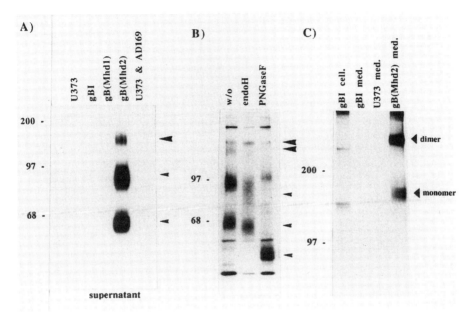

FIGURE 5. (A) Secretion into the culture medium of HCMV gB with a deletion of hd2. Supernatants of culture medium from stable transfectants expressing mutagenized gB forms with deletions for hd1 or hd2 [lanes gB(Mhd1) and gB(Mhd2), respectively] were immunoprecipitated with gB-specific monoclonal antibody from [^{35}S]methionine-labeled cultures prior to SDS-polyacrylamide gel electrophoresis (8%) and fluorography. Uninfected (U373), HCMV-infected (U373 and AD169) astrocytoma cells, and a stable transfectant expressing authentic gB were used as controls. The positions of marker proteins are indicated on the left: myosin 200 kDa, phosphorylase 97 kDa, bovine serum albumin 68 kDa. (B) Glycosidase sensitivity of the secretory gB form from gB(Mhd2). Immunoprecipitated secretory gB was left without treatment (w/o) or subjected to treatment with endoglycosidase H (lane endoH) or PNGaseF (lane PNGaseF) prior to electrophoretic separation and fluorography. (C) Oligomerization of secretory gB. Immunoprecipitated secretory gB [lane gb(Mhd2) med.] was electrophoretically separated under nonreducing conditions, precipitates of intracellular gB from a transfectant expressing authentic gB and those of culture media of the identical culture and from uninfected astrocytoma cells (lanes gBI cell., gBI med., and U373 med.) were used as controls. The filled triangles on the right indicate the positions of gB dimers and monomers.

A

B

```
NH₂-GLFGAIAGFIEGGWTGMIDGWYGYH      Influenza A
       |    |    |       |    |
NH₂-GFFGAIAGFLEGGWEGMIAGWHGTY      Influenza B
NH₂-FFGAVIGIIALGVATSAQIT           Sendai
NH₂-AVGIGALFLGFLGAAGSTMGARSM       HIV1
```

FIGURE 6. Comparative structural analysis of the C-terminal hydrophobic region of HCMV gB. (A) Alignment of the amino acid sequences of HCMV hd1 and hd2 with respect to conserved spaced glycine residues with the corresponding gB segments of the α-herpesviruses HSV,[84] VZV,[85] and PRV.[86] (B) Alignment of the N-terminal fusion peptide sequences with respect to conserved spaced glycine residues of orthomyxoviruses, HA2 of influenza A virus,[87] HA2 of influenza B virus,[88] F1 of paramyxovirus [Sendai[89]], and glycoprotein p41 of human immunodeficiency virus [HIV1[90]].

4.3. Identification of HCMV gB Cysteine Residues Responsible for Oligomerization

Formation of oligomers that are sensitive to reducing agents appears to be a relevant process for structural and functional maturation of herpesvirus gB homologues.[45,51] In the case of HCMV gB, seven cysteine residues (aa 94, 110, 185, 248, 250, 344, 391) can be deduced from the nucleotide sequence of the N-terminal cleavage product, and four (aa 506, 550, 573, 610; Fig. 7) in the luminal portion of the C-terminal cleavage product of gB, but there is presently little information about which of these residues are responsible for the formation of intra- and intermolecular disulfide bonds in the RER. So far analysis of stable human cell lines expressing HCMV gB forms in which cysteine residues residing in the luminal portion of the C-terminal cleavage product had been alternately replaced by serine showed that mutagenesis of Cys^{573} or Cys^{610} abolished oligomerization.[52] Continuation of this experimental approach should yield more detailed knowledge to devise a model of the HCMV gB complex that should eventually lead to a better understanding of the functional structure of this dominant envelope component.

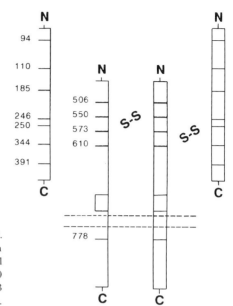

FIGURE 7. Complex formation of HCMV gB. The scheme depicts the putative participation of luminal cysteine residues of the C-terminal cleavage product at position aa 506 and aa 550 in intramolecular, and cysteine residues aa 573 and aa 610 in intermolecular disulfide linkages.

4.4. Proteolytic Cleavage of HCMV gB by a Cellular Furin-like Protease

Human CMV gB maturation involves a processing step of specific proteolytic cleavage[44,46] like that of gB homologues of a number of other members of the herpesvirus family[53,54] and unlike that of HSV gB.[44] Cleavage occurs at aa 459 (strain AD169) or aa 460 (strain Towne) and requires a dibasic amino acid consensus sequence (Lys-Arg)[44,46] that is conserved among human herpesviruses. It is noteworthy that the herpesvirus processing sites are similar to those of human immunodeficiency virus (HIV) gp160[55,56] and of the influenza virus (IV) hemagglutinin.[57] The observation that viral glycoprotein cleavage including that of HCMV gB also occurred during eukaryotic expression of the solitary authentic product[58,59] strongly suggested that cleavage resulted from a cellular enzyme activity. Recent reports indicated for HIV and IV that endoproteolytic cleavage is mediated by the cellular subtilisin-like endoprotease furin.[56,57] Involvement of this enzyme in proteolytic processing of HCMV gB as well was recently demonstrated by coexpression of the coding genes of the two products in cell cultures without endogenous furin activity[60,61] as well as by the use of a furin-specific inhibitor (decanoyl-RVKR-chloromethyl ketone) on HCMV-infected cells or gB transfectants.[61]

In most instances, processing by proteolytic cleavage appears to be essential for viral membrane proteins to function in fusion of the viral envelope to the host cell surface membrane for virus entry. Whether this notion is also valid for herpesviruses remains uncertain. Besides the fact that gB of some members of the family have no cleavage site, it was shown for bovine herpesvirus that a mutagenized uncleavable form could functionally substitute for the normal cleavable gB homologue.[62] In the case of the human herpesviruses with cleavable gB homologues, e.g., HCMV and VZV, glycoprotein deletion mutants are needed for complementation with mutagenized gB forms to elucidate this aspect.[63]

4.5. Cellular Pathway of HCMV gB into the TE

Regarding the recent observation that HCMV maturational envelopment is confined to an endocytic compartment, an intriguing question arises: By which cellular route are viral membrane glycoproteins, as essential components of this process, transported from the cellular secretory pathway into the TE? The alternatives are (1) direct intracellular vesicular transfer from the TGN to the TE or (2) transport along the exocytic pathway and exposure at the infected cell surface prior to reinternalization into the TE. Evidence in favor of the latter notion was recently obtained by the following experimental setup[40]: HCMV-infected fibroblasts were subjected to a short interval of surface biotinylation at 4°C prior to a chase at 37°C. This experimental approach would predict that extracellular virions that are released from surface-labeled cells into the chase medium should carry biotinylated envelope proteins only if these proteins had been surface components during the labeling procedure, before they were reinternalized into the TE (Fig. 1). The results, which so far are concerned only with HCMV gB, were in line with this prediction,[40] indicating that direct transfer of viral components from the TGN to the TE plays a minor role, if any, in this experimental system. The notion that viral glycoproteins have to pass the infected cell surface before they become constituents of the viral envelope may imply that, e.g., antibodies against viral membrane glycoproteins could interfere with the completion of the HCMV infectious cycle. It is of interest to evaluate whether these observations can be extended to other herpesviruses.

5. ASPECTS OF HCMV BIOLOGY RELATED TO ITS PUTATIVE ONCOGENIC POTENTIAL IN MAN

Early reports described properties of HCMV suggestive of its oncogenic potential, e.g., stimulation of host-cell DNA and RNA synthesis in productive as well as nonproductive cellular infections, transformation to malignancy *in vitro* of animal as well as human cells, and induction of tumors by transformed cells in animal models.[64–67] Furthermore, genetic sequences of HCMV with transforming capabilities were identified,[68–70] and HCMV or its products have been found in a number of human tumors, including prostatic carcinoma, adenocarcinoma of the colon, cervical carcinoma, and Kaposi's sarcoma.[71–73] The data that link HCMV to human malignancies are difficult to interpret, and it has remained debatable whether HCMV is a human carcinogen or simply a passenger in cancerous cells.[74] The recent experimental approach takes advantage of the technology available now and is aimed at the precise analysis of the virus–host-cell interaction at the molecular level.

5.1. *Trans*-Activation of Cellular Genes by HCMV

It has been consistently demonstrated that during the initial phase of cellular infection, HCMV induces transcription of RNAs specified by the heat shock protein 70 gene and by the cellular protooncogenes c-*fos*, c-*jun*, and c-*myc*, the activation of which is associated with mitogenic stimuli.[75,76] The promoters of these cellular genes appear to be *trans*-activated by HCMV immediate early proteins.[76] It was suggested that HCMV itself promotes changes in cellular gene expression to establish a favorable state of the cell cycle for permissive infection.[76,77] Activation of cellular genes involved in growth regulation and cell cycle control by HCMV is intriguing regarding its oncogenic potential; so far, however, unequivocal evidence for the involvement of defined HCMV products in loss of growth control in human cells has not been provided.

5.2. Induction of Papovavirus DNA Replication by HCMV

Like HSV,[78,79] HCMV has been shown to act as a helper virus to support the replication and extend the host cell range of papovaviruses.[80,81] These effects were apparently related either to so far unidentified HCMV gene products other than structural components or to HCMV-induced cellular functions and were dependent on papovavirus DNA replication origin as well as T antigen. Multiplication of JC virus or SV40 in human cells, which are normally nonpermissive for these agents, was suggested to reflect *trans*-activation of papovavirus genes by gene products of coinfected HCMV. These observations are presently compatible with the notion that HCMV might support papovavirus pathogenicity in humans.

5.3. Induction of Chromosome Aberrations by HCMV

Several studies have consistently shown that HCMV infection can result in a statistically significant increase in the frequency of chromosome aberrations; this effect also concerns abortively infected cells.[82,83] It was suggested that the underlying mechanism may be related to inhibition of DNA repair produced by HCMV-induced cellular DNA replication and/or *trans*-activation of cellular genes.

Taken together, there is consistent experimental evidence that HCMV–host-cell interaction may result in gene translocation, amplification, and *trans*-activation; these effects may lead to cellular dysregulation and loss of growth control, i.e., neoplastic cellular transformation. Further precise dissection of the molecular mechanisms involved in these cellular reactions may eventually answer the still unsettled question of whether or not HCMV can cause malignancies in man.

6. OUTLOOK

Our understanding of the molecular basis of HCMV–host-cell interaction and of the relationship between structure and function of HCMV constituents in viral morphogenesis, which was emphasized here, is still at its very beginning. Regarding the latter aspect, appropriate experimental systems using eukaryotic expression have to be established to study the interaction of relevant viral components, e.g., viral membrane glycoproteins and tegument proteins, at a subviral level. The observation that a compartment of cellular endocytosis is involved in HCMV maturation implicates that inhibition of cellular endocytosis should interfere with virus release. This aspect needs further examination.

With regard to practical applicability, eukaryotic recombinant virus products produced for the study of basic mechanisms of virus–host-cell interaction could in addition be used for improvement of HCMV diagnosis and possibly to help distinguish among primary, persistent, and reactivated infection, thus helping to shed light on the putative oncogenic properties of HCMV. For the essential aspect of HCMV immunobiology, which is of prime interest for differential diagnostic evaluation as well as for vaccine development, the reader is referred to an excellent review.[63]

REFERENCES

1. Gorensek, M., Stewart, R., Keys, T., McHenry, M., and Goormastic, M., 1988, A multivariate analysis of the risk of cytomegalovirus infection in heart transplant recipients, *J. Infect. Dis.* **157**:515–522.
2. Saltzman, R., Quirk, M., and Jordan, M., 1988, Disseminated cytomegalovirus infection. Molecular analysis of virus and leucocyte interaction in viremia, *J. Clin. Invest.* **81**:75–81.
3. Rapp, F., 1983, The biology of cytomegaloviruses, in: *The Herpesviruses* (B. Roizman, ed.), Plenum Press, New York, pp. 1–66.
4. Stinski, M., 1990, Cytomegalovirus and its replication, in: *Virology*, 2nd ed. (B. N. Fields and D. M. Knipe, eds.), Raven Press, New York, pp. 1959–1980.
5. Roizman, B., and Sears, A., 1990, Herpes simplex viruses and their replication, in: *Virology*, 2nd ed. (B. N. Fields and D. M. Knipe, eds.), pp. 1795–1887, Raven Press, New York.
6. Chee, M., Bankier, A., Beck, S., Bohni, R., Brown, C., Cerny, R., Horsnell, T., Hutchison C., III, Kouzarides, T., Martignetti, E., Preddie, E., Satchwell, S., Tomlinson, P., Weston, K., and Barrell, B., 1990, Analysis of the protein coding content of the sequence of human cytomegalovirus AD169, *Curr. Top. Microbiol. Immunol.* **154**:125–169.
7. Compton, T., Nepomuceno, R., and Nowlin, D., 1992, Human cytomegalovirus penetrates host cells by pH-independent fusion at the cell surface, *Virology* **191**:387–395.
8. Liu, B., and Stinski, M., 1992, Human cytomegalovirus contains a tegument protein that enhances transcription from promoters with upstream ATF and AP-1 *cis*-acting elements, *J. Virol.* **66**:4434–4444.
9. Triezenberg, S., Kingsbury, R., and McKnight, S., 1988, Functional dissection of VP16, the transactivator of herpes simplex virus immediate early gene expression, *Genes Dev.* **2**:718–729.
10. Gibson, W., 1993, Molecular biology of human cytomegalovirus, in: *Frontiers of Virology*, Vol. 2 (Y. Becker and G. Darai, eds.), *Molecular Aspects of Human Cytomegalovirus Diseases*, Springer-Verlag, Heidelberg, pp. 303–329.
11. Eggers, M., Bogner, E., Agricola, B., Kern, H., and Radsak, K., 1992, Inhibition of human cytomegalovirus maturation by brefeldin A, *J. Gen. Virol.* **73**:2679–2692.
12. Falke, D., Siegert, R., and Vogell, W., 1959, Elektronenmikroskopische Befunde zur Frage der Doppelmembranbildung des Herpes simplex Virus, *Arch. Ges. Virusforsch.* **IX**:484–496.
13. Darlington, R., and Moss, L., 1968, Herpesvirus envelopment, *J. Virol.* **2**:48–55.
14. Johnson, D. C., and Spear, P. G., 1982, Monensin inhibits the processing of herpes simplex virus glycoproteins, their transport to the cell surface, and the egress of virions from infected cells, *J. Virol.* **43**:1102–1112.

15. Campadelli-Fiume, G., Farabegoli, F., Digaeta, S., and Roizman, B., 1991, Origin of unenveloped capsids in the cytoplasm of cells infected with herpes simplex virus 1, *J. Virol.* **65**:1589–1595.
16. Whealy, M., Card, J., Meade, R., Robbins, A., and Enquist, L., 1991, Effect of brefeldin A on alphaherpesvirus membrane protein glycosylation and virus egress, *J. Virol.* **65**:1066–1081.
17. Cheung, P., Banfield, B., and Tufaro, F., 1991, Brefelldin A arrests the maturation and egress of herpes simplex virus particles during infection, *J. Virol.* **65**:1893–1904.
18. Gerace, L., and Blobel, G., 1980, The nuclear envelope lamina is reversibly depolymerized during mitosis, *Cell* **19**:277–287.
19. Weiland, F., Keil, G., Reddehase, M., and Koszinowski, U., 1986, Studies on the morphogenesis of murine cytomeglovirus, *Intervirology* **26**:192–201.
20. Radsak, K., Brücher, K., and Georgatos, S., 1991, Focal nuclear envelope lesions and specific nuclear lamin A/C dephosphorylation during infection with human cytomegalovirus, *Eur. J. Cell Biol.* **54**:299–304.
21. Bibor-Hardy, V., Suh, M., Pouchelet, and Simard, R., 1982, Modifications of the nuclear envelope of BHK cells after infection with herpes simplex virus type 1, *J. Gen. Virol.* **63**:81–94.
22. Rohsiepe, D., 1993, Untersuchungen zur Phosphorylierung der Kernlaminaproteine permissiver Kulturzellen nach Infektion mit dem Cytomegalievirus, Master's thesis in biology, Institut für Virologie, Marburg, Germany.
23. Ottaviano, Y., and Gerace, L., 1985, Phosphorylation of the nuclear lamins during interphase and mitosis, *J. Biol. Chem.* **260**:624–632.
24. Ali, M., Butcher, M., and Ghosh, H., 1987, Expression and nuclear envelope localization of biologically active fusion glycoprotein gB of herpes simplex virus in mammalian cells using cloned DNA, *Proc. Natl. Acad. Sci. USA* **84**:5675–5679.
25. Torrisi, R., Cirone, M., Pavan, A., Zompetta, C., Barile, G., Frati, L., and Faggioni, A., 1989, Localization of Epstein–Barr virus envelope glycoproteins on the inner nuclear membrane of virus-producing cells, *J. Virol.* **63**:828–832.
26. Gilbert, R., Ghosh, K., Rasile, L., and Ghosh, H., 1994, Membrane anchoring domain of herpes simplex virus glycoprotein gB is sufficient for nuclear envelope localization, *J. Virol.* **68**:2272–2285.
27. Petterson, R., 1991, Protein localization and virus assembly at intracellular membranes, *Curr. Top. Microbiol. Immunol.* **170**:67–106.
28. Radsak, K., Brücher, K., Britt, W., Shiou, H., Schneider, D., and Kollert, A., 1990, Nuclear compartmentation of glycoprotein B of human cytomegalovirus, *Virology* **177**:515–522.
29. Bogner, E., Reschke, M., Reis, B., Reis, E., Britt, W., and Radsak, K., 1992, Recognition of compartmentalized intracellular analogs of glycoprotein H of human cytomegalovirus, *Arch. Virol.* **126**:67–80.
30. Adam, S., Sterne-Marr, R., and Gerace, L., 1992, Nuclear protein import using digitonin-permeabilized cells, *Methods Enzymol.* **219**:97–110.
31. Radsak, K., and Weder, D., 1981, Effect of 2-deoxy-D-glucose on cytomegalovirus-induced DNA synthesis in human fibroblasts, *J. Gen. Virol.* **57**:33–42.
32. Weder, D., and Radsak, K., 1983, Induction of a host-specific chromatin-associated glycopolypeptide by human cytomegalovirus, *J. Gen. Virol.* **64**:2749–2761.
33. Kaiser, C. J., and Radsak, K., 1987, Inhibition by monensin of human cytomegalovirus DNA replication, *Arch. Virol.* **94**:229–245.
34. Lippicott-Schwartz, J., Donaldson, J., Schweizer, A., Berger, E., Hauri, H., Yuan, L., and Klausner, R., 1990, Microtubule-dependent retrograde transport of proteins into the ER in the presence of brefeldin A suggests an ER recycling pathway, *Cell* **60**:821–836.
35. Orci, L., Tagaya, M., Amherdt, M., Perrelet, A., Donaldson, J., Lippincott-Schwartz, J., Klausner, R., and Rothman, J., 1991, Brefeldin A, a drug that blocks secretion, prevents the assembly of non-clathrin-coated buds on Golgi cisternae, *Cell* **64**:1183–1195.
36. Severi, B., Landini, M., and Govoni, E., 1988, Human cytomegalovirus morphogenesis: An ultrastructural study of the late cytoplasmic phases, *Arch. Virol.* **98**:51–64.

37. Tooze, J., and Hollinshead, M., 1991, Tubular early endosomal networks in AtT20 cells and other cells, *J. Cell Biol.* **115**:635–653.
38. Tooze, J., and Hollinshead, M., 1992, In AtT20 and HeLa cells brefeldin A induces the fusion of tubular endosomes and changes their distribution and some of their endocytic properties, *J. Cell Biol.* **118**:813–830.
39. Tooze, J., Hollinshead, M., Reis, B., Radsak, K., and Kern, H., 1993, Progeny vaccinia and human cytomegalovirus particles utilise early endosomal cisternae for their envelops, *Eur. J. Cell Biol.* **60**:163–178.
40. Radsak, K., Eickmann, M., Mockenhaupt, T., Kern, H., Eis-Hübinger, A., and Resckke, M., 1995, Retrieval of human cytomegalovirus glycoprotein B (gpUL55) from the infected cell surface for viral envelopment, *Arch. Virol.*, under consideration.
41. Rasmussen, L., 1990, Immune response to human cytomegalovirus infection, *Curr. Top. Microbiol. Immunol.* **154**:221–254.
42. Navarro, D., Paz, P., Tugizov, S., Topp, K., La Vail, J., and Pereira, L., 1993, Glycoprotein B of human cytomegalovirus promotes virion penetration into cells, transmission of infection from cell to cell, and fusion of infected cells, *Virology* **197**:143–158.
43. Gönczöl, E., Ianacone, J., Ho, W., Starr, S., Meignier, B., and Plotkin, S., 1990, Isolated gA/gB glycoprotein complex of human cytomeglovirus envelope induces humoral and cellular immune responses in human volunteers, *Vaccine* **8**:130–136.
44. Spaete, R., Thayer, R., Probert, W., Masiarz, F., Chamberlain, S., Rasmussen, L., Merigan, T., and Pachl, C., 1988, Human cytomegalovirus strain Towne glycoprotein B is processed by proteolytic cleavage, *Virology* **167**:207–225.
45. Britt, W., and Vugler, L., 1992, Oligomerisation of the human cytomegalovirus major envelope glycoprotein complex gB (gp55–116), *J. Virol.* **66**:6747–6754.
46. Spaete, R., Saxena, A., Scott, P., Song, G., Probert, W., Britt, W., Gibson, W., Rasmussen, L., and Pachl, C., 1990, Sequence requirements for proteolytic processing of glycoprotein B of human cytomegalovirus strain Towne, *J. Virol.* **64**:2922–2931.
47. Britt, W., and Vugler, L., 1989, Processing of the gp55–116 envelope glycoprotein complex (gB) of human cytomegalovirus, *J. Virol.* **63**:403–410.
48. Rasile, L., Ghosh, K., Raviprakash, K., and Ghosh, H., 1993, Effects of deletions in the carboxy-terminal hydrophobic region of herpes simplex virus glycoprotein gB on intracellular transport and membrane anchoring, *J. Virol.* **67**:4856–4866.
49. Reschke, M., Reis, B., Nöding, K., Rohsiepe, D., Richter, A., Mockenhaupt, T., Garten, W., and Radsak, K., 1995, Constitutive expression of human cytomegalovirus glycoprotein B (gpUL55) with mutagenized carboxy-terminal hydrophobic domains, *J. Gen. Virol.* **76**:113–122.
50. White, J., 1992, Membrane fusion, *Science* **258**:917–924.
51. Ali, M., 1990, Oligomerisation of herpes simplex virus glycoprotein B occurs in the endoplasmic reticulum and 102 amino acid cytosolic domain is dispensable for dimer assembly, *Virology* **178**:588–592.
52. Eickmann, M., Reschke, M., Mockenhaupt, T., and Radsak, K., 1995, Identification of cysteine residues of human cytomegalovirus glycoprotein B (gpUL55) involved in intra- and intermolecular disulfide linkages, in preparation.
53. Montalvo, E. A., and Grose, C., 1987, Assembly and processing of the disulfide-linked varizella–zoster virus glycoprotein gpII (gp140), *J. Virol.* **61**:2877–2884.
54. Kopp, A., Blewett, E., Misra, V., and Mettenleiter, T., 1994, Proteolytic cleavage of bovine herpesvirus 1 (BHV-1) glycoprotein gB is not necessary for its function in BHV-1 or pseudorabies virus, *J. Virol.* **68**:1667–1674.
55. Brücher, K., Garten, W., Klenk, H. C., Shaw, E., and Radsak, K., 1990, Inhibition of endoproteolytic cleavage of cytomegalovirus (HCMV) glycoprotein B by palmitoyl-peptidyl-chloromethyl ketone, *Virology* **178**:617–620.
56. Hallenberger, S., Bosch, V., Angliker, H., Shaw, E., Klenk, H. D., and Garten, W., 1992, Inhibition of furin-mediated cleavage activation of HIV-1 glycoprotein gp160, *Nature* **360**:358–361.

57. Stieneke-Gröber, A., Vey, M., Angliker, H., Shaw, E., Thomas, G., Roberts, C., Klenk, H. D., and Garten, W., 1992, Influenza virus hemagglutinin with multibasic cleavage site is activated by furin, a subtilisin-like endoprotease, *EMBO J.* **11**:2407–2414.
58. Reis, B., Bogner, E., Reschke, M., Richter, A., Mockenhaupt, T., and Radsak, K., 1993, Stable constitutive expression of glycoprotein B (gpUL55) of human cytomegalovirus in permissive astrocytoma cells, *J. Gen. Virol.* **74**:1371–1379.
59. Wells, D., Vugler, L., and Britt, W., 1990, Structural and immunological characterization of human cytomegalovirus gp55–116 (gB) expressed in insect cells, *J. Gen. Virol.* **71**:873–880.
60. Vey, M., Schäfer, W., Reis, B., Ohuchi, R., Britt, W., Radsak, K., Klenk, H.-D., and Garten, W., 1993, The human subtilisin-like endoprotease furin mediates cleavage of glycoprotein B (gp55–116) of human cytomegalovirus, in: *Abstracts IXth International Congress of Virology, Glasgow*, p. 64.
61. Vey, M., Schäfer, W., Reis, B., Ohuchi, R., Britt, W., Klenk, H. D., Garten, W., and Radsak, K., 1995, Proteolytic processing of human cytomegalovirus glycoprotein B (gpUL55) is mediated by the human endoprotease furin, *Virology* **206**:746–749.
62. Blewett, E., and Misra, V., 1991, Cleavage of the bovine herpesvirus glycoprotein B is not essential for its function, *J. Gen. Virol.* **72**:2083–2090.
63. Pereira, L., Jahn, G., and Navarro, D., 1993, Proteins of human cytomegalovirus that elicit humoral immunity, in: *Frontiers of Virology*, Vol. 2, *Molecular Aspects of Human Cytomegalovirus Diseases* (Y. Becker and G. Darai, eds.), Springer-Verlag, Heidelberg, pp. 437–464.
64. St. Jeor, S., and Hutt, R., 1977, Cell DNA replication as a function in the synthesis of human cytomegalovirus, *J. Gen. Virol.* **37**:65–73.
65. Geder, L., Lausch, R., O'Neill, F., and Rapp, F., 1976, Oncogenic transformation of human embryo lung cells by human cytomegalovirus, *Science* **192**:1134–1136.
66. Geder, L., Kreider, J., and Rapp, F., 1977, Human cells transformed *in vitro* by human cytomegalovirus: Tumorigenicity in athymic mice, *J. Natl. Cancer Inst.* **58**:1003–1009.
67. Rapp, F., and Robbins, D., 1984, Cytomegalovirus and human cancer, *Birth Defects* **20**:175–192.
68. Razzaque, A., Jahan, N., McWeeney, D., Jariwalla, R., Jones, C., Brady, J., and Rosenthal, L., 1988, Localization and DNA sequence analysis of the transforming domain (mtrII) of human cytomegalovirus, *Proc. Natl. Acad. Sci. USA* **85**:5709–5713.
69. Spector, D., 1983, Human cytomegalovirus (strain AD169) contains sequences related to the avian retrovirus oncogene v-*myc*, *Proc. Natl. Acad. Sci. USA* **80**:38–39.
70. Rosenthal, L., and Choudhury, S., 1993, Potential oncogenicity of human cytomegalovirus, in: *Frontiers of Virology*, Vol. 2, *Molecular Aspects of Human Caytomegalovirus Diseases* (Y. Becker and G. Darai, eds.), Springer-Verlag, Heidelberg, pp. 412–436.
71. Huang, E., Mar, E., Boldogh, I., and Baskar, J., 1984, The oncogenicity of human cytomegalovirus, *Birth Defects* **20**:193–211.
72. Geder, L., Sanford, E., Rohner, T., and Rapp, F., 1977, Cytomegalovirus and cancer of the prostate: *In vitro* transformation of human cells, *Cancer Treat Rep.* **61**:139–146.
73. Giraldo, G., Beth, E., and Huang, E., 1980, Kaposi's sarcoma and its relationship to cytomegalovirus (CMV). III. CMV DNA and MV early antigens in Kaposi's sarcoma, *Int. J. Cancer* **26**:23–29.
74. Alford, C., and Britt, W., 1993, Cytomegalovirus, In: *The Human Herpesviruses* (B. Roizman, R. Whitley, and C. Lopez, eds.), Raven Press, New York, pp. 227–255.
75. Bolgogh, I., Abubakar, S., Deng, C., and Albrecht, T., 1991, Transcriptional activation of cellular oncogenes *fos, jun*, and *myc* by human cytomegalovirus, *J. Virol.* **65**:1568–1571.
76. Hagemeier, C., Walker, S., Sissons, P., and Sinclair, J., 1992, The 72k IE1 and 80k IE2 proteins of human cytomegalovirus independently trans-activate the c-*fos*, c-*myc* and *hsp70* promoters via basal promoter elements, *J. Gen. Virol.* **73**:2385–2393.
77. Musiani, M., and Zerbini, M., 1984, Influence of cell cycle on the efficiency of transfection with purified cytomegalovirus DNA, *Arch. Virol.* **78**:287–292.
78. Danovich, R., and Frenkel, N., 1988, Herpes simplex virus induces the replication of foreign DNA, *Mol. Cell. Biol.* **8**:3272–3281.

79. Matz, B., 1989, Herpes simplex virus causes amplification of recombinant plasmids containing simian virus 40 sequences, *J. Gen. Virol.* **70**:1347–1358.
80. Pari, G., and Stjeor, S., 1990, Effect of human cytomegalovirus on replication of SV40 origin and the expression of T antigen, *Virology* **177**:824–828.
81. Heilbronn, R., Albrecht, I., Stephan, S., Bürkle, A., and zur Hausen, H., 1993, Human cytomegalovirus induces JC virus DNA replication in human fibroblasts, *Proc. Natl. Acad. Sci. USA* **90**:11406–11410.
82. Lüleci, G., Sakizli, M., and Günalp, A., 1980, Selective chromosomal damage caused by human cytomegalovirus, *Acta Virol.* **24**:341–345.
83. Deng, C., Abubakar, S., Fons, M., Bolgogh, I., and Albrecht, T., 1992, Modulation of the frequency of human cytomegalovirus-induced chromosome aberrations by camptothecin, *Virology* **189**:397–401.
84. Pellet, P., Kousoulas, Pereira, L., and Roizman, B., 1985, Anatomy of the herpes simplex virus 1 strain F glycoprotein B gene: Primary sequence and predicted protein structure of the wild type and of monoclonal antibody-resistant mutants, *J. Virol.* **53**:243–253.
85. Keller, P., Davison, A., Lowe, R., Bennet, C., and Ellis, R., 1986, Identification and structure of the gene encoding gpII, a major glycoprotein of varicella–zoster virus, *Virology* **152**:181–191.
86. Robbins, A., Dorney, D., Wathen, M., Whealey, M., Gold, C., Watson, R., Holland, L., Weed, S., Levine, M., Glorioso, J., and Enquist, L., 1987, The pseudorabies virus gII is closely related to the gB glycoprotein gene of herpes simplex virus, *J. Virol.* **61**:2691–2701.
87. Daniels, R., Downie, J., Hay, A., Knossow, M., Skehel, J., Wang, M., and Wiley, D., 1985, Fusion mutants of the influenza virus hemagglutinin, *Cell* **40**:667–675.
88. Krystal, M., Young, J., Wilson, I., Skehel, J., and Wiley, D., 1983, Sequential mutations in hemagglutinins of influenza B virus isolates: Definition of antigenic domains, *Proc. Natl. Acad. Sci. USA* **80**:4527–4531.
89. Morrison, T., and Portner, A., 1991, Structure, function and intracellular processing of the glycoproteins of paramyxoviridae, in: *The Paramyxoviruses* (D. Kingsbury, ed.), Plenum Press, New York, pp. 347–382.
90. Gallaher, W., 1987, Detection of a fusion peptide sequence in the transmembrane protein of human immunodeficiency virus, *Cell* **50**:327–328.

17

Association of Human Herpesvirus 6 with Human Tumors

DARIO DI LUCA and RICCARDO DOLCETTI

1. INTRODUCTION

Human herpesvirus 6 (HHV-6) was originally isolated from peripheral blood mononuclear cells (PBMCs) of patients with different lymphoproliferative disorders, some of which developed in the context of acquired immunodeficiency syndrome (AIDS).[1] Since then, the virus has been isolated from children with exanthem subitum (ES), from patients with a variety of diseases but especially with AIDS or other immunologic deficits, and from healthy individuals.[2–4] The virus has a preferential tropism for $CD4^+$ T lymphocytes[5] but can also infect different T- and B-cell types as well as lymphoid and nonlymphoid cell lines,[6–9] albeit with low efficiency. HHV-6 is ubiquitous in humans, and seroprevalence in adults reaches 90%.[10] Infection occurs early in life, as shown by the fact that most children become HHV-6 seropositive by 2 years of age.

The clinical manifestation of primary infection with HHV-6 is usually represented by ES, a benign, self-limiting disease characterized by a 3-day-long cutaneous rash[2] or by febrile episodes without rash or other relevant symptoms.[11,12] Anecdotal reports suggest that lytic infection in adults may also be associated with infectious mononucleosis-like illness, acute hepatitis, chronic fatigue syndrome, afebrile lymphadenopathy, and hemophagocytic syndrome.[13] The HHV-6 also

---

DARIO DI LUCA • Institute of Microbiology, University of Ferrara, I-44100 Ferrara, Italy. RICCARDO DOLCETTI • Division of Experimental Oncology 1, Centro di Riferimento Oncologico, 33081 Aviano (PN), Italy.

DNA Tumor Viruses: Oncogenic Mechanisms, edited by Giuseppe Barbanti-Brodano *et al.* Plenum Press, New York, 1995.

plays a role in post transplant complications, where it can be associated with serious diseases such as interstitial pneumonitis and graft failure.[14–16]

Isolates of HHV-6 are divided into two variants (HHV-6A and HHV-6B), which differ in their ability to grow in T-cell lines, are clearly distinguished by DNA restriction site analysis, and show specific reactivities with some monoclonal antibodies.[17,18] The DNA sequence identity between variants ranges from 75%[19] to 96%.[18,20] The HHV-6B variant is more prevalent in the healthy population[21] and is frequently isolated from children with ES[22] as well as from immunosuppressed patients. On the other hand, until now variant A has not been detected in ES patients, with two exceptions where both variants were present[23] but may be isolated from immunosuppressed patients.

The evidence available suggests that HHV-6, like the other human herpesviruses, establishes latent infections. In fact, as many as 90% of blood samples from healthy adult donors contain small amounts of viral DNA,[21,24,25] and HHV-6 can be reactivated from macrophages of children convalescent from ES.[26] Furthermore, the observation that 4–5% of healthy adolescents and adults are positive for HHV-6-specific IgM suggests that the virus may periodically reactivate.[27] Also, HHV-6 antigens and/or DNA have been detected in salivary glands and lymph nodes,[28–30] raising the possibility that these tissues also are sites of persistent or latent infection. Infectious HHV-6 has been isolated from saliva,[31] but questions concerning the specificity of the reagents employed have been raised.[32,33] Even if infectious virus may not be frequently present in saliva, nevertheless viral DNA is detected in salivary glands and saliva by the sensitive polymerase chain reaction (PCR), suggesting that infection could possibly be transmitted by oral secretions.

A possible involvement of HHV-6 in human malignancy was originally suggested by the first isolation of the virus, since three of six viral isolates were derived from patients with lymphoma.[1] Subsequent studies have focused on the possible role of HHV-6 in human tumors, searching for the virus in clinical specimens, and studying its *in vitro* transforming potential. In this chapter we summarize the current knowledge about the association of HHV-6 with human malignancies and lymphoproliferative diseases.

2. *IN VITRO* TRANSFORMATION STUDIES

Because of the recent discovery of HHV-6, studies on its oncogenic potential are still limited in number but indicate that *in vitro* cultured cells may be transformed on transfection of HHV-6 DNA sequences. The full-length HHV-6 DNA and two distinct cloned subgenomic fragments, pZVH14 and pZVB70, respectively of 8.7 and 21 kbp, induced transformation of murine NIH3T3 cells.[34] Inoculation of the transformed cells into nude mice induced the formation of rapidly growing tumors. HHV-6 DNA corresponding to pZVH14 was detected in primary transformed cells but not in tumor-derived cell lines, suggesting that these sequences are not required to maintain the transformed state. The same two subgenomic clones induced transformation of human epidermal keratinocytes.[35] Also in this instance, the morphologically transformed cells were tumorigenic in nude mice.

Chromosomal aberrations were observed both in clones derived from the primary transfected cells and in their tumor derivatives. A smaller HHV-6 DNA fragment, Sall L of 3.9 kbp, contained within pZVB70, retained the ability to transform NIH3T3 cells.[36] Interestingly, the Sall L fragment *trans*-activates the human immunodeficiency virus (HIV) long terminal repeat (LTR). It is notable that *trans*-activating potential is often associated with oncogenic proteins encoded by other viruses, such as SV40 large T antigen, E6 and E7 of human papilloma virus (HPV), and adenovirus E1a. It should be noted that transformation has been achieved so far only in cells nonpermissive for viral replication and not in lymphoid cells, the primary target of HHV-6 infection, suggesting that inactivation of the lytic functions is important for the expression of the neoplastic potential.

3. HUMAN HERPESVIRUS 6 IN LYMPHOPROLIFERATIVE DISEASES

3.1. Non-Hodgkin's Lymphomas

The original isolation of HHV-6 from patients with lymphoproliferative diseases and the demonstration that HHV-6 DNA fragments may be oncogenic *in vitro* have stimulated the search for HHV-6 genomes in malignant lymphomas. Of the 278 non-Hodgkin's lymphomas (NHLs) studied so far by Southern blotting and reported in the literature (Table I) only five (1.8%) were shown to be HHV-6 positive, indicating that this virus may exert a direct pathogenic role in a very small proportion of NHL cases. Interestingly, of the five HHV-6-positive NHLs, two were B-cell lymphomas occurring in the context of Sjogren's syndrome (SS), and one was a T-cell lymphoma in a patient with a previous history of angioimmunoblastic lymphadenopathy (AIL).[37,38] In particular, it has been observed that the association between HHV-6 and NHLs was statistically significant in patients with SS, because two of the eight cases investigated were HHV-6 positive ($p < 0.04$ by Fisher's exact test.[37] An attractive hypothesis is that the abnormal immune responses that characterize both SS and AIL patients might favor the potential oncogenic effect of HHV-6.

Consistent with the results of Southern blot analysis, PCR-based detection methods revealed a low prevalence of HHV-6 DNA in NHL biopsies (0–12%) (Table I), the only exception being the results of Sumiyoshi *et al.*,[41] who found HHV-6 sequences in 45/76 NHL biopsies (59.2%) from Japanese patients. None of these cases, however, was HHV-6 positive at Southern blot, indicating that a limited number of HHV-6-infected cells was present. This observation suggests either that the virus was harbored by infiltrating normal cells or that viral infection took place after the expansion of the malignant clone, ruling out a possible direct involvement of HHV-6 in the development of these NHLs. The results of a recent *in situ* hybridization study performed on 45 Chinese NHL patients also seem to exclude the possibility that the eight (18%) HHV-6-positive cases observed represent the clonal expansion of lymphoid cells directly infected by HHV-6.[44] In fact, in these NHL biopsies the virus was present in only scattered neoplastic cells and/or reactive lymphocytes. Therefore, the studies published so far are consistent in indicating

TABLE I
Detection of HHV-6 DNA in Non-Hodgkin's Lymphomas

Reference and type of NHL	Technique	
	Southern blot	PCR
Jarrett et al., 1988[37]	2/53 (4%)	ND[a]
B-cell	1/35 (3%)	
T-cell	1/18 (6%)	
Josephs et al., 1988[38]	3/104 (3%)	ND
B-cell	3/82 (4%)	
T-cell	0/22	
Borisch et al., 1991[39]	ND	2/16 (13%)
Torelli et al., 1991[40]	ND	0/41
Sumiyoshi et al., 1993[41]	0/76	45/76 (59%)
B-cell		18/29 (62%)
T-cell		24/41 (59%)
Brice et al., 1993[42]		
Cutaneous T-cell	ND	1/38 (3%)
Paulus et al., 1993[43]		
Primary cerebral	ND	1/39 (3%)
Di Luca et al., 1994[21]		
B-cell	0/45	1/45 (2%)

[a]ND, not done.

that HHV-6 may be directly involved in the development of only a very small fraction of NHLs, whereas it remains to be elucidated whether the virus may indirectly contribute to lymphomagenesis. In this respect it is worth mentioning that Krueger et al.[45] have hypothesized that HHV-6 might play an indirect role in the development of some lymphoproliferative disorders because it can stimulate chronic polyclonal B-cell proliferation with consequent enlargement of the pool of cells that are at an increased risk of undergoing further transforming events. A multistep evolution of the lymphomagenic process appears particularly evident in a heterogeneous group of lymphoproliferative disorders, including peripheral T-cell lymphomas, AIL, lymphomas in patients with SS or other autoimmune diseases, lymphomas of immunocompromised patients, which Krueger et al.[45] have defined as atypical polyclonal lymphoproliferations (APL). These disorders are characterized by an increased incidence in patients with abnormal immune responses, frequent occurrence of persistent active infection by lymphotropic viruses, including HHV-6, and a high risk of evolution to overt malignant lymphomas. Nevertheless, although serologic analyses have indicated that in patients with APL HHV-6 may frequently be reactivated, the present evidence does not allow us to assess conclusively whether the prevalence and load of HHV-6 are significantly higher in APL biopsies than in reactive lymphadenopathies. The elucidation of this issue and the identification of virus-encoded antigens expressed in involved tissues are necessary to substantiate the hypothesis that HHV-6 has a contributory role in the development of these lymphoproliferative disorders.

3.2. Hodgkin's Disease

Clinical, epidemiologic, and serologic studies have suggested that Hodgkin's Disease (HD) is a heterogeneous entity that may have an infectious etiology (reviewed by Boiocchi *et al.*, Chapter 20). In particular, the bimodal age–incidence curve of HD and the different distributions of the histological subtypes in the various age groups has led to the hypothesis that different agents may be involved in the pathogenesis of this disease. An unusual host response following a late exposure to a relatively common infectious agent has been related to the development of young adult cases, the large majority of which are of the nodular sclerosis (NS) histological subtype. Conversely, the disease in older patients, which is prevalently composed of mixed cellularity (MC) cases, may have an entirely different pathogenesis. The recent demonstration of Epstein–Barr virus (EBV) DNA sequences and gene products in Reed–Sternberg cells (RSCs) (reviewed by Boiocchi *et al.*, Chapter 20) strongly indicated that EBV may be directly involved in the induction of a subset of cases. In Western countries, EBV is mainly associated with the MC subtype and with cases in older patients, supporting the hypothesis of a multifactorial etiology for HD.

In the search for other infectious agents possibly implicated in the development of HD, and particularly of the NS cases, several studies have recently attempted to assess whether HHV-6 might be involved. A possible etiological role for this virus in HD was suggested by serologic analyses that revealed a higher prevalence or titer of antibodies to HHV-6 in HD patients than in healthy controls or patients with unrelated diseases.[40,46,47] In particular, an increased HHV-6 seropositivity was observed among young adults with HD without siblings, supporting the hypothesis that a late exposure to HHV-6 consequent to a lack of social contacts might contribute to the development of HD.[46] Nevertheless, using different assays, Levine *et al.*[48] have reported that, although antibody titers to HHV-6 in sera taken from HD patients at the time of diagnosis were not significantly different from those of controls, remarkable changes in anti-HHV-6 titers may occur during the clinical course of the disease. These findings suggest that the increase in HHV-6 antibodies observed in HD patients may be related to the progression of the disease and/or to therapy rather than indicating a causal involvement of HHV-6 in the pathogenesis of HD.

The search for HHV-6 DNA sequences in HD biopsies, initially performed by Southern blotting, revealed that HHV-6 sequences are only infrequently found in HD tissues (2 of 156 published cases) as well as in benign lymphadenopathies (Tables II and III). It is worth considering that in the two HD cases positive for HHV-6 at Southern blot reported so far, the RSCs accounted for about 25% of all cellular elements,[40] whereas these cells usually represent less than 5% of the total cell population. The use of PCR-based methods of detection gave conflicting results because HHV-6 positivity ranged from 0 to 64% in the HD series investigated (Table II). Differences in the sensitivity of the PCR protocols used as well as ethnic or geographic factors may account for these discrepancies. In a recent investigation of a large series of HD samples,[21] we have detected HHV-6 DNA sequences in about one-third of cases, with no preferential distribution among the different histologi-

TABLE II
Detection of HHV-6 DNA in Hodgkin's Disease

Reference	Southern blot	PCR[a] Total	LD	MC	NS	LP
Jarrett et al., 1988[37]	0/29	ND[b]				
Josephs et al., 1988[38]	0/8	ND				
Gledhill et al., 1991[49]	0/35	ND				
Borisch et al., 1991[39]	ND	0/6				
Torelli et al., 1991[40]	2/25 (8%)	3/25 (12%)	0/1	0/10	3/10	0/4
Gompels et al., 1993[20]	ND	1/8 (13%)				
Sumiyoshi et al., 1993[41]	0/14	9/14 (64%)				
Di Luca et al., 1994[21]	0/45	13/45 (29%)	0/1	3/10	9/30	1/4

[a]LD, lymphocyte depletion; MC, mixed cellularity; NS, nodular sclerosis; LP, lymphocyte prevalence.
[b]ND, not done.

cal subtypes (Table II). In agreement with results reported by others, the positivity for HHV-6 was generally related to the presence of few viral sequences in HD lesions with no significant difference in either prevalence or viral load when compared with reactive lymphadenopathies. Characterization of HHV-6 variants showed that in HD biopsies, HHV-6B was largely prevalent over variant A,[21] indicating that the relative prevalence of the two HHV-6 variants in HD-involved tissues is not different from that detected in PBMCs from healthy donors.

Overall, these results are difficult to interpret in terms of HD pathogenesis because Southern blot and PCR do not give important information on the integrity, physical status, and transcriptional activity of the viral genome. In addition, with these methods it is not possible to identify and localize the HHV-6-infected cells. These questions are particularly relevant in the case of HD, in which only a minority of cells constitute the neoplastic component. The few *in situ* hybridization studies

TABLE III
Detection of HHV-6 DNA in Benign Lymphadenopathies

Reference	Southern blot	PCR
Jarrett et al., 1988[37]	0/10	ND[a]
Josephs et al., 1988[38]	0/31	ND
Borisch et al., 1991[39]	ND	2/4 (50%)
Sumiyoshi et al., 1993[41]	0/56	55/56 (98%)
Dolcetti et al., 1994[50]	ND	2/10 (20%)

ND, not done.

performed so far failed to reveal HHV-6 DNA sequences unambiguously within RSCs and their precursors; HHV-6 genomes were detected only in scattered lymphoid cells or macrophages.[45,51] On the other hand, it has recently been demonstrated that HD-derived cell lines are susceptible to HHV-6 infection and subsequent expression of virus-encoded antigens.[52] Therefore, the experimental evidence accumulated so far does not make it possible to establish conclusively whether HHV-6 has a direct role in the induction of a subset of HD.

However, the presence of HHV-6 in tissues involved by HD suggests that the virus might at least indirectly contribute to the disease. In fact, it has been shown that HHV-6 reactivates latent EBV in human lymphoid cell lines and up-regulates the expression of immediate-early and early EBV antigens.[53] The interactions between HHV-6 and EBV could therefore contribute by increasing the pool of EBV-infected cells, thereby enhancing the possibility of development of an EBV-related lymphoma. HHV-6 could also contribute to some pathological aspects observed in the course of HD. These patients are often affected by immunologic deficits such as reduced NK cell activity, impaired proliferative response of mononuclear cells, and increased levels of cytokines.[54] HHV-6 inhibits the proliferative response of mononuclear cells,[55] supports cytopathic infection of NK cells,[8] and up-regulates the production of interleukin 1 (IL-1) and tumor necrosis factor α (TNF-α).[13] The elucidation of these issues requires further studies and, in particular, the use of more sensitive *in situ* hybridization techniques and immunohistochemical methods suitable for the detection of HHV-6 gene products.

3.3. HIV-Associated Lymphoproliferations

HIV-infected patients develop lymphomas more frequently than the normal population, to the extent that NHL, together with Kaposi's sarcoma (KS), is the most common neoplasm observed in HIV-related immunosuppression[56] and is recognized as AIDS-defining by the Centers for Disease Control (CDC).[57] Hodgkin's disease is also frequently reported in association with HIV, and it has been suggested as a possible new AIDS-defining condition.[56] It has been shown that EBV may play a role in the pathogenesis of HIV-associated lymphoproliferative disorders; in fact, although this virus is only rarely detected in NHL arisen in HIV-seronegative individuals, about 50% of NHL from HIV-infected patients are EBV-associated.[58] Likewise, EBV is present in RSCs in about 40% of HD cases in the general population, but it is detected in the vast majority of HD from HIV-infected patients.[59] It is of interest to investigate the role of HHV-6 in HIV-associated lymphomagenesis, because the first isolations of this virus were obtained from patients with lymphoma in the course of AIDS.[60]

We have undertaken a study specifically to address this issue in comparison with the situation described for EBV[21,50] (unpublished results). We have analyzed by PCR the frequency and variant distribution of HHV-6 in nonneoplastic and malignant lymphoproliferative disorders developed in HIV-infected or HIV-negative individuals. The results, summarized in Table IV, show that HHV-6 is significantly more prevalent in lymphadenopathy syndrome (LAS) biopsies than in HIV-unrelated reactive lymphadenopathies (13/20, 65% versus 2/10, 20%; $p = 0.02$).

TABLE IV
Presence of HHV-6 and EBV DNA
in HIV-Associated and HIV-Unrelated Lymphoproliferations

Histology	HHV-6		EBV	
	HIV-seropos.	HIV-seroneg.	HIV-seropos.	HIV-seroneg.
Reactive lymphadenopathies	—	2/10 (20%)	—	4/10 (40%)
LAS	13/20 (65%)	—	14/20 (70%)	—
HD	3/10 (30%)	13/45 (29%)	10/10 (100%)	15/45 (33%)
NHL	1/17 (6%)	0/35	9/17 (53%)	1/35 (3%)

Also, the prevalence of EBV was increased in LAS (14/20, 70% versus 4/10, 40%), although the difference was not statistically significant. The prevalence of HHV-6 DNA in HD and B-cell NHL arisen in HIV-infected patients (30% and 6%, respectively) was similar to that observed in HIV-seronegative patients with the same lymphoproliferative disorders. In particular, the observation that none of these HIV-related B-cell NHL was HHV-6 positive at Southern blot analysis indicates that this virus has no direct role in the pathogenesis of these lymphomas. By contrast, EBV was significantly more associated with malignant lymphoproliferations arisen in the context of HIV infection than with the respective counterparts in HIV-negative patients, both in the case of HD (10/10, 100% versus 15/45, 33%; $p = 0.0001$) and in the case of NHL (9/17, 53% versus 1/35, 3%; $p = 0.00005$). These findings indicate that, unlike EBV,[61] the increased frequency of HHV-6 DNA detection in LAS samples does not represent a significant risk factor for the development of HHV-6-carrying malignant lymphomas in HIV-infected patients. However, HHV-6 might have an indirect role in the complex process of lymphomagenesis occurring in the context of HIV infection. In fact, the increased prevalence of HHV-6 in LAS lymph nodes, probably favored by HIV-related immunosuppression, might induce a chronic B-cell stimulation. The likelihood for these cells in active proliferation to undergo further genetic lesions would increase at each cell division. The occurrence of relevant genetic changes and the possible interaction with other oncogenic cofactors would lead to a fully transformed phenotype. The demonstration of monoclonal or oligoclonal immunoglobulin gene rearrangements in about 20% of LAS lymph nodes is in line with this hypothesis of multistep lymphomagenesis.[62] Further studies aimed at determining whether B-cell clonal expansions are more frequent in HHV-6-positive than in HHV-6-negative LAS will be helpful in elucidating the role of HHV-6 in the pathogenesis of HIV-related malignant lymphomas.

3.4. Leukemia

HHV-6 is often isolated, or high viral loads are detected, in patients with leukemia. Luka et al.[63] detected HHV-6 DNA in 14 of 16 patients with T-cell acute lymphoblastic leukemia, and in eight samples further analyzed, HHV-6-positive cells ranged between 20% and 58%. Other studies describe the reactivation of HHV-6 in patients with T-cell leukemia,[64] large granular lymphocytic leukemia,[65] and acute lymphoblastic leukemia.[66] However, these findings are too preliminary to assess whether HHV-6 has a role in the pathogenesis or may affect the clinical course of this disease.

4. ROLE OF HHV-6 IN NONLYMPHOID NEOPLASMS

HHV-6 undergoes a productive infection in PBMCs and, more generally, in several subsets of lymphoid cells, resulting in cell lysis and death. Although it is a lymphotropic virus, HHV-6 also infects epithelial and fibroblastic cells, but these systems do not support efficient viral replication.[7,67] It is possible that the neoplastic potential of the virus might be more easily expressed under conditions of nonpermissivity. In fact, as previously mentioned, *in vitro* transformation by HHV-6 has been obtained in nonpermissive cells (murine fibroblasts, human keratinocytes) but not yet in human lymphoid cells. Indeed, some recent reports have shown a significant presence of HHV-6 in nonlymphoid malignant disorders.

Analysis of human cervical intraepithelial neoplasia and squamous invasive carcinomas has recently revealed the presence of HHV-6 in 6/72 cases[67] but in none of 30 nonneoplastic cervical biopsies. Four of the six HHV-6-positive samples also contained HPV-16, the HPV type most frequently associated with cervical cancer. The potential significance of the association between HHV-6 and HPV has been shown by *in vitro* studies, indicating that HHV-6 infects cervical carcinoma cells, *trans*-activates HPV promoters, and enhances expression of the viral oncoproteins E6 and E7.[7] Furthermore, once HHV-6 infected, carcinoma cell lines showed a more aggressive tumorigenic potential on inoculation into nude mice than did the parental cell line. These studies raise the hypothesis that HHV-6 might be a cofactor in HPV-associated genital oncogenesis.

HHV-6 DNA was also detected by PCR in 7/20 (35%) KS biopsies[68]; characterization of the HHV-6 variants showed that HHV-6A accounted for the majority of positive specimens (6/7). Furthermore, HHV-6A DNA, and not HHV-6B, was detected in mesenchymal spindle cell lines derived from KS (D. Di Luca *et al.*, unpublished data), suggesting that the positive samples could not merely be explained by the presence of latently infected infiltrating lymphocytes in KS biopsies. The interest in this observation lies in the comparison between the lower occurrence of HHV-6A in the general population (less than 5%) and the fairly high prevalence in KS (30%). This difference could suggest a preferential tropism for endothelial tissue; in fact, the precise cellular origin of KS is still undetermined, but it is presumed to be from vascular or lymphatic endothelial cells. Alternatively, it could reflect viral reactivation from immunosuppression. However, it should be noted that under the influence of cytokines, endothelial cells may develop into spindle-shaped KS-like cells; because HHV-6 infection of PBMCs induces the expression of cytokines (TNF-α, IL-1β) even when viral replication is inhibited,[69] these observations suggest the opportunity to investigate further a possible contribution of HHV-6 to the pathogenesis of KS. Furthermore, HHV-6 has been detected in six of nine squamous cell carcinomas of the oral mucosa; the samples were positive both by PCR and immunohistochemical staining with HHV-6-specific monoclonal antibodies.[70] These reports are still sporadic and limited in number but raise the interesting possibility that HHV-6 might act as a cofactor in extralymphoid oncogenesis.

5. CONCLUSIONS

The available evidence on the possible pathogenetic role of HHV-6 in human tumors may be summarized as follows:

1. The virus has *in vitro* transforming potential, and the transformed cells are tumorigenic in animal systems. However, transformation has not yet been obtained in lymphoid cells, which represent the primary target of viral infection. Interestingly, viral sequences are present in some nonlymphoid neoplasms (cervical carcinoma, KS, carcinoma of the oral mucosa) whose cells do not seem to support efficient viral replication. These observations could hint at the necessity to inhibit viral lytic replication in order to allow expression of transforming potential.
2. HHV-6 is scarcely present in NHL, suggesting that the virus probably has no direct role in the development of this neoplasia. An exception is possibly represented by B-cell NHL arisen in patients with SS, where HHV-6 seems to be detected more frequently.
3. HHV-6 DNA is detected in about one-third of HD biopsies, suggesting that the virus might be associated with a subset of this disorder.
4. HHV-6 is frequently present in LAS biopsies, raising the possibility that it could contribute to subsequent development of lymphomas, a fairly common occurrence in immunodepressed individuals.

Available evidence therefore seems to indicate that HHV-6 might be associated with lymphomagenesis, but in an indirect way, either interacting with other oncogenic factors (i.e., EBV) or stimulating a chronic B-cell proliferation that may render these cells more prone to accumulate synergic damages and to undergo malignant transformation.

ACKNOWLEDGMENTS. The studies performed in our laboratories were supported by grants from Ministero della Sanità (Istituto Superiore di Sanità, AIDS Project, Rome, Italy), CNR special project "Applicazioni Cliniche della Ricerca Oncologica (ACRO)" (93.02303.PF39 and 93.02134.PF39), and Associazione Italiana per la Ricerca sul Cancro.

REFERENCES

1. Salahuddin, S. Z., Ablashi, D. V., Markham, P. D., Josephs, S. F., Sturzenegger, S., Kaplan, M., Haligan, G., Biberfeld, P., Wong-Staal, F., Kramarsky, B., and Gallo, R., 1986, Isolation of a new virus HBLV in patients with lymphoproliferative disorders, *Science* **234**:596–601.

2. Yamanishi, K., Okuno, T., Shiraki, K., Takahashi, M., Kondo, T., Asano, Y., and Kurata, T., 1988, Identification of human herpesvirus-6 as a causal agent for exanthem subitum, *Lancet* **1**: 1065–1067.
3. Downing, R. G., Sewankambo, N., Serwadda, D., Honess, R. W., Crawford, D., Jarrett, R., and Griffin, B. E., 1987, Isolation of human lymphotropic herpesviruses from Uganda, *Lancet* **2**:390.
4. Lopez, C., Pellett, P., Stewart, J., Goldsmith, C., Sanderlin, K., Black, J., Warfield, D., and Feorino, P., 1988, Characteristics of human herpesvirus 6, *J. Infect. Dis.* **157**:1271–1273.
5. Lusso, P., Markham, P. D., Tschachler, E., DiMarzo Veronese, F., Salahuddin, S. Z., Ablashi, D. V., Pawha, S., Krohn, K. J., and Gallo, R. C., 1988, In vitro cellular tropism of human B lymphotropic virus (herpesvirus-6), *J. Exp. Med.* **167**:1659–1670.
6. Ablashi, D. V., Salahuddin, S. Z., Josephs, S. F., Imam, F., Lusso, P., Gallo, R. C., Hung, C. L., Lemp, D., and Markham, P. D., 1987, HBLV (or HHV-6) in human cells, *Nature* **329**:207.
7. Chen, M., Popescu, N., Woodworth, C., Berneman, Z., Corbellino, M., Lusso, P., Ablashi, D. V., and DiPaola, J. A., 1994, Human herpesvirus 6 infects cervical epithelial cells and transactivates human papillomavirus gene expression, *J. Virol.* **68**:1173–1178.
8. Lusso, P., Malnati, M. S., Garzino-Demo, A., Crowley, R. W., Long, E. O., and Gallo, R. C., 1993, Infection of natural killer cells by human herpesvirus 6, *Nature* **362**:458–462.
9. Lusso, P., Josephs, S. F., Ablashi, D. V., Gallo, R. C., Veronese, F. D., and Markham, P. D., 1987, Diverse tropism of HBLV (human herpesvirus-6), *Lancet* **2**:743–744.
10. Parker, C. A., and Weber, J. M., 1993, An enzyme-linked immunosorbent assay for the detection of IgG and IgM antibodies to human herpesvirus type 6, *J. Virol. Methods* **41**:265–275.
11. Ward, K. N., and Gray, J. J., 1994, Primary human herpesvirus-6 infection is frequently overlooked as a cause of febrile fits in young children, *J. Med. Virol.* **42**:119–123.
12. Portolani, M., Cermelli, C., Moroni, A., Bertolani, M. F., Di Luca, D., Cassai, E., and Sabbatini, A. M., 1993, Human herpesvirus-6 infections in infants admitted to hospital, *J. Med. Virol.* **39**: 146–151.
13. Pellett, P. E., Black, J. B., and Yamamoto, M., 1992, Human herpesvirus 6: The virus and the search for its role as a human pathogen, *Adv. Virus Res.* **41**:1–52.
14. Okuno, T., Higashi, K., Shiraki, K., Yamanishi, K., Takahashi, M., Kokado, Y., Ishibashi, M., Takahara, S., Sonoda, T., Tanaka, K., Baba, K., Yabuuchi, H., and Kurata, T., 1990, Human herpesvirus 6 infection in renal transplantation, *Transplantation* **49**:519–522.
15. Drobyski, W. R., Dunne, W. M., Burd, E. M., Knox, K. K., Ash, R. C., Horowitz, M. M., Flomenberg, N., and Carrigan, D. R., 1993, Human herpesvirus-6 (HHV-6) infection in allogeneic bone marrow transplant recipients: Evidence of a marrow-suppressive role for HHV-6 *in vivo*, *J. Infect. Dis.* **167**:735–739.
16. Carrigan, D. R., Drobyski, W. R., Russler, S. K., Tapper, M. A., Knox, K. K., and Ash, R. C., 1991, Interstitial pneumonitis associated with human herpesvirus-6 infection after marrow transplantation, *Lancet* **1**:147–149.
17. Ablashi, D., Agut, H., Berneman, Z., Campadelli-Fiume, G., Carrigan, D., Ceccherini-Nelli, L., Chandran, B., Chou, S., Collandre, H., Cone, R., Dambaugh, T., Dewhurst, S., Di Luca, D., Foa-Tomasi, L., Fleckenstein, B., Frenkel, N., Gallo, R., Gompels, U., Hall, C., Jones, M., Lawrence, G., Martin, M., Montagnier, L., Neipel, F., Nicholas, J., Pellett, P., Razzaque, A., Torelli, G., Thomson, B., Salahuddin, S., Wyatt, L., and Yamanishi, K., 1993, Human herpesvirus-6 strain groups: A nomenclature, *Arch. Virol.* **129**:363–366.
18. Aubin, J. T., Collandre, H., Candotti, D., Ingrand, D., Rouzioux, C., Burgard, M., Richard, S., Huraux, J. M., and Agut, H., 1991, Several groups among human herpesvirus 6 strains can be distinguished by Southern blotting and polymerase chain reaction, *J. Clin. Microbiol.* **29**: 367–372.
19. Chou, S., and Marousek, G. I., 1994, Analysis of interstrain variation in a putative immediate-early region of human herpesvirus 6 DNA and definition of variant-specific sequences, *Virology* **198**:370–376.
20. Gompels, U. A., Carrigan, D. R., Carss, A. L., and Arno, J., 1993, Two groups of human

herpesvirus 6 identified by sequence analyses of laboratory strains and variants from Hodgkin's lymphoma and bone marrow transplant patients, *J. Gen. Virol.* **74**:613–622.
21. Di Luca, D. Dolcetti, R., Mirandola, P., De Re, V., Secchiero, P., Carbone, A., Boiocchi, M., and Cassai, E., 1994, Human herpesvirus 6: A survey of presence and variant distribution in normal peripheral lymphocytes and lymphoproliferative disorders, *J. Infect. Dis.* **170**:211–215.
22. Dewhurst, S., McIntyre, K., Schnabel, K., and Hall, C. B., 1993, Human herpesvirus 6 (HHV-6) variant B accounts for the majority of symptomatic primary HHV-6 infections in a population of U. S. infants, *J. Clin. Microbiol.* **31**:416–418.
23. Dewhurst, S., Chandran, B., McIntyre, K., Schnabel, K., and Hall, C. B., 1992, Phenotypic and genetic polymorphisms among human herpesvirus-6 isolates from North American infants, *Virology* **190**:490–493.
24. Cuende, J. I., Ruiz, J., Civeira, M. P., and Prieto, J., 1994, High prevalence of HHV-6 DNA in peripheral blood mononuclear cells of healthy individuals detected by nested PCR, *J. Med. Virol.* **43**:115–118.
25. Jarrett, R. F., Clark, D. A., Josephs, S. F., and Onions, D. E., 1990, Detection of human herpesvirus-6 DNA in peripheral blood and saliva, *J. Med. Virol.* **32**:73–76.
26. Kondo, K., Kondo, T., Okuno, T., Takahashi, M., and Yamanishi, K., 1991, Latent human herpesvirus 6 infection of human monocytes/macrophages, *J. Gen. Virol.* **72**:1401–1408.
27. Suga, S., Yoshikawa, T., Asano, Y., Nakashima, T., Yazaki, T., Fukuda, M., Kojima, S., Matsuyama, T., Ono, Y., and Oshima, S., 1992, IgM neutralizing antibody responses to human herpesvirus-6 in patients with exanthem subitum or organ transplantation, *Microbiol. Immunol.* **36**:495–506.
28. Fox, J. D., Briggs, M., Ward, P. A., and Tedder, R. S., 1990, Human herpesvirus 6 in salivary glands, *Lancet* **2**:590–593.
29. Levine, P. H., Jahan, N., Murari, P., Manak, M., and Jaffe, E. S., 1992, Detection of human herpesvirus 6 in tissues involved by sinus histiocytosis with massive lymphadenopathy (Rosai–Dorfman disease), *J. Infect. Dis.* **166**:291–295.
30. Di Luca, D., Mirandola, P., Ravaioli, T., Dolcetti, R., Frigatti, A., Bovenzi, P., Sighinolfi, L., Monini, P., and Cassai, E., 1995, HHV-6 and HHV-7 in salivary glands and shedding in saliva of healthy and HIV positive individuals, *J. Med. Virol.* **45**:462–468.
31. Levy, J. A., Ferro, F., Greenspan, D., and Lennette, E. T., 1990, Frequent isolation of HHV-6 from saliva and high seroprevalence of the virus in the population, *Lancet* **1**:1047–1050.
32. Wyatt, L. S., and Frenkel, N., 1992, Human herpesvirus 7 is a constitutive inhabitant of adult human saliva, *J. Virol.* **66**:3206–3209.
33. Black, J. B., Naoki, I., Kite-Powell, K., Zaki, S., and Pellett, P. E., 1993, Frequent isolation of human herpesvirus 7 from saliva, *Virus Res.* **29**:91–98.
34. Razzaque, A., 1990, Oncogenic potential of human herpesvirus-6 DNA *Oncogene* **5**:1365–1370.
35. Razzaque, A., Williams, O., Wang, J., and Rhim, J. S., 1993, Neoplastic transformation of immortalized human epidermal keratinocytes by two HHV-6 DNA clones, *Virology* **195**:113–120.
36. Thomson, B. J., Weindler, F. W., Gray, D., Schwaab, V., and Heilbronn, R., 1994, Human herpesvirus 6 (HHV-6) is a helper virus for adeno-associated virus type 2 (AAV-2) and the AAV-2 rep gene homologue in HHV-6 can mediate AAV-2 DNA replication and regulate gene expression, *Virology* **204**:304–311.
37. Jarrett, R. F., Gledhill, S., Qureshi, F., Crae, S. H., Madhok, R., Brown, I., Evans, I., Krajewski, A., O'Brien, C. J., Cartwright, R. A., Venables, P., and Onions, D. S., 1988, Identification of human herpesvirus 6-specific DNA sequences in two patients with non-Hodgkin's lymphoma, *Leukemia* **2**:496–502.
38. Josephs, S. F., Buchbinder, A., Streicher, H. Z., Ablashi, D. V., Salahuddin, S. Z., Guo, H. G., Wong-Staal, F., Cossman, J., Raffeld, M., Sundeen, J., Levine, P., Biggar, R., Krueger, G. R. F., Fox, R. I., and Gallo, R. C., 1988, Detection of human B-lymphotropic virus (human herpesvirus 6) sequences in B cell lymphoma tissues of three patients, *Leukemia* **2**:132–135.
39. Borisch, B., Ellinger, K., Neipel, F., Fleckenstein, B., Kirchner, T., Ott, M. M., and Mueller-

Hermelink, N. K., 1991, Lymphadenitis and lymphoproliferative lesions associated with the human herpes virus-6 (HHV-6), *Virchows Arch. B Cell Pathol.* **61:**179–187.
40. Torelli, G., Marasca, R., Luppi, M., Selleri, L., Ferrari, S., Narni, F., Mariano, M. T., Federico, M., Ceccherini, Nelli, L., Bendinelli, M., Montagnani, G., Montorsi, M., and Artusi, T., 1991, Human herpesvirus-6 in human lymphomas: identification of specific sequences in Hodgkin's lymphomas by polymerase chain reaction, *Blood* **77:**2251–2258.
41. Sumiyoshi, Y., Kikuchi, M., Ohshima, K., Takeshita, M., Eizuru, Y., and Minamishima, Y., 1993, Analysis of human herpes virus-6 genomes in lymphoid malignancy in Japan, *J. Clin. Pathol.* **46:**1137–1138.
42. Brice, S. L., Jester, J. D., Friednash, M., Golitz, L. E., Leahy, M. A., Stockert, S. S., and Weston, W. L., 1993, Examination of cutaneous T-cell lymphoma for human herpesviruses by using the polymerase chain reaction, *J. Cutan. Pathol.* **20:**304–307.
43. Paulus, W., Jellinger, K., Hallas, C., Ott, G., and Muller-Hermelink, H. K., 1993, Human herpesvirus-6 and Epstein–Barr virus genome in primary cerebral lymphomas, *Neurology* **43:**1591–1593.
44. Yin, S. Y., Ming, H. A., Jahan, N., Manak, M., Jaffe, E. S., and Levine, P. H., 1993, *In situ* hybridization detection of human herpesvirus 6 in biopsy specimens from Chinese patients with non-Hodgkin's lymphoma, *Arch. Pathol. Lab. Med.* **117:**502–506.
45. Krueger, G. R. F., Manak, M., Bourgeois, N., Ablashi, D. V., Salahuddin, S. Z., Josheps, S., Buchbinder, A., Gallo, R. C., Berthold, F., and Tesch, H., 1989, Persistent active herpes virus infection associated with atypical polyclonal lymphoproliferation (APL) and malignant lymphoma, *Anticancer Res.* **9:**1457–1476.
46. Clark, D. A., Alexander, F. E., McKinney, P. A., Roberts, P. E., O'Brien, C., Jarrett, R. F., Cartwright, R. A., and Onions, D. E., 1990, The sero-epidemiology of human herpesvirus-6 (HHV-6) from a case control study of leukaemia and lymphoma, *Int. J. Cancer* **45:**829–833.
47. Iyengar, S., Levine, P. H., Ablashi, D., Neequaye, J., and Pearson, G. R., 1991, Sero-epidemiological investigations on human herpesvirus 6 (HHV-6) infections using a newly developed early antigen assay, *Int. J. Cancer* **49:**551–557.
48. Levine, P. H., Ebbesen, P., Ablashi, D. V., Saxinger, W. C., Nordentoft, A., and Connelly, R. R., 1992, Antibodies to human herpes virus-6 and clinical course in patients with Hodgkin's disease, *Int. J. Cancer* **51:**53–57.
49. Gledhill, S., Gallagher, A., Jones, D. B., Krajewski, A. S., Alexander, F. E., Klee, E., Wright, D. H., O'Brien, C., Onions, D. E., and Jarrett, R. F., 1991, Viral involvements in Hodgkin's disease: Detection of clonal type A Epstein–Barr virus genomes in tumor samples, *Br. J. Cancer* **64:**227–232.
50. Dolcetti, R., Di Luca, D., Mirandola, P., De Vita, S., De Re, V., Carbone, A., Tirelli, U., Cassai, E., and Boiocchi, M., 1994, Frequent detection of human herpesvirus 6 DNA in HIV-associated lymphadenopathy, *Lancet* **2:**543.
51. Maeda, A., Sata, T., Enzan, H., Tanaka, K., Wakiguchi, H., Kurashige, T., Yamanishi, K., and Kurata, T., 1993, The evidence of human herpesvirus 6 infection in the lymph nodes of Hodgkin's disease, *Virchows Arch. A Pathol. Anat. Histopathol.* **423:**71–75.
52. Krueger, G. R. F., Sievert, J., Juecker, M., Tesch, H., Diehl, V., Ablashi, D. V., Balachandran, N., and Luka, J., 1992, Hodgkin's cells express human herpesvirus-6 antigens, *J. Vir. Dis.* **1:**15–23.
53. Flamand, L., Stefanescu, I., Ablashi, D. V., and Menezes, J., 1993, Activation of the Epstein–Barr virus replicative cycle by human herpesvirus 6, *J. Virol.* **67:**6768–6777.
54. Collins, R. H., 1990, The pathogenesis of Hodgkin's disease, *Blood Rev.* **4:**61–68.
55. Horvat, R. T., Parmely, M. J., and Chandran, B., 1993, Human herpesvirus 6 inhibits the proliferative responses of human peripheral blood mononuclear cells, *J. Infect. Dis.* **167:**1274–1280.
56. Milliken, S., and Boyle, M. J., 1993, Update on HIV and neoplastic disease *AIDS* **7:**S203–S209.
57. Centers for Disease Control, 1985, Revision of the case definition of acquired immunodeficiency syndrome for national reporting—United States, *Morbid. Mortal. Week. Rep.* **34:**373.

58. Boyle, M. J., Sewell, W. A. A., Sculley, T., Apolloni, A., Turner, J. J., Swanson, C. E., Penny, R., and Cooper, D. A., 1991, Subtypes of Epstein–Barr virus in HIV-associated non-Hodgkin's lymphoma, *Blood* **78:**3001–3011.
59. Uccini, S., Monardo, F., Ruco, L. P., Baroni, C. D., Faggioni, A., Agliano, A. M., Gradilone, A. M., Manzari, A., Vago, L., Costanzi, G., Carbone, A., Boiocchi, M., and De Re, V., 1989, High frequency of Epstein–Barr virus genome in HIV-positive patients with Hodgkin's disease, *Lancet* **1:**1458.
60. Salahuddin, S. Z., 1992, The discovery of human herpesvirus 6, in: *Human Herpesvirus-6* (D. V. Ablashi, G. R. F. Krueger, and S. Z. Salahuddin, eds.), Elsevier, Amsterdam, pp. 3–8.
61. Shibata, D., Weiss, L. M., Nathwani, B. N., Brynes, R. K., and Levine, A. M., 1991, Epstein–Barr virus in benign lymph node biopsies from individuals infected with the human immunodeficiency virus is associated with concurrent or subsequent development of non-Hodgkin's lymphoma, *Blood* **77:**1527–1553.
62. Pelicci, P. G., Knowles, D. M., Arlin, Z. A., Wieczorek, R., Luciw, P., Dina, D., Basilico, C., and Dalla Favera, R., 1986, Multiple monoclonal B cell expansions and c-*myc* oncogene rearrangements in acquired immunodeficiency syndrome-related lymphoproliferative disorders, *J. Exp. Med.* **164:**2049–2076.
63. Luka, J., Pirruccello, S. J., and Kersey, J. H., 1991, HHV-6 genome in T-cell acute lymphoblastic leukaemia, *Lancet* **2:**1277–1278.
64. Baurmann, H., Miclea, J. M., Ferchal, F., Gessain, A., Daniel, M. T., Guetard, D., Collandre, H., Agut, H., Castaigne, S., Rain, J. D., Montagnier, L., Perol, Y., Sigaux, F., and Degos, L., 1988, Adult T-cell leukemia associated with HTLV-1 and simultaneous infection by human immunodeficiency virus type 2 and human herpesvirus 6 in an African woman: A clinical, virologic and familial serologic study, *Am. J. Med.* **85:**853–857.
65. Tagawa, S., Mizuki, M., Onoi, U., Nakamura, Y., Nozima, J., Yoshida, H., Kondo, K., Mukai, T., Yamanishi, K., and Kitani, T., 1992, Transformation of large granular lymphocytic leukemia during the course of a reactivated human herpesvirus-6 infection, *Leukemia* **6:**465–469.
66. Yoshikawa, T., Kobayashi, I., Asano, Y., Nakashima, T., Yazaki, T., Kojima, S., and Yamada, A., 1993, Clinical features of primary human herpesvirus-6 infection in an infant with acute lymphoblastic leukemia, *Am. J. Pediatr. Hematol. Oncol.* **15:**424–426.
67. Chen, M., Wang, H., Woodworth, C. D., Lusso, P., Berneman, Z., Kingma, D., Delgado, G., and DiPaolo, J., Detection of human herpesvirus 6 and human papillomavirus 16 in cervical carcinoma, *Am. J. Pathol.* **145:**1509–1516.
68. Bovenzi, P., Mirandola, P., Secchiero, P., Strumia, R., Cassai, E., and Di Luca, D., 1993, Human herpesvirus 6 (variant A) in Kaposi's sarcoma, *Lancet* **1:**1288–1289.
69. Flamand, L., Gosselin, J., D'Addario, M., Hiscott, J., Ablashi, D. V., Gallo, R. C., and Menezes, J., 1991, Human herpesvirus 6 induces interleukin-1 beta and tumor necrosis factor alpha, but not interleukin-6, in peripheral blood mononuclear cell cultures, *J. Virol.* **65:**5105–5110.
70. Yadav, M., Chandrashekran, A., Vasudevan, D. M., and Ablashi, D. V., 1994, Frequent detection of human herpesvirus 6 in oral carcinoma, *J. Natl. Cancer Inst.* **86:**1792–1794.

18

Molecular Mechanisms of Transformation by Epstein–Barr Virus

NANCY S. SUNG and JOSEPH S. PAGANO

1. INTRODUCTION

The Epstein–Barr virus (EBV) is associated with the development of malignancies (lymphomas and carcinomas of the nasopharynx) in both lymphoid and epithelial cells. Primary infection of the oropharyngeal epithelium results in virus replication, production of virions, and ultimately the destruction of the infected cell. In contrast to the normal cytolytic consequences of epithelial infection, EBV infection in lymphocytes remains latent yet achieves rapid and efficient cell immortalization.[1,2] The EBV-infected subset of B lymphocytes in peripheral blood will grow indefinitely when explanted to culture medium. Besides the long-established association of EBV with African Burkitt's lymphoma (BL), the virus has more recently been linked to B-cell lymphomas arising in immunosuppressed persons, apparently a result of EBV-driven polyclonal lymphoproliferation unchecked by the immune response. Ironically, in BL, the disease that originally led to the discovery of EBV, the expectation of an etiological role for the virus has diminished. Although endemic African BL consistently contains EBV genomes, they are often not present in non-African tumors. Universal and more likely as a common etiological factor in BL is a chromosomal translocation placing the c-*myc* protooncogene (from chromosome 8) in a position where its expression could be influenced by its new proximity to the immunoglobulin enhancer or other Ig regulatory elements on chromosomes

---

NANCY S. SUNG and JOSEPH S. PAGANO • Departments of Microbiology and Immunology and Medicine, UNC Lineberger Comprehensive Cancer Center, University of North Carolina, School of Medicine, Chapel Hill, North Carolina 27599-7295.

DNA Tumor Viruses: Oncogenic Mechanisms, edited by Giuseppe Barbanti-Brodano *et al.* Plenum Press, New York, 1995.

2, 14, or 22.[3,4] Interestingly, EBV's association with African BL correlates with the consistent usage of a single breakpoint location on chromosome 8, in contrast to EBV-negative American lymphomas in which a variety of breakpoint locations can be found. Through stimulation of cell proliferation, EBV may help boost the likelihood of certain translocations.[5] It has been suggested that this particular structural arrangement at the breakpoint may require an EBV function to achieve deregulation of the c-myc gene. The idea that an EBV gene product might mediate translocation, although appealing, has not been sustained. An enhancer region near the c-myc gene has some functional similarities to the EBV origin of plasmid replication (ori-P), raising the possibility that, under the right circumstances, this sequence could interact with the EBV origin-binding protein, Epstein–Barr nuclear antigen 1 (EBNA-1).[4]

EBV not only dysregulates control of growth in lymphocytes primarily B-cell and rarely certain T-cell malignancies[6], but a causal role in epithelial malignancy is increasingly substantiated. Despite the fact that EBV's primary relation with epithelial cells is cytolytic, in some populations EBV-containing carcinoma can arise years later in the nearby nasopharyngeal epithelium. EBV has been detected in several tumors of epithelial origin, principally well characterized nasopharyngeal carcinoma, and also in some tumors of the parotid and salivary glands, the palatine tonsil, and the supraglottic larynx.[7–10]

Nasopharyngeal carcinoma (NPC), like BL, is a malignancy characterized by distinct patterns of geographic distribution. It is the leading cause of cancer death in southern China. Outside of southeast Asia, NPC incidence is highest among Alaskan Eskimos and in Mediterranean Africa, only rarely occurring elsewhere.[11,12] Because of its distinctive pattern of incidence, its etiology is likely to include a combination of EBV infection and dietary and environmental cofactors converging in genetically susceptible populations.

The association between EBV and NPC was first determined serologically,[13] and later the EBV genome was detected in biopsied tissues by DNA hybridization.[14,15] Passaging of the tumor in athymic mice demonstrated that the EBV DNA was present in the epithelial tumor cells as opposed to infiltrating lymphocytes.[16] Since then, EBV DNA has been found in NPC tissues from all high- and low-incidence areas, geographic areas, and ethnic groups, distinguishing it from BL, in which mainly tumors from the endemic region contain EBV.[11,17–19] In endemic regions NPC patients carry high titers of circulating IgA antibodies directed against the EBV early replicative and viral capsid antigens (EA/VCA), which can be used for early detection as well as for determining susceptibility, as titers may elevate even before the tumor is detectable.[20]

During EBV latency, as in NPC, the EBV genome exists as a chromatin-associated episome, which is the linear genome circularized via fusion of the complementary direct tandem repeats at both termini.[21–25] Heterogeneity in the number of terminal repeats within the episome is expected in polyclonal cell proliferations resulting from multiple infectious events and therefore can be used to determine the clonality of EBV-infected tissue. Analysis of the fused termini has demonstrated the monoclonality of NPC tumors, because they retain a constant rather than varying number of terminal repeats in the episomal DNA.[26] Infection with EBV is therefore likely to occur before the emergence of a malignant epithelial cell clone in NPC, suggesting that a single EBV-infected clone may be preferentially amplified. From this evidence, the universal presence of EBV genomes in NPC, the striking IgA responses, and the finding of monoclonal episomes in dysplastic tissue as well as carcinoma *in situ*, it is likely that EBV influences the initiation of this tumor.[26a,26b]

This divergent cell tropism and pathogenesis hint at a plurality of molecular mechanisms contributing to a highly adaptable viral strategy for cell transformation. With the aim of illuminating a portion of this strategy, we first briefly discuss EBV virology and then discuss in greater detail specific mechanisms the virus may use to disrupt the normal progression of cell growth and differentiation.

2. LATENT REPLICATION

The double-stranded EBV genome can assume three different structural forms (Fig. 1). During latency, it exists as a unit-length chromatin-associated episome circularized via terminal repeats.[21,21a,25] Lytic infection produces a virion-encapsidated 178- to 190-kb linear genome; the exact size varies because of deletions in different viral strains. In some cell lines, the viral genome is integrated into the cell chromosome, probably through random nonhomologous recombination at

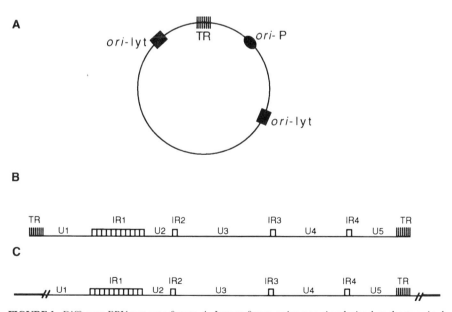

FIGURE 1. Different EBV genome forms. A. Latent form: episomes circularized at the terminal repeats (TR). B. Lytic form: unit-length linear copies. U refers to unique sequence; IR refers to internal repeated sequence. C. Integrated form: found in some cell lines; host cell sequences at the site of integration are excised and duplicated. Portions of U1 including *ori*-P may be deleted.

the termini,[27] but the existence of this viral format in primary infected tissues appears to be rare.

Inside the B lymphocyte, the EBV episome is amplified and maintained at a constant copy number.[25,28–31] Driven by host cell rather than virally encoded DNA polymerase,[32–34] episomal replication initiates at a specialized origin, *ori*-P,[35,36] once per cell cycle. The *ori*-P is comprised of two *cis*-acting elements, each of which contains multiple binding sites for the virally encoded EBNA-1 protein required for *ori*-P activation. A low-abundance protein, EBNA-1 is absolutely necessary for latent viral persistence and is therefore expressed in all cells carrying EBV DNA.

Because the EBV episome is maintained during latency at a constant copy number and is replicated coordinately early in S phase in the cell cycle, and because EBNA-1 is the only viral protein required for episomal replication, it is likely that expression or function of EBNA-1 is under cell-cycle control. In type-1 BL cells, as well as in NPC, the EBNA-1 transcript arises from Fp, a promoter in the *Bam*HI F fragment of the EBV genome[37–40] (Fig. 2). The RNA start site for Fp lies 200 bp upstream of two imperfect yet functional binding sites for the EBNA-1 protein (the Q locus) that mediate a strong autorepression of Fp.[35,41–44] Recent results show that this autorepression can be overcome by displacement of EBNA-1 from the Q locus by a member of the E2F family of cellular transcription factors.[44] The E2F transcription factor, first identified by its ability to bind to and activate the adenovirus E2 promoter,[45,46] is important in the regulation of cellular genes required for DNA synthesis and replication[47–50] as well as genes involved in cell growth.[51–54]

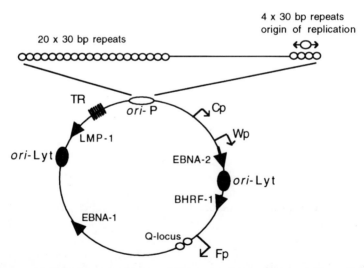

FIGURE 2. Spatial arrangement of EBNA-1 binding sites and EBNA-1 promoters. Open circles represent EBNA-1 binding sites within *ori*-P and in the Q locus. Arrows represent the three known promoters used for the EBNAs. Heavy arrowheads represent open reading frames for selected genes. An open oval represents *ori*-P; solid ovals *ori*-lyt. TR, terminal repeats; Cp, Wp, Fp, promoters for the EBNA transcripts.

During the G_1 phase of the cell cycle, E2F-like factors are inactive, existing as part of heteromeric protein complexes that include the retinoblastoma protein, [55-62] but on entry into S phase, E2F becomes dissociated and free to activate its target promoters. [53,59,63]

As more activated E2F becomes available in the cell at the G_1-S boundary, it may indirectly modulate the replication of episomes in synchrony with the cell cycle by overcoming the EBNA-1 autorepression of Fp. The newly expressed EBNA-1 arising from Fp could then activate latent episomal replication and coordinately populate the progeny cells with episomes. Once enough EBNA-1 is produced to activate the *ori*-P, to which it binds with high affinity, autorepression at the Q locus would again dominate.

3. CELL TRANSFORMATION BY EBV

The ability of EBV to immortalize B lymphocytes in culture makes it a likely candidate for involvement in the multistep process of oncogenesis. EBV immortalization is rapid and very efficient, with one infectious virus particle per cell required. At least two EBV latent gene products, Epstein–Barr nuclear antigen 2 (EBNA-2) and latent membrane protein (LMP-1), are essential for cell immortalization. Their possible roles are discussed in this section, and a summary of the roles of other latent genes is found in Table I.

3.1. Epstein–Barr Nuclear Antigen 2

That EBNA-2 is required for EBV transformation of lymphocytes is inferred from several lines of evidence. The P3HR-1 strain of EBV, in which the EBNA-2 coding region is deleted,[64,65] is unable to transform resting B cells into cell lines,[66-68] but this defect can be complemented with EBNA-2-expressing vectors.[69,70] EBNA-2, when expressed in rodent fibroblasts, reduces the serum requirements for growth,[71] and EBV-transformed cell lines carrying a wild-type EBNA-2 gene more readily produce tumors in SCID mice than do cell lines containing a mutant EBNA-2 gene.[72]

EBNA-2, encoded by the U2 region of the EBV genome, is the earliest EBV gene to be expressed after infection[73] and was first identified in latently infected lymphocyte cultures by EBV-immune human sera.[74] Extensive heterogeneity in the EBNA-2 coding region defines the two known EBV strains, with EBNA-2a (EBV-1) having 491 amino acids and EBNA-2b (EBV-2) having 443 amino acids.[74-77] EBV-1 transforms B lymphocytes much more efficiently than does EBV-2,[69,78] and it is the dominant strain present in Asian NPCs as well as in most lymphomas. EBV-2, on the other hand, is more commonly found in Alaskan NPCs and in a minority of EBV-associated lymphomas.[79]

The role of the EBNA-2 protein in EBV latency and pathogenesis is likely to center around its ability to activate both viral and cellular promoters. Several EBV latency promoters including the lymphoid-specific *Bam*HI C promoter (Cp) and promoters for the LMP-1, -2A, and -2B genes are up-regulated by EBNA-2,[80-84] as

TABLE I
Summary of EBV Latent Genes

EBV gene	Role in transformation	Function in viral life cycle	Interaction with cellular factors
EBNA-1	Establishes latent infection, essential	DNA binding protein; activates ori-P and enhancer and represses Fp	Binding to ori-P displaced by OBP,[164] binding to Q locus displaced by E2F-like factor[44]
EBNA-2	Essential but not sufficient	Transactivates Cp, LMP-1, and TP promoters; does not bind DNA	Interacts with CBF-1, up-regulates c-fgr, CD21, CD23 genes
EBNA-3A	Critical	Unknown	Unknown
EBNA-3B	Unnecessary		
EBNA-3C	Critical[165]		
EBNA-4/leader protein (LP)	Critical	Unknown	May interact with RB
LMP-1	Essential but not sufficient	Unknown	Activates bcl-2 protooncogene
Terminal protein (TP)/ LMP2A,2B	Unnecessary[166]	May block activation to lytic infection[167]	Interacts with src family tyrosine kinases[168,169]
EBERs	Unnecessary	Unknown	Interact with ribosomal protein L22,[124] La autoantigen[170]

are the cellular genes, c-fgr, vimentin, and the B-cell activation antigens CD21 and CD23.[85–88] Because of this broad range of promoter targets, it is tempting to speculate that EBNA-2, itself a transcription factor, may indirectly alter the differentiation state of infected lymphocytes by influencing the array of available cellular transcription factors.

The mechanism of EBNA-2 trans-activation may be similar to that of the adenovirus E1A and herpes simplex VP16 proteins, which do not bind DNA but activate transcription indirectly through interactions with cellular transcription factors.[89,90] Replacement of the EBNA-2 activation domain with the VP16 transactivation domain results in a chimeric protein that retains EBNA-2's transactivation and immortalization abilities.[91,92] Although the EBNA-2 protein apparently does not bind directly to DNA, it can activate transcription when fused to the DNA-binding domain of the Gal4 protein[93] and is part of protein–DNA complexes that activate enhancer-like elements near the Cp and TP-1 promoters.[94,95] An 11-bp motif, CGTGGGAAAT, lies within the EBNA-2-responsive regions of the LMP-2A, Cp, and LMP-1 promoters.[84,96,97] The EBNA-2 protein apparently can be tethered to this core consensus sequence by direct protein–protein interaction with the cellular Cp-binding factor (CBF-1) or Jκ, which mediates binding of EBNA-2 to LMP-1.[97a] These proteins are expressed in both lymphoid and epithelial cells.[95,98]

In speculation about how to account for the reported lymphoid specificity of EBNA-2 trans-activation, the p65 component of NFκB also comes to mind. Sequestered in the cytoplasm by IκB in most cell types, in B-lymphocytes p65 is released and available as a transcriptional activator.[99] Consistent with this association is the observation that EBNA-2 activation of the human immunodeficiency virus (HIV) long terminal repeat (LTR) requires κB binding sites and that EBNA-2 induces an increase in κB binding activity in transfected nuclear extracts.[100] Such an increase could be explained by either EBNA-2 up-regulation of p65 expression, an EBNA-2-induced modification of the p65 protein, or a direct interaction between EBNA-2 and p65, resulting in more efficient binding. One difficulty with this hypothesis, however, is that the activation of the HIV LTR by EBNA-2 was possible in epithelial as well as lymphoid cells. Although EBNA-2 may induce κB expression or contribute to its activation efficiency even in epithelial cells, it is still not sufficient to activate Cp in epithelial cells.[96] There is evidence that NKκB kinds to Cp in the EBNA-2 enhancer region of the promoter.[100a]

Clearly, the EBNA-2 protein, when expressed, exerts a powerful effect on the cell and may indeed divert cellular proteins such as CBF-1 and NFκB for viral use. But because EBNA-2 is not expressed in all EBV-infected tissues, it is just as clear that through intricate regulation, its expression is reserved for a specific role in the viral life cycle.

3.2. Latent Membrane Protein 1

Although not sufficient for transformation of human B lymphocytes or primary rodent cells, the EBV LMP-1 reduces the serum requirements for growth of Rat-1 and NIH3T3 cells, alters their morphology, and allows anchorage-independent growth.[101,102] These LMP-expressing Rat-1 cells are tumorigenic in nude mice.[102] Likewise, gene transfer of LMP-1 into EBV-negative BL cells results in cell clumping and increased expression of cell-adhesion molecules LFA-1, ICAM-1, and LFA-3.[83] When transferred into a squamous-cell carcinoma line, LMP-1 inhibits epithelial cell differentiation.[103] Perhaps the most compelling evidence for the necessity of LMP-1 for transformation is that EBV recombinants carrying a nonsense linker mutation within the LMP gene are unable to transform B lymphocytes,[104] although it has proven challenging to identify a discrete LMP-1 protein domain responsible for the transforming phenotype.[105]

The mRNA encoding LMP-1 contains three exons and is transcribed from the U5 portion of the EBV genome, near the right terminus.[106] In epithelial cells, LMP may be expressed from two different promoters, as two different-sized but 3' coterminal messages are detected.[107] The LMP protein has three domains: a hydrophilic amino terminus, six hydrophobic transmembrane segments, and an acidic carboxy terminus, all of which are required for transformation of Rat-1 cells.[108] An integral membrane protein, LMP-1 forms patches in the plasma membrane[109] and, by association with vimentin, is linked to the cytoskeleton.[110] This ability of LMP-1 to form membrane patches and to associate with the cytoskeleton may serve to lengthen its half-life, allowing it to act as an ion channel or as a growth-factor receptor. The recent observation that cells stably transfected with an

LMP-expressing vector have a reduced requirement for epidermal growth factor (EGF) suggests a possible interaction between LMP and the EGF receptor.[111] Another intriguing effect is LMP's ability to up-regulate expression of the *bcl*-2 protooncogene.[112] Recently LMP-1 has been shown to produce malignant changes in transgenic mice (N. Raab-Traub, personal communication).

The possibility that the EBNA-2 and LMP-1 proteins may be involved in initiation rather than maintenance of the immortalized state is inferred from the observation that they are only rarely detected in cell lines established from freshly explanted BL biopsies,[113–116] and EBNA-2 is not detected in NPC. In support of this notion is the monoclonality of the EBV genome in NPC, indicating that in these tumors EBV infection occurred prior to the outgrowth of a malignant cell clone. Once cell transformation has occurred, the EBNA-2 and LMP-1 proteins may not only be unnecessary for cell proliferation, but cells still expressing them might be preferred targets for the cytotoxic T-cell response, placing them at a growth disadvantage.[117–119]

3.3. The EBERs

The most abundant EBV transcripts in latently infected cells are two small nonpolyadenylated RNA polymerase III transcripts, EBER-1 and EBER-2 (EB-encoded RNA).[120–122] Although the precise role of the EBERs in EBV latency and cell transformation remains unclear, they exist as stable intramolecularly base-paired structures within ribonucleoprotein (RNP) complexes, which for EBER-1 includes the ribosomal L22 protein, also known as EAP (EBER-associated protein).[123–125] A possible role for EBER-1, therefore, is to sequester this protein in the nucleoplasm, resulting in a depletion of L22 from ribosome assembly during some stages of EBV infection. The resulting distortion in ribosome composition may predispose these cells to transformation. Indeed, a translocation disrupting the L22 gene has been discovered in several leukemias.[126] Because the EBERs are expressed and amplified in latently infected immortalized cells, they are a sensitive indicator of neoplastic cells and have been used for molecular diagnostics of EBV lymphomas.[127]

4. CYTOLYTIC CYCLE

Although latent EBV infection is characterized by expression of only a few genes and dependence on the host-cell replicative machinery, during lytic infection a cascade of gene activation results in the expression of many viral proteins governing the replication of viral DNA and synthesis and assembly of virions. Induction to lytic infection results in the *trans*-activation of a number of "early" promoters by the "immediate-early" gene products, Z and R. The early genes provide functions necessary for viral replication, encoding a DNA polymerase, a polymerase processivity factor, a single-stranded DNA-binding protein, and possibly primase and helicase. Accompanying the activation of the early genes are down-regulation of the latent promoters,[128] the onset of viral DNA replication, and ultimately cell lysis.

Although involvement in oncogenesis has primarily been ascribed to the EBV latent cycle gene expression consistently detected in tumor tissue, the possibility of a role for lytic cycle gene products, particularly the Z *trans*-activator, cannot be discounted. Elevated antibody titers directed against viral lytic antigens are a reliable indicator for the eventual onset of NPC. The Z *trans*-activator, though not actively expressed in BL tumor biopsies, was recently detected in a fraction of NPC cells,[129] and all these patients are seropositive for Z.

The Z protein triggers reactivation from latency by *trans*-activating an array of viral promoters containing AP-1-like Z-response elements (ZREs)[130–138] and down-regulating several EBV latency promoters.[128,139] It is very likely, however, that the timing of this reactivation is fine-tuned by cellular factors that modulate Z function. For example, Z-induced activation of EBV early promoters is inhibited by interaction between Z and the cellular p53 protein, the p65 component of NFκB, as well as members of the retinoic acid receptor family.[139–141] Because Z also inhibits p53-dependent *trans*-activation in lymphoid cells, it is possible that overexpression of Z during a period of transient viral reactivation may promote tumorigenesis by inactivating p53.[141,142] Such a mechanism is particularly conceivable in the development of EBV-associated lymphomas in immunosuppressed persons.

Because elevated antibody titers against EBV replicative antigens are prognostic for NPC development, it may be informative to search for environmental agents in the endemic areas that might intermittently reactivate EBV from latency. A very large number of extracts from plants indigenous to China have already been screened for EA induction, and positive results were obtained with 52 of 1693 species. Forty of these are distributed in the NPC endemic areas of Guangxi and Guangdong provinces in southern China,[143] and several were also shown to enhance EBV transformation of B-cells.[144]

5. EPSTEIN–BARR VIRUS INHIBITION OF APOPTOSIS

Another mechanism for the transformation of cells by EBV is abrogation of apoptosis, or programmed cell death, presumably by deregulating cell growth control. Like cell-growth transformation, apoptosis is a multistep process that can utilize multiple molecular pathways, and EBV infection may inhibit apoptosis at a number of junctures. Indeed, B cells expressing the full complement of EBV latent gene products, including all of the EBNAs and LMP-1 and -2, are resistant to induction of apoptosis.[145]

A possible mechanism for this effect is the LMP-1-mediated up-regulation of the *bcl*-2 gene, which controls cell death in mammalian cells.[112,146–152] Another viral protein, BHRF-1, expressed early in the viral lytic cycle, is loosely homologous to *bcl*-2[153] and, when expressed, can also protect cells from apoptosis.[154] Although the BHRF1 protein is not required for virus replication or transformation of lymphocytes,[155,156] its expression during reactivation of latent EBV in epithelial cells may yet contribute to the transformation of these cells, possibly leading to NPC.

Adding to these *bcl*-2-enhancing and mimicking effects is the possible Z-induced inhibition of p53 function.[141] The presence of Z in infected cells normally induces the cascade of lytic gene expression, resulting in cell lysis. If the newly expressed Z protein were tied up by p53, however, a scenario could be envisioned in which viral replication and cytolysis are inhibited, and p53 is unable to enhance apoptosis.[157,158] The result would presumably be a cell that has become more predisposed to transformation.

In seeming contrast to the idea of EBV as an inhibitor of apoptosis is the intriguing observation that EBV induces fragmentation of chromosomal DNA at the onset of lytic infection[159]; this observation was anticipated by earlier results.[160] Because this fragmentation failed to generate the nucleosome-length DNA characteristic of apoptosis, it is tempting to speculate that intermittent, abortive reactivation of EBV might create a subcellular environment that favors chromosomal translocation or deletions. In cells that are not subsequently lysed, such genetic instability could clearly contribute to cell transformation.

6. CONCLUSION

Although EBV does not have conventional oncogenes, the two key EBV genes involved in cell immortalization and transformation, EBNA-2 and LMP-1, resemble them in their actions. The effects of EBNA-2 parallel those of a nuclear oncogene in that this EBV protein, itself a transcription factor, activates the promoters of key viral and cellular genes, some of which are involved in cellular proliferation. LMP-1 resembles the class of oncogenes active at the cell membrane surface. Neither gene alone is sufficient to immortalize or transform cells. Although intact EBV readily immortalizes cells, it has not been possible to recapitulate this effect in cell culture by transfection of single or multiple EBV genes into EBV-negative cells. This suggests that an unknown factor participates in cell immortalization: either another EBV gene or a cellular oncogene or growth suppressor gene, possibly in combination with critical kinetics of cell growth. Nor has it been possible to duplicate any of the characteristic chromosomal translocations of BL by infection and immortalization of B-lymphocytes in cell culture.[161] Thus, there are lacunae in our understanding of the molecular pathogenesis of cell immmortalization by EBV and the secondary genetic events that culminate in outright transformation.

An additional EBV gene product, not itself an immortalizing gene, is critically involved in cell immortalization by the virus. That gene product is EBNA-1, which is necessary both to initiate and to maintain replication of EBV episomes in latently infected immortalized cells. Unless EBNA-1 is expressed, latent infection cannot be established, and cell immortalization can not take place. Once the episomes are present in stable copy numbers, then the latent gene products involved in cellular immortalization can be expressed. There is a potential practical offshoot of this basic EBV mechanism. In recent studies with phosphorothioated deoxynucleotide antisense oligomers to EBNA-1 mRNA, it has been possible to check the growth of EBV-immortalized lymphocyte lines by treatment over a 7-day period. Cells in the exponential growth phase treated with such antisense oligomers often stop growing entirely, presumably as the EBNA-1 protein is depleted, and episomes can therefore no longer replicate during successive cell cycles. This result has not been proven, however, because of the small number of surviving cells in the successful experiments.[162] Nevertheless, the use of antisense oligomers to EBNA-1 offers a promising conceptual approach for the treatment of both latent infection and the immortalized cell state produced by EBV (Fig. 3). Such an approach, however, would not work in cell lines harboring secondary genetic alterations such as translocations that activate c-*myc* expression.

Finally, because there is circumstantial evidence that viral reactivation may trigger the onset of NPC, the question of the utility of prophylactic antiviral therapy with an acyclovir prodrug should be considered, especially in regions endemic for NPC.[163] In addition, because retinoids are useful in prevention of other tumors of the head and neck, and retinoic acid modulates the effect of the EBV Z protein, which triggers viral reactivation,[139] a combination of these agents might be considered, guided by the appearance of IgA antibodies to EBV antigens in persons at high risk of NPC. NPC is believed to progress through EBV-positive hypercellular and dysplastic phases, which, although brief, precede the appearance of frank NPC lesions, providing the possible rationale for antiviral prophylaxis and chemoprevention.[163a] Thus, EBV presents rich opportunities for molecular pathogenetic analysis that illuminates oncogenic pathways and at the same time opportunities for early detection, diagnosis, and treatment of this and other EBV-induced malignancies.

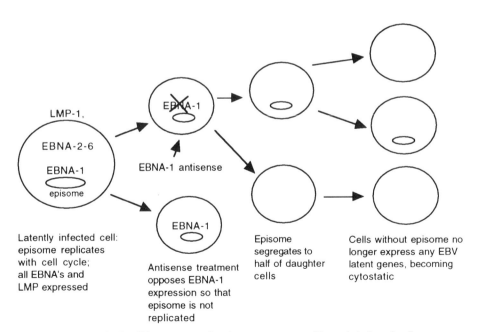

FIGURE 3. Possible outcome of antisense treatment of latently infected cells.

ACKNOWLEDGMENTS. We are indebted to Julie Wilson for help with the manuscript and to Gloria Faulkner for typing it. A portion of the work described here was supported by U.S. Public Service Grant 2-P01-CA19014-17A1 from the National Institutes of Health.

REFERENCES

1. Henderson, E., Miller, G., Robinson, J., and Heston, L., 1977, Efficiency of transformation of lymphocytes by Epstein–Barr virus, *Virology* **76:**152–163.
2. Henle, W., Diehl, B., Kohn, G., zur Hausen, H., and Henle, G., 1967, Herpes-type virus and chromosome marker in normal leukocytes after growth with irradiated Burkitt cells, *Science* **157:**1064–1065.
3. Dalla-Favera, R., Bregni, M., Erikson, J., Patterson, D., Gallo, R. C., and Croce, C., 1982, Human c-*myc* oncogene is located on the region of chromosome 8 that is translocated in Burkitt lymphoma cells, *Proc. Natl. Acad. Sci. USA* **79:**7824–7827.
4. Magrath, I., 1990, The pathogenesis of Burkitt's lymphoma, *Adv. Cancer Res.* **55:**133–270.
5. Barriga, F., Kiwanuka, J., Alvarez-Mon, M., Shiramizu, B., Huber, B., Levine, P., and Magrath, I., 1988, Significance of chromosome 8 breakpoint location in Burkitt's lymphoma: Correlation with geographical origin and association with Epstein–Barr virus. *Curr. Top. Microbiol. Immunol.* **141:**128–137.
6. Su, I. J., and Hsieh, H. C., 1992, Clinicopathological spectrum of Epstein–Barr virus associated T-cell malignancies, *Leuk. Lymphoma* **7:**47–53
7. Brichacek, B., Hirsch, I., Sibl, O., Vilikusova, E. and Vonka, V., 1983, Association of some supraglottic laryngeal carcinomas with EB virus, *Int. J. Cancer* **32:**193–197.
8. Brichacek, B., Hirsch, I., Sibl, O., Vilikusova, E. and Vonka, V., 1984, Presence of Epstein–Barr virus DNA in carcinomas of the palatine tonsil, *J. Natl. Cancer Inst.* **72:**809–815.
9. Lanier, A., Clift, S. R., Bornkamm, G., Henle, W., Goepfert, H., and Raab-Traub, N., 1991, Epstein–Barr virus and malignant lymphoepithelial lesions of the salivary gland, *Arctic Med. Res.* **50:**55–61.
10. Raab-Traub, N., Rajadurai, P., Flynn, K., and Lanier, A. P., 1991, Epstein–Barr virus infection in carcinoma of the salivary gland, *J. Virol.* **65:**7032–7036.
11. de Thé, G., 1982, Epidemiology of Epstein–Barr virus and associated diseases in: *The Herpesviruses*, Vol. 1 (B. Roizman, ed.), Plenum Press, New York, pp. 25–87.
12. Ho, J. H. C., 1990, Epidemiology of nasopharyngeal carcinoma (NPC), in: *Epstein–Barr Virus and Human Disease* (D. V. Ablashi, A. T. Huang, J. S. Pagano, G. R. Pearson, and C. S. Young, eds.), Humana Press, Clifton, NJ, pp. xli–xliv.
13. Old, L. J., Boyse, E. A., Oettgen, H. F., De Harven, E., Geering, G., Clifford, P., and Williamson, B., 1966, Precipitating antibody in human serum to an antigen present in cultured Burkitt's lymphoma cells, *Proc. Natl. Acad. Sci. USA* **56:**1699–1704.
14. Nonoyama, M., Huang, C. H., Pagano, J. S., Klein, G., and Singh, S., 1973, DNA of Epstein–Barr virus detected in tissue of Burkitt's lymphoma and nasopharyngeal carcinoma, *Proc. Natl. Acad. Sci. USA* **70:**3265–3268.
15. zur Hausen, H., Schulte-Holthauzen, H., Klein, G., Henle, G., Henle, W., Clifford, P., and Santesson, L., 1970, EBV DAN in biopsies of Burkitt tumours and anaplastic carcinomas of the nasopharynx, *Nature* **228:**1056–1058.
16. Klein, G., Giovanella, B. C., Lindahl, T., Fialkow, P. J., Singh, S., and Stehlin, J. S., 1974, Direct evidence for the presence of Epstein–Barr virus DNA and nuclear antigen in malignant epithelial cells from patients with poorly differentiated carcinoma of the nasopharynx, *Proc. Natl. Acad. Sci. USA* **71:**4737–4741.
17. Desgranges, C., Wolf, H., de Thé, G., Shanmugaratnam, K., Ellouz, R., Cammoun, N., Klein, G., and zur Hausen, H., 1975, Nasopharyngeal carcinoma X. Presence of Epstein–Barr virus genomes in epithelial cells of tumors from high and medium risk areas, *Int. J. Cancer* **16:**7–15.

18. Raab-Traub, N., Flynn, K., Pearson, G., Huang, A., Levine, P., Lanier, A., and Pagano, J., 1987, The differentiated form of nasopharyngeal carcinoma contains Epstein–Barr virus DNA, *Int. J. Cancer* **39**:25–29.
19. Zeng, Y., 1985, Seroepidemiologic studies on nasopharyngeal carcinoma in China, *Adv. Cancer Res.* **44**:121–138.
20. Henle, G., and Henle, W., 1976, Epstein–Barr virus-specific IgA serum antibodies as an outstanding feature of nasopharyngeal carcinoma, *Int. J. Cancer* **17**:1–7.
21. Dyson, J. P., and Farrell, P. J., 1985, Chromatin structure of Epstein–Barr virus. *J. Gen. Virol.* **66**:1931–1940.
21a. Shaw, J. E., Levinger, L. F., and Carter, C. W., Jr., 1979, Nucleosomal structure of Epstein–Barr virus DNA in transformed cell lines, *J. Virol.* **29**:657–665.
22. Given, D., Yee, D., Griem, K., and Kieff, E., 1979, DNA of Epstein–Barr Virus. V. Direct repeats of the ends of Epstein–Barr virus DNA, *J. Virol.* **30**:852–862.
23. Hayward, S. D., and Kieff, E., 1977, DNA of Epstein–Barr virus. II. Comparison of the molecular weights of restriction endonuclease fragments of the DNA of Epstein–Barr virus strains and identification of end fragments of the B95-8 strain, *J. Virol.* **23**:421–429.
24. Kintner, C., and Sugden, B., 1979, The structure of the termini of the DNA of Epstein–Barr virus, *Cell* **17**:661–671.
25. Nonoyama, M., and Pagano, J. S., 1972, Separation of Epstein–Barr virus DNA from large chromosomal DNA in nonvirus producing cells, *Nature* [*New Biol.*] **238**:169–171.
26. Raab-Traub, N., and Flynn, K., 1986, The structure of the termini of the Epstein–Barr virus as a marker of clonal cellular proliferation, *Cell* **47**:883–889.
26a. Raab-Traub, N., 1992, Epstein–Barr virus and nasopharyngeal carcinoma, *Semin. Cancer Biol.* **3**:297–307.
26b. Pagano, J. S., 1992, Epstein–Barr virus: Culprit or consort, *N. Engl. J. Med.* **327**:1750–1752.
27. Andersson-Anvret, M., and Lindahl, T., 1978, Integrated viral DNA sequences in Epstein–Barr virus-converted human lymphoma lines, *J. Virol.* **25**:710–718.
28. Ernberg, I., Andersson-Anvret, M., Klein, G., Lundin, L., and Killanger, D., 1977, Relationship between amount of Epstein–Barr virus-determined nuclear antigen per cell and number of EBV-DNA copies per cell, *Nature* **266**:269–271.
29. Heller, M., Dambaugh, T., and Kieff, E., 1981, Epstein–Barr virus DNA. IX. Variation among viral DNAs from producer and nonproducer infected cells, *J. Virol.* **38**:632–648.
30. Kaschka-Dierich, C., Adams, A., Lindahl, T., Bornkamm, G. W., Bjunsell, G., and Klein G., 1976, Intracellular forms of Epstein–Barr virus DNA in human tumor cells *in vivo*, *Nature* **260**:302–306.
31. Sugden, B., Phelps, M., and Domoradzki, J., 1979, Epstein–Barr virus DNA is amplified in transformed lymphocytes, *J. Virol.* **63**:590–595.
32. Adams, A., 1987, Replication of latent Epstein–Barr virus genomes in Raji cells, *J. Virol.* **61**:1743–1746.
33. Hampar, B., Tanaka, A., Nonoyama, M., and Derge, J. G., 1974, Replication of the resident repressed Epstein–Barr virus genome during the early S-phase (S-1 period) of nonproducer Raji cells, *Proc. Natl. Acad. Sci. USA* **71**:631–633.
34. Colby, B., Shaw, J., Datta, A., and Pagano, J. S., 1982, Replication of Epstein–Barr virus DNA in lymphoblastoid cells treated for extended periods with acyclovir, *Am. J. Med. Acyclovir Symp.* **73**:77–81.
35. Rawlins, D. R., Milman, G., Hayward, S. D., and Hayward, G. S., 1985, Sequence-specific DNA binding of the Epstein–Barr virus nuclear antigen (EBNA-1) to clustered sites in the plasmid maintenance region, *Cell* **42**:859–868.
36. Sugden, B., Marsh, K., and Yates, J., 1985, A vector that replicates as a plasmid and can be efficiently selected in B-lymphoblasts transformed by Epstein–Barr virus, *Mol. Cell. Biol.* **5**:410–413.
37. Kerr, B. M., Lear, A. L., Rowe, M., Croom-Carter, D., Young, L. S., Rookes, S. M., Gallimore, P. H., and Rickinson, A. B., 1992, Three transcriptionally distinct forms of Epstein–Barr virus latency in somatic cells hybrids: Cell phenotype dependence of promoter usage, *Virology* **187**:189–201.

38. Rowe, M., Lear, A. L., Croom-Carter, D., Davies, A. H., and Rickinson, A. B., 1992, Three pathways of Epstein–Barr virus gene activation from EBNA-1 positive latency in B lymphocytes, *J. Virol.* **66**:122–131.
39. Sample, J., Brooks, L., Sample, C., Young, L., Rowe, M., Gregory, C., Rickinson, A., and Kieff, E., 1991, Restricted Epstein–Barr virus protein expression in Burkitt lymphoma is due to a different Epstein–Barr nuclear antigen 1 transcriptional initiation site, *Proc. Natl. Acad. Sci. USA* **88**:6343–6347.
40. Smith, P. R., and Griffin, B. E., 1992, Transcription of the Epstein–Barr virus gene EBNA-1 from different promoters in nasopharyngeal carcinoma and B-lymphoblastoid cells, *J. Virol.* **66**:706–714.
41. Ambinder, R. F., Shah, W. A., Rawlins, D. R., Hayward, G. S., and Hayward, S. D., 1990, Definition of the sequence requirements for binding of the EBNA-1 protein to its palindromic target sites in Epstein–Barr virus DNA, *J. Virol.* **64**:2369–2379.
42. Sample, J., Henson, E. B. D., and Sample, C., 1992, The Epstein–Barr virus nuclear protein 1 promoter active in type 1 latency is autoregulated, *J. Virol.* **66**:4654–4661.
43. Sung, N., Wilson, J., and Pagano, J. S., 1993, Characterization of *cis*-acting regulatory elements of the *Bam*HI-*F* promoter of EBV, in: *Epstein–Barr Virus and Associated Diseases* (T. Tursz, J. S. Pagano, D. V. Ablashi, G. de The, G. Lenoir, and G. R. Pearson, eds.), INSERM/John Libbey, Montrouge, France, pp. 239–242.
44. Sung, N., Wilson, J., Davenport, M., Sista, N. D., and Pagano, J. S., 1994, Reciprocal regulation of the Epstein–Barr virus *Bam*HI-*F* promoter by EBNA-1 and an E2F transcription factor, *Mol. Cell. Biol.* **14**:7144–7152.
45. Kovesdi, I., Reichel, R., and Nevins, J. R., 1986, Identification of a cellular transcription factor involved in E1A transactivation, *Cell* **45**:219–228.
46. Kovesdi, I., Reichel, R., and Nevins, J. R., 1987, Role of an adenovirus E2 promoter binding factor in E1A mediated coordinate gene control, *Proc. Natl. Acad. Sci. USA* **84**:2180–2184.
47. Blake, M., and Azizkhan, J., 1989, Transcription factor E2F is required for efficient expression of the hamster dihydrofolate reductase gene *in vitro* and *in vivo*, *Mol. Cell. Biol.* **9**:4994–5002.
48. Dou, Q.-P., Fridovich-Keil, J. L., and Pardee, A. B., 1991, Inducible proteins binding to the murine thymidine kinase promoter in late G_1/S phase, *Proc. Natl. Acad. Sci. USA* **88**:1157–1161.
49. Hiebert, S. W., Blake, M., Azizkhan, J., and Nevins, J. R., 1991, Role of E2F transcription factor in E1A-mediated *trans*-activation of cellular genes, *J. Virol.* **65**:3547–3552.
50. Pearson, B. E., Nasheuer, H. P., and Wang, T. S., 1991, Human DNA polymerase a gene: Sequences controlling expression in cycling and serum-stimulated cells, *Mol. Cell. Biol.* **11**:2081–2095.
51. Dalton, S., 1992, Cell cycle regulation of the human *cdc*2 gene, *EMBO J.* **11**:1797–1804.
52. Hiebert, S. W., Lipp, M., and Nevins, J. R., 1989, E1A dependent *trans*-activation of the human MYC promoter is mediated by the E2F factor, *Proc. Natl. Acad. Sci. USA* **86**:3594–3598.
53. Mudryj, M., Devoto, S., Hiebert, S., Hunter, T., Pines, J., and Nevins, J., 1991, Cell cycle regulation of the E2F transcription factor involves an interaction with cyclin A, *Cell* **65**:1243–1253.
54. Nevins, J., 1992, E2F: A link between the Rb tumor supressor protein and viral oncoproteins, *Science* **258**:424–429.
55. Bagchi, S., Weinmann, R., and Raychaudhuri, P., 1991, The retinoblastoma protein copurifies with E2F-1, and E1A-regulated inhibitor of the transcription factor E2F, *Cell* **65**:1063–1072.
56. Bandara, L. R., Adamczewski, J. P., Hunt, T., and LaThangue, N. B., 1991, Cyclin A and the retinoblastoma gene product complex with a common transcription factor, *Nature* **352**:249–251.
57. Chellappan, S. P., Hiebert, S., Mudryj, M., Horowitz, J., and Nevins, J. R., 1991, The E2F transcription factor is a cellular target for the RB protein, *Cell* **65**:1053–1061.
58. Chittenden, T., Livingston, D. M., and Kaelin, W. G., Jr., 1991, The T/E1A-binding domain of the retinoblastoma product can interact selectively with a sequence-specific DNA-binding protein, *Cell* **65**:1073–1082.

59. Chittenden, T., Livingston, D. M., and DeCaprio, J. A., 1993, Cell cycle analysis of E2F in primary human T-cells reveals novel E2F complexes and biochemically distinct forms of free E2F, *Mol. Cell. Biol.* **13**:3975–3983.
60. Flemington, E. K., Speck, S. H., and Kaelin, W. G., Jr., 1993, E2F-1-mediated transactivation is inhibited by complex formation with the retinoblastoma susceptibility gene product. *Proc. Natl. Acad. Sci. USA* **90**:6914–6918.
61. Helin, K., Harlow, E., and Fattaey, A., 1993, Inhibition of E2F-1 transactivation by direct binding of the retinoblastoma protein, *Mol. Cell. Biol.* **16**:6501–6508.
62. Hiebert, S. W., Chellappan, S. P., Horowitz, J. M., and Nevins, J. R., 1992, The interaction of pRB with E2F inhibits the transcriptional activity of E2F, *Genes Dev.* **6**:177–185.
63. Bagchi, S., Raychaudhuri, P., and Nevins, J., 1990, Adenovirus E1A proteins can dissociate heteromeric complexes involving the E2F transcription factor: A novel mechanism for E1A transactivation, *Cell* **62**:659–669.
64. Bornkamm, G. W., Hudewentz, J., Freese, U. K., and Zimber, U., 1982, Deletion of the nontransforming Epstein–Barr virus strain P3HR1 causes fusion of the large internal repeat to the DSL region, *J. Virol.* **43**:952–968.
65. Jeang, K. T., and Hayward, S. D., 1983, Organization of the Epstein–Barr virus DNA molecule. III. Location of the P3HR1 deletion junction and characterization of the Not 1 repeat units that form part of the template for an abundant 12-O-tetradecanoylphorbol-13-acetate-induced mRNA transcript, *J. Virol.* **48**:135–148.
66. Menenzes, J., Liebold, W., and Klein, G., 1975, Biological differences between Epstein–Barr virus (EBV) strains with regard to lymphocyte transforming ability, superinfection and antigen induction, *Exp. Cell Res.* **92**:478–484.
67. Miller, G., Robinson, J., Heston, L., and Lipman, M., 1974, Differences between laboratory strains of Epstein–Barr virus based on immortalization, abortive infection, and interference, *Proc. Natl. Acad. Sci. USA* **71**:4006–4010.
68. Rabson, M., Gradoville, L., Heston, L., and Miller, G., 1982, Non-immortalizing P3J-HR1 Epstein–Barr virus: A deletion mutant of its transforming parent, Jijoye, *J. Virol.* **44**:834–844.
69. Cohen, J. I., Wang, F., Mannick, J., and Kieff, E., 1989, Epstein–Barr virus nuclear protein 2 is a key determinant of lymphocyte transformation, *Proc. Natl. Acad. Sci. USA* **86**:9558–9562.
70. Hammerschmidt, W., and Sugden, B., 1989, Genetic analysis of immortalizing functions of Epstein–Barr virus in human B-lymphocytes, *Nature* **340**:393–397.
71. Dambaugh, T., Wang, F., Hennessy, K., Woodland, E., Rickinson, A., and Kieff, E., 1986, Expression of the Epstein–Barr virus nuclear protein 2 in rodent cells, *J. Virol.* **59**:453–462.
72. Cohen, J. I., Picchio, G. R., and Mosier, D. E., 1992, Epstein–Barr virus nuclear protein 2 is a critical determinant for tumor growth in SCID mice and for transformation *in vitro*, *J. Virol.* **66**:7555–7559.
73. Alfieri, C., Birkenbach, M., and Kieff, E., 1991, Early events in Epstein–Barr virus infection of human B lymphocytes, *Virology* **181**:595–608. Erratum *Virology* **185**:946.
74. Hennessy, K., and Kieff, E., 1985, A second nuclear protein is encoded by Epstein–Barr virus in latent infection, *Science* **227**:1238–1240.
75. Dambaugh, T., Hennessey, K., Chamnankit, L., and Kieff, E., 1984, U2 region of Epstein–Barr virus DNA may encode Epstein–Barr virus nuclear antigen 2, *Proc. Natl. Acad. Sci. USA* **81**:7632–7636.
76. Mueller-Lantzsch, N., Lenoir, G.M., Sauter, M., Takaki, K., Bechet, J.M., Kuklic-Roox, C., Wunderlich, D., and Bornkamm, G.W., 1985, Identification of the coding region for a second Epstein–Barr virus nuclear antigen (EBNA-2) by transfection of cloned DNA fragments, *EMBO J.* **4**:1805–1811.
77. Rymo, L., Klein, G., and Ricksten, A., 1985, Expression of a second Epstein–Barr virus determined nuclear antigen in mouse cells after gene transfer with a cloned fragment of the viral genome, *Proc. Natl. Acad. Sci. USA* **82**:3435–3439.
78. Rickinson, A. B., Young, L. S., and Rowe, M., 1987, Influence of the Epstein–Barr virus nuclear antigen EBNA-2 on the growth phenotype of virus-transformed B cells, *J. Virol.* **61**:1310–1317.

79. Abdel-Hamid, M., Chen, J., Constantine, N., Massoud, M., and Raab-Traub, N., 1992, EBV strain variation: Geographical distribution and relation to disease state, *Virology* **190:**168–175.
80. Abbot, S. D., Rowe, M., Cadwallader, K., Ricksten, A., Gordon, J., Wang, F., Rymo, L., and Rickinson, A. B., 1990, Epstein–Barr virus nuclear antigen 2 induces expression of the virus-encoded latent membrane protein, *J. Virol.* **64:**2126–2134.
81. Fahraeus, R., Jansson, A., Ricksten, A., Sjoblom, A., and Rymo, L., 1990, Epstein–Barr virus-encoded nuclear antigen 2 activates the viral latent membrane protein promoter by modulating the activity of a negative regulatory element, *Proc. Natl. Acad. Sci. USA* **87:**7390–7394.
82. Ghosh, D., and Kieff, E., 1990, *cis*-Acting regulatory elements near the Epstein–Barr virus latent-infection membrane protein transcriptional start site, *J. Virol.* **64:**1855–1858.
83. Wang, F., Tsang, S-F., Kurilla, M. G., Cohen, J. I., and Kieff, E., 1990, Epstein–Barr virus nuclear antigen 2 transactivates latent membrane protein LMP1, *J. Virol.* **64:**3407–3416.
84. Zimber-Strobl, U., Suentzenich, K. O., Laux, G., Eick, D., Cordier, M., Calender, A., Billaud, M., Lenoir, G. M., and Bornkamm, G. W., 1991, Epstein–Barr virus nuclear antigen 2 activates transcription of the terminal protein gene, *J. Virol.* **65:**415–423.
85. Birkenbach, M., Liebowitz, D., Wang, F., Sample, J., and Kieff, E., 1989. Epstein–Barr virus latent infection membrane protein increases vimentin expression in human B-cell lines, *J. Virol.* **63:**4079–4084.
86. Calender, A., Billaud, M., Aubry, J. P., Banchereau, J., Vuillame, M., and Lenoir, G. M., 1987, Epstein–Barr virus (EBV) induces expression of B-cell activation markers on *in vitro* infection of EBV-negative B-lymphoma cells, *Proc. Natl. Acad. Sci. USA* **84:**8060–8064.
87. Knutson, J. C., 1990, The level of c-*fgr* RNA is increased by EBNA-2, an Epstein–Barr virus gene required for B-cell immortalization, *J. Virol.* **64:**2530–2536.
88. Wang, F., Gregory, C. D., Rowe, M., Rickinson, A. B., Wang, D., Birkenbach, M., Kikutani, H., Kishimoto, T., and Kieff, E., 1987, Epstein–Barr virus nuclear antigen 2 specifically induces expression of the B-cell activation antigen CD23, *Proc. Natl. Acad. Sci. USA* **84:**3452–3456.
89. Kristie, T. M., and Roizman, B., 1988, Differentiation and DNA contact points of host proteins binding at the *cis* site for virion-mediated induction of a genes of Herpes simplex virus 1, *J. Virol.* **62:**1145–1157.
90. Lillie, J. W., and Green, M. R., 1989, Transcription activation by the adenovirus E1A protein, *Nature* **338:**39–44.
91. Cohen, J. I., 1992, A region of herpes simplex virus VP16 can substitute for a transforming domain of Epstein–Barr virus nuclear protein 2, *Proc. Natl. Acad. Sci. USA* **89:**8030–8034.
92. Ling, P. D., Ryon, J. J., and Hayward, S. D., 1993, EBNA-2 of herpesvirus papio diverges significantly from the type A and type B EBNA-2 proteins of Epstein–Barr virus but retains an efficient transactivation domain with a conserved hydrophobic motif, *J. Virol.* **67:**2990–3003.
93. Cohen, J. I., Wang, F., and Kieff, E., 1991, Epstein–Barr virus nuclear protein 2 mutations define essential domains for transformation and transactivation, *J. Virol.* **65:**2545–2554.
94. Zimber-Strobl, U., Kremmer, E., Grasser, F., Marschall, G., Laux, G., and Bornkamm, G., 1993, The Epstein–Barr virus nuclear antigen 2 interacts with an EBNA-2 responsive *cis*-element of the terminal protein 1 gene promoter, *EMBO J.* **12:**167–175.
95. Ling, P., Rawlins, D. R., and Hayward, S. D., 1993, The Epstein–Barr virus immortalizing protein EBNA-2 is targeted to DNA by a cellular enhancer-binding protein. *Proc. Natl. Acad. Sci. USA* **90:**9237–9241.
96. Sung, N., Kenney, S., Gutsch, D., and Pagano, J. S., 1991, EBNA-2 transactivates a lymphoid specific enhancer in the *Bam*HI-C promoter of Epstein–Barr virus, *J. Virol.* **65:**2164–2169.
97. Tsang, S. F., Wang, F., Kenneth, M. I., and Kieff, E., 1991, Delineation of the *cis*-acting element mediating EBNA-2 transactivation of latent membrane protein expression, *J. Virol.* **65:**6765–6771.
97a. Grossman, S. R., Johannsen, E., Tong, X., Yalamanchili, R., and Kieff, E., 1994, The Epstein–Barr virus nuclear antigen 2 transactivator is directed to response elements by the Jk recombination signal binding protein, *Proc. Natl. Acad. Sci. USA* **91:**7568–7572.

98. Henkel, T., Ling, P. D., Hayward, S. D., and Peterson, M. G., 1994, Mediation of Epstein–Barr virus EBNA-2 transactivation by recombination signal-binding protein J_k, *Science* **265**:92–95.
99. Bauerle, P. A., and Baltimore, D., 1988, IkB: A specific inhibitor of the NFkB transcription factor, *Science* **242**:540–546.
100. Scala, G., Quinto, I., Ruocco, M. R., Mallardo, M., Ambrosino, C., Squitieri, B., Tassone, P., and Venuta, S., 1993, Epstein–Barr virus nuclear antigen 2 transactivates the long terminal repeat of human immunodeficiency virus type 1, *J. Virol.* **67**:2853–2861.
100a. Sista, N. D., Barry, C. M., Sung, N. S., and Pagano, J. S., 1994, Negative regulation of the Epstein–Barr Virus BCRF1 promoter by BZLF1 is mediated through DNA binding, in preparation.
101. Baichwal, V. R., and Sugden, B., 1988, Transformation of Balb/3T3 cells by the BNLF-1 gene of Epstein–Barr virus, *Oncogene* **2**:461–467.
102. Wang, D., Liebowitz, D., and Kieff, E., 1985, An EBV membrane protein expressed in immortalized lymphocytes transforms established rodent cells, *Cell* **43**:831–840.
103. Dawson, C. W., Rickinson, A. B., and Young, L. S., 1990, Epstein–Barr virus latent membrane protein inhibits human epithelial cell differentiation, *Nature*, **344**:777–780.
104. Kaye, K. M., Izumi, K. I., and Kieff, E., 1993, Epstein–Barr virus latent membrane protein 1 is essential for B-lymphocyte growth transformation, *Proc. Natl. Acad. Sci. USA* **90**:9150–9154.
105. Izumi, K. M., Kaye, M. K., and Kieff, E., 1994, Epstein–Barr virus recombinant molecular genetic analysis of the LMP-1 amino-terminal cytoplasmic domain reveals a probable structural role, with no component essential for primary B-lymphocyte growth transformation, *J. Virol.* **68**:4369–4376.
106. van Santen, V., Cheung, A., and Kieff, E., 1981, Epstein–Barr virus RNA VII: Size and direction of transcription of virus-specified cytoplasmic RNAs in a transformed cell line, *Proc. Natl. Acad. Sci. USA* **78**:1930–1934.
107. Gilligan, K., Sato, H., Rajadurai, P., Busson, P., Young, L., Rickinson, A., Tursz, T., and Raab-Traub, N., 1990, Novel transcription from the EBV terminal *Eco*RI fragment, DIJhet, in a nasopharyngeal carcinoma, *J. Virol.* **64**:4948–4956.
108. Moorthy, R. K., and Thorley-Lawson, D. A., 1993, All three domains of the Epstein–Barr virus-encoded latent membrane protein LMP-1 are required for transformation of Rat-1 fibroblasts, *J. Virol.* **67**:1638–1646.
109. Liebowitz, D., Wang, D., and Kieff, E., 1986, Orientation and patching of the latent membrane protein encoded by Epstein–Barr virus, *J. Virol.* **58**:233–237.
110. Liebowitz, D., Kopan, R., Fuchs, E., Sample, J., and Kieff, E., 1987, An Epstein–Barr virus transforming protein associates with vimentin in lymphocytes. *Mol. Cell. Biol.* **7**:2299–2308.
111. Miller, W., Earp, H., and Raab-Traub, N., 1995, The Epstein–Barr virus latent membrane protein 1 induces expression of the epidermal growth factor receptor, *J. Virol.* **69**:4390–4398.
112. Henderson, S., Rowe, M., Gregory, C., Croom-Carter, D., Wang, F., Longnecker, R., Kieff, E., and Rickinson, A., 1991, Induction of *bcl*-2 expression by Epstein–Barr virus latent membrane protein 1 protects infected B-cells from programmed cell death, *Cell* **65**:1107–1115.
113. Masucci, M. G., Torsteinsdottir, S., Columbani, J., Brautbar, C., Klein, E., and Klein, G., 1987, Down-regulation of class I HLA antigens and of the Epstein–Barr virus-encoded latent membrane protein in Burkitt lymphoma lines, *Proc. Natl. Acad. Sci. USA* **84**:4567–4571.
114. Modrow, S., and Wolf, H., 1986, Characterization of two related Epstein–Barr virus-encoded membrane proteins that are differentially expressed in Burkitt lymphoma and in *in-vitro*-transformed cell lines, *Proc. Natl. Acad. Sci. USA* **83**:5703–5707.
115. Rowe, D. T., Rowe, M., Evan, G. I., Wallace, L. E., Farrell, P. J., and Rickinson, A. B., 1986, Restricted expression of EBV latent genes and T-lymphocyte-detected membrane antigen in Burkitt's lymphoma cells, *EMBO J.* **5**:2599–2607.
116. Rowe, M., Rowe, D. T., Gregory, C. D., Young, L. S., Farrell, P. J., Rupani, H., and Rickinson, A. B., 1987, Differences in B cell growth phenotype reflect novel patterns of Epstein–Barr virus latent gene expression in Burkitt's lymphoma cells, *EMBO J.* **6**:2743–2751.

117. Gregory, C. D., Murray, R. J., Edwards, C. F., and Rickinson, A. B., 1988, Down-regulation of cell adhesion molecules LFA-3 and ICAM-1 in Epstein–Barr virus-positive Burkitt's lymphoma underlies tumour cell escape from virus-specific T-cell surveillance, *J. Exp. Med.* **167:**1811–1824.
118. Knutson, J. C., and Sugden, B., 1989, Immortalization of B lymphocytes by Epstein–Barr virus: What does the virus contribute to the cell? in: *Advances in Viral Oncology.* Vol. 8 (G. Klein, ed.), Raven Press, New York, pp. 151–172.
119. Moss, D. J., Misko, I. S., Burrows, S. R., Burman, K., MaCarthy, R., and Sculley, T. B., 1988, Cytotoxic T-cell clones discriminate between A- and B-type Epstein–Barr virus transformants, *Nature* **331:**719–721.
120. Arrand, J. R., and Rymo, L., 1982, Characterization of the major Epstein–Barr virus-specific RNA in Burkitt lymphoma-derived cells, *J. Virol.* **41:**376–389.
121. Howe, J. G., and Shu, M. D., 1989, Epstein–Barr virus small RNA (EBER) genes: Unique transcription units that combine RNA polymerase II and III promoter elements, *Cell* **57:** 825–834.
122. Jat, P., and Arrand, J. R., 1982, *In vitro* transcription of two Epstein-Barr virus specified small RNA molecules, *Nucleic Acids Res.* **10:**3407–3425.
123. Toczyski, D., and Steitz, J. A., 1991, EAP, a highly conserved cellular protein associated with Epstein–Barr virus small RNAs (EBERs), *EMBO J.* **10:**459–466.
124. Toczyski, D., and Steitz, J. A., 1993, The cellular RNA-binding protein EAP recognizes a conserved stem-loop in the Epstein–Barr virus small RNA EBER-1, *Mol. Cell. Biol.* **13:**703–710.
125. Toczyski, D. P., Matera, A. G., Ward, D. C., and Steitz, J. A., 1994. The Epstein–Barr virus (EBV) small RNA EBER 1 binds and relocalizes ribosomal protein L22 in EBV-infected human B-lymphocytes, *Proc. Natl. Acad. Sci. USA* **91:**3463–3467.
126. Nucifora, G., Begy, C. R., Erickson, P., Drabkin, H. A., and Rowley, J. D., 1993, The 3;21 translocation in myelodysplasia results in a fusion transcript between the AML1 gene and the gene for EAP, a highly conserved protein associated with the Epstein–Barr virus small RNA EBER 1. *Proc. Natl. Acad. Sci. USA* **90:**7784–7788.
127. Rhandawa, P. S., Jaffe, R., Demetris, A. J., Nalesnik, M., Starzl, T. E., Chen, Y. Y., and Weiss, L. M., 1992, Expression of Epstein–Barr virus-encoded small RNA (by the EBER-1 gene) in liver specimens from transplant recipients with post-transplantation lymphoproliferative disease, *N. Engl. J. Med.* **327:**1710–1714.
128. Kenney, S., Kamine, J., Holley-Guthrie, E., Lin, J-C., Mar, E-C., and Pagano, J., 1989, The Epstein–Barr virus (EBV) BZLF1 immediate early gene product differentially affects latent versus productive EBV promoters, *J. Virol.* **63:**1729–1736.
129. Cochet, C., Martel-Renoir, D., Grunewald, V., Bosq, J., Cochet, G., Schwaab, G., Bernaudin J. F., and Joab, I., 1993, Expression of the Epstein–Barr virus immediate-early gene, BZLF-1, in nasopharyngeal carcinoma tumor cells, *Virology* **197:**358–365.
130. Chang, Y. N., Dong, D. L. Y., Hayward, G. S., and Hayward, S. D., 1990, The Epstein–Barr virus *Zta* transactivator: A member of the b-*ZIP* family with unique DNA binding specificity and a dimerization domain that lacks the characteristic heptad leucine zipper motif, *J. Virol.* **64:**3358–3369.
131. Chavrier, P., Gruffat, H., Chevallier-Graco, A., Buisson, M., and Sergeant, A., 1989, The Epstein–Barr virus (EBV) early promoter DR contains a *cis*-acting element responsive to the EBV transactivator EB1 and an enhancer with constitutive and inducible activities, *J. Virol.* **63:**607–614.
132. Cox, M. A., Leahy, J., and Hardwick, J. M., 1990, An enhancer within the divergent promoter of Epstein–Barr virus responds synergistically to the R and Z transactivators, *J. Virol.* **64:**313–321.
133. Holley-Guthrie, E. A., Quinlivan, E. B., Mar, E. C., and Kenney, S., 1990, The Epstein–Barr virus (EBV) BMRF-1 promoter for early antigen (EA-D) is regulated by the EBV transactivators BRLF-1 and BZLF-1 in a cell-specific manner. *J. Virol.* **64:**3753–3759.
134. Lieberman, P. M., Hardwick, J. M., and Hayward, S. D., 1989, Responsiveness of the Epstein–

Barr virus Not1 repeat promoter to the Z transactivator is mediated in a cell-type-specific manner by two independent signal regions, *J. Virol.* **63**:3040–3050.
135. Lieberman, P. M., Hardwick, J. M., Sample, J., Hayward, G. S., and Hayward, S. D., 1990, The *Zta* transactivator involved in induction of lytic cycle gene expression in Epstein–Barr virus-infected lymphocytes binds to both AP-1 and ZRE sites in target promoter and enhancer regions, *J. Virol.* **64**:1143–1155.
136. Lieberman, P. M., and Berk, A. J., 1990, *In vitro* transcriptional activation, dimerization, and DNA binding specificity of the Epstein–Barr virus *Zta* protein, *J. Virol.* **64**:2560–2568.
137. Rooney, C., Howe, J. G., Speck, S. H., and Miller, G., 1989, Influences of Burkitt's lymphoma B-cells and primary B-cells on latent gene exopression by the nonimmortalizing P3J-HR1 strain of Epstein–Barr virus, *J. Virol.* **63**:1531–1539.
138. Rooney, C. M., Rowe, D. T., Ragot, T., and Farrell, P. J., 1989, The spliced BZLF-1 gene of Epstein–Barr virus (EBV) transactivates an early promoter and induces the virus lytic cycle, *J. Virol.* **63**:3109–3116.
139. Sista, N. D., Pagano, J. S., Liao, W., and Kenney, S., 1993, Retinoic acid is a negative regulator of the Epstein–Barr virus protein (BZLF-1) that mediates disruption of latent infection, *Proc. Natl. Acad. Sci. USA* **90**:3894–3898.
140. Gutsch, D., Holley-Guthrie, E., Zhang, Q., Stein, B., Blanar, M., Baldwin, A., and Kenney, S., 1994, The bZIP transactivator, BZLF-1, of EBV functionally and physically interacts with the p65 subunit of NFkB, *Mol. Cell. Biol.* **14**:1939–1948.
141. Zhang, Q., Gutsch, D., and Kenney, S., 1994, Functional and physical interaction between p53 and BZLF-1: Implications for Epstein–Barr virus latency, *Mol. Cell. Biol.* **14**:1929–1938.
142. Pietenpol, J. A., Tokino, T., Thiagalingam, S., El-Deiry, W. S., Kinzler, K. W., and Bogelstein, B., 1994, Sequence-specific transcriptional activation is essential for growth suppression by p53, *Proc. Natl. Acad. Sci. USA* **91**:1998–2002.
143. Zeng, Y., Zhong, J. M., Ye, S. Q., Ni, Z. Y., Miao, X. Q., Mo, Y. K., and Li, Z. L., 1994, Screening of Epstein–Barr virus early antigen expression inducers from Chinese medicinal herbs and plants, *Biomed. Environ. Sci.* **7**:50–55.
144. Hu, Y., and Zeng, Y., 1985, [Enhanced transformation of human lymphocytes by Chinese herbs], *Chin. J. Oncol.* **7**:417–419 (in Chinese).
145. Gregory, C. D., Dive, C., Henderson, S., Smith, C. A., Williams, G. T., Gordon, J., and Rickinson, A. B., 1991, Activation of Epstein–Barr virus latent genes protects human B cells from death by apoptosis, *Nature* **349**:612–614.
146. McDonnell, T. J., Deane, N., Platt, F. M., Nunez, C., Jaeger, U., McKaern, J. P., and Korsmeyer, S. J., 1989, *bcl*-2-immunoglobulin transgenic mice demonstrate extended B cell survival and follicular lymphoproliferation, *Cell* **57**:79–88.
147. Vaux, D. L., Cory, S., and Adams, J. M., 1988, *Bcl*-2 gene promotes haemopoietic cell survival and cooperates with c-*myc* to immortalize pre-B cells, *Nature* **335**:440–442.
148. Garcia, I., Martinou, I., Tsujimoto, Y., and Martinou, J. C., 1992. Prevention of programmed cell death of sympathetic neurons by the *bcl*-2 proto-oncogene, *Science* **258**:302–304.
149. Bissonnette, R. P., Echeverry, F., Mahboubi, A., and Green, D. R., 1992, Apoptotic cell death induced by c-*myc* is inhibited by *bcl*-2, *Nature* **359**:552–554.
150. Fanidi, A., Harrington, E. A., and Evans, G. I., 1992, Cooperative interaction between c-*myc* and *bcl*-2 proto-oncogenes, *Nature* **359**:554–556.
151. Sentman, C. L., Shutter, J. R., Hockenberry, D., Kanagawa, O., and Korsmeyer, S. J., 1991, *bcl*-2 inhibits multiple forms of apoptosis but not negative selection in thymocytes, *Cell* **67**:879–888.
152. Levine, B., Huang, Q., Isaacs, J. T., Reed, J. C., Griffin, B. E., and Hardwick, J. M., 1993, Conversion of lytic to persistent alphavirus infection by the *bcl*-2 cellular oncogene, *Nature* **361**:739–742.
153. Pearson, G., Luka, J., Petti, L., Sample, J., Birkenbach, M., Braun, D., and Kieff, E., 1987, Identification of an Epstein–Barr virus early gene encoding a second component of the restricted early antigen complex, *Virology* **160**:151–161.

154. Henderson, S., Huen, D., Rowe, M., Dawson, C., Johnson, G., and Rickinson, A., 1993, Epstein–Barr virus-coded BHRF1 protein, a viral homologue of *bcl*-2, protects human B-cells from programmed cell death, *Proc. Natl. Acad. Sci. USA* **90:**8479–8483.
155. Marchini, A., Tomkinson, B., Cohen, J. I., and Kieff, E., 1991, BHRF1, the Epstein–Barr virus gene with homology to Bcl2, is dispensable for B-lymphocyte transformation and virus replication, *J. Virol.* **65:**5991–6000.
156. Lee, M. A., and Yates, J. L., 1992, BHRF1 of Epstein–Barr virus, which is homologous to human proto-oncogene *bcl*2, is not essential for transformation of B cells or for virus replication *in vitro*, *J. Virol.* **66:**1899–1906.
157. Lowe, S. W., Ruley, H. E., Jacks, T., and Housman, D. E., 1993, p53-dependent apoptosis modulates the cytotoxicity of anticancer agents, *Cell* **74:**957–967.
158. Lowe, S. W., Jacks, T., Housman, D. E., and Ruley, H. E., 1994, Abrogation of oncogene-associated apoptosis allows transformation of p53-deficient cells, *Proc. Natl. Acad. Sci. USA* **91:**2026–2030.
159. Kawanishi, M., 1993, Epstein–Barr virus induces fragmentation of chromosomal DNA during lytic infection, *J. Virol.* **67:**7654–7658.
160. Nonoyama, M., and Pagano, J. S., 1972, Replication of viral DNA and breakdown of cellular DNA in Epstein–Barr virus infection, *J. Virol.* **9:**714–716.
161. Klein, C., and Raab-Traub, N., 1987, Human neonatal lymphocytes immortalized after microinjection of Epstein–Barr virus DNA, *J. Virol.* **61:**1552–1558.
162. Pagano, J. S., Jimenez, G., Sung, N. S., Raab-Traub, N., and Lin, J. C., 1992, Epstein–Barr virus latency and cell immortalization as targets for antisense oligomers, *Ann. N. Y. Acad. Sci.* **660:** 107–116.
163. Pagano, J. S., 1995, Epstein–Barr virus: Therapy of active and latent infection. In: *Antiviral Chemotherapy* (D. J. Jeffries and E. DeClerq, eds.), John Wiley & Sons, Chichester, pp. 155–195.
163a. Pathmanathan, R., Prasad, U., Sadler, R., Flynn, K., and Raab-Traub, N., 1995, Clonal proliferations of cells infected with Epstein–Barr virus in preinvasive lesions related to nasopharyngeal carcinoma, *N. Engl. J. Med.* **333:**693–698.
164. Zhang, S., and Nonoyama, M., 1994, The cellular proteins that bind specifically to the Epstein–Barr virus origin of plasmid DNA replication belong to a gene family, *Proc. Natl. Acad. Sci. USA* **91:**2843–2847.
165. Tomkinson, B., Robertson, E., and Kieff, E., 1993, Epstein–Barr virus nuclear proteins EBNA-3A and EBNA-3C are essential for B-lymphocyte growth transformation. *J. Virol.* **67:**2014–2025.
166. Longnecker, R., Miller, C. L., Miao, X. Q., Tomkinson, B., and Kieff, E., 1993, The last seven transmembrane and carboxy terminal cytoplasmic domains of Epstein–Barr virus latent membrane 2 (LMP-2) are dispensible for lympocyte infection and growth transformation *in vitro*, *J. Virol.* **67:**2006–2013.
167. Miller, C. L., Lee, J. H., Kieff, E., and Longnecker, R., 1994, An integral membrane protein (LMP2) blocks reactivation of Epstein–Barr virus from latency following surface immunoglobulin crosslinking, *Proc. Natl. Acad. Sci. USA* **91:**772–776.
168. Longnecker, R., Druker, B., Roberts, T. M., and Kieff, E., 1991, An Epstein–Barr virus protein associated with cell growth transformation interacts with tyrosine kinase, *J. Virol.* **65:**3681–3692.
169. Burkhardt, A. L., Bolen, J. B., Kieff, E., and Longnecker, R., 1992, An Epstein–Barr virus transformation-associated membrane protein interacts with *src*-family α tyrosine kinases, *J. Virol.* **66:**5161–5167.
170. Lerner, M. R., Andrews, N. C., Miller, G., and Steitz, J., 1981, Two small RNAs encoded by Epstein–Barr virus and complexes with protein are precipitated by antibodies from patients with systemic lupus erythematosus, *Proc. Natl. Acad. Sci. USA* **78:**805–809.

19

Epstein–Barr Virus
Mechanisms of Oncogenesis

LAYLA KARIMI and DOROTHY H. CRAWFORD

1. INTRODUCTION

Epstein–Barr virus (EBV), like all human herpesviruses, infects almost all of the world's population with an asymptomatic or trivial primary infection followed by lifelong persistence. Despite the ubiquitous nature of EBV and its potent *in vitro* immortalizing ability, the persistent infection is only rarely associated with disease. However, when the balance between the virus and the cell, and/or the infected cell and the immune system, is disturbed, serious consequences may ensue. In particular, EBV is associated with a variety of malignancies in which it appears to have an essential etiological role. However, it is now clear that the etiology of these diseases is extremely complex and that EBV is one among a series of essential factors required for the eventual outgrowth of the malignant cell. Several of these factors have now been clarified at the molecular level, although there remain many facets yet to be elucidated. The purpose of this chapter is to review EBV-associated tumors with emphasis on the oncogenic mechanisms involved.

2. THE VIRUS

Like other herpesviruses, EBV has a protein core wrapped with DNA, a nucleocapsid composed of 162 capsomeres that form an eicosahedral protein shell, and an outer envelope containing glycoprotein spikes. The EBV genome is approximately 172 kbp, linear, double-stranded DNA with a coding potential of 100–200 proteins. Restriction endonuclease maps have been derived,[1] and the entire genome

---

LAYLA KARIMI and DOROTHY H. CRAWFORD • Department of Clinical Sciences, London School of Hygiene and Tropical Medicine, London WC1E 7HT, England.

DNA Tumor Viruses: Oncogenic Mechanisms, edited by Giuseppe Barbanti-Brodano *et al.* Plenum Press, New York, 1995.

cloned[2] and sequenced.[3] The nomenclature for the promoters and open reading frames within the viral genome has been assigned using the abbreviated name of the *Bam*HI restriction fragment[3] (Fig. 1, Table I).

Two types of EBV exist, A and B (also called 1 and 2), which differ mainly in the *Bam*HI WYH region of the viral genome coding for the EBV nuclear antigen (EBNA) 2 protein.[4] Differences also occur in the EBNA-3, -4, and -6 coding regions and in the small EBV-encoded RNAs (EBERs)[5,6] (Table I). These differences are reflected in biological diversity, with A-type viruses being more efficient at *in vitro* immortalization of B lymphocytes than B-type viruses (see below).[7] Type-A EBV predominates worldwide, with the B type being rare in Western societies but more common in African communities[8] and in the setting of immunosuppression.[9] However, no specific disease associations have been noted.

Multiple copies of the EBV genome exist as closed circular episomes in the nucleus of the infected cell, and viral DNA replication occurs during the S phase of the cell cycle with transfer of EBV DNA to the daughter cells leaving the viral genome copy number relatively constant.

3. EPSTEIN–BARR VIRUS INFECTION *IN VITRO*

The EBV has a restricted host range in which man is its natural host and certain nonhuman primates can be infected experimentally.[10] The virus infects B lymphocytes *in vitro*,[11] resulting in B-cell immortalization and the generation of continuously proliferating lymphoblastoid cell lines (LCL).[12] The virus can also infect squamous epithelial cells *in vitro*,[13] resulting in lytic infection with the release of virus particles and cell death. However, this process is very inefficient, and the nonavailability of cell lines permissive for virus replication has made the study of virus replication and individual gene functions by the generation of mutants difficult. In contrast, immortalization of B lymphocytes by EBV has been studied in great detail. This is because of the ease with which large numbers of immortalized cells can be generated for study and the assumption that the genes involved in the immortalization process are important in EBV-induced tumorigenesis. Limiting

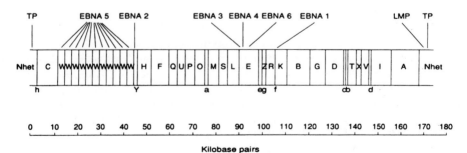

FIGURE 1. Linear representation of the EBV genome showing the *Bam*HI restriction fragments (A–Z) and the locations of open reading frames for the latent viral genes. The EBNA-5 is coded for by the *Bam*HI W fragments, which are repeated a variable number of times in the genome.

TABLE I
EBV Latent Genes[a]

ORF	Gene product	Alternative nomenclature	MW (kd)	Cellular location	Required for immortalization	Suggested function/property	Transactivation
BKRF1	EBNA-1	—	65–85	Nucleus	Yes	Plasmid maintenance	Cp
BYRF1	EBNA-2	—	86	Nucleus	Yes	Viral oncogene	CP, LMP, TP, CD23, Fgr, bcl2
BLRF3-BERF1	EBNA-3	EBNA3A	140–157	Nucleus	Yes	NK	
BERF2a-BERF2b	EBNA-4	EBNA3B	148–180	Nucleus	No	NK	
BWRF1	EBNA-5	EBNA leader protein	20–130	Nucleus	No	Binds Rb, p53	
BERF3-BERF4	EBNA-6	EBNA3c	160	Nucleus	Yes	Potential viral oncogene	LMP, CD21, CD23, vimentin
BNLF1	LMP	LMP1	58–63	Membrane	Yes	Viral oncogene	bcl2, NFkβ, CD23, vimentin
BARF1/BNRF1	TP-1	LMP2A	53	Membrane	No	Prevents Ca$^+$ mobilization	
BNRF1	TP-2	LMP2B	40	Membrane	No	NK	
BCRF1	EBER-1,2	—		Cytoplasm/nucleus	No	Regulation of PKR activity	
BARF0	BamHI A transcripts	—		Cytoplasm	NK	NK	

[a] Abbreviations: ORF, open reading frame, abbreviated using the Bam HI (B) fragment (A-Z)m LF or RF for left or rightwards, ORF followed by the number of the gene in a particular ORF; MW, molecular weight; NK, not known; PKR, double-stranded RNA protein kinase; EBNA, EB virus nuclear antigen; LMP, latent membrane protein; TP, terminal protein; P, promoter.

dilution analysis reveals that around 1–10% of infected B lymphocytes are induced to immortalize,[14] indicating that EBV is the most potent known immortalizing agent for mammalian cells. The susceptible population is the small resting B cell expressing surface IgM.[15]

Entry of the virus into B lymphocytes occurs through a ligand–receptor interaction between the major viral envelope glycoprotein gp340 and the cell surface complement receptor molecule CR2 (CD21).[16] CR2 is expressed on all mature circulating lymphocytes and forms part of a signal transduction complex including CD19, TAPA1, and Leu 13.[17] It is postulated that the binding of EBV to CR2 induces initial activating signals, but these have not yet been clearly defined. Once bound, the viral envelope fuses with the cell membrane, releasing the capsid into cytoplasm. The viral genome rapidly enters the cell nucleus, where it circularizes to form an episome that later amplifies to give multiple copies per cell.

In the first 72 hr following infection, cellular changes occur that are coincident with, and dependent on, a sequentially ordered cascade of viral gene expression and that eventually lead to proliferation and immortalization of the cell (Fig. 2). A limited number of latent viral genes are expressed during this process that are assumed to be necessary for driving cell activation and proliferation and at the same time inhibiting further cellular differentiation and suppressing viral DNA replication. These gene products include six nuclear antigens (EBNA-1 to -6), three latent membrane proteins (LMP), terminal proteins (TP) 1 and 2, and two small EB-encoded RNAs (EBER-1 and -2), which are transcribed but not translated. The characteristics of these gene products and their alternative nomenclature are listed in Table I. The EBNAs 2, 3, and 6 and LMP have been shown by analysis of

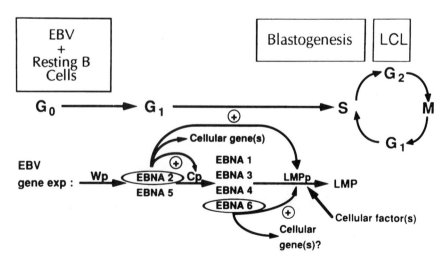

FIGURE 2. Diagrammatic summary of viral and cellular gene expression during EBV infection of primary B lymphocytes. The activation and proliferation of B lymphocytes by EBV is orchestrated by the EBNA-2 protein, which *trans*-activates viral and cellular genes and also mediates the promoter switch from Wp to Cp culminating in full latent viral gene expression and immortalization. (Courtesy of Dr. M. J. Allday, St Mary's Hospital Medical School, London, UK.)

recombinant mutant viruses to be essential for immortalization,[18–21] whereas EBNA-4, -5, TP-1, -2, and EBER-1 and -2 are not.[18,22,25] In addition to these proteins, complex transcripts arising from the *Bam*HI A open reading frame have been demonstrated in LCL, but no protein products have yet been identified.[26,27]

During the early phase of infection (12–24 hr), the Wp promoter in the *Bam*HI W repeat sequence (Fig. 3) is activated, initiating long primary transcripts that are alternatively spliced to give rise to individual mRNAs for the EBNA proteins. EBNA-2 and EBNA-5 are the first proteins to be detected at around 12 hr postinfection.[28] The function of EBNA-5 is unknown, and it is not essential for immortalization.[18] However, the findings that the protein colocalizes with the retinoblastoma gene product Rb[29] and binds to Rb and the tumor suppressor gene product p53 in *in vitro* systems[30] and that B cells infected with EBNA-5 deletion mutants grow very poorly[18] suggest its involvement in the control of cellular proliferation.

EBNA-2 is a transcription factor with an acidic activation domain that is essential for immortalization.[18,19] The protein clearly plays a pivotal role in the immortalization process, causing a switch in promoter usage from Wp to the alternative EBV latent promoter Cp (Fig. 3), with the consequent expression of EBNAs 1–6.[31] Coincident *trans*-activation of the LMP and TP promotors is also EBNA-2-mediated.[32,33] In addition, EBNA-2 up-regulates certain cellular genes including the oncogenes c-*fgr*[34] and *bcl*2[35] and the B-cell activation molecules CD21 and CD23.[36] CD23 is a protein that is invariably expressed on EBV-

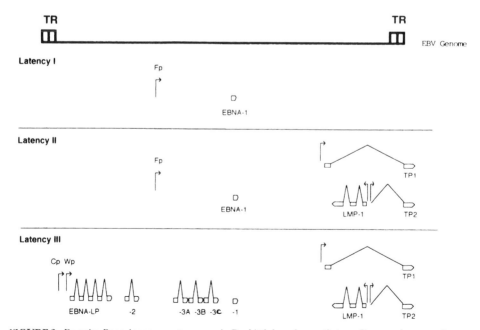

FIGURE 3. Epstein–Barr virus promoter usage in Burkitt's lymphoma (latency I), nasopharyngeal carcinoma and Hodgkin's lymphoma (latency II), and lymphoblastoid cell lines (latency III). (Courtesy of Dr. L. A. Brooks, London School of Hygiene and Tropical Medicine, London, UK.)

immortalized B cells,[37] acts as a low-affinity receptor for IgE,[38] and, once cleaved from the cell surface, may act as an autocrine B-cell growth factor.[39] EBNA-2 does not bind directly to DNA but interacts with a variety of cellular factors to form an activation complex. Jk and PU.1 are both DNA-binding proteins that have been shown by specific mutagenesis of their binding sites to convey EBNA-2 responsiveness. Jk (also called CBF1) is a ubiquitous factor that binds to the EBNA-2 response elements upstream of Cp, CD23, and LMP, all of which contain one or more Jk binding site. This binding probably causes repression that is relieved by the binding of EBNA-2. The expression of PU.1 protein is restricted to B cells, macrophages, and red cell precursors, and the finding that PU.1 is essential for EBNA-2-mediated activation of the LMP promoter explains the restriction of EBNA-2 *trans*-activation of LMP to the B-cell type. Several other DNA binding factors, as well as factors binding to the EBNA-2 acidic domain, have been identified that also enhance its activity.[40–42]

As a result of EBNA-2 *trans*-activation, full latent viral gene expression is achieved by 48–72 hr postinfection.[28,43] The roles of the other nuclear proteins in the immortalization process are unclear. EBNA-1 binds to the latent viral origin of replication (*ori*-P) and is a requirement for maintenance of the viral episome and synchronization of its replication with cell division.[44] EBNA-6 is a nuclear protein that contains a leucine zipper upper domain. It cooperates with EBNA-2 in the *trans*-activation of LMP[45] and also induces the cellular genes CD21[36] and vimentin,[45] probably as a result of the up-regulation of LMP (see below) (Table I).

Fundamental changes in cellular phenotype occur at 48 hr postinfection, and these have been attributed primarily to the effect of LMP on B-cell activation, moving the cell through the phases of the cell cycle. This protein, when expressed under a heterologous promoter in Rat I or NIH3T3 rodent fibroblast cell lines, acts as a classical viral oncogene in *in vitro* assays, inducing loss of contact inhibition and anchorage dependence, altered cell morphology, and tumorigenicity in nude mice.[46] In B-lymphocyte cell lines, the protein causes an increase in cell size, expression of the activation markers CD23β, CD40, and the transferrin receptor and the cell adhesion molecules (CAMs) CD11a, CD54, CD57, and vimentin, and increase in steady-state intracellular free calcium levels.[47] The LMP, like EBNA-2, induces expression of the oncogene *bcl*2,[48] the product of which is an inhibitor of the apoptotic response (programmed cell death) to growth factor withdrawal. Thus, LMP-induced *bcl*2 expression correlates with resistance to apoptosis induced by low serum concentration and ionophores in B-cell lines. The mechanism of LMP induction of cellular genes and the signaling pathways involved are unclear, although LMP induction of the HIV-1 LTR has been shown to be mediated by activation of the transcription factor NF-kB.[49] Because NF-kB activation is necessary for many of the LMP-induced phenotypic changes, it is likely that this transcription factor, perhaps in concert with other factors, mediates these changes.

The predicted protein structure of LMP indicates a short cytoplasmic amino terminus separated from a large cytoplasmic carboxy-terminal domain by six hydrophobic segments with five membrane-spanning units forming three external loops. The carboxy-terminal domain is the major site of phorphorylation.[50] Deletion mutants and site-directed mutagenesis to the sites of phosphorylation have been used to determine the domains necessary for activation of B cells and transformation of rodent fibroblasts. The results are contradictory.[51,52] It is reported that the carboxy-terminal cytoplasmic domain is required for activation of NF-kB and B-cell surface phenotypic changes[53]; the amino-terminal cytoplasmic and transmembrane domains, but not the carboxy-terminal domain, are required for transformation of Balb 3T3 fibroblasts,[51] whereas all three domains are required for transformation of rat fibroblasts.[52] The similarity between the structure of LMP and that of a growth factor receptor or ion channel has led to speculation on the mode of action of LMP; however, to date these functions have not been demonstrated.

The TP open reading frame is located at the two ends of the linear EBV genome and is completed only when the genome has circularized to form an episome.[54] TP-1 and -2 are produced by alternative splicing of the mRNA. Although little is known about the function of TP-1 and -2 because of the lack of specific antibodies, TP-1 associates with, and is phosphorylated by, the B-lymphocyte *src* family of tyrosine kinases.[55] This interaction inhibits calcium mobilization following stimulation of latently EBV-infected B lymphocytes and may thereby prevent fortuitous activation of the viral lytic cycle, which is known to be linked to B-cell activation/differentiation processes.[56]

By 72 hr postinfection, EBV-infected cells showing full latent viral gene expression have progressed through G_1, S, and G_2 phases of the cell cycle and are entering mitosis. These cells then continue to cycle indefinitely, giving rise to an immortalized LCL. Their phenotype is similar to that of antigen-stimulated lymphoblasts with expression of the B-cell activation markers typical of this stage of B-cell differentiation (such as CD23, CD39, CD30, CD70, ki24) as well as the CAMs, CD11a, CD54, CD58. The LCL cells are immortal but cytogenetically normal. They are not, by definition, transformed cells because they do not form colonies in soft agar or give rise to tumors in nude mice.[57] These characteristics differ from EBV-carrying cell lines derived from Burkitt's lymphoma (BL) biopsy material (see later).

Within a LCL, very few cells (0–5%) switch from a latent to a lytic type of infection with the expression of the lytic cycle antigens, the production of virus particles, and cell death. This switch is linked to the state of differentiation of the B cell, with the more differentiated cells of a plasma cell phenotype being permissive for viral replication.[58] The process can be induced by agents such as the tumor promoter 12-O-tetradecanoyl phorbol-13-acetate (TPA),[59] *n*-butyrate,[60] and antibodies to IgM,[61] which induce B-cell differentiation. This treatment activates the promoter of the lytic cycle immediate early gene BZLF1, which contains TPA-response elements.[62] BZLF1 protein then induces the sequential expression of early, nonstructural, followed by late structural, genes by *trans*-activating early gene promotors.[63]

4. EPSTEIN–BARR VIRUS INFECTION *IN VIVO*

The EBV is efficiently transmitted from one individual to another through shedding from lytically infected squamous epithelial cells in the oropharynx into secretions. Primary infection usually occurs during the first few years of life, when it is asymptomatic. However, if infection is delayed until adolescence, it causes infectious mononucleosis (IM) in around 50% of cases.[64] In IM, EBV replicates in pharyngeal epithelial cells, and from this lymphoid-rich site infiltrating B lymphocytes become infected, enter the circulation, and are disseminated throughout the body.[65] It is postulated that these B cells, which support full latent viral gene expression,[66] proliferate to form a pool of long-lived virus-carrying B cells. However, once cell-mediated immune mechanisms are active, virus replication in epithelial cells and carriage in B cells drop to low levels. The dramatic T-cell response seen in IM is the cause of most of the symptoms of IM; thus, IM is an immunopathological disease. The disease is self-limiting, usually lasting 3–6 weeks.

As with other members of the *Herpesviridae*, EBV has the capacity for lifelong persistence in the infected host. Whether primary infection is asymptomatic or manifests as IM, almost all individuals are persistently infected and shed infectious virus at low levels into the buccal fluid and carry around one in a million virus-infected B cells.[67] This persistent infection is controlled by cellular immune mechanisms, which are maintained at easily detectable levels in normal seropositive individuals.[68] Cytotoxic T cells (CTL) specific for all the latent viral gene products with the exception of EBNA-1 have been detected[69] and are assumed to be the major control mechanism. Antibodies to the latent and lytic cycle antigens also persist for life, and those to the membrane antigen gp340 are neutralizing. In the light of these findings it is generally believed that lifelong viral persistence must require the establishment of a latent type of infection in a cell in which the highly immunogenic latent gene products are not expressed.[70] The exact site of this infection, however, remains a point of controversy. The basal cells of the oropharyngeal squamous epithelium have been suggested as one site of latency.[71] The alternative site is the long-lived B lymphocyte. In both these cell types, experimental evidence suggests that EBV gene expression is dependent on the differentiation status of the cell.[58,72,73] In addition, both arise from self-renewing stem cells, which could form a reservoir of virus-carrying cells with very restricted viral gene expression constantly giving rise to differentiating daughter cells with the potential for full viral gene expression and induction of the lytic cycle. In the search for such a cell, several groups have detected circulating B cells with a restricted viral gene expression, although the exact extent of this remains controversial.[66,74]

5. EPSTEIN–BARR VIRUS-ASSOCIATED MALIGNANCIES

The EBV is associated with a spectrum of malignancies of lymphoid and epithelial cell origin. The lymphoid malignancies include BL, lymphoproliferative disease, and lymphoma in immunocompromised individuals and Hodgkin's lymphoma.[75–77] The classical EBV-associated epithelial cell tumor, undifferentiated nasopharyngeal carcinoma (NPC), is a tumor of the squamous epithelium of the nasopharynx.[78] In addition to these malignancies, the virus has recently been associated with a varying number of cases of other tumor types including peripheral T-cell lymphoma,[79,80] gastric carcinoma,[81] salivary gland tumors,[82] and most recently with tumors of smooth muscle in the immunocompromised.[83,84] The role

5.1. Epstein–Barr Virus in the Immunocompromised Host

5.1.1. EBV-Associated Lymphoproliferative Disease

In individuals with defects in cell-mediated immune mechanisms, the balance between EBV load and immune control is tipped such that there is increased virus production in the oropharynx, increased numbers of circulating EBV-infected B cells, and a decrease in detectable circulating EBV-specific CTL.[85] These findings are reflected in increased levels of antibodies to virus lytic gene products, but no overt disease occurs in the majority of individuals. However, in certain groups of immunocompromised patients, such as organ transplant recipients and AIDS patients, an increased incidence of lymphoma has been recorded,[86] and most of these are EBV associated.[87] These tumors form a histologically heterogeneous group, termed B-lymphoproliferative disease (BLPD), which generally resemble large-cell lymphoma.[88] AIDS patients also have an increased incidence of BL (see below).[89] The BLPD may be polyclonal, oligoclonal, or monoclonal, and tumors have been recorded that progress from poly- to monoclonality over time. In around 50% of cases occurring in transplant recipients, there is evidence of a recent primary EBV infection.[90]

Examination of the viral and cellular phenotype of BLPD reveals a picture very similar to that seen in LCLs (see Section 3). The majority of cells in most tumors express EBNA 1–6 and LMP (Table II) as well as CAMs and B-cell activation markers.[87] This viral phenotype, with expression of the viral oncogenes EBNA-2 and LMP, strongly suggests that EBV plays an essential role in the evolution of the tumor. In addition, all the known viral targets and essential accessory molecules for CTL recognition (CAMs, HLA class I) are expressed, indicating that these cells would be recognized by a normally functioning immune system. In this context, it is noteworthy that the treatment of choice for BLPD in organ transplant recipients is a reduction in immunosuppressive drugs, which, in the majority of cases, leads to

TABLE II
Latent Viral Gene Expression in EBV-Associated Tumors[a]

| Tumor | EBNA | | | | | | LMP | TP | | EBERS | BamHI-A | Promoter usage |
	1	2	3	4	5	6		1	2			
BL	+	–	–	–	–	–	–	–	–	+	+	Fp
BLPD	+	+	+	+	+	+	+	?	?	+	+?	Cp/Wp
NPC	+	–	–	–	–	–	±	+	±	+	+	Fp
HL	+	–	–	–	–	–	++	+	+	+	+	Fp

[a]Abbreviations: BL, Burkitt's lymphoma; BLPD, B lymphoproliferative disease; NPC, nasopharyngeal carcinoma; HL, Hodgkin's lymphoma; EBNA, EB virual nuclear antigen; LMP, latent membrane protein; TP, terminal protein.

tumor regression.[91] Despite this form of treatment, however, the disease often progresses or recurs, and BLPD carries a high mortality.[92] Recently, treatment with *in vitro*-grown autologous, EBV-specific CTL has been successful in both preventing and treating BLPD.[93] These forms of immunotherapy underline the importance of the immune deficiency and EBV in the evolution of this disease.

It is still unclear whether EBV infection of B cells in the context of severe immunosuppression is alone sufficient to cause BLPD or whether other cofactors are required. The facts that only a minority of patients (up to 10%) develop BLPD and that, although there are many infected B cells in the body, the tumors are often solitary and monoclonal, suggest that other factors, either genetic or epigenetic, are required. Although consistent cytogenetic abnormalities have not been demonstrated, the number of tumors studied is low. It would be of interest to analyze these tumors for the lymphoma-associated translocations involving *bcl*2, *bcl*6, as well as mutations of the tumor suppressor genes p16, p15, and p106.[94] On the other hand, the unusual sites (brain, gut, transplant organ) in which BLPD is commonly found suggest that the virus-carrying cells may have stringent growth factor requirements that depend on local soluble factor production in areas of chronic antigenic stimulation such as the gut and transplanted organ and/or that are enhanced by the immunologic deficit in the brain. Here again, detailed molecular studies are required to define these factors more precisely.

5.1.2. · *X-Linked Lymphoproliferative Syndrome*

X-linked lymphoproliferative syndrome (XLPS) is a rare complication of primary EBV infection that is invariably fatal. The genetic defect is carried on the X chromosome, and the syndrome is therefore inherited throught the female line and affects 50% of male offspring. The affected children remain well until they are infected by EBV, when they usually die within 2–3 weeks of liver and/or bone marrow failure. All tissues are infiltrated by EBV-infected polyclonal B cells and $CD8^+$ T cells. Occasionally patients survive the initial EBV infection and die of an EBV-associated lymphoma 2–3 years later. The phenotype of the EBV-infected cells in the XLPS and in the associated lymphoma is identical to that described for BLPD[87] (above). Occasionally XLPS is associated with hemophagocytic syndrome. Although the genetic abnormality has been localized on the X chromosome, the basic immunologic abnormality in XLPS has not been identified.

5.2. Burkitt's Lymphoma

Burkitt's lymphoma is well known as an EBV-associated malignancy; it is the tumor from which the virus was first isolated in 1964.[95] Two geographically distinct forms of BL are now recognized.[96] The first form, endemic (e)BL, is the most frequent childhood cancer in the areas of equatorial Africa and the coastal plains of Papua–New Guinea, a geographic distribution that correlates with that of holoendemic malaria.[97] In eBL cases that occur in these high-incidence areas, around 97% of tumors carry EBV DNA. The second form, sporadic (s)BL, occurs throughout the world at low incidence, representing less than 3% of childhood cancers. In some areas such as North America and Europe, 15–20% of these tumors contain

EBV DNA, but in others such as North Africa, up to 80% are associated with the virus.[98] EBV-positive and -negative BL have now also been recognized in AIDS patients.[99]

Burkitt's lymphoma usually occurs in extranodal sites, commonly in the jaw, but is often multifocal with involvement of the central nervous system, gut, ovary, thyroid, and breast in eBL and in abdominal sites and bone marrow in sBL. The two forms of BL are histologically indistinguishable and characteristically show a "starry sky" picture microscopically with a medium-sized, noncleaved cell morphology. The cells are clonal B lymphocytes and have the phenotype of germinal center cells expressing the markers CD10 and CD77 as well as surface or cytoplasmic IgM.[100]

5.2.1. Epstein–Barr Virus Gene Expression in BL

Evidence for an etiological association between EBV and eBL is still indirect; however, the finding of EBV in 97% of tumors argues strongly in favor of its essential role in BL development. Furthermore, analysis of EBV terminal repeat sequences, which vary in number in individual isolates, shows EBV of a single clonal type in BL, suggesting the involvement of the virus prior to expansion of the malignant cell.[101] This argues against a simple passenger role for EBV. Studies analyzing EBV gene expression in BL on tumor biopsy material have revealed that, unlike LCL, where the full 11 EBV latent genes are expressed, in BL it is restricted to EBNA-1[102] (Table II). EBER-1 and -2 RNAs and a low level of BamHI A transcripts can also be detected.[103] The lack of EBNA 2–6 expression is probably a result of inactivation of the latent promotors Cp and Wp, which have been shown to be methylated in BL.[104,105] Activation of another promoter, Fp, located in the BamHI F fragment of the EBV genome, occurs,[106] and the resultant transcript only splices to EBNA-1 (Fig. 3). The lack of LMP and TP expression in BL cells may be caused, at least in part, by the lack of EBNA-2-induced activation of their promoters. Because EBNA-1, which is required for maintenance of the viral episomes, has not, to date, been shown to have a growth-promoting function, and the known viral oncogenes EBNA-2 and LMP are not expressed in BL, it has been suggested that BL is derived from a cell that initially expressed the latent viral genes known to have growth-promoting potential. In this senario, these latent gene products (EBNA 2, LMP) are required only early in the evolution of BL, and later immune selection would lead to the outgrowth of a clone expressing only EBNA-1 because this is the only latent viral protein not recognized by CTL (see Section 4). As a consequence of this restricted gene expression in BL, expression of CAMs (induced by LMP) is low or undetectable, thereby compounding the lack of T-cell recognition of the tumor cells.

Recent findings on BL biopsy material and BL cell lines suggest that BL cells from some tumors may contain integrated EBV genomes[107] or subgenomic fragments[108] either alone or in addition to multiple viral episomes. Although the integration site is not constant in these cases, it is possible that transcripts from the junction regions of viral and cellular DNA code for fusion proteins with growth-promoting ability,[109] although these have not been demonstrated. In addition, the presence of integrated subgenomic fragments or deletion mutants of EBV with loss of episomal DNA may account for the apparent absence of EBV in EBV-negative BL. In this regard, it has recently been demonstrated that EBV infection of B cells induces the RAG1 gene, the product of which is a site-specific DNA recombinase that is necessary for Ig gene rearrangement in B cells (J. Sixbey, personal communication). It is therefore theoretically possible that this protein facilitates homologous recombination between virus and cellular sequences, causing viral integration of viral sequences. Furthermore, it is possible that EBV, by inducing RAG1, enhances the likelihood of a c-myc translocation occurring in an EBV-infected cell (see below).

Lymphoma cell lines derived from BL material are relatively easy to establish and maintain in culture; however, studies of viral gene expression in these cell lines have revealed a more heterogenic pattern than in BL biopsy material, particularly in those cell lines that have been in culture for several years. These patterns have been characterized into three groups: I, II, and III.[110,111] Group I are cultured BL biopsy cells, which initially grow as single small round cells in suspension and are very susceptible to apoptosis induced by low serum concentration. They show a latency I pattern of gene expression identical to that of the tumor with only EBNA-1 protein and EBER-1, -2, and BamHI A transcripts detected (Fig. 3). These cell lines grow as colonies in soft agar and form tumors in nude mice. With time this phenotype usually drifts to a group III BL with a latency III pattern of gene expression that is identical to the full latent viral gene expression seen in LCLs (Fig. 3). This drift is associated with demethylation of Cp promoter sequences and can be enhanced by the addition of the demethylating compound 5-azacytadine.[112] The B-cell activation markers and CAMs expressed on LCL cells are also expressed on BL cells showing the latency III pattern. The cells grow in tight clump in suspension and are relatively resistant to apoptosis. Latency II (seen in NPC, Section 5.4, and HL, Section 5.3) is characterized by EBNA-1 expressed from Fp and LMP and TP-1 and -2 expression (Fig. 3). In some BL cell lines EBNA-2 is not expressed because of a deletion in the EBNA-2 coding sequences, but in others the viral genome is integrated into the cellular DNA.

5.2.2. Alteration of Cell Growth Regulatory Genes in BL

Reciprocal chromosomal translocations involving the c-myc oncogene and the Ig gene loci are consistent features of BL whether or not it is EBV associated.[96] One of three translocations occurs, the most common being a translocation of c-myc on chromosome 8 to the Ig heavy chain gene region on chromosome 14 (8:14). Translocations of c-myc to the kappa (2:8) or lambda light chain genes (8:22) on chromosomes 2 and 22, respectively, also occur. Such translocations bring Ig promoter/enhancer sequences into the vicinity of the c-myc gene, resulting in its deregulation. Translocation breakpoints vary between the two types of BL, generally being close to the c-myc gene in sBL and distant from it in eBL.[113] The Ig gene breakpoints also vary, being at the switch region in sBL and at the J, V, or D region in the eBL.[114] This observation led to the hypothesis that the translocations of eBL occur as the result of errors in VDJ joining in an early bone marrow B stem cell whereas those of sBL occur in a germinal center B cell at the time of Ig isotype switching. Both these translocation types occur in AIDS BL.

The c-myc gene product is a nuclear protein that dimerizes with the MAX protein to form a DNA-binding complex that is involved in cell cycle control,[115] and translocations involving the c-*myc* gene locus clearly play a privotal role in the evolution of BL. However, experiments in which the c-*myc* gene under a heterologous promoter has been transfected into LCL cells show that *myc* deregulation does not induce a BL phenotype,[116] and the cell lines generally remain nontumorigenic in nude mice.[117] In addition, although mice transgenic for c-*myc* linked to the Eμ Ig enhancer develop B-cell tumors (plasmacytoma), these tumors do not occur in all mice and are monoclonal, indicating their origin from a few, rare precursor cells.[118] These two findings argue in favor of additional factors being required for BL development, and in this regard point mutations in the c-*myc* gene have been detected that may enhance deregulation.[119]

Loss of function of tumor suppressor genes such as p53 and Rb are a common feature of a wide variety of tumor types, in particular the virus-associated tumors in which DNA tumor viruses such as SV40, adenovirus, and human papillomavirus code for oncoproteins that bind to, and inactivate, these gene products (see other chapters in the volume). No latent EBV gene products have been shown to bind p53 or Rb *in vivo*, and, although EBNA-5 binding to Rb and p53 *in vitro* may have *in vivo* significance (see Section 3), the absence of EBNA-5 expression in BL excludes this as an oncogenic mechanism in this case. However, mutations of p53 with loss of suppressor function have been reported in both eBL and sBL[120] as well as AIDS-associated BL, and these are likely to be important in the evolution of the disease.

5.2.3. Pathogenesis of BL

Like most malignancies, BL is a multifactorial tumor with a multistep evolution. Any hypothesis put forward to explain its development must take into account the oncogene/tumor suppressor gene abnormalities, the EBV association, the geographic restriction, and the recent appearance of BL in AIDS patients.

The geographic restriction of eBL to equatorial Africa was first noted by Burkitt,[97] who suggested on epidemiologic grounds that holoendemic malaria was a cofactor in the development of the tumor—an association that is upheld today. However, the exact role of malaria infection in BL development is still much debated. Malaria is known to cause T-cell immunosuppression[121] and polyclonal B-cell activation.[122] Both of these changes are also a consequence of human immunodeficiency virus (HIV) infection, where progressive loss of $CD4^+$ T cells causes immunodeficiency and HIV-coded envelope glycoproteins directly activate B cells.[123] In both malaria and HIV infections, circulating EBV-specific CTL activity is decreased or absent,[124,125] and, as a consequence of this, the levels of circulating EBV-carrying B cells are raised.[125,126]

A unifying hypothesis has been formulated to incorporate all the known features of BL into a multistep senario for the evolution of the tumor.[127] Polyclonal activation of B cells as a result of a chronic malaria or HIV infection increases the numbers of B cells in the body undergoing Ig gene rearrangements and/or Ig class switching and thereby increases the chance of translocation and deregulation of the c-*myc* gene occurring during these events. Inappropriate expression of c-*myc* in the absence of a growth signal required for cell survival may induce apoptosis. However, combined with LMP-induced *bcl*2, expressing this may enhance cell survival when many early B cells and germinal center B cells die from apoptosis. Immunosuppression promotes the survival of EBV-infected B cells despite the expression of viral growth-promoting gene products. These two mechanisms, acting in parallel, would thus increase the chance of EBV infection and c-*myc* translocation coinciding in the same cell. Although the sequence of these events remains to be elucidated, the available evidence suggests that EBV infection and c-*myc* translocation are early events that are crucial to tumor expansion, whereas p53 mutation is a later consequence of rapid proliferation of the malignant cells and confers an added growth advantage to the clone.

5.3. Hodgkin's Lymphoma

Hodgkin's lymphoma (HL) is a nodal tumor of unknown etiology. Four histological types exist: lymphocyte-predominant, mixed cellularity, lymphocyte-depleted, and nodular-sclerosing. The tumor tissue characteristically has a low number (up to 2%) of the multinucleate malignant cells: Hodgkin–Reed–Sternberg (HRS) cells in a background of normal reactive lymphocytes. The origin of HRS cells remains obscure—they variably show B, T, or "null" cell phenotype but consistently express the IL-2 receptor, CD25, and CD30.

It has long been recognized that many of the seroepidemiologic features of HL suggest a viral etiology, in particular the biphasic age distribution with peaks in young adulthood and the elderly.[128] An association with EBV was suggested by the increased incidence of HL following IM[129] and the high antibody titers to EBV antigens found in HL sera.[130] However, it was not until the development of sensitive detection techniques that an association was definitely established by the direct demonstration of monoclonal EBV episomes and viral gene expression in HRS cells in a subset of tumors.[131–134] The frequency of EBV association appears to vary with geographic location and with HL type. Thus, around 50% of HL in United States[135] and Europe[136] contain EBV DNA, whereas an association of 90–100% is reported from Peru[137] and Honduras.[135] Type A EBV is more prevelant in HL than type B, reflecting the normal distribution; however, in the AIDS setting, HL often contains EBV of the B type.[138] Although all four types of HL have been reported to carry EBV DNA, whether the HRS cells show features of B, T, or null cells, the closest association is with the most aggressive mixed cellularity/lymphocyte-depleted types (>80%), but this figure is around 30% for nodular sclerosing HL, and only the occasional lymphocyte-predominant type contains EBV DNA.[139] The EBV association also correlates with the site of origin of HL. Those tumors arising in the cervical lymph nodes more commonly contain EBV than those arising in other body sites.[140]

Examination of HL biopsy material by a combination of polymerase chain reaction (PCR) and *in situ* hybridization reveals a latency II pattern (Fig. 3) with EBNA-1, initiating from Fp, as well as LMP, TP-1, -2, and EBER-1 and -2 expression in HRS cells[134,141] (Table II). In a few HL (around 6%), a small number of cells express BZLF1, the *trans*-activator of the EBV lytic cycle; however, because other lytic cycle antigens, EA, MA, and VCA, are not detected, this is interpreted as the triggering of an abortive lytic infection in a few cells, the remainder being tightly latent.[142] Because LMP is the only viral oncogene expressed in HL, attention has focused on defining a growth-promoting role for this protein in the HRS cell type. Analysis of *bcl*2 expression shows no up-regulation in those HRS cells expressing LMP, thus ruling out a mechanism by which LMP prolongs cell survival by inducing *bcl*2 and thereby a resistance to apoptosis.[143] In around 30% of HL, polymorphisms of the LMP gene have been identified[144] that were first recognized in NPC (see below). These result from deletions and mutations of the gene, but their frequency in the normal population has not been determined, and their significance remains unclear.

In summary, no consistent differences in cellular phenotype, cytokine production, or disease outcome have been demonstrated between EBV- and non-EBV-associated HL, and the role of EBV in the malignant process remains to be elucidated.

5.4. Nasopharyngeal Carcinoma

Nasopharyngeal carcinoma is a malignancy of the stratified squamous epithelium of the nasopharynx. This tumor is the most consistently EBV-associated tumor, although genetic and environmental factors also play a role in its development. Nasopharyngeal carcinoma has been classified by the World Health Organization (WHO) into three categories according to the stage of cellular differentiation: WHO 1, keratinizing squamous cell carcinomas; WHO 2, nonkeratinizing carcinomas; and WHO 3, undifferentiated carcinomas. The NPC shows a remarkable degree of geographic variations, having a high incidence in Southern China, Hong Kong, Singapore, and Southeast Asia, an intermediate incidence in North and East Africa, Alaska, Greenland, Iceland, and the Mediterranean rim, and a low incidence in Europe and North America.[78] Regardless of geographic origin, EBV is consistently associated with WHO 2 and 3 carcinomas. This association was initially demonstrated by detection of elevated serum IgG and IgA antibodies to EBV early antigens and viral capsid antigens,[145] and was consolidated by the detection of the viral genome and viral-coded antigens in the malignant epithelial cells.[146,147] The demonstration of monoclonality of the viral DNA by the detection of clonal, fused EBV terminal fragments indicates that the malignancy has arisen from the clonal expansion of a single EBV-infected progenitor cell.[101]

The association of EBV with WHO 1 NPC is still controversial. Early studies found no association[148]; however, in a recent large study on biopsy material from over 100 WHO 1 NPC tumors and nine premalignant lesions, EBV was found in all cases regardless of the stage of differentiation.[149]

Type A EBV is predominant in Southern China in both healthy Chinese carriers and NPC biopsy material. However, a number of alterations in the EBV genome involving the repeat sequences have been identified in isolates from both of these sources. The predominant strain is characterized by deletion of a *Bam*HI site between the *W1* and *I1* fragments, which gives a characteristic restriction enzyme digest pattern.[150,151] In addition, numerous sequence changes in the BNLF-1 gene encoding LMP have been noted in clones from NPC tumors of

Chinese origin when compared with that of the prototype laboratory strain (B95-8).[152] The biological differences between these virus strains, and their relationship to the etiology of NPC, are not clear.

Studies on NPC have been hampered by the small amount of available material and the difficulty in growing cell lines from NPC biopsy material. To date only five NPC epithelial tumor cell lines have been reported, and, with one exception, these cell lines are EBV-genome negative.[153] A second approach, which was originally used to eliminate infiltrating nonmalignant lymphocytes, is the growth of NPC biopsy material in athymic nude mice.[154] Again, this technique has been successful with only a small minority of tumor biopsies, and to date four transplantable NPC lines have been developed (designated C15, C17, C18, and C19).[155] C15 was derived from a primary tumor, and C17, C18, and C19 from metastatic tissue. This has provided abundant and homogeneous NPC material for phenotypic, cellular, and molecular analyses.

5.4.1. Epstein–Barr Virus Gene Expression in NPC

Analysis of EBV gene expression in NPC biopsies and nude mouse-grown tumor material has identified a latency II pattern of expression[156] that is distinct from BL (latency I) (see Section 5.2.1) and LCLs (latency III) (see Section 3) but similar to HL (see Section 5) Fig. 3, Table II). The NPC cells consistently express EBNA-1 driven from the Fp promoter and EBER-1 and -2 RNA. In addition, in the majority of cases, the cells also express LMP and TP-1 and -2, albeit at low levels, often detectable only by reverse-transcription PCR.[156] Northern blot and PCR studies have also detected expression of latent transcripts that run through the *Bam*HI A region of the EBV genome, but no gene product has been identified.[27,156] In one study the expression of BZLF1 protein was detected in a few cells of undifferentiated NPC,[157] although replicative virus has not been identified.

The expression of LMP in NPC is of interest because of its proven ability to act as an oncogene in *in vitro* assays (see Section 3).[46] When expressed under a heterologous promoter in the SV40-immortalized, nonturmorigenic, keratinocyte cell line, Rhek-1, LMP induces morphological transformation accompanied by down-regulation of expression of cytokeratins that are associated with epithelial cell differentiation.[158] In the stratified squamous epithelial cell line SCC12, LMP has been reported to inhibit the cellular capacity for terminal differentiation and, in *in vitro* culture, gives rise to a disorganized multilayered epithelial structure lacking tight desmosomal junctions. Increased levels of surface expression of the CAM, CD54, and the proliferation-linked membrane marker CD40 were also noted.[159] The expression of LMP in transgenic mice *in vivo* resulted in a hyperplastic condition of the epidermis with aberrant expression of the hyperproliferative keratin 6 in the suprabasal layers of the epithelium.[160] Taken together, these data strongly suggest that LMP plays a role in predisposing the nasopharyngeal epithelium to malignancy.

The functional significance of the sequence divergence between the prototype B95-8 (B)LMP and LMP of Chinese NPC (C) origin has been explored by transfecting the genes into Rhek-1 cells and comparing their growth *in vitro* and tumorigenicity in SCID mice. It was demonstrated that clones expressing CLMP formed invasive tumors more often than BLMP-expressing clones.[161] However, because in other epithelial cell lines BLMP has been shown to induce tumorigenicity,[162] these findings may be a reflection of the properties of the cell background rather than the LMP molecule.

5.4.2. Alteration of Cell Growth Regulation in NPC

In contrast to BL, our knowledge about growth regulation of the malignant epithelial cells in NPC is minimal. Expression of the protooncogene *bcl*2 in NPC has been investigated because it is induced by LMP. However, although most NPC studied expressed high levels of *bcl*2, no correlation was found with LMP expression or with the detection of EBV DNA. In addition, *bcl*2 expression was detected in basal epithelial cells from the normal nasopharynx, suggesting a role in normal keratinocyte differentiation.[163] p53 gene mutations have not been identified in primary NPC tumors, although they have been detected in a few metastatic lesions and appear at high frequency in nude-mouse-grown material.[164,165] This suggests that p53 mutation is not involved in the initial clonal outgrowth of NPC but, if present, confers a growth advantage. Other workers, using immunocytochemical methods, report overexpression of p53 in a proportion of EBV-positive NPC and suggest that this results from stabilization of p53 protein by interaction with viral protein(s).[166] However, the only latent viral protein previously shown to interact with p53 *in vitro*, EBNA-5, is not expressed in NPC.

Other factors influencing the growth of the malignant epithelial cells in EBV-positive NPC are immunosurveillance mechanisms, which, although not studied in detail in NPC patients, are assumed to be normal. Because in most cases the tumor cell express LMP, which is a known T-cell target, these cells should be recognized by immune T cells. The tumor tissue is heavily infiltrated with lymphoid cells, mainly $CD4^+$ and $CD8^+$ T cells,[167] but it is unclear whether these are antigen-specific T cells or nonspecific T cells attracted to the site by tumor cell cytokine production. The malignant epithelial cells have been shown to express HLA class I and II[167] and the CAM CD54, which are required for T-cell recognition, although expression of CAM CD58 is minimal.[168] Tumor outgrowth may therefore result from the down-regulation of LMP expression to levels below that required to stimulate T cells or to mutate LMP sequences, giving rise to unrecognized target peptides (see above).

5.4.3. Pathogenesis of NPC

As with BL, NPC develops through a multistep process, in t his case involving the interplay of EBV with genetic and/or environmental cofactors. A genetic predisposition for NPC was first implied by the marked geographic restriction and family clustering. Thus, the incidence of NPC in southern China is 100-fold higher than that in most European countries, reflecting the high frequency in Mongoloid races and low frequency in Caucasians. This high incidence of NPC is maintained for one or two generations in emigres from high- to low-incidence areas. Genetic analysis on Chinese populations reveals links with specific HLA haplotypes: HLA A2, BW46, and the antigen B17 confering a twofold relative risk.[169] More recent studies on clustering of NPC in Chinese families suggests that a susceptibility gene that is closely linked to the HLA locus confers a relative risk of approximately 21.[170]

Studies on environmental factors involved in the etiology of NPC have concentrated on dietary components, particularly traditional salted fish, which forms the staple diet for young children in high-incidence areas. Large epidemiologic investigations indicate that the ingestion of these foodstuffs is a risk factor for NPC,[171,172] and this is backed up by the experimental induction of tumors in Wistar rats fed on this material.[173] In addition, extracts of salted fish have been found to contain N-nitroso compounds with mutagenic properties,[174] and more recently these food extracts have been shown to induce EBV lytic-cycle antigen expression in latently infected B-cell lines.[175] Other workers have concentrated on studying the *Euphorbiaceae* and *Thymelaeaceae* plant families, which are indigenous to southern China and contain phorbol esters, known inducers of the EBV lytic cycle *in vitro*.[176]

Despite all these studies, our knowledge of the evolution of NPC remains fragmentary, and the links among genetic, environmental, and viral factors in NPC remain to be elucidated.

6. CONCLUSIONS

The EBV is an oncogenic virus which plays a major role in the evolution of a variety of human tumors. Since the discovery of the virus in 1964, a large body of work has contributed to our understanding of EBV/B-lymphocyte interactions and taught us not only about tumorigenesis but also about normal B-cell activation/differentiation processes. We are now in a position to be able to postulate the series of events required to give rise to EBV-associated B-cell tumors and to suggest intervention strategies. More recently, the virus has been associated with a wider spectrum of diseases and tumors from a bewildering variety of tissue and cellular origins. In many of these situations we have no clear picture of the role of the virus, if any, in the etiology of the disease. It will require detailed molecular analysis based on our current knowledge of the virus to elucidate these disease processes.

REFERENCES

1. Given, D., and Kieff, E., 1978, DNA of Epstein–Barr virus. IV Linkage map for restriction enzyme fragments of the B95-8 and Wai strains of EBV, *J. Virol.* **21**:524–542.
2. Dambaugh, T., Beisel, C., Hummel, M., King, W., Fennewald, S., Cheung, A., Heller, M., Raab-Traub, N., and Kieff E., 1980, EBV DNA. VII Molecular cloning and detailed mapping of EBV (B95-8) DNA, *Proc. Natl. Acad. Sci. USA* **77**:2999–3003.
3. Baer, R., Bankier, A. T., Biggin, M. D., Deininger, P. L., Farrell, P. J., Gibson, T. J., Hatfull, G., Hudson, G. S., Satchwell, S. C., Seguin, C., Tuffnell, P. S., and Barrell, B. G., 1984, DNA sequence and expression of the B95-8 Epstein–Barr virus genome, *Nature* **310**:207–211.
4. Dambaugh, T., Wang, F., Hennessy, K., Woodland, E., Rickinson, A., and Kieff, E., 1986, Expression of the Epstein–Barr virus nuclear protein 2 in rodent cells, *J. Virol.* **59**:453–462.

5. Arrand, J. R., Young, L. S., and Tugwood, J. D., 1989, Two families of sequences in the small RNA-encoding region of Epstein–Barr virus (EBV) correlate with EBV types A and B, *J. Virol.* **63:**983–986.
6. Rowe, M., Young, L. S., Cadwallader, K., Petti, L., and Kieff, E., 1989, Distinction between Epstein–Barr virus type A (EBNA 2A) and type B (EBNA 2B) isolates extends to the EBNA 3 family of nuclear proteins, *J. Virol.* **63:**1031–1039.
7. Rickinson, A. B., Young, L. S., and Rowe, M., 1987, Influence of the Epstein–Barr virus nuclear antigen EBNA2 on the growth phenotype of virus-transformed B cells, *J. Virol.* **61:**1310–1317.
8. Zimber, U., Adldinger, H. K., Lenoir, G. M., Vuillauma, M., Knebel-Doeberitz, M. V., Laux, G., Desgranges, C., Wittmann, P., Freese, U.-K., Schneider, U., and Bornkamm, G. W., 1986, Geographical prevalence of two types of Epstein–Barr virus, *Virology* **154:**56–66.
9. Sculley, T. B., Apolloni, A., Hurren, L., Moss, D. J., and Cooper, D. A., 1990, Coinfection with A- and B-type Epstein–Barr virus in human immunodeficiency virus-positive subjects, *J. Infect. Dis.* **162:**643–648.
10. Frank, A., Ambiman, W., and Miller, G., 1976, Epstein–Barr virus and non-human primates: Natural and experimental infection, *Adv. Cancer Res.* **23:**171–210.
11. Pattengale, P. K., Smith, R. W., and Gerber, P., 1973, Selective transformation of B lymphocytes by EB virus, *Lancet* **2:**93–94.
12. Pope, J. H., Horne, M. K., and Scott, W., 1968, Transformation of foetal human leucocytes *in vitro* by filtrates of a human leukaemia cell line containing herpes-like virus, *Int. J. Cancer* **3:**857–866.
13. Sixbey, J. W., Vesterinen, E. H., Nedrud, J. G., Raab-Traub, N., Walton, L. A., and Pagano, J. S., 1983, Replication of Epstein–Barr virus in human epithelial cells infected *in vitro, Nature* **306:**480–483.
14. Tosato, G., Blaese, M. R., and Yarchoan, R., 1985, Relationship between immunoglobulin production and immortalization by Epstein–Barr virus, *J. Immunol.* **135:**959–964.
15. Steel, C. M., Philipson, J., Arthur, E., Gardiner, S. E., Newton, M. S., and McIntosh, R. V., 1977, Possibility of EB virus preferentially transforming a subpopulation of human B lymphocytes, *Nature* **270:**729–731.
16. Nemerow, G. R., Wolfert, R., McNaughton, M. E., and Cooper, N. R., 1985, Identification and characterisation of the Epstein–Barr virus receptor on human B lymphocytes and its relationship to the C3d complement receptor (CR2), *J. Virol.* **55:**347–351.
17. Bradbury, L. E., Kansas, G. S., Levy, S., Evans, R. L., and Tedder, T. T., 1992, The CD19/CD21 signal transducing complex of human B lymphocytes includes the target of anti proliferative antibody-1 and Leu-13 molecules, *J. Immunol.* **149:**2841–2850.
18. Hammerschmidt, W., and Sugden, B., 1989, Genetic analysis of immortalising functions of Epstein–Barr virus in human B lymphocytes, *Nature* **340:**393–397.
19. Cohen, J. I., Wang, F., Mannick, J., and Kieff, E., 1989, Epstein–Barr virus nuclear protein 2 is a key determinant of lymphocyte transformation, *Proc. Natl. Acad. Sci. USA* **86:**9558–9562.
20. Tomkinson, B., Robertson, E., and Kieff, E., 1993, Epstein–Barr virus nuclear proteins EBNA-3A and EBNA-3C are essential for B-lymphocyte growth transformation, *J. Virol.* **67:**2014–2025.
21. Kaye, K. M., Izumi, K. M., and Kieff, E., 1993, Epstein–Barr virus latent membrane protein 1 is essential for B lymphocyte growth transformation, *Proc. Natl. Acad. Sci. USA* **90:**9150–9154.
22. Tomkinson, B., and Kieff, E., 1992, Use of second-site homologous recombination to demonstrate that Epstein–Barr virus nuclear protein 3B is not important for lymphocyte infection or growth transformation *in vitro, J. Virol.* **66:**2893–2903.
23. Longnecker, R., Miller, C. L., Tomkinson, B., Miao, X-Q., and Kieff, E., 1993, Deletion of DNA encoding the first five transmembrane domains of Epstein–Barr virus latent membrane proteins 2A and 2B, *J. Virol.* **67:**5068–5074.
24. Kim, O.-J., and Yates, J. L., 1993, Mutants of Epstein–Barr virus with a selective marker

disrupting the TP gene transform B cells and replicate normally in culture, *J. Virol.* **67:**7634–7640.
25. Swaninathan, S., Tomkinson, B., and Kieff, E., 1991, Recombinant Epstein–Barr virus with small RNA (EBER) genes deleted transforms lymphocytes and replicates *in vitro*, *Proc. Natl. Acad. Sci. USA* **88:**15446–1550.
26. Karran, L., Gao, Y., Smith, P. R., and Griffin, B. E., 1992, Expression of a family of complementary-strand transcripts in Epstein–Barr virus-infected cells, *Proc. Natl. Acad. Sci. USA* **89:**8058–5062.
27. Smith, P., Gao, Y., Karran, L., Jones, M. D., Snudden, D., and Griffin, B. E., 1993, Complex nature of the major viral polyadenylated transcripts in Epstein–Barr virus-associated tumors, *J. Virol.* **67:**3217–3225.
28. Alfieri, C., Birkenbach, M., and Kieff, E., 1991, Early events in Epstein–Barr virus infection of human B lymphocytes, *Virology* **181:**595–608.
29. Jiang, W.-Q., Szekely, L., Wendel-Hausen, V., Ringertz, N., Klein, G., and Rosen, A., 1991, Colocalisation of the retinobastoma protein and the Epstein–Barr virus-encoded nuclear antigen EBNA-5, *Exp. Cell Res.* **197:**314–318.
30. Szekely, L., Selivanova, G., Magnusson, K. P., Klein, G., and Wiman, K. G., 1993, EBNA-5, an Epstein–Barr virus-encoded nuclear antigen, binds to the retinoblastoma and p53 proteins, *Proc. Natl. Acad. Sci. USA* **90:**5455–5459.
31. Woisetschlaeger, M., Jin, X. W., Yandava, C. N., Furmanski, L. A., Strominger, J. L., and Speck, S. H., 1991, Role for the Epstein–Barr virus nuclear antigen 2 in viral promoter switching during initial stages of infection, *Proc. Natl. Acad. Sci. USA* **88:**3942–3946.
32. Fåhraeus, R., Jansson, A., Ricksten, A., Sjoblom, A., and Rymo, L., 1990, Epstein–Barr virus-encoded nuclear antigen 2 actives the viral latent membrane protein promoter by modulating the activity of a negative regulatory element, *Proc. Natl. Acad. Sci. USA* **87:**7390–7394.
33. Zimber-Strobl, U., Kremmer, E., Grässer, F., Marschall, G., Laux, G., and Bornkamm, G. W., 1993, The Epstein–Barr virus nuclear antigen 2 interacts with an EBNA2 responsive *cis*-element of the terminal protein 1 gene promoter, *EMBO J.* **12:**167–175.
34. Knutson, J. C., 1990, The level of c-*fgr* RNA is increased by EBNA2, an Epstein–Barr virus gene required for B cell immortalisation, *J. Virol.* **64:**2530–2536.
35. Finke, J., Fritzen, R., Ternes, P., Trivedi, P., Bross, K. J., Lange, W., Mertelsmann, R., and Dölken, G., 1992, Expression of *bcl*-2 in Burkitt's lymphoma cell lines: Induction by latent Epstein–Barr virus genes, *Blood* **80:**459–469.
36. Wang, F., Gregory, C., Sample, C., Rowe, M., Liebowitz, D., Murray, R., Rickinson, A. G., and Kieff, E., 1990, Epstein–Barr virus latent membrane protein (LMP1) and nuclear protein 2 and 3C are effectors of phenotypic changes in B lymphocytes: EBNA-2 and LMP1 cooperatively induce CD23, *J. Virol.* **64:**2309–2318.
37. Thorley-Lawson, D. A., and Mann, K. P., 1985, Early events in Epstein–Barr virus infection provide a model for B cell activation. *J. Exp. Med.* **162:**45–59.
38. Conrad, D. H., 1990, The low affinity receptor for IgE, *Annu. Rev. Immunol.* **8:**623.
39. Swenderman, S., and Thorley-Lawson, D. A., 1987, The activation antigen BLAST-2, when shed, is an autocrine BCGF for normal and transformed B cells, *EMBO J.* **6:**1637–1642.
40. Ling, P. D., Hsieh, J. J.-D., Ruf, I. K., Rawlins, D. R., and Hayward, S. D., 1994, EBNA-2 upregulation of Epstein–Barr virus latency promoters and the cellular CD23 promoter utilizes a common targeting intermediate, CBF1, *J. Virol.* **68:**5375–5383.
41. Waltzer, L., Logeat, F., Brou, C., Israel, A., Sergeant, A., and Manet, E., 1994, The human Jk recombination signal sequence binding protein (RBP-Jk) targets the Epstein–Barr virus EBNA 2 protein to its DNA responsive elements, *EMBO J.* **13:**5633–5638.
42. Johannesen, E., Koh, E., Mosialos, G., Tong, X., Keiff, E., and Grossman, S. R., 1995, Epstein–Barr virus nuclear protein 2 transactivation of the latent membrane protein 1 promoter is mediated by Jk and PU1, *J. Virol.* **69:**253–262.

43. Allday, M. J., Crawford, D. H., and Griffin, B. E., 1989, Epstein–Barr virus latent gene expression during the initiation of B cell immortalisation, *J. Gen. Virol.* **70:**1755–1764.
44. Yates, J. L., Warren, N., and Sugden, B., 1985, Stable replication of plasmids derived from Epstein–Barr virus in mammalian cells, *Nature* **313:**812–815.
45. Allday, M. J., Crawford, D. H., and Thomas, J. A., 1993, Epstein–Barr virus (EBV) nuclear antigen 6 induces expression of the EBV latent membrane protein and an activated phenotype in Raji cells, *J. Gen. Virol.* **74:**361–369.
46. Wang, D., Liebowitz, D., and Kieff, E., 1985, An EBV membrane protein expressed in immortalized lymphocytes transforms established rodent cells, *Cell* **43:**831–840.
47. Wang, D., Liebowitz, D., Wang, F., Gregory, C., Rickinson, A., Larson, R., Springer, T., and Kieff, E., 1988, Epstein–Barr virus latent infection membrane protein alters the human B-lymphocyte phenotype: Deletion of the amino terminus abolishes activity, *J. Virol.* **62:**4137–4184.
48. Henderson, S., Rowe, M., Gregory, C., Croom-Carter, D., Wang, F., Longnecker, R., Kieff, E., and Rickinson, A., 1991, Induction of *bcl*-2 expression by Epstein–Barr virus latent membrane protein 1 protects infected B cells from programmed cell death, *Cell* **65:**1107–1115.
49. Hammarskjold, M.-L., and Simurdo, M. C., 1992, Epstein–Barr virus latent membrane protein *trans*-activates the human immunodeficiency virus type I long terminal repeat through induction of NF-kB activity, *J. Virol.* **66:**6496–6501.
50. Moorthy, R., and Thorley-Lawson, D. A., 1990, Processing of the Epstein–Barr virus-encoded latent membrane protein P63/LMP, *J. Virol.* **64:**829–837.
51. Baichwal, V. R., and Sugden, B., 1989, The multiple membrane-spanning segments of the BNLF-1 oncogene from Epstein–Barr virus are required for transformation, *Oncogene* **4:**67–74.
52. Moorthy, R. K., and Thorley-Lawson, D. A., 1993, All three domains of the Epstein–Barr virus-encoded latent membrane protein LMP-1 are required for transformation of Rat-1 fibroblasts, *J. Virol.* **67:**1638–1646.
53. Huen, D. S., Henderson, S. A., Croom-Carter, D., and Rowe, M., The Epstein–Barr virus latent membrane protein-1 (LMP 1) mediates activation of NK-kB and cell surface phenotype via two effector regions in its carboxy-terminal cytoplasmic domain, *Oncogene* **10**(3):549–560.
54. Laux, G., Perricaudet, M., and Farrell, P. J., 1988, A spliced Epstein–Barr virus gene expressed in immortalized lymphocytes is created by circularization of the linear viral genome, *EMBO J.* **7:**769–774.
55. Longnecker, R., Druker, B., Roberts, T., and Kieff, E., 1991, An Epstein–Barr virus protein associated with cell growth transformation interacts with a tyrosine kinase, *J. Virol.* **65:**3681–3692.
56. Miller, C. L., Lee, J. H., Kieff, E., and Longnecker, R., 1994, An integral membrane protein (LMP2) blocks reactivation of Epstein–Barr virus from latency following surface immunoglobulin crosslinking, *Proc. Natl. Acad. Sci. USA* **91:**772–776.
57. Nilsson, K., Klein, G., Henle, W., and Henle, G., 1971, The establishment of lymphoblastoid lines from adult and foetal human lymphoid cells and its dependence on EBV, *Int. J. Cancer* **8:**443–450.
58. Crawford, D. H., and Ando, I., 1986, EB virus induction is associated with B cell maturation, *Immunology* **59:**405–409.
59. Zur Hausen, H., O'Neill, F. J., and Freese, U. K., 1978, Persisting oncogenic herpesvirus induced by the tumour promoter TPA, *Nature* **272:**373–375.
60. Luka, J., Kallin, B., and Klein, G., 1979, Induction of the Epstein–Barr virus (EBV) cycle in latently infected cells by *n*-butyrate, *Virology* **94:**228–231.
61. Tovey, M. G., Lenoir, G., and Begon-Lours, J., 1978, Activation of latent Epstein–Barr virus by antibody to human IgM, *Nature* **276:**270–272.
62. Flemington, E., and Speck, S. H., 1990, Identification of phorbol ester response elements in the promotor of Epstein–Barr virus putative lytic switch gene BZLF1, *J. Virol.* **64:**1217–1226.

63. Grogan, E., Jenson, H., Countryman, J., Heston, L., Gradoville, L., and Miller, G., 1987, Transfection of a rearranged viral DNA fragment, WZhet, stably converts latent Epstein–Barr virus infection to productive infection in lymphoid cells, *Proc. Natl. Acad. Sci. USA* **84**:1332–1336.
64. Niederman, J. C., Evans, A. S., Subrahmanyan, L., and McCollum, R. W., 1970, Prevalence, incidence and persistence of EB virus antibody in young adults, *N. Engl. J. Med.* **282**:361–365.
65. Klein, G., Svedmyr, E., Jondal, U., and Persson, P. O., 1976, EBV-determined nuclear antigens (EBNA)-positive cells in the peripheral blood of infectious mononucleosis patients, *Int. J. Cancer* **17**:21–26.
66. Tierney, R. J., Steven, N., Young, L. S., and Rickinson, A. B., 1994, Epstein–Barr virus latency in blood mononuclear cells: Analysis of viral gene transcription during primary infection and in the carrier state, *J. Virol.* **68**:7374–7385.
67. Lam, K. M.-C., Whittle, H., Grzywacz, M., and Crawford, D. H., 1994, Epstein–Barr virus carrying B cells are large, surface IgM, IgD bearing cells in normal individuals and acute malaria patients, *Immunology* **82**:383–388.
68. Moss, D. J., Rickinson, A. B., and Pope, J. H., 1978, Long-term T-cell-mediated immunity to Epstein–Barr virus in man. I. Complete regression of virus-induced transformation in cultures of seropositive donor leucocytes, *Int. J. Cancer* **22**:662–668.
69. Murray, R. J., Kurilla, M. G., Brooks, J. M., Thomas, W. A., Rowe, M., Kieff, E., and Rickinson, A. B., 1992, Identification of target antigens for the human cytotoxic T cell response to Epstein–Barr virus (EBV): Implication for the immune control of EBV-positive malignancies. *J. Exp. Med.* **176**:157–168.
70. Klein, G., 1989, Viral latency and transformation: The strategy of Epstein–Barr virus, *Cell* **58**:5–8.
71. Allday, M. J., and Crawford, D. H., 1988, The role of epithelium in EBV persistence and the pathogenesis of B cell tumours, *Lancet* **1**:855–858.
72. Sixbey, J. W., 1989, Epstein–Barr virus and epithelial cells, *Adv. Viral Oncol.* **8**:187–202.
73. Li, Q. X., Young, L. S., Niedobitek, G., Dawson, C. W., Birkenbach, M., Wang, F., and Richinson, A. B., 1992, Epstein–Barr virus infection and replication in a human epithelial cell system, *Nature* **356**:347–350.
74. Qu, L., and Rowe, D. T., 1992, Epstein–Barr virus latent gene expression in uncultured peripheral blood lymphocytes, *J. Virol.* **66**:3715–3724.
75. Farrell, P. J., and Sinclair, A. J., 1994, Burkitt's lymphoma, in: *Viruses and Cancer*, Vol. 51 (A. Mison, J. Neil, and M. McCrae, eds.), Cambridge University Press, Cambridge, pp. 101–121.
76. Thomas, J. A., Allday, M. J., and Crawford, D. H., 1991, Epstein–Barr virus-associated lymphoproliferative disorders in immunocompromised individuals, *Adv. Cancer Res.* **57**:329–380.
77. Herbst, H., and Niedobitek, G., 1994, Epstein–Barr virus in Hodgkin's disease, *Epstein–Barr Virus Rep.* **1**:31–35.
78. Raab-Traub, N., 1992, The Epstein–Barr virus and nasopharyngeal carcinoma, in: *Seminars in Cancer Biology*, Vol. 3 (A. B. Rickinson, ed.), Saunders Scientific Publications/Academic Press, London, pp. 297–307.
79. Thomas, J. A., Cotter, F., Hanby, A. M., Long, L. Q., Morgan, P. R., Bramble, B., and Bailey, B. M. W., 1993, Epstein–Barr virus-related oral T-cell lymphoma associated with human immunodeficiency virus immunosuppression, *Blood* **81**:3350–3356.
80. Jones, J. F., Shurin, S., Abramowsky, C., Tubbs, R. R., Sciotto, C. G., Wahl, R., Sands, J., Gottman, D., Katz, B. Z., and Sklar, J., 1988, T-cell lymphomas containing Epstein–Barr viral DNA in patients with chronic Epstein–Barr virus infections, *N. Engl. J. Med.* **318**:733–741.
81. Imai, S., Koizumi, S., Sujuira, M., Tokunaja, M., Uemura, Y., Yamamoto, N., Tanaka, S., Sato, E., and Osato, T., 1994, Gastric carcinoma: Monoclonal epithelium malignant cells expressing Epstein–Barr virus latent infection protein, *Proc. Natl. Acad. Sci. USA* **91**:9131–9135.
82. Raab-Traub, N., Rajadurai, P., Flynn, K., and Lanier, A. P., 1991, Epstein–Barr virus infection in carcinoma of the salivary glands, *J. Virol.* **65**(12):7032–7036.

83. McClain, K. L., Leach, C. T., Jenson, H. B., Joshi, V. V., Pollock, B. H., Parmley, R. T., DiCarlo, F. J., Gould Chadwick, E., and Murphy, S. B., 1995, Association of Epstein–Barr virus with leiomyosarcoma in young people with AIDS, *N. Engl. J. Med.* **332:**12–18.
84. Lee, E. S., Locker, J., Nalesnik, M., Reyes, J., Jaffe, R., Alashari, M., Nour, B., Tzakis, A., and Dickman, P. S., 1995, The association of Epstein–Barr virus with smooth-muscle tumors occurring after organ transplantation, *N. Engl. J. Med.* **332:**19–25.
85. Crawford, D. H., Sweny, P., Edwards, J. M. B., Janossy, G., and Hoffbrand, A. V., 1981, Long-term T-cell-mediated immunity to Epstein–Barr virus in renal-allograft recipients receiving cyclosporin A, *Lancet* **1:**10–12.
86. Opelz, G., and Henderson, R., 1993, Incidence of non-Hodgkin lymphoma in kidney and heart transplant recipients, *Lancet* **342:**1514–1516.
87. Thomas, J. A., Hotchin, N., Allday, M. J., Amlot, P., Rose, M., Yacoub, M., and Crawford, D. H., 1990, Immunohistology of Epstein–Barr virus associated antigens in B cell disorders from immunocompromised individuals, *Transplantation* **49:**944–953.
88. Frizzera, G., 1987, The clinico-pathological expressions of Epstein–Barr virus infection in lymphoid tissues, *Virchows Arch. [B]* **53:**1–12.
89. Beral, V., Peterman, T., Berkelman, R., and Jaffe, H., 1991, AIDS-associated non-Hodgkin lymphoma, *Lancet* **337:**805–809.
90. Ho, M., Jaffe, R., Miller, G., Breinig, M. K., Dummer, J. S., Makowka, L., Atchison, R. W., Karrer, F., Nalesnik, M. A., and Starzl, T. E., 1988, The frequency of Epstein–Barr virus infection and associated lymphoproliferative syndrome after transplantation and its manifestations in children, *Transplantation* **45:**719–727.
91. Starzl, T. E., Porter, K. A., Iwatsuki, S., Rosenthal, J. T., Shaw, B. W., Atchison, R. W., Nalesnik, M. A., Ho, M., Griffith, B. P., Hakala, T. R., Hardesty, R. L., Jaffe, R., and Bahnson, H. T., 1984, Reversibility of lymphomas and lymphoproliferative lesions developing under cyclosporin-steroid therapy, *Lancet* **1:**583–587.
92. Armitage, J. M., Kormos, R. L., Stuart, S., Fricker, F. J., Griffith, B. P., Nalesnik, M., Hardesty, R. L., and Dummer, J. S., 1991, Posttransplant lymphoproliferative disease in thoracic organ transplant patients: Ten years of cyclosporine-based immunosuppression, *J. Heart Lung Transplant* **10:**877–887.
93. Rooney, C. M., Smith, C. A., Ng, C. Y. C., Loftin, S., Li, C., Krance, R. A., Brenner, M. K., and Heslop, H. E., 1995, Use of gene-modified virus-specific T lymphocytes to control Epstein–Barr virus-related lymphoproliferation, *Lancet* **345:**9–12.
94. Hartwell, L. H., and Castan, M. B., 1994, Cell cycle control and cancer, *Science* **266:**1821–1828.
95. Epstein, M. A., and Barr, Y. M., 1964, Virus particles in cultured lymphoblasts from Burkitt's lymphoma, *Lancet* **1:**702–703.
96. Magrath, I., 1990, The pathogenesis of Burkitt's lymphoma, *Adv. Cancer Res.* **55:**130–270.
97. Burkitt, D., 1962, A children's cancer dependent on climatic factors, *Nature* **194:**232–234.
98. Epstein, M. A., and Achong, B. G., 1979, The relationship of the virus to Burkitt's lymphoma, in: *The Epstein–Barr Virus*, (M. A. Epstein, and B. G. Achong, eds.), Springer, Berlin, pp. 321–337.
99. Subar, M., Neri, A., Inghirami, G., Knowles, D. M., and Dalla-Favera, R., 1988, Frequent c-*myc* oncogenes activation and infrequent presence of Epstein–Barr virus genome in AIDS-associated lymphoma, *Blood* **72:**667–671.
100. Gregory, C. D., Edwards, C. F., Milner, A., Wiels, J., Lipinski, M., Rowe, M., Tursz, T., and Rickinson, A. B., 1988, Isolation of a normal B cell subset with a Burkitt-like phenotype and transformation *in vitro* with Epstein–Barr virus, *Int. J. Cancer* **42:**213–220.
101. Raab-Traub, N., and Flynn, K., 1986, The structure of the termini of the Epstein–Barr virus as a marker of clonal cellular proliferation, *Cell* **47:**883–889.
102. Rowe, D. T., Rowe, M., Evans, G. I., Wallace, L. E., Farrell, P. J., and Rickinson, A. B., 1986, Restricted expression of EBV latent genes and T-lymphocyte-detected membrane antigen in Burkitt's lymphoma cells, *EMBO J.* **5:**2599–2607.

103. Brooks, L. A., Lear, A. L., Young, L. S., and Rickinson, A. B., 1993, Transcripts from the Epstein–Barr virus *Bam*HI A fragment are detectable in all three forms of virus latency, *J. Virol.* **67**:3182–3190.
104. Allday, M. J., Kundu, D., Finerty, S., and Griffin, B. E., 1990, CpG methylation of viral DNA in EBV-associated tumours, *Int. J. Cancer* **45**:1125–1130.
105. Altiok, E., Minarovits, J., Hu, L. F., Contrerasbrodin, B., Klein, G., and Ernberg, I., 1992, Host-cell-phenotype-dependent control of the BCR2/BWR1 promoter complex regulates the expression of Epstein–Barr virus nuclear antigens 2-6, *Proc. Natl. Acad. Sci. USA* **89**:905–909.
106. Sample, J., Brooks, L., Sample, C., Young, L., Rowe, M., Gregory, C., Rickinson, A., and Kieff, E., 1991, Restricted Epstein–Barr virus protein expression in Burkitt lymphoma is due to a different Epstein–Barr virus nuclear antigen 1 transcriptional initiation site, *Proc. Natl. Acad. Sci. USA* **88**:6343–6347.
107. Delecluse, H.-J., Bartnizke, S., Hammerschmidt, W., Bullerdiek, J., and Bornkamm, G. W., 1993, Episomal and integrated copies of Epstein–Barr virus coexist in Burkitt lymphoma cell lines, *J. Virol.* **67**:1292–1299.
108. Sixbey, J. W., 1994, Defective Epstein–Barr virus and viral disease processes, *Epstein–Barr Virus Rep.* **1**:133–137.
109. Zimber-Strobl, U., Suentzenich, K.-O., Laux, G., Eick, D., Cordier, M., Calender, A., Billaud, M., Lenoir, G. M., and Bornkamm, G. W., 1991, Epstein–Barr virus nuclear antigen 2 activates transcription of the terminal protein gene, *J. Virol.* **65**:415–423.
110. Rowe, M., Lear, A. L., Croom-Carter, D., Davies, A. H., and Rickinson, A. B., 1992, Three pathways of Epstein–Barr virus gene activation from EBNA1-positive latency in B lymphocytes, *J. Virol.* **66**:122–131.
111. Rowe, M., Rowe, D. T., Gregory, C. D., Young, L. S., Farrell, P. J., Rupani, H., and Rickinson, A. B., 1987, Differences in B cell growth phenotype reflect novel patterns of Epstein–Barr virus latent gene expression in Burkitt's lymphoma cells, *EMBO J.* **6**:2743–2751.
112. Masucci, M. G., Contreras-Salazar, B., Ragnar, E., Falk, K., Minarovits, J., Ernberg, I., and Klein, G., 1989, 5-Azacytidine upregulates the expression of Epstein–Barr virus nuclear antigen 2(EBNA2) through EBNA6 and latent membrane protein in the Burkitt's lymphoma line Rael, *J. Virol.* **63**:3135–3141.
113. Shiramizu, B., Barriga, F., and Neequaye, J., 1991, Patterns of chromosomal breakpoint locations in Burkitt's lymphoma: Relevance to geography and Epstein–Barr virus association, *Blood* **77**:1516–1526.
114. Neri, A., Barriga, F., Knowles, D. M., McGrath, I. T., and Dalla-Favera, R., 1988, Different regions of the immunoglobulin heavy-chain locus are involved in chromosomal translocations in distinct pathogenetic forms of Burkitt lymphoma, *Proc. Natl. Acad. Sci. USA* **85**:2748–2752.
115. Kretzner, L., Blackwood, E. M., and Eisenman, R. N., 1992, Myc and max proteins possess distinct transcriptional activities, *Nature* **359**:426–429.
116. Lombardi, L., Newcomb, E. W., and Dalla-Favera, R., 1989, Pathogenesis of Burkitt lymphoma: Expression of an activated c-myc oncogene causes the tumorigenic conversion of EBV-infected human B lymphoblasts, *Cell* **59**:161–170.
117. Hotchin, N. A., Allday, M. J., and Crawford, D. H., 1990, Deregulated c-*myc* expression in Epstein–Barr virus-immortalised B-cells induces altered growth properties and surface phenotype but not tumorigenicity, *Int. J. Cancer* **45**:566–571.
118. Langdon, W. Y., Harris, A. W., Cory, S., and Adams, J. M., 1986, The c-*myc* oncogene perturbs B lymphocyte development in Eμ-*myc* transgenic mice, *Cell* **47**:11–18.
119. Rabbitts, T. H., Hamlyn, P. H., and Baer, R., 1983, Altered nucleotide sequences of a translocated c-*myc* gene in Burkitt lymphoma, *Nature* **306**:760–765.
120. Vousden, K. H., Crook, T., and Farrell, P. J., 1993, Biological activities of p53 mutants in Burkitt's lymphoma cells, *J. Gen. Virol.* **74**:803–810.
121. Ho, M., Webster, H. K., Looareesuwan, S., Supanaranond, W., Phillips, R. E., Chanthavanich, P., and Warrell, D. A., 1986, Antigen-specific immunosuppression in human malaria due to *P. falciparum*, *J. Infect. Dis.* **153**:763.

122. Greenwood, B. M., Oduloju, A. J., and Platts-Mills, T. A. E., 1979, Partial characterisation of a malaria mitogen, *Trans. R. Soc. Trop. Med. Hyg.* **73:**178.
123. Clifford-Lane, H., Masur, H., Edgar, L. C., Whalen, G., Rook, A. H., and Fauci, A., 1983, Abnormalities of B cell activation and immunoregulation in patients with the acquired immunodeficiency syndrome, *N. Engl. J. Med.* **309:**453–458.
124. Whittle, H. C., Brown, J., Marsh, K., Greenwood, B. M., Seidelin, P., Tighe, H., and Wedderburn, L., 1984, T-cell control of Epstein–Barr virus-infected B cells is lost during *P. falciparum* malaria, *Nature* **312:**449–450.
125. Birx, D. L., Redfield, R. R., and Tosato, G., 1986, Defective regulation of Epstein–Barr virus infection in patients with acquired immunodeficiency syndrome (AIDS) or AIDS-related disorders, *N. Engl. J. Med.* **314:**874–879.
126. Lam, K. M.-C., Syed, N., Whittle, H., and Crawford, D. H., 1991, Circulating Epstein–Barr virus-carrying B cells in acute malaria, *Lancet* **337:**876–878.
127. Lenoir, G., and Bornkamm, G., 1987, Burkitt's lymphoma, a human cancer model for the study of the multistep development of cancer: Proposal for a new scenario, *Adv. Viral Oncol.* **6:**173–206.
128. MacMahon, B., 1957, Epidemiological evidence of the nature of Hodgkin's disease, *Cancer* **10:**1045–1054.
129. Rosdahl, N., Larsen, S. O., and Clemmesen, J., 1974, Hodgkin's disease in patients with previous infectious mononucleosis: 30 years' experience, *Br. Med. J.* **2:**243–256.
130. Mueller, N., 1991, An epidemiologist's view of the new molecular biology findings in Hodgkin's disease, *Ann. Oncol.* **2:**23–28.
131. Weiss, L. M., Strikler, J. G., Warnke, R. A., Purtilo, D. T., and Sklar, J., 1987, Epstein–Barr viral DNA in tissues of Hodgkin's disease, *Am. J. Pathol.* **129:**86–91.
132. Weiss, L. M., Movahed, L. A., Roger, B. S., Warnke, R. A., and Sklar, J., 1989, Detection of Epstein–Barr viral genomes in Reed–Sternberg cells of Hodgkin's disease. *N. Engl. J. Med.* **320:**502–506.
133. Anagnostopoulos, I., Herbst, H., Niedobitek, G., and Stein, H., 1989, Demonstration of monoclonal EBV genomes in Hodgkin's disease and KI-1-positive anaplastic large cell lymphoma in combined Southern blot and *in situ* hybridization. *Blood* **74:**810–816.
134. Pallesen, G., Hamilton-Dutoit, S. J., Rowe, M., and Young, L. S., 1991, Expression of Epstein–Barr virus latent gene products in tumour cells of Hodgkin's disease, *Lancet* **337:**320–322.
135. Ambinder, R. F., Browing, P. J., Lorenzana, I., Leventhal, B. G., Cozenza, H., Mann, R. B., MacMahon, E. M. E., Medina, R., Cardona, V., Grufferman, S., Olshan, A., Levin, A., Petersen, E. A., Blattner, W., and Levine, P. H., 1993, Epstein–Barr virus and childhood Hodgkin's disease in Honduras and the United States, *Blood* **81:**462–467.
136. Murray, P. G., Young, L. S., Rowe, M., and Crocker, J., 1992, Immunohistochemical demonstration of the Epstein–Barr virus-encoded latent membrane protein in paraffin sections of Hodgkin's disease, *J. Pathol.* **166:**1–5.
137. Chang, K. L., Albujar, P. F., Chen, Y. Y., Johnson, R. M., and Weiss, L. M., 1993, High prevalence of Epstein–Barr virus in the Reed–Sternberg cells of Hodgkin's disease occuring in Peru, *Blood* **81:**496–501.
138. Boyle, M. J., Vasak, E., Tschuchnigg, M., Turner, J. J., Sculley, T., Penny, R., Cooper, D. A., Tindall, B., and Sewell, W. A., 1993, Subtypes of Epstein–Barr virus (EBV) in Hodgkin's disease: Association between B-type EBV and immunocompromise, *Blood* **81:**468–474.
139. Herbst, H., Steinbrecher, E., Niedobitek, G., Young, L. S., Brooks, L., Müller-Lantzsch, N., and Stein, H., 1992, Distribution and phenotype of Epstein–Barr virus-harboring cells in Hodgkin's disease, *Blood* **80:**484–491.
140. O'Grady, J., Stewart, S., Elton, R. A., and Krajweski, A. S., 1994, EBV and Hogkin's disease, *Lancet* **343:**265–266.
141. Deacon, E. M., Pallesen, G., Niedobitek, G., Crocker, J., Brooks, L., Rickinson, A. B., and Young, L. S., 1993, Epstein–Barr virus and Hodgkin's disease: Transcriptional analysis of virus latency in the malignant cells, *J. Exp. Med.* **177:**339–349.
142. Pallesen, G., Sandvej, K., Hamilton-Dutoit, S. J., Rowe, M., and Young, L. S., 1991, Activation of Epstein–Barr virus replication in Hodgkin and Reed–Sternberg cells, *Blood* **78:**1162–1165.

143. Armstrong, A. A., Gallager, A., Krayewski, A. S., Jones, D. B., Wilkins, B. S., Onions, D. E., and Jarret, R. F., 1992, The expression of the EBV latent membrane protein (LMP-1) is independent of CD23 and *bcl*-2 in Reed-Sternberg cells in Hodgkin's disease, *Histopathology* **21**:72-73.
144. Knecht, H., Bachmann, E., Joske, D. J. L., Sahli, R., Emery-Goodman, A., Casanova, J.-L., Zilic, M., Bachmann, F., and Odermatt, B. F., 1993, Molecular analysis of the LMP (latent membrane protein) oncogene in Hodgkin's disease, *Leukemia* **7**:580-585.
145. Henle, W., and Henle, G., 1976, Epstein-Barr virus-specific IgA serum antibodies as an outstanding feature of nasopharyngeal carcinoma, *Int. J. Cancer* **17**:1-7.
146. Wolf, H., Zur Hausen, H., and Becker, V., 1973, EB virus genomes in epithelial nasopharyngeal carcinoma cells, *Nature [New Biol.]* **244**:245-247.
147. Huang, D. P., Ho, J. H. C., Henle, W., and Henle, G., 1974, Demonstration of Epstein-Barr virus associated nuclear antigen in nasopharyngeal carcinoma cells from fresh biopsies, *Int. J. Cancer* **14**:580-588.
148. Klein, G., Giovanella, B. C., Landahl, T., Fialkow, P. J., Singh, S., and Stehlin, J., 1974, Direct evidence for the presence of Epstein-Barr virus DNA and nuclear antigen in malignant epithelial cells from patients with anaplastic carcinoma of the nasopharynx, *Proc. Natl. Acad. Sci. USA* **71**:4737-4741.
149. Pathmanathan, K., Prasad, U., Flynn, K., Chandrika, M., and Raab-Traub, N., 1994, EBV infection in nasopharyngeal carcinoma and premalignant lesions: Gene expression and strain variation, in: *Epstein-Barr Virus and Associated Diseases*, (J. Pagano & A. Rickinson, eds.), Cold Spring Harbor Laboratory, Cold Spring Harbor, NY, p. 238.
150. Lung, M. L., Chang, R. S., Huang, M. L., Guo, H.-Y., Choy, D., Sham, J., Tsao, S. Y., Cheng, P., and Ng, M. H., 1990, Epstein-Barr virus genotypes associated with nasopharyngeal carcinoma in Southern China, *Virology* **185**:44-53.
151. Lung, M. L., Lam, W.P., Sham, J., Choy, D., Young-Sheng, Z., Guo, H.-Y., and Ng. M. H., 1991, Detection and prevelence of the "f" variant of Epstein-Barr virus in Southern China, *Virology* **185**:67-71.
152. Hu, L.-F., Zabarovsky, E. R., Chen, F., Cao, S.-L., Ernberg, I., Klein, G., and Winberg, G., 1991, Isolation and sequencing of the Epstein-Barr virus BNLF-1 gene (LMP1) from a Chinese nasopharyngeal carcinoma, *J. Gen. Virol.* **72**:2399-2409.
153. Yao, K., Zhang, H.-Y., Zhu, H.-C., Wang, F.-X., Li, G.-Y., Wen, D.-S., Li, Y.-P., Tsai, C.-H., A., and Glaser, R., 1990, Establishment and characterization of two epithelial tumour cell lines (HNE-1 and HONE-1) latently infected with Epstein-Barr virus and derived from nasopharyngeal carcinoma, *Int. J. Cancer* **45**:83-89.
154. Klein, G., Giovanella, B. C., Lindahl, T., Fialkow, P. J., Singh, S., and Stehlin, J., 1974, Direct evidence for the presence of Epstein-Barr virus DNA and nuclear antigen in malignant epithelial cells from patients with poorly differentiated carcinoma of the nasopharynx, *Proc. Natl. Acad. Sci. USA* **71**:4737-4741.
155. Busson, P., Ganem, G., Flores, P., Mugneret, F., Clausse, B., Caillou, B., Braham, K., Wakasugi, H., Lipinski, M., and Tursz, T., 1988, Establishment and characterization of three transplantable EBV-containing nasopharyngeal carcinomas, *Int. J. Cancer* **42**:599-606.
156. Brooks, L., Yao, Q. Y., Rickinson, A. B., and Young, L. S., 1992, Epstein-Barr virus latent gene transcription in nasopharyngeal carcinoma cells: coexpression of EBNA1, LMP1, and LMP2 transcripts, *J. Virol.* **66**:2689-2697.
157. Cochet, C., Martel-Renoir, D., Grunewald, V., Bosq, J., Cochet, G., Schwaab, G. Bernaudin, J.-F., and Joab, I., 1993, Expression of the Epstein-Barr virus immediate early gene, BZLF1, in nasopharyngeal carcinoma tumor cells, *Virology* **197**:358-365.
158. Fahraeus, R., Rymo, L., Rhim, J. S., and Klein, G., 1990, Morphological transformation of human keratinocytes expressing the LMP gene of Epstein-Barr virus, *Nature* **345**:447-449.
159. Dawson, C. W., Rickinson, A. B., and Young, L. S., 1990, Epstein-Barr virus latent membrane protein inhibits human epithelial cell differentiation, *Nature* **344**:777-780.
160. Wilson, J. B., Weinberg, W., Johnson, R., Yuspa, S., and Levine, A. J., 1990, Expression of the BNLF-1 oncogene of Epstein-Barr virus in the skin of transgenic mice induces hyperplasia and aberrant expression of keratin 6, *Cell* **61**:1315-1327.

161. Hu, L-F., Chen, F., Zheng, X., Ernberg, I., Cao, S.-L., Christensson, B., Klein, G., and Winberg, G., 1993, Clonability and tumorigenicity of human epithelial cells expressing the EBV encoded membrane protein LMP1, *Oncogene* **8:**1575–1583.
162. Nicholson, L., Hopwood, P., Johannessen, I., and Crawford, D. H., unpublished observation.
163. Lu, Q.-L., Elia, G., Lucas, S., and Thomas, J. A., 1993, Bcl-2 proto-oncogene expression in Epstein–Barr virus-associated nasopharyngeal carcinoma, *Int. J. Cancer* **53:**29–35.
164. Effert, P., McCoy, R., Abdel-Hamid, M., Flynn, K., Zhang, Q., Busson, P., Tursz, T., Liu, E., and Raab-Traub, N., 1992, Alterations of the p53 gene in nasopharyngeal carcinoma, *J. Virol.* **66:** 3768–3775.
165. Sun, Y., Hegamyer, G., Cheng, Y.-J., Hildesheim, A., Chen, J.-Y., Chen, I.-H., Cao, Y., Yao, K.-T., and Colburn, N. H., 1992, An infrequent point mutation of the p53 gene in human nasopharyngeal carcinoma, *Proc. Natl. Acad. Sci. USA* **89:**6516–6520.
166. Niedobitek, G., Agathanggelou, A., Barber, P., Smallman, L., Jones, E., and Young, L., 1993, p53 overexpression and Epstein–Barr virus infection in undifferentiated and squamous cell nasopharyngeal carcinomas, *J. Pathol.* **170:**457–461.
167. Thomas, J. A., Iliescu, V., Crawford, D. H., Ellouz, R., Cammoun, M., and De-Thé, G., 1984, Expression of HLA-DR antigens in nasopharyngeal carcinoma: An immunohistological analysis of the tumour cells and infiltrating lymphocytes, *Int. J. Cancer* **33:**813–819.
168. Busson, P., Zhang, Q., Guillon, J.-M., Gregory, C. D., Young, L. S., Clausse, B., Lipinksi, M., Rickinson, A. B., and Tursz, T., 1992, Elevated expression ICAM1 (CD45) and minimal expression of LFA3 (CD58) in Epstein–Barr virus-positive nasopharyngeal carcinoma cells, *Int. J. Cancer* **50:**863–867.
169. Simons, M. J., Wee, G. B., Chan, S. H., Shanmugaratnam, K., Day, N. E., and De-Thé, G. B., 1975, Probable identification of an HLA-A second-locus antigen associated with a high risk of nasopharyngeal carcinoma, *Lancet* **1:**142–143.
170. Lu, S.-J., Day, N.E., Degos, L., Lepage, V., Wang, P.-C., Chan, S.-H., Simons, M., McKnight, B., Easton, D., Zeng, Y., and De Thé, G., 1990, Linkage of a nasopharyngeal carcinoma susceptibility locus to the HLA region, *Nature* **346:**470–471.
171. Geser, A., Charney, N., Day, N. E., Ho, H. C., and De-Thé, G., 1978, Environmental features in the etiology of nasopharyngeal carcinoma, in: *Nasopharyngeal Carcinoma: Etiology and Control* (G. De-Thé and Y. Ito, eds.) IARC Scientific Publications no. 20, Lyon, pp. 213–229.
172. Anderson, E. N. Jr., Anderson, M. L., and Ho, H. C., 1978, Environmental backgrounds of young Chinese nasopharyngeal carcinoma patients, in: *Nasopharyngeal Carcinoma: Etiology and Control* (G. De-Thé and Y. Ito, eds.) IARC Scientific Publications no. 20, Lyon, pp. 231–239.
173. Huang, D. P., Ho, J. H. C., and Saw, D., and Teogh, T. B., 1978, Carcinoma of the nasal and paranasal regions in rats fed Cantonese salted marine fish, in: *Nasopharyngeal Carcinoma: Etiology and Control* (G. De-Thé and Y. Ito, eds.) IARC Scientific Publications no. 20, Lyon, pp. 315–328.
174. Huang, D. P., Ho, J. H. C., Webb, K. S., Wood, B. J., and Gough, T. A., 1981, Volatile nitrosamines in salt-preserved fish before and after cooking, *Fed. Cosmet. Toxicol.* **19:**167–171.
175. Bouvier, G., Polack, A., Traub, B., Bornkamm, G. W., Ohshima, H., Bartsch, H., and De-Thé, G., 1988, Food extracts from high risk areas for NPC induce an EBV early promoter, in: *Epstein–Barr Virus and Human Disease* (D. V. Ablashi, A. Faggioni, G. R. F. Krueger, J. S. Pagano, and G. R. Pearson, eds.) Humana Press, Clifton, New Jersey, pp. 501–504.
176. Ito, Y., 1986, Vegetable activators of the viral genome and the causation of Burkitt's lymphoma and nasopharyngeal carcinoma, in: *The Epstein–Barr Virus* (M. A. Epstein, and B. G. Achong, eds.), William Heineman, London, pp. 207–236.

20

Association of Epstein–Barr Virus with Hodgkin's Disease

MAURO BOIOCCHI, RICCARDO DOLCETTI, VALLI DE RE, ANTONINO CARBONE, and ANNUNZIATA GLOGHINI

1. INTRODUCTION

The nature of Hodgkin's disease (HD) and the origin of its characteristic cells, the Reed–Sternberg (RS) cells, are still obscure in spite of the abundant literature on this disorder. In recent years, the concept that HD is a malignant lymphoma has become widely accepted among pathologists, clinicians, and medical researchers working on this disease. However, the evidence accumulated so far with regard to the standard parameters defining neoplastic proliferation, such as clonality, chromosomal abnormality, and transplantability, is still controversial. A further enigma concerning HD as a neoplastic lymphoproliferative disease is the extremely low percentage of putative neoplastic cells (RS cells and their variants) occurring in the polymorphic histopathological setting that characterizes the disease. In contrast, well-recognized neoplastic lymphoproliferative diseases, the so-called non-Hodgkin's lymphomas (NHLs), usually show monomorphic histological patterns and are constituted predominantly of monoclonal or, more rarely, oligoclonal cell populations.

The presence of typical RS cells in an appropriate cellular background is required for the pathological diagnosis of HD. The bizarre features of RS cells, which are surrounded by apparently "normal" cells of the background, have indicated that these cells are the most likely candidates for malignant cells of the

---

MAURO BOIOCCHI, RICCARDO DOLCETTI, and VALLI DE RE • Division of Experimental Oncology 1, Centro di Riferimento Oncologico, 33081 Aviano (PN), Italy.   ANTONINO CARBONE and ANNUNZIATA GLOGHINI • Division of Pathology, Centro di Riferimento Oncologico, 33081 Aviano (PN), Italy.

DNA Tumor Viruses: Oncogenic Mechanisms, edited by Giuseppe Barbanti-Brodano *et al.* Plenum Press, New York, 1995.

disease. However, attempts to characterize RS cells morphologically, immunocytochemically, and immunogenetically have yielded conflicting results.[1-7] Among the cell components of the hematopoietic system, so far there is no cell type that has not been suspected of being the normal counterpart of RS cells.

Finally, the heterogeneity of the clinical [8,9] and histological features[10-14] of HD suggests that this disease includes a group of etiopathogenetically related, but not identical, clinicopathological entities. Such heterogeneity might be dependent on the differentiation stage of the presumed "HD neoplastic cells" (RS cells and their precursors) [15,16] or, alternatively, on the existence of several biologically related diseases, each with a different etiopathogenesis.[17-19]

2. EPIDEMIOLOGIC DATA

Although its malignant nature is widely accepted, HD shows features very similar to those of an infectious disease. A bimodal age–incidence curve has been described for HD, with an early peak comprising cases from young adults (aged 15–34 years) and a second peak occurring in older individuals.[20] The first peak mainly includes HD cases of the nodular sclerosis (NS) subtype, whereas the incidence of the mixed cellularity (MC) histotype increases with age. These findings suggest that HD is a heterogeneous disease and that different etiologies are probably responsible for cases of different age groups or histotypes.

It has been proposed that HD cases of young adults may arise as an unusual and late host response to an infectious agent that is widespread in all communities. In support of this hypothesis are epidemiologic data indicating that factors that would favor late infection, such as early birth order, small sibship size, high maternal education, and low-density housing in childhood, are associated with an increased risk of HD in young adults.[21-23] In addition, the report of local spatial clustering of young HD cases, particularly of the NS histotype, further strengthens the possibility that the transmission of an infectious agent, probably a virus, is involved in the development of these cases.[24] In older HD patients, the same virus may be reactivated from its latent state as a consequence of an age-related decline in immunosurveillance, or, alternatively, these cases have a completely different etiopathogenesis.

A possible pathogenetic role of Epstein–Barr virus (EBV) was originally suggested by the observation that individuals with a history of mononucleosis (IM) have a two to four times increased risk of developing HD.[25-27] Furthermore, most HD patients have abnormally high levels of antibodies against EBV antigens at disease presentation.[28,29] These findings, however, may simply be the consequence of the immune dysfunction of these patients, which would favor the reactivation of latent infections. To evaluate such a possibility, a prospective study on sera obtained several years before the onset of HD has been carried out. In these samples, antibodies to VCA, EBNA, and to the diffuse form of EA were significantly elevated, suggesting that a chronic EBV reactivation may occur over a long period of time before the onset of overt disease.[30] These findings have renewed the interest in the possible pathogenetic association of EBV with HD.

3. DETECTION OF EBV IN HD SAMPLES

A variety of different techniques at the DNA, RNA, and protein levels have been used to investigate the presence of EBV in HD specimens. Although early studies failed to detect the EBV genome in pathological tissues from HD patients, with the advent of cloned viral probes and blot hybridization, EBV DNA has recently been detected in about a third of HD cases.[31-34] For screening purposes, the hybridization of Southern blots with the *Bam*HI W fragment of the B-95-8 EBV strain, specific for the large internal repeat of the virus (Fig. 1), has been particularly useful.[35] This fragment, in fact, is usually reiterated ten times, thus permitting coverage of approximately 20% of an individual EBV genome. Moreover, the use of probes for the unique sequences adjacent to the 3' and 5' terminal repeats of EBV allowed the analysis of the configuration of the viral genome termini (Fig. 2).[36] The EBV genome encapsidated in the virion is a double-stranded linear molecule with homologous direct tandem repeats of approximately 500 bp at each terminus serving as cohesive sites for the circularization. Following viral infection of the host cell, the linear termini join to form covalently closed episomal DNA. Because both the number of terminal repeats at the ends of the linear EBV genome and the extent of terminal sequence overlapping during episome formation are variable, the molecular configuration of fused termini after circularization varies for each episome. In fact, because of different numbers of terminal repetitive sequences enclosed in independently originated episomes, the length of each EBV genome is remarkably variable. Multiple identical viral episomes that have the same fused termini are maintained in the progeny of each infected cell, and the number of the terminal repeats in a particular episome represents a constant clonal marker for the virus and consequently for the infected cell.[36] Differences in the configuration of EBV fused termini can be detected as differently sized fragments at Southern blot analysis. By such an approach it has been clearly demonstrated that in EBV-positive HD cases, the viral genome is in episomal form and generally monoclonal, indicating that the EBV-infected cell population has arisen from a single cell.[31-34,37]

Because of the limited sensitivity of Southern blotting, the search for EBV

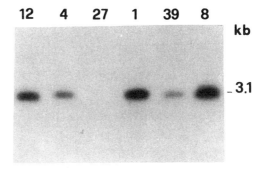

FIGURE 1. Southern blot autoradiogram showing EBV genomic DNA in pathological tissues from HD patients. Lane numbers indicate different HD cases. DNAs were digested with the *Bam*HI restriction endonuclease and hybridized with the *Bam*HI W EBV probe; 3.1 kb indicates the size of the *Bam*HI W EBV internal repeat. Lane 27 contains the DNA of an EBV-negative HD.

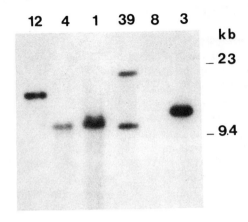

FIGURE 2. Southern blot autoradiogram showing the analyses of the fused terminal portion of the EBV genome. DNAs were digested with the *Bam*HI restriction endonuclease and hybridized with the *Xho*I 1.9-kb probe representing unique EBV sequences adjacent to the 3′ terminal repeats. Lane numbers indicate different HD cases. Case 39 contains two EBV-infected cellular clones, whereas case 8 is an EBV-negative HD case. Dimensions (kilobases) are from *Hind*III-digested λ DNA fragments (not shown).

genomes in HD biopsy specimens was extended using the polymerase chain reaction (PCR) technique, which showed EBV DNA sequences in a considerably larger proportion of cases, ranging between 40% and 80%, depending on the sensitivity of the method used.[38–41] However, these results may be confusing because it cannot be ruled out that PCR positivity may derive from normal bystander cells, which are not directly involved in the pathogenesis of the disease.[42,43] Therefore, PCR amplification may be used as a primary screening method to detect EBV-negative HD cases, whereas EBV positivity has to be confirmed by other analytical methods, namely, those allowing the identification of EBV-infected cells. *In situ* hybridization (for either genomic DNA or EBV-encoded mRNAs such as EBERs) provided direct evidence that in EBV-positive cases, the virus is prevalently found in RS cells and their precursors, further supporting the possible etiopathogenetic role of the virus in HD (Fig. 3).[32,39,44,45] Finally, immunostaining with antibodies against EBV-encoded proteins, such as latent membrane protein 1 (LMP-1), was successfully used to detect EBV-positive cases and also to ascertain that RS cells and their precursors harbor and actively transcribe the viral genome (Fig. 4).[17,46]

In a recent review of the available data concerning the detection of EBV by different techniques, Drexler[4] reported that, in a total of 928 HD cases investigated up to 1992, EBV positivity was ~19% by genomic DNA *in situ* hybridization, ~34% by either RNA *in situ* hybridization or Southern blot analysis of genomic DNA, ~45% by immunostaining with LMP-1 antibodies, and 56% by PCR amplification. In conclusion, although the frequency of EBV positivity varied as a function of both the series studied and the analytical method used, it is undoubted that a large proportion of HD cases contains a clonal cell population, derived from a single EBV-infected cell, that expresses specific viral proteins. Such a clonal population is identifiable with the RS cells and their mononucleated precursors.

The existence of divergent sequences in the genomic region encoding the EBV EBNA-2 and EBNA-3 antigens defines two different EBV strains.[47–49] Type-A EBV has a more efficient *in vitro* transforming activity than type-B strain.[50] In addition, type-A EBV is largely widespread in healthy Western populations, which only rarely

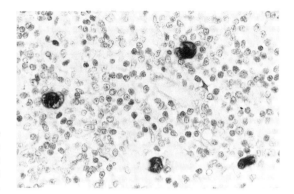

FIGURE 3. EBER *in situ* hybridization performed in an EBV-positive HD case. Signal is present as dense grains over the nuclei of RS cells (nuclear fast red counterstain; ×400).

harbor type-B EBV in peripheral blood.[51] In contrast, type-B virus has been observed in a large percentage of peripheral lymphocytes from immunocompromised patients, in human immunodeficiency virus (HIV-1)-associated NHL, and in endemic Burkitt's lymphoma.[52–54] Recent studies aimed at characterizing the subtype of EBV involved in HD from HIV-1-negative patients reported that type-A virus is predominant in EBV-associated cases, with frequencies ranging from 78% to 100%.[52,55,56] Interestingly, we detected type-B EBV in five of ten HD samples from HIV-1-positive patients,[55] indicating that in the context of HIV-1-induced immuno-

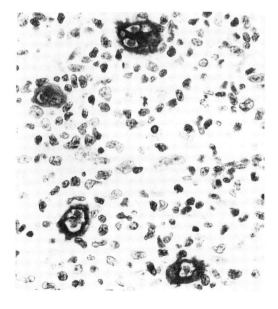

FIGURE 4. A paraffin section from an HD specimen exhibiting Hodgkin and RS cells with strong cytoplasmic staining for the EBV-encoded antigen LMP-1 (Bouin-fixed, paraffin-embedded section, alkaline-phosphatase–anti-alkaline-phosphatase immunostaining method; hematoxylin counterstain; ×630).

depression, both EBV subtypes seem to be equally capable of contributing to the development of HD, consistently with reported findings concerning NHL.

4. EPSTEIN–BARR VIRUS ASSOCIATION WITH SPECIFIC HD HISTOLOGICAL SUBTYPES

Although HD is considered a single disease and a distinct clinical entity, clinical and pathological experience over the past two decades and recent advances in immunohistochemistry and gene rearrangement studies have indicated that HD exhibits significant heterogeneity.[8–14] At present, three HD groups may be identified[57]: (1) a major subset including the NS subtype and the "cellular-phase nodular sclerosis" (CPNS) variant showing "lacunar" RS cells with no evidence of sclerosis[58,59]; (2) a subset encompassing almost the entire spectrum of the non-NS type. This subset mainly includes the MC subtype along with a smaller group of related cases with features of the "diffuse lymphocyte predominance" (DLP) and "lymphocyte depletion" (LD) subtypes; and (3) a minor subset including the "nodular lymphocyte predominance" (NLP) subtype showing clinical and immunohistochemical characteristics different from those of the other HD subtypes, suggesting that NLP should probably be classified as a low-grade NHL.[60,61] These HD subsets may reflect different evolutive forms of a single disease or, alternatively, may represent separate, albeit closely related, diseases, perhaps with different etiopathogenesis.

We recently found[41] a higher association of the EBV genome with the MC HD subtype (10 of 15 positive cases, 67%) than with the NLP (0 of 5 cases) or NS subtypes (12 of 46 cases, 26%) ($p < 0.01$, χ^2 test). The frequency of EBV-positive MC HD subtypes in our series was consistent with that recently found by Herbst et al.[38] in a large series of West Germany HD cases. In contrast, we found a remarkably lower percentage of EBV-positive NS HD cases (26% versus 61% in Herbst series). A similar higher association of EBV with the MC HD subtype (96%) than with the NS (32%) and LP (10%) subtypes was also found in a Danish HD series by immunohistochemical detection of LMP-1.[46] In our single-institution experience, immunostaining with anti-LMP-1 antibodies showed EBV positivity in 27 of 104 HD cases (26%); in accordance with the study by Pallesen et al.,[46] a higher association of EBV with the MC HD subtype (56%) than with the NS (16%) and LP (0%) subtypes was detected.[62] The meaning of such a differential association of EBV with NS HD in different series is presently not understood. It is possible that in different human populations a specific etiopathogenetic stimulus, such as EBV infection, might cause differential pathogenetic pressure, resulting in the induction of distinct forms of a single disease. Alternatively, similar pathological entities might have a different etiopathogenesis in distinct human populations. It is worth noting that in our NS HD series, the EBV genome showed a higher association with cellular-phase NS (CPNS) (five of six EBV-positive cases, 83%) than with typical NS HD cases (7 of 40 EBV-positive cases, 17.5%). Such a dichotomic distribution of EBV between CPNS and typical NS HD subsets may play an important role in resolving the controversy over whether CPNS should be included in the NS subtype. In fact,

CPNS shows the lacunar cell typical of the NS subtype but lacks sclerosis and displays clinical features and survival rates more closely related to MC HD.[8,9] A stricter association of EBV with the MC subtype is also suggested by data demonstrating that EBV is prevalently found in the pediatric and older-age patient groups,[63,64] in which the MC subtype is the more frequent form of HD,[65–67] as well as by the very high frequency of EBV-positive HD in HIV-1-infected patients,[68–70] in whom the NS and NLP subtypes are by far less frequent than the MC HD subtype.[62,71]

In conclusion, much experimental data suggest that the EBV genome is primarily associated with the MC HD subset. Thus, the proposed subdivision of HD into two main groups, the MC and NS subsets, may find a further validation from virological findings. The high association of EBV with MC HD cases, in the context of the previously demonstrated monoclonal and episomal nature of the EBV genome in HD samples,[31–34,37] suggests a possible etiopathogenetic role for the virus in the development of this HD subset.

5. OTHER VIRUSES

The possibility that other herpesviruses might be pathogenetically associated with HD has been investigated by various approaches. With regard to cytomegalovirus (CMV), serologic studies showed that in HD patients, anti-CMV antibody titers were not significantly higher than those of control groups.[30] In addition, several investigations failed to show the presence of CMV DNA sequences in HD pathological tissues.[31,72,73]

A possible etiological role for another member of the human herpesvirus family, the human herpesvirus type 6 (HHV-6), has been suggested by investigations reporting that sera obtained from HD patients had a higher prevalence or titer of antibody to this virus.[16,18,74] Nevertheless, only limited and conflicting data as to the presence of HHV-6 genomes in HD samples are presently available. In fact, although HHV-6 DNA was detected in 12% to 30% of HD biopsy specimens using Southern blotting or PCR analysis,[16,18,75,76] other studies performed with similar approaches yielded negative results.[63,77] (For a review see Chapter 17.) A more extensive and detailed analysis, including *in situ* hybridization with HHV-6-specific probes, is needed to define the role of HHV-6 infection in HD.

Finally, no human T-cell leukemia virus I (HTLV-1) DNA was detected in a series of 12 HD biopsy specimens.[78]

6. PERSISTENCE OF EBV IN THE COURSE OF HD

It is highly unlikely that the presence of EBV genome in HD-involved tissues reflects the expansion of EBV-infected lymphoblastoid B-cell clones because of the reduced immunocompetence associated with the disease. In fact, the detection of a single EBV episome in each case of HD is a generalized finding in every study reported so far.[31–34,37] These studies also included those analyzing HD cases arising in HIV-1-infected patients,[67] whose profound immunodeficiency is thought to be a strong predisposing factor for the expansion of different EBV-infected lymphoblastoid cell clones. In situ hybridization studies[32,39,44,45] localized EBV genome in RS cells and their mononucleated precursors. This association was also consistently verified when multiple HD biopsy samples, either synchronous or metachronous, were analyzed.[72] Moreover, we have recently reported that the persistence of EBV in multiple metachronous localizations of the disease is consequent on the expansion of a unique EBV-infected cellular clone with morphological features of RS cells (Fig. 5).[37] These findings suggest that such a cell population plays a functional role in the pathological tissue and, therefore, that EBV is likely to be involved in the pathogenesis of EBV-positive HD cases.

7. PATTERN OF EBV GENE EXPRESSION IN HD

In EBV-infected cells, the viral episome is not merely a silent element; depending on the differentiative and functional status of the host cells, peculiar interactions between the viral and cellular genome occur that lead to differential expression of latent viral genes. Such a latent viral gene expression may affect the

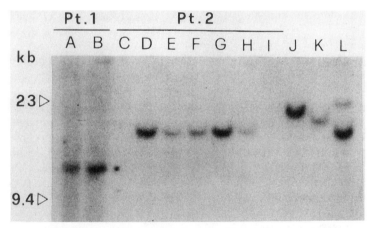

FIGURE 5. Southern blot autoradiogram demonstrating the persistence of the same EBV-infected cellular clone in multiple metachronous localizations of the disease. The analysis of EBV fused termini was performed by hybridizing BamHI-digested DNAs with a 4.1-kb EcoRI fragment corresponding to a unique EBV sequence adjacent to the 5'-terminal repeats. Lanes A and B contain DNA from two metachronous lymph-node localizations of HD obtained from patient 1. Lanes C to I contain DNAs from tissues of patient 2: a left cervical lymph node with persistent generalized lymphadenopathy (lane C), a biopsy sample of a right cervical lymph node with HD (lane D), two superficial (lanes E and F) and two abdominal (lanes G and H) involved lymph nodes obtained at autopsy, and a cerebellar autopsy sample without neoplastic infiltration (lane I). Lanes J, K, and L contain EBV-positive HD cases introduced as controls. Sizes are in kilobases. Hybridization of the same DNAs with a probe specific for unique genomic sequences contiguous to the 3'-terminal repeat of the virus gave superimposable results.

properties of the infected cell and its interaction with the immune system of the host.

Based on the analysis of EBV latent protein expression in cell culture systems and in EBV-associated tumors, three different patterns of viral gene expression (called latency) have been identified: latency I, observed in Burkitt lymphomas (BL) *in vivo* and in BL-derived cell lines that retain *in vitro* the original tumor phenotype, in which only the virus-encoded nuclear antigen EBNA-1 is expressed[79,80]; latency II, characteristic of nasopharyngeal carcinoma (NPC) cells that express EBNA-1 and LMP-1 both *in vivo* and *in vitro*[81–83]; and latency III, detected in immunoblastic B-cell NHL arising in immunosuppressed patients and in EBV-transformed lymphoblastoid B-cell lines (LCLs) *in vitro*, which expresses the full spectrum of EBNA proteins (EBNA 1–6) together with three membrane proteins (LMP-1, -2A, and -2B).[82] Recent studies have confirmed that these different forms of EBV latency are transcriptionally distinct.[85,86]

Studies on the expression of EBV latent viral genes in HD have clearly demonstrated that RS cells are EBNA-1-positive, EBNA-2-negative, and LMP-1-positive, thus displaying a pattern of viral latency superimposable on that of NPC.[17,46] Moreover, Deacon et al.,[87] using a PCR-based amplification of viral mRNA, confirmed that also at the transcriptional level, and in particular in terms of usage of specific viral promoters, the pattern of gene expression in EBV-positive HD is almost identical to that of NPC cells. These data are in contrast with the possibility that the EBV-positive cellular clone detected in about 40% of HD cases may correspond to the expansion of LCL-like cells, which are characterized by a different pattern of viral gene expression (latency III) and usage of different viral promoters.

The observation that a similar pattern of EBV gene expression is constantly found in RS cells and their precursors suggests that the LMP-1 protein may play an important role in the pathogenesis of HD. LMP-1 expression is probably of particular relevance in the light of the putative role of the protein in EBV-induced transformation of both human and animal cells. In fact, LMP-1 expression is crucial for B-lymphocyte immortalization,[88] whereas LMP-1 transfection in rodent fibroblastoid cells is able to induce some features of the transformed phenotype, such as reduction of serum requirement for growth, contact inhibition, and anchorage dependence.[89] In B lymphocytes, LMP-1 can also induce many of the phenotypic changes associated with EBV infection, including increased homotypic adhesion and up-regulation of adhesion molecules (LFA-1, ICAM-1, LFA-3) and B-cell activation markers (CD23, CD71).[90,91] Recent genetic analysis performed in primary B lymphocytes has shown that the carboxy terminus of LMP-1 is involved in cell growth and transformation.[92] In fact, the first 44 amino acids of this protein domain interact with a cellular protein(s)[93] related to the very recently described putative effectors of tumor necrosis factor receptor (TNFR) signaling.[94] The interaction of LMP-1 with TNFR signaling pathways may be important in enabling EBV-infected cells to evade cellular host defense mechanisms in latent or lytic EBV infection or to block the induction of cell death mediated by TNFR.[95] The identification in clinically and histologically aggressive HD and HD-related lymphoproliferative disorders (i.e., angioimmunoblastic lymphadenopathy, peripheral T-cell lymphoma, etc.)[96] of mutational hot spots within the carboxy terminal region of the LMP-1 oncogene identical to those originally described in Asian NPC[97,98] further supports the role of the LMP-1 carboxyl terminal domain in cell transformation or immortalization. Transfectans expressing such LMP-1 deletion mutants were, in fact, readily tumorigenic when compared with the wild-type gene following inoculation in nude and SCID mice.[98,99] Moreover, it has been reported that LMP-1 expression increases the stimulatory capacity of EBV-negative B-cell lines in allogeneic mixed lymphocyte cultures.[100] An intriguing aspect is that LMP-1 can provide target epitopes for EBV-specific cytotoxic T lymphocytes.[101,102] Therefore, it is reasonable to assume that the expression of LMP-1 by RS cells may reflect the general impairment of immunosurveillance that is characteristic of HD patients. Alternatively, it may be the consequence of a more limited, selective dysfunction affecting the EBV-specific cytotoxic T lymphocytes able to recognize and kill LMP-1-positive cells. We have recently observed a class I restricted EBV-specific cytotoxic response in tumor-infiltrating lymphocytes (TILs) from three EBV-negative HD cases, whereas TILs from six EBV-associated HD cases were either noncytotoxic or exerted a LAK-type cytotoxic activity. These results, together with the finding that the precursors of EBV-specific cytotoxic T lymphocytes were present in the peripheral blood of one patient with an EBV-positive HD, suggest that local inhibition of EBV-specific cytotoxic responses may be involved in the pathogenesis of EBV-associated HD cases.[103]

By analyzing the expression of the BZLF1 protein, which controls the switch between EBV latency and replication, subsequent studies have investigated the possibility that activation of the EBV productive cycle may occur *in vivo* in EBV-positive HD cases. Only 6% of the cases studied showed RS cells positive for BZLF1.[104] In these cases, however, no evidence of early antigen (EA) or virus capsid antigen (VCA) expression was obtained,[104] indicating that EBV infection in RS cells is usually latent, and when activation of the productive cycle occurs, this usually results in abortion of virus production.

8. PATTERN OF EBV GENE EXPRESSION IN HD-RELATED LYMPHOPROLIFERATIVE DISORDERS

A broad group of reactive-appearing lymphoproliferative disorders, such as CD30-positive anaplastic large-cell lymphomas, T-cell-rich B-cell lymphomas, peripheral T-cell lymphomas, and angioimmunoblastic lymphadenopathy, share an intriguing characteristic with HD: the admixture in the pathological tissue of a presumed neoplastic clonal cell population with a polyclonal reactive-like T-cell component, a feature also common to NPC. These reactive-like lymphoproliferative diseases are also frequently associated with EBV. In these cases the virus infects, in clonal and episomal form, the presumed neoplastic cellular clone, which often expresses LMP-1 in the absence of the EBNA-2 viral antigen.[57,105–109] We recently found that the frequency of LMP-1 expression in a series of HD cases (71% in patients with and 21% in patients without HIV-1 infection) was close to that found in CD30-positive anaplastic large-cell lymphomas (72% in patients with and 24% in patients without HIV-1 infection).[109] Moreover, all these reactive-like, EBV-positive lymphoproliferative disorders share the activation marker CD30 with EBV-positive HD.[57,105-108]

As supposed for EBV-positive HD, in these pathological processes, LMP-1 expression may be the drive behind the clonal proliferation of EBV-infected neoplastic cells and may account for the florid T-reactive component. In fact, LMP-1 has strong transforming potentialities,[88-91] and it is an extremely effective stimulator of the T-cell immunologic response.[100] It is tempting to speculate that, on the basis of the impressive similarities in virological, immunologic, and histopathological properties they show, EBV-positive HD and the other EBV-positive, reactive-like lymphoproliferative disorders represent a continuum of diseases with a common pathogenetic mechanism. However, further studies are required to elucidate this issue.

9. CONCLUDING REMARKS

Hodgkin's disease and HD-related, reactive-like lymphoproliferative disorders share progressive, invasive, and diffuse properties with monomorphous NHLs but display a peculiar characteristic that differentiates them biologically from the latter lymphomas. This consists of the coexistence, within the pathological tissue, of a polyclonal normal-appearing, presumed reactive cellular component, mainly composed of T lymphocytes, with a clonal cell population that constitutes a minority of the pathological mass. In EBV-positive HD cases, the clonal cell population is identified by the presence of the viral episome, the monoclonality of which indicates that the expansion of a single EBV-infected cell has occurred. The RS cells and their mononuclear precursors belong to the progeny of this presumed HD neoplastic cell population, as confirmed by their constant association with the viral genome and expression of viral gene products. However, the genotypic and phenotypic characteristics of RS cells vary from case to case and within a single case.[4,110,111] This heterogeneity of marker expression, which is very unusual for a typical neoplastic cell population, may reflect a possible heterogeneity of origin of RS cells, both among different cases and within each single case. These findings markedly contrast with the well-defined monoclonality of the EBV-infected cell population and may be explained by assuming that RS cells are heterokaryons induced by the fusion of the EBV-infected mononucleated cell with different cellular partners, possibly consequent on virus-mediated cytopathic effects.[112-116] On this ground, the cell population that sustains the disease may be identifiable with the EBV-infected cell mononucleated precursors of RS cells.

A very puzzling question regarding HD and related diseases is how and why different cell populations coexist within the pathological tissue, in an almost constant qualitative and quantitative pattern, in all synchronous and metachronous localizations of the disease even for a long period of time. In particular, it is unclear why the proliferating, "neoplastic" EBV-infected clone, in spite of its capability to recruit a considerable "reactive" cell component, presumably consequent on the expression of some viral antigens, neither escapes the control of the immune system evolving to a monomorphous NHL nor is eradicated by the immune response of the host. These questions might be explained by assuming that the clone that sustains the disease is not composed of fully transformed cells, able therefore to grow autonomously, but rather is the expansion of a cell population that has been immortalized by the action of EBV-encoded gene products in EBV-positive HD cases or as a consequence of infection by other viruses or genetic alterations of the cell in EBV-negative cases. In any case, the proliferating clone would be dependent on still unidentified exogenous factors for cell replication. In this context, the reactive component of the disease, evoked by viral or cell-encoded antigenicities, may favor the growth of the clonal component by secretion of biological mediators. Thus, in the proposed scenario, the clonal and the reactive-like populations would be interconnected by "biological interactions," which determine and maintain the pathological process. It is worth considering that "biological interactions" similar to those that are supposed to occur in HD have been clearly demonstrated to play a pivotal role in the induction and maintenance, both *in vivo* and *in vitro*, of the composite lymphoproliferative diseases that spontaneously arise in the SJL mouse strain, an experimental model that strongly mimics HD and T-cell-rich B-cell lymphomas.[117–126] Finally, it has been shown that contact-dependent interactions of RS cells with cytokine-producing $CD4^+$ T lymphocytes are mediated in part by the CD40 molecule, a member of the nerve growth factor receptor family showing a significant homology to the HD-associated antigen CD30.[127]

Recognition of the nature of the biological interactions occurring between the polyclonal T-cell population and the clonal cells in HD, as well as the role EBV plays in determining such interactions, may have important clinical implications in the development of future management strategies.

ACKNOWLEDGMENTS. This work was supported by CNR, Progetto Finalizzato "Applicazioni cliniche della ricerca oncologica" (contract 92.02324.PF39), and by the Associazione Italiana per la Ricerca sul Cancro (AIRC).

REFERENCES

1. Angel, C. A., Warford, A., Campbell, A. C., Pringle, J. H., and Lauder, I., 1987, The immunohistology of Hodgkin's disease—Reed–Sternberg cells and their variants, *J. Pathol.* **153**:21–30.
2. Diehl, V., von Kalle, C., Fonatsch, C., Tesch, H., Jücker, M., and Schaadt, M., 1990, The cell of origin in Hodgkin's disease, *Semin. Oncol.* **17**:660–672.
3. Drexler, H. G., and Leber, B. F., 1988, The nature of the Hodgkin's cell, *Blut* **56**:135–137.
4. Drexler, H. G., 1992, Recent results on the biology of Hodgkin and Reed–Sternberg cells. I. Biopsy material, *Leuk. Lymphoma* **8**:283–313.
5. Ford, R. J., 1988, Hodgkin's disease in 1987—is history repeating itself? *Hematol. Oncol.* **6**:201–204.
6. Griesser, H., and Mak, T. W., 1988, Immunogenotyping in Hodgkin's disease, *Hematol. Oncol.* **6**:239–245.
7. Jones, D. B., 1987, The histogenesis of the Reed–Sternberg cell and its mononuclear counterparts, *J. Pathol.* **151**:191–195.

8. Colby, T. V., Hoppe, R. T., and Warnke, R. A., 1981, Hodgkin's disease: A clinicopathologic study of 659 cases, *Cancer* **49:**1848–1858.
9. Mac Lennan, K. A., Bennett, M. H., Tu, A., Hudson, B. V., Easterling, J., Hudson, G. V., and Jelliffe, A. M., 1989, Relationship of histopathologic features to survival and relapse in nodular sclerosing Hodgkin's disease. A study of 1659 patients, *Cancer* **64:**1686–1693.
10. Anastasi, J., and Variakojis, D., 1988, Heterogeneity in Hodgkin's disease: No simple answer for a complex disorder, *Hum. Pathol.* **19:**1251–1254.
11. Chittal, S. M., Caveriviere, P., Schwarting, R., Gerdes, J., Al Saati, T., Rigal-Huguet, F., Stein, H., and Delsol, G., 1988, Monoclonal antibodies in the diagnosis of Hodgkin's disease. The search for a rational panel, *Am. J. Surg. Pathol.* **12:**9–21.
12. Griesser, H., Feller, A. C., Mak, T. W., and Lennert, K., 1987, Clonal rearrangements of T-cell receptor and immunoglobulin genes and immunophenotypic antigen expression in different subclasses of Hodgkin's disease, *Int. J. Cancer* **40:**157–160.
13. Kadin, M. E., Muramoto, L., and Said, J., 1988, Expression of T-cell antigens on Reed–Sternberg cells in a subset of patients with nodular sclerosing and mixed cellularity Hodgkin's disease, *Am. J. Pathol.* **130:**345–353.
14. Knowles, D. M. II, Neri, A., Pelicci, P. G., Burke, J. S., Wu, A., Winberg, C. D., Sheibani, K., and Dalla Favera, R., 1986, Immunoglobulin and T-cell receptor β-chain gene rearrangement analysis of Hodgkin's disease: Implications for lineage determination and differential diagnosis, *Proc. Natl. Acad. Sci. USA* **83:**7942–7946.
15. Leoncini, L., Del Vecchio, M. T., Kraft, R., Megha, T., Barbini, P., Cevenini, G., Poggi, S., Pileri, S., Tosi, P., and Cottier, H., 1990, Hodgkin's disease and CD30-positive anaplastic large cell lymphomas—A continuous spectrum of malignant disorders. A quantitative morphometric and immunohistologic study, *Am. J. Pathol.* **137:**1047–1057.
16. Krueger, G. R. F., Manak, M., Bourgeois, N., Ablashi, D. V., Salahuddin, S. Z., Josephs, S. S., Buchbinder, A., Gallo, R. C., Berthold, F., and Tesch, H., 1989, Persistent active herpes virus infection associated with atypical polyclonal lymphoproliferation (APL) and malignant lymphoma, *Anticancer Res.* **9:**1457–1476.
17. Herbst, H., Dallenbach, F., Hummel, M., Niedobitek, G., Pileri, S., Müller-Lantzsch, N., and Stein, H., 1991, Epstein–Barr virus latent membrane protein expression in Hodgkin and Reed–Sternberg cells, *Proc. Natl. Acad. Sci. USA* **88:**4766–4770.
18. Torelli, G., Marasca, R., Luppi, M., Selleri, L., Ferrari, S., Narni, F., Mariano, M. T., Federico, M., Ceccherini-Nelli, L., Bendinelli, M., Montagnani, G., Montorsi, M., and Artusi, T., 1991, Human herpes virus-6 in human lymphomas: Identification of specific sequences in Hodgkin's lymphomas by polymerase chain reaction, *Blood* **77:**2251–2258.
19. Gutensohn, N. M., and Cole, P., 1977, Epidemiology of Hodgkin's disease in the young, *Int. J. Cancer* **19:**595–604.
20. Gutensohn, N. M., and Cole, P., 1980, Epidemiology of Hodgkin's disease, *Semin. Oncol.* **7:**92–102.
21. Vianna, N. J., Greenwald, P., and Davies, J. N. P., 1971, Extended epidemic of Hodgkin's disease in high-school students, *Lancet* **1:**1209–1210.
22. Gutensohn, N. M., and Cole, P., 1981, Childhood social environment and Hodgkin's disease, *N. Engl. J. Med.* **304:**135–140.
23. Gutensohn, N. M., 1982, Social class and age at diagnosis of Hodgkin's disease: New epidemiological evidence for the "two-disease hypothesis," *Cancer Treat. Rep.* **66:**689–695.
24. Alexander, F. E., 1990, Clustering and Hodgkin's disease, *Br. J. Cancer* **62:**708–711.
25. Rosdahl, N., Larsen, S. O., and Clemmesen, J., 1974, Hodgkin's disease in patients with previous infectious mononucleosis. 30 years' experience, *Br. Med. J.* **2:**253–256.
26. Muñoz, N., Davidson, R. J. L., Witthoff, B., Ericsson, J. E., and De-The, G., 1978, Infectious mononucleosis and Hodgkin's disease, *Int. J. Cancer* **22:**10–13.
27. Kvaale, G., Hoiby, E. A., and Pedersen, E., 1979, Hodgkin's disease in patients with previous infectious mononucleosis, *Cancer* **23:**593–597.

28. Johansson, B., Klein, G., Henle, W., and Henle, G., 1970, Epstein–Barr virus (EBV)-associated antibody patterns in malignant lymphoma and leukemia. I. Hodgkin's disease, *Int. J. Cancer* **6:**450–462.
29. Levine, P. H., Ablashi, D. V., Berard, C. V., Carbone, P. P., Waggoner, D. E., and Malan, L., 1971, Elevated antibody titers to Epstein–Barr virus in Hodgkin's disease, *Cancer* **27:**416–421.
30. Mueller, N., Evans, A., Harris, N. L., Comnstock, G. W., Jellum, E., Magnus, K., Orentreich, N., Polk, B. F., and Vogelman, J., 1989, Hodgkin's disease and Epstein–Barr virus. Altered antibody pattern before diagnosis, *N. Engl. J. Med.* **320:**689–695.
31. Weiss, L. M., Strickler, J. G., Warnke, R. A., Purtilo, D. T., and Sklar, J., 1987, Epstein–Barr viral DNA in tissues of Hodgkin's disease, *Am. J. Pathol.* **129:**86–91.
32. Anagnostopoulos, I., Herbst, H., Niedobitek, G., and Stein, H., 1989, Demonstration of monoclonal EBV genomes in Hodgkin's disease and KI-1-positive anaplastic large cell lymphoma by combined Southern blot and *in situ* hybridization, *Blood* **74:**810–816.
33. Boiocchi, M., Carbone, A., De Re, V., and Dolcetti, R., 1989, Is the Epstein–Barr virus involved in Hodgkin's disease? *Tumori* **75:**345–350.
34. Staal, S. P., Ambinder, R., Beschorner, W. E., Hayward, G. S., and Mann, R., 1989, A survey of Epstein–Barr virus DNA in lymphoid tissue. Frequent detection in Hodgkin's disease, *Am. J. Clin. Pathol.* **91:**1–5.
35. Polach, A., Hartl, G., Zimber, U., Frese, U. K., Lauv, G., Takaki, M., Hohn, B., Gissman, L., and Bornkamm, G. W., 1984, A complete set of overlapping cosmid clones of M-ABA virus derived from nasopharyngeal carcinoma and its similarity to other Epstein–Barr virus isolated, *Gene* **27:**279–288.
36. Raab-Traub, N., and Flynn, K., 1986, The structure of the termini of the Epstein–Barr virus as a marker of clonal cellular proliferation, *Cell* **47:**883–889.
37. Boiocchi, M., Dolcetti, R., De Re, V., Gloghini, A., and Carbone, A., 1993, Demonstration of a unique Epstein–Barr virus-positive cellular clone in metachronous multiple localizations of Hodgkin's disease, *Am. J. Pathol.* **142:**33–38.
38. Herbst, H., Niedobitek, G., Kneba, M., Hummel, M., Finn, T., Anagnostopoulos, I., Bergholz, M., Krieger, G., and Stein, H., 1990, High incidence of Epstein–Barr virus genomes in Hodgkin's disease, *Am. J. Pathol.* **137:**13–18.
39. Uhara, H., Sato, Y., Mukai, K., Akao, I., Matsuno, Y., Furuya, S., Hoshikawa, T., Shimosato, Y., and Saida, T., 1990, Detection of Epstein–Barr virus DNA in Reed–Sternberg cells of Hodgkin's disease using the polymerase chain reaction and in situ hybridization, *Jpn. J. Cancer Res.* **81:**272–278.
40. Knecht, H., Odermatt, B. F., Bachmann, E., Teixeira, S., Sahli, R., Hayoz, D., Heitz, P., and Bachmann, F., 1991, Frequent detection of Epstein–Barr virus DNA by the polymerase chain reaction in lymph node biopsies from patients with Hodgkin's disease without genomic evidence of B- or T-cell clonality, *Blood* **78:**760–767.
41. Boiocchi, M., De Re, V., Dolcetti, R., Carbone, A., Scarpa, A., and Menestrina, F., 1992, Association of Epstein–Barr virus genome with mixed cellularity and cellular phase nodular sclerosis Hodgkin's disease subtypes, *Ann. Oncol.* **3:**307–310.
42. Masih, A., Weisenburger, D., Duggan, M., Armitage, J., Bashir, R., Mitchell, D., Wickert, R., and Purtilo, D. T., 1991, Epstein–Barr viral genome in lymph nodes from patients with Hodgkin's disease may not be specific to Reed–Sternberg cells, *Am. J. Pathol.* **139:**37–43.
43. Armstrong, A. A., Weiss, L. M., Gallagher, A., Jones, D. B., Krajewski, A. S., Angus, B., Brown, G., Jack, A. S., Wilkins, B. S., Onions, D. E., and Jarrett, R. F., 1992, Criteria for the definition of Epstein–Barr virus association in Hodgkin's disease, *Leukemia* **6:**869–874.
44. Weiss, L. M., Movahed, L. A., Warnke, R. A., and Sklar, J., 1989, Detection of Epstein–Barr viral genomes in Reed–Sternberg cells of Hodgkin's disease, *N. Engl. J. Med.* **320:**502–506.
45. Brousset, P., Chittal, S., Schlaifer, D., Icart, J., Payen, C., Rigal-Huguet, F., Voigt, J. J., and Delsol, G., 1991, Detection of Epstein–Barr virus messenger RNA in Reed–Sternberg cells of Hodgkin's disease by *in situ* hybridization with biotinylated probes on specially processed modified acetone methyl benzoate xylene (ModAMeX) sections, *Blood* **77:**1781–1786.

46. Pallesen, G., Hamilton-Dutoit, S. J., Rowe, M., and Young, L. S., 1991, Expression of Epstein–Barr virus latent gene products in tumor cells of Hodgkin's disease, *Lancet* **337**:320–322.
47. Adldinger, H. K., Delius, H., Freese, U. K., Clarke, J., and Bornkamm, G. W., 1985, A putative transforming gene of Jijoye virus differs from that of Epstein–Barr virus prototypes, *Virology* **141**:221–234.
48. Rowe, M., Young, L. S., Cadwallader, K., Petti, E., Kieff, E., and Rickinson, A. B., 1989, Distinction between Epstein–Barr virus type A (EBNA 2A) and type B (EBNA 2B) isolates extend to the EBNA 3 family of nuclear proteins, *J. Virol.* **63**:1031–1039.
49. Sculley, T. B., Apolloni, A., Stumm, R., Moss, D. J., Müeller-Lantzsch, N., Misko, I. S., and Cooper, D. A., 1989, Expression of Epstein–Barr virus nuclear antigens 3, 4, and 6 are altered in cell lines containing B-type virus, *Virology* **171**:401–408.
50. Rickinson, A. B., Young, L. S., and Rowe, M., 1987, Influence of the Epstein–Barr virus nuclear antigen EBNA 2 on the growth phenotype of virus-transformed B-cells, *J. Virol.* **61**: 1310–1317.
51. Sixbey, J. W., Shirley, P., Chesney, P. J., Buntin, D. M., and Resnick, L., 1989, Detection of a second widespread strain of Epstein–Barr virus, *Lancet* **2**:761–765.
52. Boyle, M. J., Sewell, W. A., Sculley, T. B., Apolloni, A., Turner, J. J., Swanson, C. E., Penny, R., and Cooper, D. A., 1991, Subtypes of Epstein–Barr virus in human immunodeficiency virus-associated non-Hodgkin's lymphoma, *Blood* **78**:3004–3011.
53. Goldschmidts, W. L., Bhatia, K., Franklin Johnson, J., Akar, N., Gutierrez, M. I., Shibata, D., Carolan, M., Levine, A., and Magrath, I. T., 1992, Epstein–Barr virus genotypes in AIDS-associated lymphomas are similar to those in endemic Burkitt's lymphomas, *Leukemia* **6**: 875–878.
54. Young, L. S., Yao, Q. Y., Rooney, C. M., Sculley, T. B., Moss, D. J., Rupani, H., Laux, G., Bornkamm, G. W., and Rickinson, A. B., 1987, New type B isolates of Epstein–Barr virus from Burkitt lymphoma and from normal individuals in endemic areas, *J. Gen. Virol.* **68**:2853–2862.
55. De Re, V., Boiocchi, M., De Vita, S., Dolcetti, R., Gloghini, A., Uccini, S., Baroni, C., Scarpa, A., Cattoretti G., and Carbone, A., 1993, Subtypes of Epstein–Barr virus in HIV-1-associated and HIV-1-unrelated Hodgkin's disease cases, *Int. J. Cancer* **54**:895–898.
56. Boyle, M. J., Vasak, E., Tschuchnigg, M., Turner, J. J., Sculley, T., Penny, R., Cooper, D. A., Tindall, B., and Sewell, W. A., 1993, Subtypes of Epstein–Barr virus (EBV) in Hodgkin's disease: Association between B-type EBV and immunocompromise, *Blood* **81**:468–474.
57. Diebold, J., and Audowin, J., 1989, Maladie de Hodgkin. Une ou plusieurs maladies? *Ann. Pathol.* **9**:86–91.
58. Kadin, M. E., Glatstein, E., and Dorfman, R. F., 1971, Clinicopathologic studies of 117 untreated patients subjected to laparotomy for the staging of Hodgkin's disease, *Cancer* **27**: 1277–1294.
59. Strum, S. B., and Rappaport, H., 1971, Interrelations of the histologic types of Hodgkin's disease, *Arch. Pathol.* **91**:127–134.
60. Hansmann, M., Stein, H., Fellbaum, C., Hui, P. K., Parwarescg, M. R., and Lennert, K., 1989, Nodular paragranuloma can transform into high grade malignant lymphoma of B type, *Hum. Pathol.* **20**:1169–1175.
61. Chittal, S. M., Alard, C., Rossi, J. F., Al Saati, T., LeTourneau, A., Diebold, J., and Delsol, G., 1990, Further phenotypic evidence that nodular, lymphocyte-predominant Hodgkin's disease is a large B-cell lymphoma in evolution, *Am. J. Surg. Pathol.* **14**:1024–1035.
62. Tirelli, U., Errante, D., Dolcetti, R., Gloghini, A., Serraino, D., Vaccher, E., Franceschi, S., Boiocchi, M., and Carbone, A., 1995, Hodgkin's disease and HIV infection: Clinico-pathological and virological features of 114 patients from the Italian cooperative group on AIDS and tumors, *J. Clin. Oncol.* **13**:1758–1767.
63. Gledhill, S., Gallagher, A., Jones, D. B., Krajewski, A. S., Alexander, F. E., Klee, E., Wright, D. H., O'Brien, C., Onions, D. E., and Jarrett, R. F., 1991, Viral involvement in Hodgkin's disease: Detection of clonal type A Epstein–Barr virus genomes in tumor samples, *Br. J. Cancer* **64**: 227–232.

64. Armstrong, A. A., Alexander, F. E., Pinto Paes, A., Morad, N. A., Gallagher, A., Krajewski, A. S., Jones, D. B., Angus, B., Adams, J., Cartwright, R. A., Onions, D. E., and Jarrett, R. F., 1993, Association of Epstein–Barr virus with pediatric Hodgkin's disease, *Am. J. Pathol.* **142:**1683–1688.
65. Correa, P., and O'Conor, G. T., 1971, Epidemiologic patterns of Hodgkin's Disease, *Int. J. Cancer* **8:**192–201.
66. McKinney, P. A., Alexander, F. E., Ricketts, T. J., Williams, J., and Cartwright, R. A., 1989, A specialist leukemia/lymphoma registry in the UK. Part I: Incidence and geographical distribution of Hodgkin's disease, *Br. J. Cancer* **60:**942–947.
67. Glaser, S. L., and Swartz, W. G., 1990, Time trends in Hodgkin's disease incidence. The role of diagnostic accuracy, *Cancer* **66:**2196–2204.
68. Boiocchi, M., De Re, V., Gloghini, A., Vaccher, E., Dolcetti, R., Marzotto, A., Bertola, G., and Carbone, A., 1993, High incidence of monoclonal EBV episomes in Hodgkin's disease and anaplastic large-cell Ki-1-positive lymphomas in HIV-1-positive patients, *Int. J. Cancer* **54:**53–59.
69. Uccini, S., Monardo, F., Ruco, L. P., Baroni, C. D., Faggioni, A., Agliano, A. M., Gradilone, A. M., Manzari, A., Vago, L., Costanzi, G., Carbone, A., Boiocchi, M., and De Re, V., 1989, High frequency of Epstein–Barr virus genome in HIV-positive patients with Hodgkin's disease, *Lancet* **1:**1458.
70. Herndier, B. G., Sanchez, H. C., Chang, K. L., Chen, Y. Y., and Weiss, L. M., 1993, High prevalence of Epstein–Barr virus in the Reed–Sternberg cells of HIV-associated Hodgkin's disease, *Am. J. Pathol.* **142:**1073–1079.
71. Carbone, A., Tirelli, U., Vaccher, E., Volpe, R., Gloghini, A., Bertola, G., De Re, V., Rossi, C., Boiocchi, M., and Monfardini, S., 1991, A clinicopathologic study of lymphoid neoplasias associated with human immunodeficiency virus infection in Italy, *Cancer* **68:**842–852.
72. Coates, P. J., Slavin, G., and D'Ardenne, A. J., 1991, Persistence of Epstein–Barr virus in Reed–Sternberg cells throughout the course of Hodgkin's disease, *J. Pathol.* **164:**291–297.
73. Samoszuk, M., and Ravel, J., 1991, Frequent detection of Epstein–Barr viral deoxyribonucleic acid and absence of cytomegalovirus deoxyribonucleic acid in Hodgkin's disease and acquired immunodeficiency syndrome-related Hodgkin's disease, *Lab. Invest.* **65:**631–636.
74. Clark, D. A., Alexander, F. E., McKinney, P. A., Roberts, B. E., O'Brien, C., Jarrett, R. F., Cartwright, R. A., and Onions, D. E., 1990, The seroepidemiology of human herpesvirus-6 (HHV-6) from a case-control study of leukemia and lymphoma, *Int. J. Cancer* **45:**829–833.
75. Gompels, U. A., Carrigan, D. R., Carss, A. L., and Arno, J., 1993, Two groups of human herpesvirus 6 identified by sequence analyses of laboratory strains and variants from Hodgkin's lymphoma and bone marrow transplant patients, *J. Gen. Virol.* **74:**613–622.
76. Di Luca, D., Dolcetti, R., Mirandola, P., De Re, V., Secchiero, P., Carbone, A., Boiocchi, M., and Cassai, E., 1994, Human herpesvirus 6: A survey of presence and variant distribution in normal peripheral lymphocytes and lymphoproliferative disorders, *J. Infect. Dis.* **170:**211–215.
77. Jarrett, R. F., Gledhill, S., Qureshi, F., Crae, S. H., Madhok, R., Brown, I., Evans, I., Krajewski, A., O'Brien, C. J., Cartwright, R. A., Venables, P., and Onions, D. E., 1988, Identification of human herpesvirus 6-specific DNA sequences in two patients with non-Hodgkin's lymphoma, *Leukemia* **2:**496–502.
78. Tesch, H., Jücker, M., Krönke, M., and Diehl, V., 1991, Analysis of HTLV-1 in Hodgkin's disease, *Leuk. Lymphoma* **4:**371–374.
79. Gregory, C. D., Rowe, M., and Rickinson, A. B., 1990, Different Epstein–Barr virus–B-cell interactions in phenotypically distinct clones of a Burkitt lymphoma cell line, *J. Gen. Virol.* **71:**1481–1495.
80. Rowe, M., Rowe, D. T., Gregory, C. D., Young, L. S., Farrell, P. J., Rupani, H., and Rickinson, A. B., 1987, Differences in B-cell growth phenotype reflect novel patterns of Epstein–Barr virus latent gene expression in Burkitt's lymphoma cells, *EMBO J.* **6:**2743–2751.
81. Fahraeus, R., Fu, H. L., Ernberg, I., Finke, J., Rowe, M., Klein, G., Falk, K., Nilsson, E., Yadav,

M., Busson, P., Tursz, T., and Kallin, B., 1988, Expression of Epstein–Barr virus-encoded proteins in nasopharyngeal carcinoma, *Int. J. Cancer* **42**:329–338.
82. Young, L. S., Dawson, C. W., Clark, D., Rupani, H., Busson, P., Tursz, T., Johnson, A., and Rickinson, A. B., 1988, Epstein–Barr virus gene expression in nasopharyngeal carcinoma, *J. Gen. Virol.* **69**:1051–1065.
83. Brooks, L., Yao, Q. Y., Rickinson, A. B., and Young, L. S., 1992, Epstein–Barr virus latent gene transcription in nasopharyngeal carcinoma cells: Coexpression of EBNA1, LMP1 and LMP2 transcripts, *J. Virol.* **66**:2689–2697.
84. Kieff, E., and Liebowitz, D., 1990, Epstein–Barr virus and its replication, in: *Virology* (B. N. Fields and D. M. Knife, eds.), Raven Press, New York, pp. 1889–1920.
85. Brooks, L. A., Lear, A. L., Young, L. S., and Rickinson, A. B., 1993, Transcripts from the Epstein–Barr virus *Bam*HI A fragment are detectable in all three forms of virus latency, *J. Virol.* **67**:3182–3190.
86. Smith, P. R., Gao, Y., Karran, L., Jones, M. D., Snudden, D., and Griffin, B. E., 1993, Complex nature of the major viral polyadenylated transcripts in Epstein–Barr virus-associated tumors, *J. Virol.* **67**:3217–3225.
87. Deacon, E. M., Pallesen, G., Niedobitek, G., Crocker, J., Brooks, L., Rickinson, A. B., and Young, L. S., 1993, Epstein–Barr virus and Hodgkin's disease: Transcriptional analysis of virus latency in the malignant cells, *J. Exp. Med.* **177**:339–349.
88. Hennessy, K., Fennewald, S., Hummel, M., Cole, T., and Kieff, E., 1984, A membrane protein encoded by Epstein–Barr virus in latent growth-transforming infection. *Proc. Natl. Acad. Sci. USA* **81**:7207–7211.
89. Wang, D., Liebowitz, D., and Kieff, E., 1985, An EBV membrane protein expressed in immortalized lymphocytes transforms established rodent cells, *Cell* **43**:831–840.
90. Wang, D., Liebowitz, D., Wang, F., Gregory, C., Rickinson, A. B., Larson, R., Springer, R., and Kieff, E., 1988, Epstein–Barr virus latent infection membrane protein alters the human B-lymphocyte phenotype: Deletion of the amino terminus abolishes activity, *J. Virol.* **62**:4173–4184.
91. Wang, F., Gregory, C., Sample, C., Rowe, M., Liebowitz, D., Murray, R., Rickinson, A. B., and Kieff, E., 1990, Epstein–Barr virus latent membrane protein (LMP-1) and nuclear proteins 2 and 3C are effectors of phenotypic changes in B-lymphocytes: EBNA2 and LMP-1 cooperatively induce CD23, *J. Virol.* **64**:2309–2318.
92. Kaye, K. M., Izumi, K. M., Mosialos, G., and Kieff, E., 1995, The Epstein–Barr virus LMP1 cytoplasmic carboxyl terminus is essential for B-lymphocyte transformation; fibroblast co-cultivation complements a critical function within the terminal 155 residues, *J. Virol.* **69**:675–683.
93. Masialos, G., Birkeubach, M., Yalamachini, R., Van Arsdale, T., Ware, C., and Kieff, E., 1995, The Epstein–Barr virus transforming protein LMP1 engages signaling proteins for the tumor necrosis factor receptor family, *Cell* **80**:389–399.
94. Rothe, M., Wang, S. C., Henzel, W. J., and Coeddel, D. V., 1994, A novel family of putative signal transducers associated with the cytoplasmic domain of the 75 kDa tumor necrosis factor receptor, *Cell* **78**:681–692.
95. Tartaglia, L. A., Rothe, M., Hu, Y. F., and Goeddel, D. V., 1993, Tumor necrosis factor's cytotoxic activity is signaled by the p55 TNF receptor, *Cell* **73**:213–216.
96. Knecht, H., Bachmann, E., Brousset, P., Rothenberger, S., Einsele, H., Lestou, V. S., Delsol, G., Bachmann, F., Ambros, P. F., and Odermatt, B. F., 1995, Mutational hot spots within the carboxyl terminal region of the LMP1 oncogene of Epstein–Barr virus are frequent in lymphoproliferative disorders, *Oncogene* **10**:523–528.
97. Hu, L. F., Zabarovsky, E. R., Chen, F., Cao, S. L., Ernberg, I., Klein, G., and Winberg, G., 1991, Isolation and sequencing of the Epstein–Barr virus BNLF-1 gene (LMP1) from a Chinese nasopharyngeal carcinoma, *J. Gen. Virol.* **72**:2399–2409.
98. Chen, M. L., Tsai, C. N., Liang, C. L., Shu, C. H., Hang, C. R., Sulitzeanu, D., Liu S. T., and

Chang, Y. S., 1992, Cloning and characterization of the latent membrane protein (LMP) of a specific Epstein-Barr virus variant derived from the nasopharyngeal carcinoma in the Taiwanese population, *Oncogene* **7:**2131-2140.
99. Hu, L. F., Chen, F., Zheng, X., Ernberg, I., Cao, S. L., Christensson, B., Klein, G., and Winberg, G., 1993, Clonability and tumorigenicity of human epithelial cells expressing the EBV encoded membrane protein LMP1, *Oncogene* **8:**1575-1583.
100. Cuomo, L., Trivedi, P., Wang, F., Winberg, G., Klein, G., and Masucci, M. G., 1990, Expression of the Epstein-Barr virus (EBV)-encoded membrane antigen (LMP) increases the stimulatory capacity of EBV-negative B-lymphoma lines in allogeneic mixed lymphocyte cultures, *Eur. J. Immunol.* **20:**2293-2299.
101. Murray, R. J., Wang, D., Young, L. S., Wang, F., Rowe, M., Kieff, E., and Rickinson, A. B., 1988, Epstein-Barr virus-specific cytotoxic T-cell recognition of transfectants expressing the virus-coded latent membrane protein LMP, *J. Virol.* **62:**3747-3755.
102. Khanna, R., Burrows, S. R., Kurilla, M. G., Jacob, C. A., Misko, I. S., Sculley, T. B., Kieff, E., and Moss, D. J., 1992, Localization of Epstein-Barr virus cytotoxic T cell epitopes using recombinant vaccinia: Implications for vaccine development, *J. Exp. Med.* **176:**169-176.
103. Frisan, T., Sjöberg, J., Dolcetti, R., Boiocchi, M., De Re, V., Carbone, A., Brautbar, C., Battat, S., Biberfeld, P., Eckman, M., Öst, Å., Christensson B., Sundström, C., Björkholm, M., Pisa, P., and Masucci, M. G., 1995, Local suppression of Epstein-Barr virus (EBV) specific cytotoxicity in biopsies of EBV positive Hodgkin's disease, *Blood* **86:**1493-1501.
104. Pallesen, G., Sandvej, K., Hamilton-Dutoit, S. J., Rowe, M., and Young, L. S., 1991, Activation of Epstein-Barr virus replication in Hodgkin and Reed-Sternberg cells, *Blood* **78:**1162-1165.
105. Herbst, H., Dallenbach, F., Hummel, M., Niedobitek, G., Finn, T., Young, L. S., Rowe, M., Müller-Lantzsch, N., and Stein, H., 1991, Epstein-Barr virus DNA and latent gene products in Ki-1 (CD30)-positive large cell lymphomas, *Blood* **78:**2666-2673.
106. Loke, S. L., Ho, F., Srivastava, G., Fu, K. H., Leung, B., and Liang, R., 1992, Clonal Epstein-Barr virus genome in T-cell-rich lymphomas of B or probable B lineage, *Am. J. Pathol.* **140:**981-989.
107. Chen, C., Sadler, R. H., Walling, D. M., Su, I., Hsieh, H., and Raab-Traub, N., 1993, Epstein-Barr virus (EBV) gene expression in EBV-positive peripheral T-cell lymphomas, *J. Virol.* **67:**6303-6308.
108. Dolcetti, R., Carbone, A., Zagonel, V., De Re, V., Gloghini, A., Frisan, T., and Boiocchi, M., 1993, Type 2 Epstein-Barr virus genome and latent membrane protein-1 expression in a T-cell rich lymphoma of probable B-cell origin, *Am. J. Clin. Pathol.* **100:**541-549.
109. Carbone, A., Gloghini, A., Volpe, R., Boiocchi, M., Tirelli, U., and the Italian Cooperative group on AIDS and Tumors, 1994, High frequency of Epstein-Barr virus latent membrane protein-1 expression in acquired immunodeficiency syndrome-related Ki-1 (CD30)-positive anaplastic large-cell lymphomas, *Am. J. Clin. Pathol.* **101:**768-772.
110. Chu, W. S., Abbondanzo, S. L., and Frizzera, G., 1992, Inconsistency of the immunophenotype of Reed-Sternberg cells in simultaneous and consecutive specimens from the same patients. A paraffin section evaluation in 56 patients, *Am. J. Pathol.* **141:**11-17.
111. Trümper, L. H., Brady, G., Bagg, A., Gray, D., Loke, S. L., Griesser, H., Wagmann, R., Braziel, R., Gascoyne, R. D., Vicini, S., Iscove, N. N., Cossman, J., and Mak, T. W., 1993, Single cell analysis of Hodgkin and Reed-Sternberg cells: Molecular heterogeneity of gene expression and p53 mutations, *Blood* **81:**3097-3115.
112. Bayliss, G. J., and Wolf, H., 1980, Epstein-Barr virus induced cell fusion, *Nature* **287:**164-165.
113. Linder, J., and Purtilo, D. T., 1984, Infectious mononucleosis and complications in immune deficiency and cancer, in: *Immune Deficiency and Cancer: Epstein-Barr Virus and Lymphoproliferative Malignancies* (D. T. Purtilo, ed.), Plenum Medical, New York, pp. 11-36.
114. Lukes, R. J., Tindle, B. H., and Parker, J. W., 1969, Reed-Sternberg-like cells in infectious mononucleosis, *Lancet* **2:**1003-1004.
115. Tindle, B. H., Parker, J. W., and Lukes, R. J., 1972, Reed-Sternberg cells in infectious mononucleosis? *Am. J. Clin. Pathol.* **58:**607-617.

116. Strum, S. B., Park, J. K., and Rappaport, H., 1970, Observation of cells resembling Sternberg–Reed cells in condition other than Hodgkin's disease, *Cancer* **26:**176–190.
117. Katz, I. R., Chapman-Alexander, J., Jacobson, E. B., Lerman, S. P., and Thorbecke, G. J., 1981, Growth of SJL/J-derived transplantable reticulum cell sarcoma as related to its ability to induce T-cell proliferation in the host. III. Studies on thymectomized and congenitally athymic SJL mice, *Cell. Immunol.* **65:**84–92.
118. Katz, J. D., Lebow, L. T., and Bonavida, B., 1989, The *in vivo* depletion of Vβ17a$^+$ T-cells results in the inhibition of reticulum cell sarcoma growth in SJL/J mice. Evidence for the use of anticlonotypic antibody therapy in the control of malignancy, *J. Immunol.* **143:**1387–1395.
119. De Kruiff, R. H., Brown, P. H., Thorbecke, G. J., and Ponzio, N. M., 1985, Characterization of SJL cell clones responsive to syngeneic lymphoma (RCS): RCS-specific clones are stimulated by activated B-cells, *J. Immunol.* **135:**3581–3586.
120. Ponzio, N. M., Lerman, S. P., Chapman, J. M., and Thorbecke, G. J., 1977, Properties of reticulum cell sarcomas in SJL/J mice. IV. Minimal development of cytotoxic cells despite marked proliferation to syngeneic RCS *in vivo* and *in vitro*, *Cell. Immunol.* **32:**10–22.
121. Lasky, J. L., Ponzio, N. M., and Thorbecke, G. J., 1988, Characterization and growth factor requirements of SJL lymphomas. I. Development of a B-cell growth factor-dependent *in vitro* cell line, cRCS-X, *J. Immunol.* **140:**679–687.
122. Lasky, J. L., and Thorbecke, G. J., 1989, Characterization and growth factor requirements of SJL lymphomas. II. Interleukin 5-dependence of the *in vitro* cell line cRCS-X, and influence of other cytokines, *Eur. J. Immunol.* **19:**365–371.
123. Nakauchi, H., Osada, H., Yagita, H., and Okamura, K., 1987, Molecular evidence that SJL reticulum cell sarcomas are derived from pre-B cell. Clonal rearrangement of heavy chain but not light chain immunoglobulin genes, *J. Immunol.* **139:**2803–2809.
124. Haran-Ghera, N., Katler, M., and Meshores, A., 1967, Studies on leukemia development in the SJL/J strain of mice, *J. Natl. Cancer Inst.* **39:**653–661.
125. Siegler, R., and Rich, M. A., 1968, Pathogenesis of reticulum cell sarcoma in mice, *J. Natl. Cancer Inst.* **41:**125–143.
126. Bolocchi, M., Dolcetti, R., and Carbone, A., 1992, Pathogenesis of human reactive-appearing non-monomorphous malignant lymphoproliferative disorders: A hypothesis, *Tumori* **78:**221–227.
127. Carbone, A., Gloghini, A., Gattei, V., Aldinucci, D., Degan, M., De Paoli, P., Zagonel, V., and Pinto, A., 1994, Expression of functional CD40 antigen on Reed–Sternberg cells and Hodgkin's disease cell lines, *Blood* **85:**780–789.

21

The Development of Epstein–Barr Virus Vaccines

ANDREW J. MORGAN

1. THE DEMAND FOR AN EPSTEIN–BARR VIRUS VACCINE

Epstein–Barr virus (EBV) is one of the handful of human herpesviruses and is of great significance because of its oncogenic potential in humans [reviewed by Liebowitz and Kieff[1,2] and Miller[3]]. Infection with EBV usually occurs in early childhood with no clinical consequences, and humoral and cell-mediated immune responses develop that control virus replication and the number of infected cells present in the individual. The growing number of clinical conditions, including Burkitt's lymphoma (BL) and nasopharyngeal carcinoma (NPC), now associated with this virus has warranted the development of vaccines to prevent or modify EBV infection, a thesis first articulated by M. A. Epstein in 1976.[4]

The range of host cells that can be infected by EBV is very small, and infection principally occurs in B lymphocytes by way of the CD21 complement receptor binding the virus through its major viral envelope glycoprotein, gp350.[5,6] Certain epithelial cells carry this receptor and can also be infected by the virus.[7,8] The location of these cells is predominantly in the oropharyngeal epithelium[9] but also in the uterine cervix[10] and male genital tract.[11] Productive viral replication occurs in oropharyngeal epithelial cells, which somehow evade mucosal immune responses, and shedding of the virus into the buccal fluid provides the route by which the virus can be transmitted to another individual. At present, it is not clear what cell acts as a reservoir for EBV during its life-long persistent infection. In a normal seropositive individual, only one cell in 10^5 to 10^6 of the total B-cell population is infected with EBV,[12] and only a vanishingly small quantity of these appar-

---

ANDREW J. MORGAN • Department of Pathology and Microbiology, School of Medical Sciences, University of Bristol, Bristol BS8 1TD, England.

DNA Tumor Viruses: Oncogenic Mechanisms, edited by Giuseppe Barbanti-Brodano *et al.* Plenum Press, New York, 1995.

ently enter the productive replication cycle and generate infectious virus. Is the oropharyngeal epithelium a reservoir that provides infectious virus that can then infect B lymphocytes traveling through that zone, or do latently infected B lymphocytes switch into productive replication when in transit through this region?

The population of latently infected B lymphocytes is clearly regulated by cell-mediated immune responses,[13] at least in part by recognition of certain EBV latent antigens. It appears that the expression of certain of these antigens is necessary for them to be targets for the cellular immune system because the expression of Epstein–Barr nuclear antigen 1 (EBNA-1) alone is not enough.[14,15] It has been argued that these infected cells, which have their EBV latent gene expression restricted to EBNA-1, as resting B cells may themselves to be the reservoir for lifelong persistent infection. The EBV is classified as type 1 or type 2[16] on the basis of sequence differences in EBNA-2 and EBNA-3a, -3b, and -3c,[17] and the two types are readily distinguished by polymerase chain reaction methods. Studies have been carried out in bone marrow transplantation to ask whether the transplant recipients retain the original EBV type, accommodate both the indigenous and donor EBV types, or whether the indigenous is replaced by the donor type. In the limited number of studies that have been carried out, it is clear that the indigenous EBV type is replaced by the virus of the donor.[18] These data suggest that bone marrow B cells may themselves act as long-term reservoirs of a single EBV type.

Following the discovery of EBV[19] as a new herpesvirus in biopsies of BL from East Africa in 1964, it became apparent that a strong association existed between the virus and endemic BL. The arguments for there being a causal link between the virus and this lymphoma are discussed at length elsewhere.[1–3,20] The principal criteria supporting the causative hypothesis are that (1) antibodies against the virus capsid antigen are predictive of the later occurrence of BL, (2) the virus genome is found in 90% of cases, and (3) the virus can induce malignant lymphoma in certain New World primates.[21–24] Similarly, it became clear that undifferentiated NPC, a major cancer in southeast Asia and southern China, was also associated with the virus. The virus is always found in undifferentiated NPC tumor cells,[24,26] and a serologic indicator of the onset of the disease is the rapid rise of serum IgA antibodies against virus capsid antigen[27] in a large proportion of affected individuals. Nasopharyngeal carcinoma is a major world health problem, with the number of new cases being reported each year in excess of 80,000.[28]

More recently, it has been observed that a large proportion of Hodgkin's lymphomas[29–33] contain EBV and express a limited number of EBV latent antigens, and there are also serologic data that link delayed EBV infection to an increased risk of developing certain Hodgkin's lymphomas.[34,35] It is also well known that a proportion of the lymphomas that occur in immunosuppressed patients with AIDS or who are undergoing organ transplantation are in fact EBV lymphomas.[36–38]

Ninety-five percent or more of the world's population are infected with EBV for life and, except for a proportion of those living in the industrial West, became infected in early childhood. The incidence of BL and NPC is low in the West, and in those parts of the world where they are found, they occur in well-defined geographic regions. Thus, EBV cannot be the only factor in the causation of these diseases but may be an essential one. Other factors in the causation of BL include holoendemic malaria,[39–41] but cofactors in the causation of NPC remain unidentified[42] with the possible exception of the consumption of salted fish during childhood in the southern Chinese. People with certain HLA types are more predisposed than others to contract NPC, and susceptibility to the disease may be linked to a particular gene locus.[43]

There have been two prime motivating factors for research into the development of EBV vaccines. The first is to achieve the goal of a cheap and effective vaccine that will reduce the incidence of, or eliminate entirely, one or more of the above diseases. In particular, such a vaccine will be aimed at NPC, which is the most common cancer in men and the second most common in women in southern China and therefore afflicts a great many people. A second motivation has been a purely scientific one, as there are some fundamental questions about the biology of EBV that cannot be answered other than by preventing natural infection by vaccination. For example, vaccination offers the only means of obtaining a direct and unequivocal proof that EBV is a causative agent in the human cancers mentioned above. This result could be obtained within 7 or 8 years in the population of susceptible African children who will develop endemic BL.

Unfortunately, diseases that affect only the Third World and have little impact in the West do not attract the research and development resources that are ideally needed. The likelihood of obtaining a return on the investment required by the commercial pharmaceutical sector is greatly reduced, as Third World countries cannot afford to pay the cost of vaccines made in the Western drug company environment. However, EBV causes infectious mononucleosis (IM), otherwise more commonly known as glandular fever[44] or the "kissing disease," as it probably often arises from osculatory contact in young adults. In those individuals in whom primary infection is delayed until adolescence, there is a 50% chance of the individual contracting IM when the infection does occur. The high incidence of IM in the West may reflect higher hygiene standards: IM is almost unknown in the Third World, where infection takes place during the first year of life. In all but a very few cases, IM resolves within weeks or months at the most, and the subject will display the subclinical immunologic characteristics of a normal seropositive individual with a lifelong persistent infection.

Until recently it had been taken for granted that the cost of developing an EBV vaccine was not justified simply to prevent IM because the vast majority of cases were thought to have no long-term consequences, although the patient may have been quite unwell for a long period of time. Furthermore, a very good case indeed has to be made to justify the vaccination of otherwise perfectly healthy young children. A proper cost–benefit analysis was needed to answer this question, and this task has recently been attempted by Evans.[45] This study indicates that an effective EBV vaccine that could prevent IM would pay for itself in a relatively short time from savings made in health care costs. The incidence of IM in the United States is 65 per 100,000 or 170,000 cases per year in a population of 243 million; 15–20% of cases require hospital stays of 4.5 days each on average, the total cost being in the region of US$100 million in any one year. The loss of time at work and physicians' fees can only add to this figure.

In predicting the development cost of an EBV vaccine, some assumptions have to be made, and it is not possible, at this stage, to predict the cost per dose of the candidate EBV vaccines. In the Evans study, it was assumed that the cost of a gp350 subunit EBV vaccine would be in the region of US$40 per dose, which would give rise to an overall cost of $480 million if 4 million children were given three immunizations. In addition, who can say what the cost will be of the malignancies associated with EBV infection that might be preventable by vaccination? On the face of it, there is a powerful case for the commercial development of an EBV vaccine for the reasons given above. The incidence of IM in the Unites States is greater than all other reportable diseases except gonorrhea, and these include AIDS, tuberculosis, measles, and hepatitis, although these latter diseases, of course, are likely to have much more serious consequences when compared to IM in the vast majority of the population. It is to be hoped that the increased perception of the importance of IM and its cost to Western health services along with the apparent link with Hodgkin's lymphoma, which is a relatively common group of tumors in the West, will provide a strong incentive for the commercial development of an EBV vaccine. In turn, this would promote the development of viable products for use in the Third World to prevent NPC.

2. SELECTION OF AN EBV VACCINE MOLECULE

In some respects, the development of herpes simplex virus (HSV) vaccines has followed a similar path to that of EBV vaccines. The observations that passively administered monoclonal antibodies against one or more HSV envelope glycoproteins provided immunity to HSV infection under certain circumstances in laboratory animals single out herpesvirus glycoproteins as being good subunit vaccine candidates.[46] In the early days, that immunization could prevent herpesvirus infection at all was demonstrated for Marek's disease of chickens[47] and *Herpes saimiri* in nonhuman primates.[48] Because it is not possible to use attenuated or killed EBV variants as vaccines because of the oncogenic potential of the virus, the first prototype EBV vaccine was based on a purified viral envelope glycoprotein. Once the rationale for developing a subunit vaccine had been established, the task of identifying components of the EBV envelope was pursued because, at that time, it was presumed that protective immunity would be provided by virus-neutralizing antibodies directed against external viral envelope components. A correlation was observed between the levels of *in vitro* virus-neutralization activity of some sera and reactivity of antibodies from the same sera against the so-called membrane antigens (MA) on productively infected lymphocytes.[49,50] When these same virus-neutralizing antibodies were used to immunoprecipitate viral glycoproteins from lysates of infected and radiolabeled cells, the viral MA was found to consist of three principal glycoprotein components.[51-55] The largest and most abundant of the three are gp350 and gp220, the latter being a spliced variant of the former that is encoded by the same open reading frame (ORF) BLLF1.[56] The third component, found in much lower abundance, is gp85, a viral envelope glycoprotein that plays a role in fusion between the host cell and viral envelopes.[57-59] The potential role of this molecule in an EBV vaccine has not yet been investigated, but the potential for gp350 as a subunit vaccine has been evaluated at some length and will shortly enter human trials.

The very first subunit vaccine was based on gp350 purified from bulk cultures of B958 virus-infected cells[60] that had been induced to increase the proportion of cells involved in productive infection by addition of sodium butyrate and phorbol esters.[61] The purified but denatured antigen made by preparative SDS-PAGE was renatured by the removal of SDS in the presence of urea, subsequent removal of urea by dialysis, and incorporation into artificial liposomes. This vaccine formulation induced virus-neutralizing antibodies in mice[60] and induced a protective immune response in the cottontop tamarin model of EBV lymphoma[62] (see below).

A number of interesting properties of gp350 have emerged since the first studies showed it to be a protective immunogen in an animal model of EBV-induced lymphoma. The gp350 glycoprotein is the virus ligand that binds to its host cell complement component receptor CD21.[5-7] It contains up to 50% carbohydrate, much of which is O-linked.[63,64] In view of examples of human immunodeficiency virus and influenza in which the capacity to generate antigenic variants poses serious problems in vaccine design, the question arose as to whether wild-type variation in the gp350 antigen was significant and whether this would render a conventional vaccine ineffective if it were based on a single laboratory strain. The sequences of gp350 genes have been compared from several type 1 and type 2 laboratory virus isolates, and no significant differences have been found. Also, a panel of monoclonal antibodies recognized the gp350s from all these sources equally well.[65]

At least 20 ORFs in the EBV genome could potentially code for glycoproteins that would have some N-linked sugars,[66] but with the exception of the EBV counterpart of gB coded for by BALF4,[67,68] the product of BILF2,[69] and gp85, which is the gH analogue,[57-59] none have been identified or characterized, let alone evaluated, for incorporation into any vaccine. Several EBV glycoprotein products might ultimately need to be incorporated into any vaccine formulation.

As mentioned above, the selection of a virus molecule for use as a subunit vaccine was originally based on an assumption that may turn out to have been unjustified in terms of what immune responses will be required in protection against primary virus infection. The assumption was that protective immunity would be provided by virus-neutralizing antibodies directed at glycoproteins on the surface of the virus. This is clearly not the case in the tamarin model (see below) and is unlikely to be the only immunologic criterion for protection against EBV infection in humans. In addition, it can be argued that it might not even be necessary to immunize with glycoproteins that retain their native conformations if cytotoxic T cells (CTL) were to provide the essential protective immune responses. In any event, final judgment of these issues must await information on the correlates of protective immunity in humans, some of which will become available on the completion and evaluation of the first human trials.

3. AN ANIMAL MODEL OF EBV-INDUCED LYMPHOMA

As far as possible, it is essential to determine the safety and efficacy of an experimental vaccine by testing in an animal model of the virus infection in question, and preferably in an animal model in which the virus induces a similar or identical disease to that which it causes in humans. This ideal is almost never achieved, and the inevitable shortcomings of animal models must be taken into account when they are used to predict the outcome of vaccinations in humans. Inoculation of EBV into the common marmoset (*Callithrix jacchus*) gives rise to a poorly defined mononucleosis-like syndrome.[70,71] It was thought at one time that this model might be useful, as a very few animals, at least, appeared to support a persistent infection, but this phenomenon has not been reproducible. However, when EBV is inoculated in sufficient quantities into cottontop tamarins (*Saguinus oedipus oedipus*) or the owl monkey (*Aotus trivirgatus*), B cell lymphomas develop rapidly.[21–24] Studies have concentrated on the tamarins since the owl monkeys that were originally used were found to have different karyotypes in different individuals.[72] The malignant lymphoma induced in the tamarin by EBV injection has been closely studied, and the tumors, which arise invariably within 2 to 3 weeks, bear a striking resemblance to those that have a high incidence in human immunosuppressed organ graft patients. The tumors arise independently at different sites in lymphoid tissues and have been shown to be mono- or oligoclonal in origin.[21] Furthermore, as is the case in EBV lymphomas in immunodepressed patients, but in contrast to BL and NPC tumors, all of the known latent viral antigens are expressed.[73]

This animal model has been used to test the efficacy of the variety of EBV vaccines that have been developed. In one sense, the model is a good test for efficacy of an EBV vaccine because protection can be achieved against a massive tumorigenic dose of virus that is injected intraperitoneally. This severe challenge should be compared with the mode of infection that normally occurs in humans, where a very small quantity of virus is transmitted orally. However, one shortcoming of the model is that the tamarin has never been shown to be infected by the oral route and, unlike humans, appears not to be able to sustain a persistent infection, at least not in the same way as in humans. This animal also has a very restricted major histocompatibility complex class I polymorphism.[74] The significance of the tamarin model in EBV research might be greatly enhanced if evidence of persistent EBV infection could be obtained. Until very recently, it was assumed that the tamarin could not sustain persistent EBV infection because measurable antibody responses to the virus capsid antigen (VCA) tailed off and became undetectable within 18 months to 2 years of inoculation. However, recently, by use of polymerase chain reaction methods and immunohistochemical staining, small numbers of EBV-positive B lymphocytes have been detected in animals that had been immunized and challenged with a lymphomagenic dose of EBV but were protected against tumor induction. Similarly, virus-infected B lymphocytes were also detected in those animals that had not been immunized and had received a tumorigenic challenge but in which the tumors spontaneously regressed and the animal survived. Both of these observations were made only several months after the challenge dose of virus was given, and further studies will be needed to demonstrate viral persistence.[75] All the gp350 vaccination experiments carried out to date in the cottontop tamarin are summarized in Table I.

4. NATURAL PRODUCT gp350 SUBUNIT VACCINES

The first demonstration that gp350 was an effective subunit vaccine in the tamarin lymphoma model was obtained using material isolated from very large bulk cultures of cells infected with the B958 laboratory isolate of EBV and induced to productive infection with sodium butyrate and phorbol esters.[61] Only very small quantities of protein were isolated, and this material was purified using SDS-PAGE followed by elution from the gel, removal of SDS, and renaturation. These very small quantities of protein were incorporated into artificial liposomes made from phosphatidylcholine. Despite the very small quantities of protein and the crude adjuvant, complete protective immunity was induced in tamarins against the tumorigenic dose of EBV.[62] This study gave the first definitive evidence that a vaccine incorporating the major envelope glycoprotein gp350 could prevent malignant lymphoma experimentally induced by inoculation with a large dose of virus. This experiment was a landmark in the development of EBV vaccines.

In order to simplify the purification procedure and to prepare larger quantities of gp350 for vaccine studies, monoclonal antibody affinity chromatography was employed. Milligram quantities of natural product gp350 were obtained from

TABLE I
gp350 Vaccination Experiments Carried Out in the Cottontop Tamarin

Immunogen	Adjuvant	Animals protected	Neutralizing antibodies	Reference
Natural product gp350 isolated by SDS-PAGE	Liposomes	1/4[a]	Yes	62
Natural product gp350 isolated by SDS-PAGE	Liposomes	2/2	Yes	62
Natural product gp350 isolated by monoclonal antibody	Liposomes	0/4	Yes	76
Natural product gp350 isolated by ion exchange	ISCOMs	4/4	Yes	79
Natural product gp350 isolated by ion exchange	SAF-1	4/4	Yes	81
BPV gp350 isolated by ion exchange	SAF-1	3/4[b]	Yes	100
BPV gp350 isolated by ion exchange	Alum	3/5	Yes	101
WR vaccinia gp350	None	3/4	No	113
Wyeth vaccinia gp350	None	0/4	No	113
Replication-defective adenovirus gp350	None	4/4	No	109

[a]The protected animal was the only animal in which an immune response could be detected.
[b]The single unprotected animal developed severe ulcerative colitis during the experiment.

purified plasma membrane preparations obtained from large bulk cultures of EBV-infected cells that had been chemically induced to increase productive virus infection. Following incorporation into artificial liposomes and immunization of tamarins, high titers of virus-neutralizing antibodies were generated as in the previous experiment. However, unlike the previous experiment, none of the immunized animals were protected against the lymphomagenic virus challenge, and all developed malignant lymphoma within 2 to 3 weeks.[76] These results have never been satisfactorily explained, although it has been put forward that important protective immunogenic sites are destroyed by elution of gp350 from monoclonal antibody affinity columns at pH 11.5. This is not at all consistent with the fact that gp350 purified in SDS-PAGE followed by exposure to urea, removal of SDS, and renaturation is a protective immunogen. It is possible that the monoclonal antibody selected a subpopulation of gp350 that is somehow insufficient to induce a protective immune response. A third explanation suggested that denaturation may be an important step in altering the immunologic profile of the gp350 molecule. The knowledge that protective immunity in this animal model using gp350 can be induced without specific antibody production (see below) may also give a clue. It may be that the induction of virus-neutralizing antibodies specific for gp350 are not central in a protective immune response, and the failure of a monoclonal antibody affinity-purified gp350 reflects a failure of this material under the particular circumstances of that experiment to induce cell-mediated immunity.

A new purification procedure for gp350 was developed using anion-exchange chromatography.[77] The relatively high negative charge carried by gp350 at neutral pH probably results from the large amounts of sialic acid present.[63] An ion-exchange purification is possible from detergent lysates of infected cells at pH 7.2 in a low-salt piperazine buffer. Under these conditions, gp350 and virtually nothing else binds to a Mono Q anion-exchange column and is eluted at pH 5.2. The gp350 prepared in this way, after having been incorporated into immunostimulating complexes (ISCOMs)[78,79] or into the Syntex aduvant formulation (SAF-1),[80,81] induces protective immunity in tamarins against the standard lymphomagenic dose of virus along with high titers of virus-neutralizing antibodies. Protective immune responses were obtained with a relatively low dose of antigen when SAF-1 or ISCOMs were used, this dosage being 5 µg or less.

Although purified natural product gp350 has been available for some time, and sufficient could have been prepared for initial human trials, the product was rejected for a number of reasons. First, although the natural product gp350 could have been prepared for small-scale phase I and phase II human trials, ultimately the preparation method would have been entirely unsuitable for large-scale production. Second, because the cells from which the gp350 was purified contained EBV, there would always be a risk of contamination of the product with viral DNA. In view of the oncogenic potential of EBV DNA, this would be an unacceptable risk. Third, in order to maximize the yield of natural product gp350 from bulk cultures of B958 cells, it was necessary to include phorbol esters and sodium butyrate to induce as many cells as possible to productive viral replication. The use of phorbol esters on any scale to prepare materials for human use is unacceptable because of their tumor-promoting properties.

5. RECOMBINANT gp350 SUBUNIT VACCINES

To overcome the problems of yield, purity, and presence of EBV DNA, efforts were made to express the gp350 gene in bacteria, yeast, and eukaryotic cells. The complete nucleotide sequence of the B958 EBV isolate was determined some years ago,[66] and the gene coding for gp350 and gp220, its spliced companion coming from the same ORF, has long been mapped. Parts of the gp350 gene have been expressed in bacteria,[56,82–86] and the complete gene in yeast.[87] However, glycosylation does not occur in bacteria, and different carbohydrates from those found on mammalian cells are added when the gene is expressed in yeast.[88] At that time, emphasis was on producing gp350 that could induce virus-neutralizing antibodies because it was believed that this would be the key immunologic parameter required to obtain protective immunity against the virus. Although the bacterial product was recognized by antibodies from normal seropositives, none of these antibodies were virus-neutralizing.[83,84] Similarly, the yeast gp350 product was not sufficiently antigenically similar, presumably because of yeast glycosylation, and this approach was discontinued. However, it is likely that cell-mediated immune responses are going to be important in protective immunity in humans as they are known to be in protection against EBV-induced lymphoma in the tamarin (see below). Because tertiary structure and glycosylation are unlikely to impede the generation of appropriate T-cell responses, both the bacterial and yeast products should be reexamined as candidate vaccines, but this has not yet occurred. A truncated version of the gp350 has been expressed in a baculovirus system, but this product has not been characterized in terms of its ability to induce virus-neutralizing antibodies nor in terms of its ability to induce protective immunity in the tamarin lymphoma model[89] and may be an equally good vaccine candidate.

The gp350 gene has now been expressed in a number of mammalian cell expression systems where glycosylation and posttranslational modifications occur that are closely similar to those found on the natural product.[90–94] It has not been possible to distinguish between these products and the natural product gp350 in terms of their ability to induce virus-neutralizing antibodies, bind a range of monoclonal antibodies, and, in some studies, to stimulate the proliferation of gp350-specific T cells.[91,92,95,96] In some cases, the membrane anchor sequence has been removed from the gene, allowing secretion of the expressed eukaryotic product into the culture medium.[91,92,94] This approach offers major advantages in the large-scale preparation of a defined product that is relatively easy to purify. A bovine papillomavirus (BPV) expression system and a Chinese hamster ovary (CHO) cell system have been adapted to produce a secreted gp350 product in large quantities.[91,92] Again, the secreted product is indistinguishable from the authentic gp350 in immunologic terms. Another major benefit from the use of this system is to guarantee the absence of potentially oncogenic EBV DNA, although it will certainly be necessary to demonstrate formally the absence of EBV DNA and acceptable levels of BPV DNA.

From a regulatory point of view, the BPV grown in C127 mouse fibroblasts is less than ideal. Papillomaviruses have an oncogenic potential, and the C127 cell line itself is tumorigenic in nude mice. The fact that the gp350 product made in this system is secreted and highly purified should be set against this. Furthermore, there is a precedent for the use of this BPV cell line for the production of human growth hormone for commercial human use.[97]

6. CHOICE OF ADJUVANT

Most proteins or glycoproteins are weakly immunogenic when inoculated alone into animals, and gp350 is no different in this respect. An adjuvant is invariably required to stimulate the immune response to the antigen except when it is presented as part of a live virus or other vehicle, in which case, a whole range of the natural immune responses are triggered. A variety of effective immunologic adjuvants are now available, and new ones appear from time to time. No doubt they do not all work in the same way and are certainly not equally effective. The commercial potential of a safe and effective adjuvant for use in humans is very great, and this field has become extremely competitive, with the big drug companies having a strong interest. As a result, it has been very difficult to establish just what results have been obtained by various groups evaluating their own adjuvants, where the precise composition or formulation is often confidential.

Adjuvants comprise a small immunostimulatory molecule carried in a delivery vehicle that reduces to a minimum the dispersal and dilution of the immunostimulant and antigen. Presumably, this allows the effective presentation of antigen and immunostimulator to the immune system by either being in the right place at the right time or allowing the cells of the immune system to focus their activities in concert at one location. Antigen presentation, targeting, and delayed decomposition may all be important functions of the delivery vehicle. The mode of action of the small immunoactivator molecule may include lymphokine production, antigen processing, mitogenicity, and up-regulation of HLA expression. Probably a key element in the action of adjuvants is their influence on the differentiation of $CD4^+$ T cells into T_H1 or T_H2 subsets following vaccination, and presumably the adjuvant can influence the ratio of the two types. T_H1 cells produce interleukin 2 and interferon γ and enhance cell-mediated immune responses, whereas T_H2 cells produce interleukins 4, 5, and 10 and augment humoral immune responses.[98] Recently, it has been shown that interleukin 12 can substitute for certain bacterial adjuvants in enhancing cell-mediated immune responses, at least against *Leishmania major*.[99] Protection studies in the tamarin lymphoma model with natural product gp350 subunit vaccines have so far been confined to the use of artificial liposomes,[62,60] threonyl muramyl dipeptide used in pluronic block copolymers (SAF-1),[81] and ISCOMs.[79]

The gp350 produced in the BPV system induces complete protective immunity in the tamarin lymphoma model when used with SAF-1.[100] In order for it to be used in conjunction with ISCOMs, a new coupling technique is employed because BPV gp350 has no hydrophobic membrane anchor sequence, and it is this hydrophobic sequence that enables gp350 ISCOMs to be created with natural product gp350. The coupling methodology for the BPV product and ISCOMs that have chemically defined pure components is well advanced.

The original objective was to focus attention on the use of the latter powerful adjuvants, and this strategy was based on the incorrect assumption that one of these adjuvants would be licensed for human use well before the time of writing of this chapter. Although their use, or use of derivatives of them, in humans is likely to be some time in the future, it cannot be said when. There have been grounds, therefore, for returning to evaluate aluminum salts as an adjuvant in the tamarin lymphoma model because only adjuvants of this type are licensed for use in humans. As there is a pressing need to carry out some human trials as soon as possible, the alum adjuvant was tested in the tamarin lymphoma model with BPV gp350.[101] The results of these experiments clearly show that alum/gp350 formulations are capable of inducing protective immunity in the tamarin against EBV-induced lymphoma, although they may be less effective than SAF-1. It is worth remembering that the challenge with a lymphomagenic dose of virus represents an extreme and completely unphysiological event and, in some respects, a very stringent test of the efficacy of any vaccine. Three out of five animals immunized with alum/gp350 were protected against lymphoma, so it could be that alum/gp350 will be sufficiently effective in humans. Again, it must stated that we do not know the correlates of human immune protection against EBV, and so there is little to be gained by speculating on which adjuvant formulation is likely to give the best result before some human trials have taken place. It is noteworthy that in the *Callithrix jucchus* model of an EBV-induced mononucleosis-like syndrome, protective immunity was obtained using gp350 in alum.[70,101a]

7. LIVE VIRUS VECTOR RECOMBINANTS

The disadvantages often cited against the use of subunit vaccines is that they generate poor and sometimes inappropriate responses, although the new generation of adjuvants mentioned above should overcome this objection. Any failure to induce a broad-ranging and powerful immune response has to be set against the advantages of using biologically dead material of absolutely defined composition.[102] The choice may depend on a variety of factors that will include what immune responses are required to be induced. One assumes that the induction of both T- and B-cell responses is necessary, along with some memory function. It is known that both ISCOMs[103] and SAF-1 induce T-cell responses and memory.[104] Aluminum salts also allow the induction of certain cell-mediated immune responses.[105]

Many of the conventional and very successful vaccines used in recent times have been attenuated live virus vaccines.[106] Many recombinant viruses have now been made expressing one or more important vaccine molecules, but none has so far been used in humans. Recombinant vaccinia expressing rabies glycoprotein has been used to immunize wild animals in Europe and the United States by impregnating food bait.[107] The concerns raised by the use of recombinant live virus are, first, the possibility, remote or otherwise, that the recombinant virus may mutate in the wild or recombine again with wild-type virus and become pathogenic once more, and second, depending on the recombinant virus in question, the inevitable presence side effects in a fraction of the population where some individuals are unusually susceptible even to attenuated strains because they have a defective immune system. Much effort has been devoted to producing several live recombinant virus vaccine vectors that minimize these concerns, and this goal is now well within reach if it has not already been achieved.

The protein gp350 has been expressed in vaccinia,[108] adenovirus,[109] and varicella.[110] The gp350 gene could be expressed in a canarypox vector[111] or in a vaccinia recombinant derived from the Copenhagen strain, which has been specifically attenuated by the removal or inactivation of individual genes.[112] Recombinant vaccinias expressing gp350 have been derived from both the WR laboratory strain and the Wyeth vaccine strain, and both have been tested in the tamarin lymphoma model.[113] When the WR strain recombinant was used, protective immunity was induced in three out of four animals, but no protective immunity was afforded to those animals immunized with the Wyeth strain derivative. Both groups of animals responded to vaccination with high levels of antibody against vaccinia proteins, but no antibodies against gp350 could be detected in any animal. Antibody levels against vaccinia proteins were substantially lower where the Wyeth derivative was used, and presumably this is a reflection of the degree to which the strain has been attenuated. The key observation in these experiments was that the WR strain derivative gave protective immunity in the absence of antibodies to gp350. Clearly, protective immunity in this case is provided by some form of cell-mediated immune response. When protective immunity was induced in the tamarin lymphoma model using replication-defective adenovirus expressing gp350 (see below), antibodies against gp350 were induced but had no capacity to neutralize EBV *in vitro*.[109] More work needs to be done in developing effective vaccinia recombinants that strike the correct balance between attenuation and immunogenicity. Progress to this end should be made with the Copenhagen strain vaccinia derivative[112] and canarypox recombinants[111] mentioned above.

A number of recombinant adenoviruses expressing foreign antigens have now been constructed.[114–118] Recombinant adenovirus expressing gp350 has been made and recently tested in the tamarin model.[109] These recombinants were based on human adenovirus type 5, which had been well characterized. Replication-defective recombinants were made in which the E1 region is deleted, and the virus can be propagated only in a helper cell line that provides the deleted E1 function.[119] Adenoviruses have the capacity to down-regulate HLA expression in the cells they are infecting, but the E3 region, which is responsible for this, has also been deleted in these replication-defective recombinants. Several features of adenovirus have, quite apart from the above, made them attractive for vaccine delivery. First, adenoviruses types 4 and 7 have already been used on a large scale in the U.S. Armed Forces to prevent respiratory disease, and they have a good safety record.[120,121] Second, in this case, the adenovirus was given by oral administration following encapsulation. Mucosal immunity is induced in the respiratory tract, although primary immune contact is in the gut lymphoid tissue.

The Oka varicella–zoster virus vaccine strain has been used to make recombinants expressing gp350.[110] These recombinants failed to induce protective immunity in the tamarin lymphoma model, and it is again suggested that these recombinants are too attenuated in this species. Antibodies were generated against varicella proteins but not against gp350 except in one tamarin where the levels were very low (R. Lowe, S. Finerty, and A. J. Morgan, unpublished observations). The future use of this latter recombinant in humans cannot yet be dismissed because, as has been stated above, the criteria that have to be met in the induction of protective immunity in humans are not yet known.

Probably the most significant results obtained using recombinant virus expressing gp350 are those in which the Chinese vaccinia strain (tien tan) was used to vaccinate a small group of both seronegative and seropositive children in southern China.[122,123] It was reported that antibody levels to gp350 were raised in those subjects who were already seropositive and were induced in those who were seronegative at the beginning of the trial. Six of nine vaccinated children who were seronegative for EBV at the time of vaccination remain seronegative at the time of writing, which is 3 years since vaccination. Reports on the progress of these Chinese children will be awaited with keen interest, particularly with respect to the prevention of, or delay in, infection of those who were EBV-seronegative at the time of vaccination.

8. CELL-MEDIATED IMMUNE RESPONSES TO gp350

The use of the WR strain vaccinia recombinants expressing gp350 in the tamarin lymphoma model revealed that protective immunity could be obtained in the absence of serum antibody specific for gp350. This result suggests that in this case at least, protection was provided by cell-mediated immune responses. Certainly, other examples of vaccinia- and adenovirus-induced protection against viral infection are recorded where specific antibody against the inserted gene product has been absent.[124,125]

Although studies on cellular immunity to EBV have been extensive, these have almost invariably centred on the 11 latent viral antigens including the EBNAs and the latent membrane proteins (LMPs).[13] However, four studies have been carried out on cellular immune responses to gp350 in humans. First, it is now known that normal seropositive individuals possess T-helper cells specific for gp350,[95,96] and gp340/ISCOMs stimulate the production of T cells that prevent EBV-induced transformation *in vitro*.[126] Recently, it was shown that gp350 can be a target antigen in HLA class II-restricted CD4 T-cell recognition and cytotoxicity.[127] In this case, the target cells are EBV-transformed B-cell lines coated with the infectious virus they have made themselves, which carries gp350. The gp350 is the major EBV envelope glycoprotein and allows the binding, at least, of virus to potential host cells expressing CD21, the complement receptor. It is difficult to imagine how gp350 can be a target for T-cell recognition, as so few infected B lymphocytes in the circulation in humans express this glycoprotein because they remain only latently infected. However, CTL target sensitivity may be such that only very small, undetectable quantities of gp350 peptides are needed. B lymphocytes passing in the region of the oropharynx, where infectious virus is present, presumably bind virus at this point and could become susceptible to HLA class II-restricted recognition.

The mechanism of cell-mediated immune protection in the tamarin lymphoma model is also very difficult to explain if a protective function for antibody is excluded. Because protection is gp350-specific, it must be assumed that on inoculation of the vaccinated animal with live virus, protective immune responses are directed against newly infected B cells coated with EBV shortly after inoculation with the lymphomagenic dose of EBV through an MHC class II mechanism. Because the tumor cells themselves induced in the system do not express detectable quantities of lytic cycle genes, it seems unlikely that they would be the target of the protective immune response unless very small quantities of gp350 are made and are sufficient to allow recognition by gp350-specific T cells.

Other mechanisms of protection may exist, and if it is possible to generate protective immunity in humans by vaccination using gp350, it may be by quite a different mechanism from the one adopted by the tamarin in the face of a massive dose of virus introduced by injection. Antibody responses may well be important in humans, although the virus persists while encountering high titers of virus-neutralizing antibody to gp350 in many cases. In the tamarin model, it appears that gp350-specific antibodies, whether they be virus-neutralizing or not, are not involved in protective immunity, because, as discussed above, protection is achieved using vaccinia recombinants where no antibodies are made.[113] Protection is also achieved when the antibodies made are not virus-neutralizing following immunization with adenovirus recombinants,[109] and protective immunity is not always induced when high titers of the neutralizing antibody against gp350 are generated.[76] In the *Callithrix jacchus* model of the EBV-induced mononucleosis-like syndrome, protective immunity did not correlate with the presence of serum virus-neutralizing antibodies following immunization with gp350 produced in transfected Vero cells and injected with alum.[101a]

In certain circumstances, cells expressing gp350 can be good targets for antibody-dependent cellular cytotoxicity (ADCC).[128] The ADCC activity is readily detected in sera from rabbits immunized with whole EBV but has never been detected in sera from tamarins or rabbits vaccinated with purified gp350 (R. Pither, A. J. Morgan, and J. Menezes, unpublished observations).

Tamarins mount cell-mediated immune responses following inoculation with a lymphomagenic dose of EBV, and this has been measured in terms of a regression in the outgrowth *in vitro* of autologous EBV-infected B cells in the presence of peripheral blood lymphocytes from tamarins in which the EBV-induced tumors have spontaneously regressed.[129] The capacity to cause regression in the outgrowth *in vitro* of autologous EBV-infected B-cell lines could be induced by reinoculation of these tamarins with a sublymphomagenic dose of virus 18 months after the first challenge. More recently, EBV-specific CTL have been identified in these restimulated animals (M. Shooshstari, A. D. Wilson, and A. J. Morgan, unpublished observations).

Demonstrating the presence of gp350-specific T cells in vaccinated animals that can proliferate in response to gp350 *in vitro* is an immediate goal. In addition, efforts are being made to detect gp350-specific CTL, but here the problem is creating an effective target cell. Live recombinant vaccinia virus expressing gp350 could be used to infect autologous fibroblasts or lymphocytes as has been done for some of the latent antigens.[14,15] The C127 cell line in which the BPV gp350 expression system is present is a mouse fibroblast line, and autologous mouse laboratory strains exist. Vaccination of these mice could allow the detection of CTL activity against C127 cells expressing gp350.

9. T- AND B-CELL EPITOPES ON THE gp350 MOLECULE

The generation of a panel of monoclonal antibodies some years ago[130] allowed the compartmentalization of gp350 B-cell epitopes into six groups. Some monoclonal antibodies in this panel will inhibit the binding of some of the others, but where one antibody does not inhibit the binding of another, they were clearly binding to separate epitopes. These six epitope groups have not been mapped on the intact molecule, but some epitopes have been mapped using bacterial fusion proteins.[82–86] The mapping of particular immune functions to different regions of the molecule could be important and could ultimately allow the design of a synthetic peptide or a recombinant-derived vaccine of predetermined immunologic specificity. The immunologic profile of gp350 is certainly complex, and most, if not all, of those epitopes associated with virus-neutralizing antibodies are discontinuous and are dependent on conformation and the tertiary structure of the molecule. A possible exception to this is a linear epitope in the amino-terminal region between residues 316 and 327. This epitope is recognized by antibodies from rabbits immunized either with gp350 bacterial fusion proteins containing the sequence or with intact virus or gp350 ISCOMs.[83]

The initial approach adopted in this laboratory to the identification of both B- and T-cell epitopes was to express overlapping fragments of the gene in bacteria as β-galactosidase fusion proteins.[82–85] The ability of antibodies against gp350, made in vaccinated animals and present in normal human sera, to bind to the various fragments of gp350 expressed as bacterial fusion proteins was determined in Western blotting. This procedure allows the detection of linear epitopes only, and certainly a number of immunodominant epitopes of the linear kind were recognized by antisera from normal human seropositives. None of these epitopes were able to bind virus-neutralizing antibodies. Clear differences in recognition patterns between normal human seropositives and sera from immunized animals were seen. These data might simply reflect species differences in immune responses but may also be the reflection of different immune responses given following natural infection when compared to vaccination with a purified molecule. This would explain why neutralizing antibodies against linear epitopes were found in vaccinated rabbit sera but not in sera from naturally infected humans. There are no methods known as yet for identifying or reconstructing discontinuous B-cell epitopes, although the use of random-sequence synthetic peptides or random peptide sequences in phage libraries should be pursued.[131–133]

The gp350 bacterial fusion proteins have also been used to detect T-cell epitopes and, in some cases, to map them. It would be expected that although the bacterial products do not retain the native information of gp350, nor do they carry their correct carbohydrate complement, they should still have the capacity to be recognized by T cells following MHC class I or class II presentation. Bacterial fusion proteins were screened for their ability to induce proliferation in previously isolated gp350 specific T-cell clones from normal seropositives.[85] A large number of potential T-cell epitopes have been tentatively identified using an algorithm for this purpose, but only a few epitopes have been located in the amino-terminal region to date.[83–85] The mapping and characterization of gp350 T-cell epitopes should be pursued but will be a major task. Given the difficulty in identifying appropriate epitopes and the inevitable variation in epitopes for different HLA types, the development of synthetic peptide EBV vaccines seems to be remote at present.

10. EPSTEIN–BARR VIRUS LATENT ANTIGEN VACCINES

Infection of B cells by EBV is latent and switches to productive infection, where infectious virus is generated, in only a small number of cells and under certain circumstances. The EBV-infected B cells resemble B cells driven to proliferate by cytokines, interleukins 10 and 4, antigens, or by CD40 cross-linking[134,135] Proliferation may also be enhanced by an autocrine loop consisting of CD23 that is induced during latent infection. The net result of the above factors is that EBV-infected B lymphocytes are transformed to immortalized cell lines able to grow indefinitely. The capacity to immortalize B-cells has proved to be an invaluable asset in analyzing latent gene expression because transformed cell lines can be grown indefinitely, are readily available, and are easily made. Eleven EBV genes are expressed in latent infection, and apart from two abundant small viral RNAs known as EBERs 1 and 2, there are six EBNAs (1, 2, 3a, 3b, 3c, and the leader protein, LP). In addition, three LMPs (1, 2A and 2B) are expressed.[1,2]

A completely different approach to EBV vaccination has begun to emerge during the past few years. This approach is based on the knowledge that $CD8^+$ cells are responsible for limiting the number of EBV-infected B lymphocytes in the circulation of normal seropositive individuals. Could the induction of these immune cells prior to primary infection be effective in preventing or modifying primary infection? The central problem remains the same with this approach as with the approach using gp350. How does the virus persist despite the presence of what should be effective cellular and humoral immune responses? Normal seropositive individuals contain CTLs that are specific for EBNA-3a and EBNA-3c and may also have activity against LMP-2 and EBNA-2. Activities against EBNA-1, EBNA-LP, or LMP-1 are low or undetectable. An important observation from these studies was that a large proportion of the total CTL activity in normal seropositive individuals could not be accounted for by the EBV latent genes.[14,15] It remains to identify the target antigens for this proportion of the CTL population. A number of epitopes that are recognized by CTLs have been located in EBNA-3. The view taken by Moss and others[14,15] is that these epitopes could be used in a vaccine to elicit T-cell memory, which can then be activated to produce EBNA-3-specific CTL. Because the target epitope varies among HLA types, several synthetic peptides corresponding to a number of different epitopes will have to be incorporated in such a vaccine. A relatively small number of CTL epitopes would be able to elicit an immune response in the vast majority of the population. Phase 1 trials are planned of an EBNA-3a peptide, FLRGRAYGL, which is restricted through the HLA B8 allele.[14] The ability to induce CTLs specific for cells expressing one or more EBV latent genes raises the possibility of inducing CTLs to kill EBV tumor cells. An example of this reported recently shows that BL cells can be made susceptible to killing by CTLs by sensitization with added latent antigen synthetic peptides.[136]

11. CONCLUSIONS

The major EBV envelope glycoprotein gp350, when expressed as a genetically engineered product in either the BPV[92] or CHO[91] cell expression systems, should now be evaluated in human trials as soon as possible. The efficacy of gp350 subunit vaccines in the tamarin lymphoma model does not necessarily give any indication of the outcome of vaccination in humans, given that humans sustain a lifelong persistent infection and are infected by the oral route. Set against this, protective immunity is achieved in the tamarin under extreme conditions where a massive dose of EBV is introduced by injection. In some respects, therefore, the tamarin model could be a stringent test for vaccine efficacy. Until human trials are carried out, there are many important questions that must remain unanswered. The problems fall into two categories. First, can vaccination be used to prevent EBV infection in humans? This approach is based on a number of assumptions about the life cycle of EBV itself. Like other herpesviruses, EBV persists throughout life in the face of substantial and continuing immune responses from the host, which include virus-neutralizing antibodies and control of the infected B-cell population by CTLs. It can be argued that induction of these immune responses, by vaccination before primary infection can take place, will not prevent primary infection, as the virus has evolved mechanisms for bypassing the natural immune responses.

The second principal problem is not specific to the EBV but concerns all modern vaccines, and that is designing vaccines that are acceptable for human use but retain sufficient potency to induce the appropriate immune responses. Choices must be made between subunit vaccines based on gp350 and adjuvants and recombinant virus vectors expressing the gp350 gene. Unfortunately, it is necessary to make these choices without having a sound rational basis for doing so. What kind of immune responses will be induced, and will they include gp350-specific CTLs and/or virus-neutralizing antibodies? Will the induction of systemic immunity be sufficient to provide protection at the mucosal interface that is the point of infection in humans? Is it possible that the induction of mucosal immunity in the form of IgA antibodies will enhance infection by EBV? It is possible that the choice of adjuvant in human trials will have a profound effect on the outcome, and so it will be essential not only to evaluate conventional alum formulations but also to test ISCOMs and others. Whether the use of these latter adjuvants, which are not yet licensed for human use, will be possible in the near future is unknown at present.

The results of a small-scale human trial in Southern China using the tien tan gp350 vaccinia recombinant indicate that protection against primary EBV infection can be achieved for at least a limited period. In this trial, six children out of nine who were vaccinated with the recombinant remain seronegative with respect to EBV 30 months after vaccination at the time of writing.[122,123] It will be many years before it is known whether infection that occurs in vaccinated individuals is of a different quality in the long term from infection in nonvaccinated individuals. The difficulties with assessing this trial fall into the categories mentioned above and depend on our assumptions about the biology and life cycle of EBV and the efficacy of the vaccine preparation. Different immune responses can be elicited using different recombinant viruses and by following different immunization protocols. These issues can be resolved only by further human trials with a range of recombinant virus vectors and subunit vaccines. Human trials for the BPV gp350 product in several adjuvant formulations are planned to take place in the near future.

ACKNOWLEDGMENTS. The author thanks the many contributors to the field of EBV vaccines for their cooperation and discussions and the European Commission DG XII Biotech Programme (grant CT930105) and the Cancer Research Campaign of the U.K. for financial support.

REFERENCES

1. Liebowitz, D., and Kieff, E., 1990, Replication of Epstein–Barr virus, in: *Fields Virology* (B. Fields and D. Knipe, eds.), Raven Press, New York, pp. 1889–1920.
2. Liebowitz, D., and Kieff, E., 1993, Epstein–Barr virus, in: *The Human Herpesviruses* (B. Roizman, R. J. Whitley, and C. Lopez, eds.), Raven Press, New York, pp. 107–172.
3. Miller, G., 1990, The Epstein–Barr virus, in: *Fields Virology*, (B. Fields and D. Knipe, eds.), Raven Press, New York, pp. 1921–1958.
4. Epstein, M. A., 1976, Epstein–Barr virus—Is it time to develop a vaccine program? *J. Natl. Cancer Inst.* **56:**697–700.
5. Tanner, J., Whang, Y., Sample, J., Sears, A., and Kieff, E., 1988, Soluble gp350/220 and deletion mutant glycoproteins block Epstein–Barr virus adsorption to lymphocytes, *J. Virol.* **62:**4452–4464.
6. Moore, M. D., DiScipio, R. G., Cooper, N. R., and Nemerow, G. R., 1989, Hydrodynamic, electron microscopic, and ligand-binding analysis of the Epstein–Barr virus/C3dg receptor (CR2), *J. Biol. Chem.* **264:**20576–20582.
7. Birkenbach, M., Tong, X., Bradbury, L. E., Tedder, T., and Kieff, E., 1992, Characterization of an Epstein–Barr virus receptor on human epithelial cells, *J. Exp. Med.* **176:**1405–1414.
8. Sixbey, J. W., Vesterinen, E. H., Nedrud, J. G., Raab-Traub, N., Walton, L. A., and Pagano, J. S., 1983, Replication of Epstein–Barr virus in human epithelial cells infected *in vitro*, *Nature* **306:**480–483.
9. Sixbey, J. W., Nedrud, J. G., Raab-Traub, N., Hanes, R. A., and Pagano, J. S., 1984, Epstein–Barr virus replication in oropharyngeal epithelial cells, *N. Engl. J. Med.* **310:**1225–1230.
10. Sixbey, J. W., Lemon, S. M., and Pagano, J. S., 1986, A second site for Epstein–Barr virus shedding: The uterine cervix, *Lancet* **2:**1122–1124.
11. Israele, V., Shirley, P., and Sixbey, J. W., 1991, Excretion of the Epstein–Barr virus from the genital tract of men, *J. Infect. Dis.* **163:**1341–1343.
12. Tosato, G., and Blaese, R. M., 1985, Epstein–Barr virus infection and immunoregulation in man, *Adv. Immunol.* **37:**99–149.
13. Rickinson, A. B., Gregory, C. D., Murray, R. J., Ulaeto, D. O., and Rowe, M., 1989, Cell

mediated immunity to Epstein–Barr virus and the pathogenesis of virus-associated B cell lymphomas, in: *Immune Responses, Virus Infections and Disease* (N. J. Dimmock and P. D. Minor, eds.), IRL Press, Oxford, pp. 59–83.
14. Khanna, R., Burrows, S. R., Kurilla, M. G., Jacob, C. A., Misko, I. S., Sculley, T. B., Kieff, E., and Moss, D. J., 1992, Localization of Epstein–Barr virus cytotoxic T cell epitopes using recombinant vaccinia: Implications for vaccine development, *J. Exp. Med.* **176:**169–176.
15. Murray, R. J., Kurilla, M. G., Brooks, J. M., Thomas, W. A., Rowe, M., Kieff, E., and Rickinson, A. B., 1992, Identification of target antigens for the human cytotoxic T cell response to Epstein–Barr virus (EBV): Implications for the immune control of EBV-positive malignancies, *J. Exp. Med.* **176:**157–168.
16. Zimber U., Adldinger H. K., Lenoir G. M., Vuillaume, M., Knebel-Doeberitz, M. V., Laux, G., Desgranges, C., Wittmann, P., Freese, U. K., and Schneider, U., 1986, Geographical prevalence of two types of Epstein–Barr virus. *Virology* **154:**56–66.
17. Sample, J., Young, L., Martin, B., Chatmou, T., Kieff, E., Rickinson, A., and Kieff, E., 1990, Epstein–Barr virus types 1 and 2 differ in their BNA-3a, EBNA-3b, and EBNA-3c genes, *J. Virol.* **64:**4084–4092.
18. Gratama J. W., Oosterveer M. A., Lepoutre J. M., Van Rood, J. J., Zwaan, F. E., Vossen, J. M., Kapsenberg, J. G., Richel, D., Klein, G., and Ernberg, I., 1990, Serological and molecular studies of Epstein–Barr virus infection in allogeneic marrow graft recipients, *Transplantation* **49:**725–730.
19. Epstein, M. A., Achong, B. G., and Barr, Y. M., 1964, Virus particles in cultured lymphoblasts from Burkitt's lymphoma, *Lancet* **1:**702–703.
20. Magrath, I., 1990, The pathogenesis of Burkitt's lymphoma, *Adv. Cancer Res.* **55:**133–270.
21. Cleary, M. L., Epstein, M. A., Finerty, S., Dorfman, R. F., Bornkamm, G. W., Kirkwood, J. K., Morgan, A. J., and Sklar, J., 1985, Individual tumors of multifocal EB virus-induced malignant lymphomas in tamarins arise from different B-cell clones, *Science* **228:**722–724.
22. Miller, G., Shope, T., Coope, D., Waters, L., Pagano, J., Bornkamm, G. W., and Henlé, W., 1977, Lymphoma in cotton-top marmosets after inoculation with Epstein–Barr virus: Tumor incidence, histologic spectrum, antibody responses, demonstration of viral DNA, and characterization of viruses. *J. Exp. Med.* **145:**948–967.
23. Miller, G., 1979, Experimental carcinogenicity by the virus *in vivo*, in: *The Epstein–Barr Virus* (M. A. Epstein and B. G. Achong, eds.), Springer, Berlin, pp. 351–372.
24. Shope, T., Dechairo, D., and Miller, G., 1973, Malignant lymphoma in cottontop marmosets after inoculation with Epstein–Barr virus, *Proc. Natl. Acad. Sci. USA* **70:**2487–2491.
25. Henlé, W., Henlé, G., Ho, H.-C., Burtin, P., Cachin, Y., Clifford, P., de Schryver, A., de-Thé, G., Diehl, V., and Klein, G., 1970, Antibodies to Epstein–Barr virus in nasopharyngeal carcinoma, other head and neck neoplasms, and control groups, *J. Natl. Cancer Inst.* **44:**225–231.
26. Zur Hausen, H., Schulte-Holthausen, H., Klein, G., Henlé, W., Henlé, G., Clifford, P., and Santesson, L., 1970, EBV DNA in biopsies of Burkitt tumors and anaplastic carcinomas of the nasopharynx, *Nature* **228:**1056–1058.
27. Zeng, Y., 1985, Seroepidemiological studies on nasopharyngeal carcinoma in China, *Adv. Cancer Res.* **44:**121–138.
28. Parkin, D. M., Stjemsward, J., and Muir, C. S., 1984, Estimates for the worldwide frequency of twelve major cancers, *Bull. WHO* **62:**163–182.
29. Herbst, H., Niedobitek, G., Kneba, M., Hummel, M., Finn, T., Anagnostopoulos, I., Bergholz, M., Krieger, G., and Stein, H., 1990, High incidence of Epstein–Barr virus genomes in Hodgkin's disease, *Am. J. Pathol.* **137:**13–18.
30. Herbst, H., Stein, H., and Niedobitek, G., 1993, Epstein–Barr virus and CD30$^+$ malignant lymphomas, *Crit. Rev. Oncol.* **4:**191–239.
31. Pallesen, G., Hamilton Dutoit, S. J., Rowe, M., and Young, L. S., 1991, Expression of Epstein–Barr virus latent gene products in tumour cells of Hodgkin's disease, *Lancet* **337:**320–322.
32. Stein, H., Herbst, H., Anagnostopoulos, I., Niedobitek, G., Dallenbach, F., and Kratzsch,

H. C., 1991, The nature of Hodgkin and Reed–Sternberg cells, their association with EBV, and their relationship to anaplastic large-cell lymphoma, *Ann. Oncol.* **2**:33–38.
33. Wu, T. C., Mann, R. B., Charache, P., Hayward, S. D., Staal, S., Lambe, B., and Ambinder, R. F., 1990, Detection of EBV gene expression in Reed–Sternberg cells of Hodgkin's disease, *Int. J. Cancer* **46**:801–804.
34. Ambinder, R. F., Browning, P. J., Lorenzana, I., Leventhal, B. G., Cosenza, H., Mann, R. B., MacMahon, E. M., Medina, R., Cardona, V., and Grufferman, S., 1993, Epstein–Barr virus and childhood Hodgkin's disease in Honduras and the United States, *Blood* **81**:462–467.
35. Mueller, N., Evans, A., Harris, N. L., Comstock, G. W., Jellum, E., Magnus, K., Orentreich, N., Polk, B. F., and Vogelman, J., 1989, Hodgkin's disease and Epstein–Barr virus. Altered antibody pattern before diagnosis, *N. Engl. J. Med.* **320**:689–695.
36. Hanto, D. W., Frizzera, G., Gajl-Peczalska, K. J., Sakamoto, K., Purtilo, D. T., Balfour, H. H., Jr., Simmons, R. L., and Najarian, J. S., 1982, Epstein–Barr virus-induced B-cell lymphoma after renal transplantation: Acyclovir therapy and transition from polyclonal to monoclonal B-cell proliferation, *N. Engl. J. Med.* **306**:913–918.
37. Purtilo, D. T., Tatsumi, E., Manolov, G., Manolova, Y., Harada, S., Lipscomb, H., and Krueger, G., 1985, Epstein–Barr virus as an etiological agent in the pathogenesis of lymphoproliferative and aproliferative diseases in immune deficient patients, *Int. Rev. Exp. Pathol.* **27**:113–183.
38. Thomas, J. A., and Crawford, D. H., 1989, Epstein–Barr virus associated B-cell lymphomas in AIDS and after organ transplantation, *Lancet* **1**:1075–1076.
39. Facer, C. A., and Playfair, J. H., 1989, Malaria, Epstein–Barr virus, and the genesis of lymphomas, *Adv. Cancer Res.* **53**:33–72.
40. Whittle, H. C., Brown, J., Marsh, K., Blackman, M., Jobe, O., and Shenton, F., 1990, The effects of *Plasmodium falciparum* malaria on immune control of B lymphocytes in Gambian children, *Clin. Exp. Immunol.* **80**:213–218.
41. Whittle, H. C., Brown, J., Marsh, K., Greenwood, B. M., Seidelin, P., Tighe, H., and Wedderburn, L., 1984, T-cell control of Epstein–Barr virus-infected B cells is lost during *P. falciparum* malaria, *Nature* **312**:449–450.
42. Chen, C. J., Liang, K. Y., Chang, Y. S., Wang, Y. F., Hsieh, T., Hsu, M. M., Chen, J. Y., and Liu, M. Y., 1990, Multiple risk factors of nasopharyngeal carcinoma: Epstein–Barr virus, malarial infection, cigarette smoking and familial tendency, *Anticancer Res.* **10**:547–553.
43. Lu, S., Day, N. E., Degos, L., Lepage, V., Wang, P.-C., Chan, S.-H., Simons, M., McKnight, B., Easton, D., Zeng, Yi, and de-Thé, G., 1990, Linkage of a nasopharyngeal carcinoma susceptibility locus to the HLA region, *Nature* **346**:470–471.
44. Henlé, G., Henlé, W., and Diehl, V., 1968, Relation of Burkitt's tumor-associated herpes-type virus to infectious mononucleosis, *Proc. Natl. Acad. Sci. USA* **59**:94–101.
45. Evans, A. S., 1993, Epstein–Barr vaccine: Use in infectious mononucleosis, in: *The Epstein–Barr Virus and Associated Diseases* (T. Tursz, J. S. Pagano, D. V. Ablashi, G. de Thé, G. Lenoir, and G. R. Pearson, eds.) J. Libbey, London/INSERM, Paris, pp. 593–598.
46. Burke, R. L., 1993, Current developments in HSV vaccines, *Semin. Virol.* **4**:187–197.
47. Kaaden, O. R., and Dietzschold, B., 1974, Alterations of the immunological specificity of plasma membranes of cells infected with Marek's disease and turkey herpes viruses, *J. Gen. Virol.* **25**:1–10.
48. Laufs, R., and Steinke, H., 1975, Vaccination of non-human primates against malignant lymphoma, *Nature* **253**:71–72.
49. De Schryver, A., Klein, G., Hewetson, J., Rocchi, G., Henlé, W., Henlé, G., Moss, D. J., and Pope, J. H., 1974, Comparison of EBV neutralization tests based on abortive infection or transformation of lymphoid cells and their relation to membrane reactive antibodies (anti MA), *Int. J. Cancer* **13**:353–362.
50. Pearson, G., Dewey, F., Klein, G., Henlé, G., and Henlé, W., 1970, Relation between neutralization of Epstein–Barr virus and antibodies to cell-membrane antigens induced by the virus, *J. Natl. Cancer Inst.* **45**:989–995.

51. Hoffman, G. J., Lazarowitz, S. G., and Hayward, S. D., 1980, Monoclonal antibody against a 250,000-dalton glycoprotein of Epstein–Barr virus identifies a membrane antigen and a neutralizing antigen, *Proc. Natl. Acad. Sci. USA* **77**:2979–2983.
52. North, J. R., Morgan, A. J., and Epstein, M. A., 1980, Observations on the EBV envelope and virus-determined membrane antigen (MA) polypeptides, *Int. J. Cancer* **26**:231–240.
53. Qualtière, L. F., and Pearson, G. R., 1980, Radioimmune precipitation study comparing the Epstein–Barr virus membrane antigens expressed on P_3HR-1 virus-superinfected Raji cells to those expressed on cells in a B95-8 virus-transformed producer culture activated with tumor-promoting agent (TPA), *Virology* **102**:360–369.
54. Qualtière, L. F., and Pearson, G. R., 1979, Epstein–Barr virus-induced membrane antigens: Immunochemical characterisation of Triton X100 solubilized viral membrane antigens from EBV-superinfected Raji cells, *Int. J. Cancer* **23**:808–817.
55. Thorley-Lawson, D. A., and Geilinger, K., 1980, Monoclonal antibodies against the major glycoprotein (gp350/220) of Epstein–Barr virus neutralize infectivity, *Proc. Natl. Acad. Sci. USA* **77**:5307–5311.
56. Beisel, C., Tanner, J., Matsuo, T., Thorley-Lawson, D., Kezdy, F., and Kieff, E., 1985, Two major outer envelope glycoproteins of Epstein–Barr virus are encoded by the same gene, *J. Virol.* **54**:665–674.
57. Haddad, R. S., and Hutt-Fletcher, L. M., 1989, Depletion of glycoprotein gp85 from virosomes made with Epstein–Barr virus proteins abolishes their ability to fuse with virus receptor-bearing cells, *J. Virol.* **63**:4998–5005.
58. Heineman, T., Gong, M., Sample, J., and Kieff, E., 1988, Identification of the Epstein–Barr virus gp85 gene, *J. Virol.* **62**:1101–1107.
59. Miller, N., and Hutt-Fletcher, L. M., 1988, A monoclonal antibody to glycoprotein gp85 inhibits fusion but not attachment of Epstein–Barr virus, *J. Virol.* **62**:2366–2372.
60. North, J. R., Morgan, A. J., Thompson, J. L., and Epstein, M. A., 1982, Purified Epstein–Barr virus M_r 340,000 glycoprotein induces potent virus-neutralizing antibodies when incorporated in liposomes, *Proc. Natl. Acad. Sci. USA* **79**:7504–7508.
61. Morgan, A. J., North, J. R., and Epstein, M. A., 1983, Purification and properties of the gp340 component of Epstein–Barr virus membrane antigen in an immunogenic form, *J. Gen. Virol.* **64**:455–460.
62. Epstein, M. A., Morgan, A. J., Finerty, S., Randle, B. J., and Kirkwood, J. K., 1985, Protection of cottontop tamarins against Epstein–Barr virus-induced malignant lymphoma by a prototype subunit vaccine, *Nature* **318**:287–289.
63. Morgan, A. J., Smith, A. R., Barker, R. N., and Epstein, M. A., 1984, A structural investigation of the Epstein–Barr (EB) virus membrane antigen glycoprotein, gp340, *J. Gen. Virol.* **65**:397–404.
64. Serafini-Cessi, F., Malagolini, N., Nanni, M., Dall'Olio, F., Campadelli-Fiume, G., Tanner, J., and Kieff, E., 1989, Characterization of N- and O-linked oligosaccharides of glycoprotein 350 from Epstein–Barr virus, *Virology* **170**:1–10.
65. Lees, J. F., Arrand, J. E., Pepper, S. D., Stewart, J. P., Mackett, M., and Arrand, J. R., 1993, The Epstein–Barr virus candidate vaccine antigen gp340/220 is highly conserved between virus types A and B, *Virology* **195**:578–586.
66. Baer, R., Bankier, A. T., Biggin, M. D., Deininger, P. L., Farrell, P. J., Gibson, T. J., Hatfull, G., Hudson, G. S., Satchwell, S. C., Seguin, C., Tuffrrell, P. S., and Barrell, B. G., 1984, DNA sequence and expression of the B95-8 Epstein–Barr virus genome, *Nature* **310**:207–211.
67. Gong, M., and Kieff, E., 1990, Intracellular trafficking of two major Epstein–Barr virus glycoproteins, gp350/220 and gp110, *J. Virol.* **64**:1507–1516.
68. Gong, M., Ooka, T., Matsuo, T., and Kieff, E., 1987, Epstein–Barr virus glycoprotein homologous to herpes simplex virus gB, *J. Virol.* **61**:499–508.
69. Mackett, M., Conway, M. J., Arrand, J. R., Haddad, R. S., and Hutt-Fletcher, L. M., 1990, Characterization and expression of a glycoprotein encoded by the Epstein–Barr virus *Bam*HI *I* fragment, *J. Virol.* **64**:2545–2552.

70. Emini, E. A., Luka, J., Armstrong, M. E., Banker, F. S., Provost, P. J., and Pearson, G. R., 1986, Establishment and characterization of a chronic infectious mononucleosislike syndrome in common marmosets, *J. Med. Virol.* **18:**369–379.
71. Wedderburn, N., Edwards, J. M. B., Desgranges, C., Fontaine, C., Cohen, B., de Thé, G., 1984, Infectious mononucleosis-like response in common marmosets infected with Epstein–Barr virus, *J. Infect. Dis.* **150:**878–882.
72. Ma, N. S. F., 1981, Chromosome evolution in the owl monkey, Aotus, *Am. J. Phys. Anthropol.* **54:**293–303.
73. Young, L. S., Finerty, S., Brooks, L., Scullion, F., Rickinson, A. B., and Morgan, A. J., 1989, Epstein–Barr virus gene expression in malignant lymphomas induced by experimental virus infection of the cottontop tamarin, *J. Virol.* **63:**1967–1974.
74. Watkins, D. I., Hodi, F. S., and Letvin, N. L., 1988, A primate species with a limited major histocompatibility complex class I polymorphism, *Proc. Natl. Acad. Sci. USA* **85:**7714–7718.
75. Niedobitek, G., Agathanggelou, A., Finerty, S., Tierney, R., Jones, E. L., Watkins, P., Morgan, A. J., Young, L. S., and Rooney, N., 1994, Latent Epstein–Barr virus infection in cottontop tamarins: A possible model for EBV infection in humans, *Am. J. Pathol.* **145:**969–978.
76. Epstein, M. A., Randle, B. J., Finerty, S., and Kirkwood, J. K., 1986, Not all potently neutralizing, vaccine-induced antibodies to Epstein–Barr virus ensure protection of susceptible experimental animals, *Clin. Exp. Immunol.* **63:**485–490.
77. David, E. M., and Morgan, A. J., 1988, Efficient purification of Epstein–Barr virus membrane antigen gp340 by fast protein liquid chromatography, *J. Immunol. Methods* **108:**231–236.
78. Morein, B., Sundquist, B., Höglund, S., Dalsgaard, K., and Osterhaus, A., 1984, ISCOM, a novel structure for antigenic presentation of membrane proteins from enveloped viruses, *Nature* **308:**457–460.
79. Morgan, A. J., Finerty, S., Lovgren, K., Scullion, F. T., and Morein, B., 1988, Prevention of Epstein–Barr (EB) virus-induced lymphoma in cottontop tamarins by vaccination with the EB virus envelope glycoprotein gp340 incorporated into immune-stimulating complexes, *J. Gen. Virol.* **69:**2093–2096.
80. Allison, A. C., and Byers, N. E., 1986, An adjuvant formulation that selectively elicits the formation of antibodies of protective isotype and cell-mediated immunity, *J. Immunol. Methods* **95:**157–168.
81. Morgan, A. J., Allison, A. C., Finerty, S., Scullion, F. T., Byars, N. E., and Epstein, M. A., 1989, Validation of a first-generation Epstein–Barr virus vaccine preparation suitable for human use, *J. Med. Virol.* **29:**74–78.
82. Pither, R. J., Zhang, C. X., Wallace, L. E., Rickinson, A. B., and Morgan, A. J., 1991, Mapping of B and T cell epitopes on the Epstein–Barr major envelope glycoprotein gp340, in: *Vaccines 91* (R. A. Lerner, H. Ginsberg, R. M. Chanock, and F. Brown, eds.), Cold Spring Harbor, New York, pp. 197–201.
83. Pither, R. J., Nolan, L., Tarlton, J., Walford, J., and Morgan, A. J., 1992, Distribution of epitopes within the amino acid sequence of the Epstein–Barr virus major envelope glycoprotein, gp340, recognized by hyperimmune rabbit sera, *J. Gen. Virol.* **73:**1409–1415.
84. Pither, R. J., Zhang, C. X., Shiels, C., Tarlton, J., Finerty, S., and Morgan, A. J., 1992, Mapping of B-cell epitopes on the polypeptide chain of the Epstein–Barr virus major envelope glycoprotein and candidate vaccine molecule gp340, *J. Virol.* **66:**1246–1251.
85. Pither, R. J., Zhang, C. X., Wallace, L. E., Rickinson, A. B., and Morgan, A. J., 1991, Mapping of B and T cell epitopes on the Epstein–Barr receptor ligand gp340: A candidate subunit vaccine, in: *Epstein–Barr Virus and Human Disease 1990* (D. V. Ablashi, A. T. Huang, A. S. Pagano, J. R. Pearson, and C. S. Yang, eds.) Humana, Clifton, New Jersey, pp. 207–211.
86. Zhang, P. F., Klutch, M., Armstrong, G., Qualtiere, L., Pearson, G., and Marcus–Sekura, C. J., 1991, Mapping of the epitopes of Epstein–Barr virus gp350 using monoclonal antibodies and recombinant proteins expressed in *Escherichia coli* defines three antigenic determinants, *J. Gen. Virol.* **72:**2747–2755.

87. Schultz, L. D., Tanner, J., Hofmann, K. J., Emini, E. A., Condra, J. H., Jones, R. E., Kieff, E., and Ellis, R. W., 1987, Expression and secretion in yeast of a 400-kDa envelope glycoprotein derived from Epstein–Barr virus, *Gene* **54:**113–123.
88. Emini, E. A., Schleif, W. A., Armstrong, M. E., Silberklang, M., Schultz, L. D., Lehman, D., Maigetter, R. Z., Qualtiere, L. F., Pearson, G. R., and Ellis, R. W., Antigenic analysis of the Epstein–Barr virus major membrane antigen (gp350/220) expressed in yeast and mammalian cells: Implications for the development of a subunit vaccine, *Virology* **166:**387–393.
89. Nuebling, C. M., Buck, M., Boos, H., von Deimling, A., and Mueller Lantzsch, N., 1992, Expression of Epstein–Barr virus membrane antigen gp350/220 in *E. coli* and in insect cells, *Virology* **191:**443–447.
90. Conway, M., Morgan, A. J., and Mackett, M., 1988, Expression of Epstein–Barr virus membrane antigen gp340/220 in mouse fibroblasts using a bovine papilloma virus vector, *J. Gen. Virol.* **70:**729–734.
91. Hessing, M., van Schijndel, H. B., van Grunsven, W. M., Wolf, H., and Middeldorp, J., 1992, Purification and quantification of recombinant Epstein–Barr viral glycoproteins gp350/220 from Chinese hamster ovary cells, *J. Chromatogr.* **599:**267–272.
92. Madej, M., Conway, M. J., Morgan, A. J., Sweet, J., Wallace, L., Arrand, J., and Mackett, M., 1992, Purification and characterisation of Epstein–Barr virus gp340/220 produced by a bovine papilloma virus vector system, *Vaccine* **10:**777–782.
93. Motz, M., Deby, G., Jilg, W., and Wolf, H., 1986, Expression of the Epstein–Barr virus major membrane proteins in Chinese hamster ovary cells, *Gene* **44:**353–359.
94. Whang, Y., Silberklang, M., Morgan, A., Munshi, S., Lenny, A. B., Ellis, R. W., and Kieff, E., 1987, Expression of the Epstein–Barr virus gp350/220 gene in rodent and primate cells, *J. Virol.* **61:**1796–1807.
95. Ulaeto, D., Wallace, L., Morgan, A. J., Morein, B., and Rickinson, A. B., 1988, *In vitro* T cell responses to a candidate Epstein–Barr virus vaccine: Human CD4+ T-cell clones specific for the major envelope glycoprotein gp340, *Eur. J. Immunol.* **18:**1689–1697.
96. Wallace, L. E., Wright, J., Ulaeto, D. O., Morgan, A. J., and Rickinson, A. B., 1991, Identification of two T-cell epitopes of the candidate Epstein–Barr virus vaccine glycoprotein gp340 recognized by CF4+ T-cell clones, *J. Virol.* **65:**3821–3828.
97. Carter, M. J., Facklam, T. J., Long, P. C., and Scotland, R. A., 1988, Are continuous cell lines safe as substrates for human drugs and biologics? A case study with human growth hormone, *Dev. Biol. Stand.* **70:**101–107.
98. Mossman, T. R., and Coffman, R. L., 1989, Heterogeneity of cytokine secretion patterns and functions of helper T cells, *Adv. Immunol.* **46:**111–147.
99. Afonso, L. C. C., Scharton, T. M., Vieira, L. Q., Wysocka, M., Trinchieri, G., and Scott, P., 1994, The adjuvant effect of interleukin-12 against *Leishmania major*, *Science* **263:**235–237.
100. Finerty, S., Tarlton, J., Mackett, M., Conway, M., Arrand, J. R., Watkins, P. E., and Morgan, A. J., 1992, Protective immunization against Epstein–Barr virus-induced disease in cottontop tamarins using the virus envelope glycoprotein gp340 produced from a bovine papillomavirus expression vector, *J. Gen. Virol.* **73:**449–453.
101. Finerty, S., Mackett, M., Arrand, J. R., Watkins, P. E., Tarlton, J., and Morgan, A. J., 1994, Immunisation of cottontop tamarins and rabbits with a candidate vaccine against the Epstein–Barr virus based on the major viral envelope glycoprotein gp340 and alum, *Vaccine* **12:**1180–1184.
101a. Emini, E. A., Schleif, W. A., Silberklang, M., Lehman, D., and Ellis, R. W., 1989, Vero cell-expressed Epstein–Barr virus (EBV) gp350/220 protects marmosets from EBV challenge, *J. Med. Virol.* **27:**120–123.
102. Murphy, F. A., 1989, Vaccinia-vectored vaccines, risks and benefits, *Res. Virol.* **140:**463–491.
103. Takahashi, H., Takeshita, T., Morein, B., Putney, S., Germain, R. N., and Berzofsky, J. A., 1990, Induction of CD8+ cytotoxic T cells glycoprotein by immunisation with purified HIV-1 envelope protein in ISCOMS, *Nature* **344:**873–875.

104. Byars, N. E., Nakano, G., Welch, M., Lehman, D., and Allison, A. C., 1991, Improvement of hepatitis B vaccine by the use of a new adjuvant, *Vaccine* **9**:309–318.
105. Dillon, S. B., Demuth, S. G., Schneider, M. A., Weston, C. B., Jones, C. S., Young, J. F., Scott, M., Bhatnaghar, S., LoCastro, S., and Hanna, N., 1992, Induction of Class I MHC-restricted CTL in mice by a recombinant influenza vaccine in aluminium hydroxide adjuvant, *Vaccine* **10**:309–318.
106. Murphy, R., and Chanock, R. M., 1990, Immunisation against viruses, in: *Fields Virology* (B. Fields and D. Knipe, eds.), Raven Press, New York, pp. 469–502.
107. Brochier, B., Languet, B., Blancou, J., Thomas, I., Kieny, M. P., Costy, F., Desmettre, P., and Pastoret, P.-P., 1989, Use of recombinant vaccinia–rabies virus for oral protection of wildlife against rabies: Inocuity to several non target bait-consuming species, *J. Wildl. Dis.* **25**:540.
108. Mackett, M., and Arrand, J. A., 1985, Recombinant vaccinia virus induces neutralising antibodies in rabbits against Epstein–Barr virus membrane antigen gp340, *EMBO J.* **4**:3229–3234.
109. Ragot, T., Finerty, S., Watkins, P. E., Perricaudet, M., and Morgan, A. J., 1993, Replication-defective recombinant adenovirus expressing the Epstein–Barr virus (EBV) envelope glycoprotein gp340/220 induces protective immunity against EBV-induced lymphomas in the cottontop tamarin, *J. Gen. Virol.* **74**:501–507.
110. Lowe, R. S., Keller, P. M., Keech, B. Y., Davison, A. J., Whang, Y., Morgan, A. J., Kieff, E., and Ellis, R. W., 1987, Varicella zoster virus as a live vector for the expression of foreign genes, *Proc. Natl. Acad. Sci. USA* **84**:3896–3900.
111. Taylor, J., Weinberg, R., Tartaglia, J., Richardson, C., Alkhatib, G., Briedis, D., Appel, M., Norton, E., and Paoletti, E., 1992, Nonreplicating viral vectors as potential vaccines, recombinant canarypox virus expressing measles virus fusion (F) and hemagglutinin (HA) glycoproteins, *Virology* **187**:321–328.
112. Tartaglia, J., Perkus, M. E., Taylor, J., Norton, E. K., Audonnet, J.-C., Cox, W. I., Davis, S. W., Van der Hoeven, J., Meignier, B., Riviere, M., Languet, B., and Paoletti, E., 1992, NYVAC: A highly attenuated strain of vaccinia virus, *Virology* **188**:217–232.
113. Morgan, A. J., Mackett, M., Finerty, S., Arrand, J. R., Scullion, F. T., and Epstein, M. A., 1988, Recombinant vaccinia virus expressing Epstein–Barr virus glycoprotein gp340 protects cottontop tamarins against EBV-induced malignant lymphomas, *J. Med. Virol.* **25**:189–195.
114. Alkhatib, G., and Briedis, D. J., 1988, High level eucaryotic *in vivo* expression of biologically active measles virus haemagglutinin by using an adenovirus type 5 helper-free vector system, *J. Virol.* **62**:2718–2727.
115. Ballay, A., Levrero, M., Buendia, M. A., Tiollais, P., and Perricaudet, M., 1985, *In vitro* and *in vivo* synthesis of the hepatitis B surface antigen and of the receptor for polymerised human serum albumin from recombinant human adenovirus, *EMBO J.* **4**:3861–3865.
116. Dewar, R. L., Natarajan, V., Vasudevachari, M. B., and Salzman, N. P., 1989, Synthesis and processing of human immunodeficiency virus type 1 envelope proteins encoded by a recombinant human adenovirus, *J. Virol.* **63**:129–136.
117. Eloit, M., Gilardi-Hebeystreit, P., Toma, B., and Perricaudet, M., 1990, Construction of a defective adenovirus expressing the pseudorabies virus glycoprotein gp50 and its use as a live vaccine, *J. Gen. Virol.* **71**:2425–2431.
118. Johnson, D. C., Ghosh-Choudery, G., Smiley, J. R., Fallis, L., and Graham, F. L., 1988, Abundant expression of herpes simplex virus glycoprotein gB using an adenovirus vector, *Virology* **164**:1–14.
119. Graham, F. L., Smiley, J., Russell, W. C., and Nairn, R., 1977, Characteristics of a human cell line transformed by DNA from human adenovirus type 5, *J. Gen. Virol.* **36**:59–72.
120. Top, F. H., Jr., Grossman, R. A., Bartelloni, P. J., Segal, H. E., Dudding, B. A., Russell, P. K., and Büscher, E. L., 1971, Immunization with live types 7 and 4 adenovirus vaccines. I. Safety, infectivity, antigenicity, and potency of adenovirus type 7 vaccine in humans, *J. Infect. Dis.* **124**:148–154.

121. Top, F. H., Jr., Büscher, E. L., Bancroft, W. H., and Russell, P. K., 1971, Immunization with live types 7 and 4 adenovirus vaccines. II. Antibody response and protective effect against acute respiratory disease due to adenovirus type 7, *J. Infect. Dis.* **124**:155–160.
122. Gu, S., Huang, T., Ruan, L., Miao, Y., Lu, H., Chu, C. M., Motz, M., and Wolf, H., 1993, On the first EBV vaccine trial in humans using recombinant vaccinia virus expressing the major membrane antigen, In: *Epstein–Barr Virus and Associated Disease* (T. Tursz, J. S. Pagano, D. V. Ablashi, G. de Thé, G. Lenoir, and G. R. Pearson, eds.), Libbey, London/INSERM, Paris, pp. 579–584.
123. Gu, S., Huang, T., Miao, Y., Ruan, L., Zhao, Y., Han, C., Xiao, Y., Zhu, J., and Wolf, H., 1991, A preliminary study on the immunogenicity in rabbits and in human volunteers of a recombinant vaccinia virus expressing Epstein–Barr virus membrane antigen, *Chin. Med. Sci. J.* **6**:241–243.
124. Moss, B., 1991, Vaccinia virus: A tool for research and vaccine development, *Science* **252**:1662–1667.
125. Moss, B., Smith, G. L., Gerin, J. L., and Purcell, R. H., 1984, Live recombinant vaccinia virus protects chimpanzees against hepatitis B, *Nature* **311**:67–69.
126. Bejarano, M. T., Masucci, M. G., Morgan, A. J., Morein, B., Klein, G., and Klein, E., 1990, Epstein–Bar virus (EBV) antigens processed and presented by B cells, B blasts and macrophages trigger T-cell-mediated inhibition of EBV-induced B-cell transformation, *J. Virol.* **64**:1398–1401.
127. Lee, S. P., Wallace, L. E., Mackett, M., Arrand, J. R., Searle, P. F., Rowe, M., and Rickinson, A. B., 1993, MHC class II-restricted presentation of endogenously synthesized antigen: Epstein–Barr virus transformed B cell lines can present the viral glycoprotein gp340 by two distinct pathways, *Int. Immunol.* **5**:451–460.
128. Khyatti, M., Patel, P. C., Stefanescu, I., and Menezes, J., 1991, Epstein–Barr virus (EBV) glycoprotein gp350 expressed on transfected cells resistant to natural killer cell activity serves as a target antigen for EBV-specific antibody-dependent cellular cytotoxicity, *J. Virol.* **65**:996–1001.
129. Finerty, S., Scullion, F. T., and Morgan, A. J., 1988, Demonstration *in vitro* of cell-mediated immunity to Epstein–Barr virus in cottontop tamarins, *Clin. Exp. Immunol.* **73**:181–185.
130. Qualtière, L. F., Decoteau, J. F., and Hassan Nasr-el-Din, M., 1987, Epitope mapping of the major Epstein–Barr virus outer envelope glycoprotein gp350/220, *J. Gen. Virol.* **68**:535–543.
131. Cwirla, S. E., Peters, E. A., Barrett, R. W., and Dower, W. J., 1990, Peptides on phage: A vast library of peptides for identifying ligands, *Proc. Natl. Acad. Sci. USA* **87**:6378–6382.
132. Geysen, H. M., Rodda, S. J., and Mason, T. J., 1986, A priori delineation of a peptide which mimics a discontinuous antigenic determinant, *Mol. Immunol.* **23**:709–715.
133. Scott, J. K., and Smith, G. P., 1990, Searching for peptide ligands with an epitope library, *Science* **249**:386–390.
134. Banchereau, J., and Rousset, F., 1991, Growing human B lymphocytes in the CD40 system, *Nature* **353**:678–679.
135. Banchereau, J., de Paoli, P., Valle, A., Garcia, E., and Rousset, F., 1991, Long-term human B cell lines dependant on interleukin 4 and antibody to CD40, *Science* **251**:70–72.
136. Khanna, R., Burrows, S. R., Suhrbier, A., Jacob, C.A., Griffin, H., Misko, I. S., Sculley, T. B., Rowe, M., Rickinson, A. B., and Moss, D. J., 1993, EBV peptide epitope sensitization restores human cytotoxic T cell recognition of Burkitt's lymphoma cells. Evidence for a critical role for ICAM-2, *J. Immunol.* **150**:5154–5162.

Index

Acquired immunodeficiency syndrome (AIDS)
  B-lymphoproliferative disease in, 355–356, 396
  Burkitt's lymphoma in, 357, 358, 359
  cervical cancer in, 101
  HHV-6 in, 313, 319–320
  Hodgkin's disease in, 319–320, 360
  lymphadenopathy syndrome in, 319–320
  non-Hodgkin's lymphoma in, 319–320
  See also Human immunodeficiency virus
Adenoviruses
  immune repression by, 207–214
  persistent infection, 212–214
  transformation by, 195–196
    cell cycle deregulation, 199–202, 207
    repression of differentiation, 203–204, 207
  for vaccine delivery, 406, 408
  See also E1A proteins; E1B proteins
Aflatoxin B1, 174, 185
AIDS: *See* Acquired immunodeficiency syndrome
ALV: *See* Avian leukosis virus
Antisense treatment, of EBV infection, 336–337
Apoptosis
  induction by p53, 34, 35, 140–141
  inhibition by adenovirus E1B, 204, 206–207
  inhibition by EBV, 335–336, 352
AP-1 transcription factor, HSV induction of, 269
Asbestos, mesothelioma and, 82, 83
Atherosclerosis, herpesvirus and, 235
Autocrine secretion, of lymphokines, 247
Avian leukosis virus (ALV), 228–229

BK: *See* Polyomaviruses, human
B-lymphoproliferative disease, 355–356, 396

BPV: *See* Papillomaviruses, bovine
Burkitt's lymphoma
  c-myc alterations in, 327–328, 358–360
  EBV gene expression in, 357–358
  EBV subtypes in, 379
  etiological hypothesis, 359–360, 396
  forms of, 356–357
  pathology, 357
  tumor suppressor genes in, 359

Carcinogens, chemical
  aflatoxin B1, 174, 185
  as BPV cofactors, 112–114, 116–118
  as CRPV cofactors, 92
  DNA amplification by, 268
  as EBV cofactors, 335, 364
  as HPV cofactors, 288
  as HSV cofactors, 270
Cervical cancer
  HHV-6 association with, 102, 321
  HPV association with, 101–103, 123, 281–282, 288
    chromosome alterations, 99–101
    immune response, 144, 161–162
    integration into host genome, 94, 128, 135, 143, 287
    pRb and, 139
    primary infection, 158, 159–160
    retinoids and, 143
    viral subtypes, 124–125
  HPV-negative, 143–144, 160
  HSV association with
    epidemiologic evidence, 102, 281, 284–285, 288
    mouse model, 270
    viral antigens in tumors, 285

Cervical cancer (cont.)
　HSV association with (cont.)
　　viral DNA sequences in tumors, 271, 281, 285–287
　　in vitro evidence, 281, 283
　in immunosuppressed patients, 161
　keratinocyte differentiation and, 98–99
　risk factors, 101–103, 288
　See also Genital cancers
Chemotherapeutic agents, p53 and, 34
Chromosome damage
　by bracken fern chemicals, 113–114, 117
　by HCMV, 307
　by HPV, 99–101
　by HSV, 283
　in liver cancer, 180, 184–186
　role in oncogenesis, 15–16
　by T antigen
　　of BK, 56
　　in fibroblast model, 17–22
　　molecular mechanism, 22–23
　　of SV40, 15–26
Cirrhosis, pathophysiology of, 173
Cottontail rabbit papillomavirus (CRPV), 91–92
CRPV: See Cottontail rabbit papillomavirus
Cyclins
　p53 inhibition of, 205–206
　pRb regulation of, 30–31, 139, 202
Cytokines, in chronic hepatitis, 174, 175
Cytomegalovirus, human (HCMV)
　chromosome damage by, 307
　glycoproteins of envelope
　　alternative theories, 297
　　hydrophobic domain, 301, 303
　　inhibitors of, 298–299
　　oligomerization, 304
　　proteolytic cleavage, 305–306
　　structure and processing, 300–301, 303
　　transport to nuclear membrane, 297, 298
　　transport to tubular endosome, 297, 306
　herpesvirus characteristics, 295
　Hodgkin's disease and, 381
　oncogenic potential, 306–307
　as papovavirus cofactor, 307
　reproductive cycle, 295–297
　　maturational budding, 299–300
　　transport budding, 297–299
　retrovirus interaction with, 229
　trans-activation of cellular genes, 307

Differentiation
　control by adenovirus E1A, 203–204
　in HPV-containing cells, 98–99

DNA amplification
　by chemical carcinogens, 268
　by HSV, 268, 283
　p53 and, 141
　premature termination and, 232
DNA repair, p53 and, 34, 97–98, 140–141, 204–205
Ductin, BPV E8 binding of, 116

E1A proteins, of adenoviruses
　homology with HPV E7, 130
　homology with T antigen, 29, 31, 35–36
　p53 and, 34
　p107 and, 31, 202
　p130 and, 202
　p300 and, 35–36, 203–204
　pRb and, 138, 199–202
　repression of MHC class I antigens, 207–214
　structures, 196–198
　trans-activating functions, 199
　transformation role, 131, 196, 204, 207
E1B proteins, of adenoviruses
　apoptosis inhibition by, 204, 206–207
　p53 and, 204–207
　transformation role, 196, 204, 206–207
E2F transcription factor
　in EBV latent replication, 330–331
　functions, 138–139
　pRb binding to, 30, 31
　　E1A protein and, 199–202
　　E7 protein and, 96–97, 138–139
E6 and E7 proteins: See Papillomaviruses, bovine; Papillomaviruses, human
EBNA-1
　in Hodgkin's disease, 360, 383
　as immortalization factor, 336–337
　in nasopharyngeal carcinoma, 362
EBNA-2
　in B-cell immortalization, 350–352
　in B-lymphoproliferative disease, 355
　in Burkitt's lymphoma, 357, 358
　in transformation, 331–333, 334, 336
EBV: See Epstein–Barr virus
Endoplasmic reticulum, HBV protein accumulation in, 180
Epithelioma, hamster polyomavirus
　natural infections, 1, 3–4, 5–6
　in transgenic mice, 6–7
Epithelium
　difficulty in modeling, 145
　normal differentiation, 141–142
Epstein–Barr virus (EBV)
　apoptosis inhibition by, 335–336, 352

INDEX 423

Epstein–Barr virus (EBV) (cont.)
  episomes
    replication, 329–331
    structure, 328, 329, 377
  genome organization, 347–348
  in HD-related lymphoproliferative disorders, 384–385
  HHV-6 interactions with, 319
  in HIV-associated lymphoproliferative disorders, 319–320, 355–356, 396
  homology with H. saimiri, 241
  immortalization of B cells, 336, 348, 350–353
  integration into host genome, 329–330
  latent replication, 329–331
    EBNA-1 in, 336–337
    host cells for, 395–396
    small RNAs in, 334, 350–351
  lytic infection, 334–336, 348
    in oncogenesis, 335, 336
    switch from latent infection, 353
  malignant spectrum, 354–355
  primary infection of humans, 353–354, 376, 395–396
  strains of, 331, 348, 378–380
  structure, 347
  transformation by
    chemical enhancers, 335
    EBNA-2 in, 331–333, 334, 336
    LMP-1 in, 333–334, 336
    small RNAs in, 334
    unknown factor, 336
  treatment possibilities, 336–337
  vaccines, 395–419
    adjuvant choice, 404–405
    animal model for testing, 400–401
    cell-mediated responses, 407–409
    demand for, 395–398
    latent antigen vaccines, 410–411
    live virus recombinants, 405–406
    natural subunit vaccines, 401–402
    recombinant subunit vaccines, 403–404
    selection of vaccine molecule, 398–399
    T- and B-cell epitopes, 409–410
    unsolved problems, 411–412
  See also Burkitt's lymphoma; Hodgkin's disease; Nasopharyngeal carcinoma
Exanthem subitum, 313, 314

Firones, 268
c-fos, HSV-2 induction of, 269
Free radicals, in chronic hepatitis, 174–175

GADD45, 34, 205
Gap junctions, BPV4 E8 protein and, 116

Gene amplification: See DNA amplification
Genital cancers
  epidemiology, 284–285
  HPV association with, 281–282
    in males, 92, 101, 160
    viral types, 92, 124–125
    vulvar carcinoma, 285, 286
  HSV association with, 281, 284, 285–287
  multistep etiology, 288
  See also Cervical cancer
Genital warts, vaccines for, 166, 167
Ground squirrel hepatitis virus (GSHV)
  liver cancer and, 172
  myc genes and, 182, 184
Growth factor receptors
  in alimentary cancer explants, 117
  BPV E5 protein and, 98, 136
  EBV LMP-1 as, 333–334, 353
  in HPV-immortalized keratinocytes, 98, 142–143
  HSV ICP10 phosphokinase, 263–266
  transformation mechanism of, 265–266
GSHV: See Ground squirrel hepatitis virus

HaPV: See Polyomavirus, hamster
HBV: See Hepatitis B virus
HCMV: See Cytomegalovirus, human
Hepatitis B, chronic
  mutagenic agents in, 174–175
  pathophysiology, 172–174
Hepatitis B virus (HBV)
  in chronic hepatitis, 172–173
    aflatoxin activation and, 174
    cytokines and, 174
    peroxide and, 175
  endoplasmic reticulum alterations, 180
  epidemiology, 171–172
  genome organization, 175
  integration into host genome, 176, 179, 180–181
  in liver cancer etiology, 172, 173, 186
  replication, 176–177
  rodent models, 182–184
  X trans-activator
    interactions with host proteins, 177–178, 185
    potential oncogenicity, 178–179, 181
Hepatocellular carcinoma
  epidemiology, 171–172
  etiological factors, 172, 173, 186
  genetic alterations in, 184–186
  HBV DNA integration in, 176, 179, 180–181
  rodent models, 182–184
  TGF-α in, 174

Herpes simplex viruses (HSV)
  activation of cellular genes, 269, 282, 283
  activation of DNA synthesis, 268
  activation of endogenous viruses, 229, 269, 283
  chemical cocarcinogens, 270
  gene amplification by, 268, 283
  homology with cellular DNA, 269–270
  ICP10 PK transforming factor
    as appropriated cellular gene, 266
    expression of RR and PK domains, 266–267, 282
    in genital cancers, 285
    as growth factor receptor, 263–266
    novel catalytic motifs, 261–263
    ras activation by, 265–266
    as transcription factor, 269, 283
    verification of kinase activity, 260–261, 282
  immortalization genes, 259–260
  integration into host genome, 269, 287
  mutagenesis by, 257–258, 267–268, 283
  transformation by
    hit-and-run mechanism, 254, 257–258, 282
    ICP10 role, 263–267, 282, 288
    with inactivated virus, 253–255
    multistep process, 258–259, 270–271, 288
    with viral DNA sequences, 255–257, 282–283
  See also Cervical cancer; Genital cancers
Herpesvirus, avian: *See* Marek's disease virus
Herpesvirus 6, human (HHV-6)
  in cervical cancer, 102, 321
  EBV interactions with, 319
  epidemiology, 313
  in HIV-associated lymphoproliferations, 313, 319–320
  in Hodgkin's disease, 317–319, 381
  immunologic deficits from, 319
  latent infection, 314
  in leukemia, 320–321
  in non-Hodgkin's lymphomas, 315–316, 320
  in nonlymphoid neoplasms, 102, 321–322
  primary infection, 313–314
  transformation *in vitro,* 314–315
  variants, 314
Herpesvirus of turkey (HVT), 227, 228
Herpesvirus saimiri (HVS)
  genome organization, 240–242
  homology with cellular genes, 241–242, 243

Herpesvirus saimiri (HVS) (*cont.*)
  immortalization of T cells, 239–240
  transformation of T cells
    genome alterations, 242–244
    lymphokines and, 245, 246, 247–248
    possible mechanisms, 248–249
    SCOL oncogene, 244–245, 247–249
    small RNAs in, 245–246
HHV-6: *See* Herpesvirus 6, human
Hit-and-run mechanism
  by BK, 56
  by BPV4, 115–116
  by HSV, 254, 257–258, 282
  by SV40 T antigen, 76
HIV: *See* Human immunodeficiency virus
Hodgkin's disease
  EBV in
    detection of virus, 377–380
    epidemiologic data, 360, 376
    HD subtypes and, 360, 380–381
    inhibition of cytotoxic responses, 384
    monoclonal episomes, 377, 378
    persistence of virus, 381–382
    viral gene expression, 360–361, 382–384
    viral subtypes and, 360, 378–380
  epidemiology, 376
  HIV association with, 319–320, 360, 379–380, 381–382
  non-EBV viruses and, 381
    HHV-6, 317–319
  pathology, 375–376
  subtypes, 360, 376, 380–381
  unsolved problems, 375, 385–386
Hodgkin's-disease-related lymphoproliferative disorders, 384–385
HPV: *See* Papillomaviruses, human
HSV: *See* Herpes simplex viruses
HTLV-l: *See* Human T-cell leukemia virus I
Human immunodeficiency virus (HIV)
  Burkitt's lymphoma and, 359
  genital cancers and, 101, 102
  HHV-6 in lymphomas with, 319–320
  Hodgkin's disease and, 319, 379–380, 381–382
  non-Hodgkin's lymphoma and, 319, 379
  trans-activation by herpesviruses, 229
  See also Acquired immunodeficiency syndrome
Human T-cell leukemia virus I (HTLV-1)
  absence in Hodgkin's disease, 381
  trans-activation by herpesviruses, 229
HVS: *See Herpesvirus saimiri*
HVT: *See* Herpesvirus of turkey

IGF: *See* Insulin-like growth factor
Immune deficiency
 cervical cancer and, 161
 from dietary cofactors of BPV, 112–113, 114
 EBV malignancies and, 319–320, 335, 355–356, 396
 human polyomaviruses and, 52–53, 60
 See also Human immunodeficiency virus
Immune response
 cell-mediated immune mechanism, 162
 in chronic hepatitis, 173–174, 175
 to HPV, 144, 161–162
 repression by adenovirus E1A, 207–214
 See also Vaccines
Immunotherapy
 for B-lymphoproliferative disease, 356
 against SV40 T antigen, 86
Infectious mononucleosis
 Hodgkin's disease and, 376
 as primary EBV infection, 353–354
 vaccines, 397–398
Insulin-like growth factor (IGF)
 in HPV-immortalized keratinocytes, 142
 large T antigen and, 76
Interferons
 in chronic hepatitis, 174
 stimulation of class I antigens, 208
Interleukin-2
 in chronic hepatitis, 174
 in HVS-transformed T cells, 239–240, 247–248
Interleukin-4
 in HVS-transformed T cells, 245, 246, 247–248

JC: *See* Polyomaviruses, human
c-*jun*, HSV induction of, 269

Kaposi's sarcoma
 cytomegalovirus association with, 60
 HHV-6 association with, 60, 321–322
 HPV association with, 60, 125
 polyomaviruses association with, 59–60
Keratinocytes
 with HPV
  differentiation, 94, 98–99
  immortalization, 95–98, 142–143
  pRb binding, 139
 in normal epithelium, 142
 organotypic raft cultures, 94, 98, 132
17kT antigen of SV40, 18

Large T antigen
 chromosome damage by
  with BK virus, 56

Large T antigen (*cont.*)
 chromosome damage by (*cont.*)
  in fibroblast model, 17–22
  molecular mechanism, 22–23
  with SV40 virus, 15–26
 of hamster polyomavirus, 7, 9, 10–11
 homology with E1A and E7, 29, 130, 200
 of human polyomaviruses, 37, 52, 56–57
 of murine polyomavirus, 10, 37–38
 replication functions, 28
 of SV40
  chromosome damage by, 15–26
  functional domains, 28, 36–37
  homology with polyomavirus T antigen, 52
  in mesotheliomas, 83, 85
  p53 and, 22–23, 28, 31–35, 36, 39, 76
  p107 and, 31, 35, 76
  p300 and, 28, 35–36, 39, 76
  pRb and, 28–31, 36, 39, 76, 200
  transformation role, 39, 75–76
 tumorigenesis and, 22
Latent viruses, criteria for oncogenicity of, 62
Leukemia
 hamster polyomavirus and, 1, 4
 HHV-6 in, 320–321
Liver cancer: *See* Hepatocellular carcinoma
LMP-1
 in B-cell immortalization, 350–353
 in B-lymphoproliferative disease, 355
 in Burkitt's lymphoma, 357, 360
 in HD-related lymphoproliferative disorders, 384–385
 in Hodgkin's disease, 360–361, 383–384
 in nasopharyngeal carcinoma, 362–363
 in transformation, 333–334, 336
LPV: *See* Lymphotropic papavovirus
Lymphomas
 animal model with EBV, 400–401
 B-lymphoproliferative disease, 355–356, 396
 hamster polyomavirus and, 1, 4–6
 Herpesvirus saimiri and, 239
 HIV-associated, 319–320, 355, 359, 379–380, 381–382, 396
 Hodgkin's-related, 384–385
 in immunosuppressed persons, 319–320, 335, 355–356, 396
 Marek's disease, 227, 228, 234
 See also Burkitt's lymphoma; Hodgkin's disease; Non-hodgkin's lymphoma
Lymphoproliferative syndrome, X-linked, 356
Lymphotropic papavovirus (LPV), 3, 5

Marek's disease virus (MDV), 227–238
 ALV interaction with, 228–229

Marek's disease virus (MDV) (cont.)
  gene expression
    in latently infected cells, 233–234
    in lytically infected cells, 229–233
  genome organization, 228
  integration into host genome, 233
  vaccines, 227, 229
mdm2, p53 and, 32, 33, 34, 206
MDV: See Marek's disease virus
Mesotheliomas
  in hamsters
    from t antigen deletion mutants, 79
    from wild-type SV40, 77–78
  in humans
    increased incidence, 81, 82
    pathogenesis, 81–83
    polio vaccines and, 80
    SV40 association, 61, 82–86
MHC class I antigens
  function of, 207–208
  repression by E1A, 207–214
Middle T antigen
  cooperation with adenovirus E1A, 196
  of hamster polyomavirus, 2–3
  of murine polyomavirus, 11, 37–38, 39
Mononucleosis: See Infectious mononucleosis
myb gene, HPV deregulation of, 139
myc genes
  activation by middle T antigen, 38
  BK cooperation with, 55
  in Burkitt's lymphoma, 327–328, 358–360
  in HPV-transformed cells, 100, 143
  in HSV-transformed cells, 269
  in liver cancer, 179, 180, 182–184
  mechanism of action, 358–359
  WHV and GSHV cooperation with, 182–184

Nasopharyngeal carcinoma
  cellular types, 361
  EBV gene expression in, 335, 362–363
  EBV in tissues, 328–329, 361–362, 396
  epidemiology, 328, 361, 363–364, 396
  etiology, 335, 363–364, 396–397
  possible therapies, 337
  primary EBV infection, 327
Nitrosamines, in chronic hepatitis, 175
Non-Hodgkin's lymphoma
  EBV in, 319–320, 383
  HHV-6 in, 315–316, 320
  in HIV-infected patients, 319–320

Oncogenes, definition of, 253
Oncogenicity
  criteria for, 62

Oncogenicity (cont.)
  multistage process, 111
  variables affecting, 124

p21, 34, 140
p53
  adenovirus E1A and, 34
  adenovirus E1B and, 204–207
  apoptosis induced by, 34, 35, 140–141
  BPV4 and, 114, 115
  in Burkitt's lymphoma, 359, 360
  cell cycle control by, 16, 31–34
  in cervical cancer, 143–144
  DNA repair and, 34, 97–98, 140–141, 204–205
  EBV inactivation of, 335, 336
  HBV X protein and, 178, 185
  HPV E6 protein and, 97, 133–134, 135, 140–141
  large T antigen and
    of human polyomaviruses, 37, 56
    of murine polymomavirus, 38
    of SV40, 22–23, 28, 31–35, 36, 39, 76
  in liver cancer, 172, 184–185
  mechanisms for inactivation of, 32
  in mesotheliomas, 85
  in nasopharyngeal carcinoma, 363
  trans-activation by, 32–34, 204–206
p107
  adenovirus E1A and, 31, 202
  in HPV, 139
  T antigen and
    of JC, 56
    of SV40, 31, 35, 76
p130
  adenovirus E1A and, 202
  SV40 T antigen and, 31, 76
p300
  adenovirus E1A and, 35–36, 203–204
  SV40 T antigen and, 28, 35–36, 39, 76
p400
  SV40 T antigen and, 76
Papillomaviruses, bovine (BPV)
  carcinogenic mechanism, 115–116
  dietary carcinogenic cofactors
    bracken carcinogens and mutagens, 113–114
    bracken hypothesis, 112–113
    quercetin transformation mechanism, 116–118
  E proteins of, 115–116, 128, 136–137
  genome organization, 114–115
  types of, 112
  vaccines, 166

INDEX

Papillomaviruses, human (HPV)
  capsid proteins, 137–138, 144
  chromosome damage by, 99–101
  classification, 92, 94, 124–125
  E1 protein, 136
  E2 protein, 136–137
  E4 protein, 137
  E5 protein, 135–136
  E6 protein, 132–135
    p53 binding to, 97, 133–134, 135, 140–141
    as target for vaccines, 160–161
  E7 protein, 129–132
    E6 transcription and, 133
    p107 binding to, 139
    pRb binding to, 96–97, 130–131, 138–139
    RNA binding to, 140
    as target for vaccines, 160–161
  genome organization, 125–126, 158
  growth factors and, 98, 142–143
  immortalization of keratinocytes, 95–98, 100, 137, 142
  immune response to, 144, 158, 161–162
  integration into host genome
    in cervical cancer, 94, 100–101, 128, 135, 143, 287
    in immortalized keratinocytes, 95–96, 137
  keratinocyte differentiation and, 94, 98–99
  limitations of current research, 145
  natural infection, 158–160
  raft cultures, 94, 98, 132
  species specificity, 91, 93–94
  tissue specificity, 91, 94, 125, 128
  transcription regulation, 126–129
  transformation process, 160–161
  vaccines
    current trials, 166–167
    delivery, 164–165
    immune response, 144, 158, 161–162
    points of intervention, 157
    prophylactic, 162–163
    therapeutic, 163–164
    validation of efficacy, 166
  See also Cervical cancer; Genital cancers
Papovaviruses
  classification, 92–93
  HCMV cooperation with, 307
  stimulation of quiescent cells, 76
  See also Polyomaviruses; Simian virus 40
Peroxides, in chronic hepatitis, 174–175
Polio vaccine, SV40 contamination of, 60, 61, 62, 79–80
Polyomavirus, hamster (HaPV), 1–14
  epitheliomas, 1, 3–4
  genome organization, 1–3

Polyomavirus, hamster (HaPV) (cont.)
  lymphotropism, 1, 4–6
  replication *in vitro*, 7, 9
  transformation *in vitro*, 9–11
  transgenic mice from, 6–7
Polyomavirus, murine (Py)
  homology with HaPV, 2–3
  is not lymphotropic, 5
  T antigens of, 10–11, 37–38, 39
Polyomaviruses, human
  general characteristics, 51–52
  in human tissues, 57–60
  natural infection, 52–53
  oncogenicity, 53–54, 61–63
  transformation *in vitro*, 37, 54–57
pRb
  adenovirus E1A and, 138, 199–202
  binding motif of viral proteins, 28, 29, 200
  BPV4 E7 protein and, 116
  HPV E7 protein and, 96–97, 130–131, 138–139
  large T antigen and
    of hamster polyomavirus, 10–11
    of human polyomaviruses, 37, 56
    of murine polyomavirus, 10, 37–38
    of SV40, 28–31, 36, 39, 76, 200
  in liver cancer, 185
  phosphorylation cycle, 29–30, 34, 200
Premature termination of transcripts, 232
Protooncogenes, definition of, 253
Py: *See* Polyomavirus, murine

Quercetin, 114, 116–118

Radiation
  p53 response to UV, 33, 34, 141
  X-rays as cofactor with HPVs, 92
ras genes
  activation by adenovirus E1A, 34, 196
  activation by HSV ICP10 PK, 265–266
  activation by middle T antigen, 38, 196
  in bovine carcinomas, 114
  in cervical cancer, 102–103
  cooperation with BK, 54, 55
  cooperation with BPV4, 116, 117
  cooperation with HPV, 34, 130, 131, 135
  p53 and, 34, 35
Rb protein: *See* pRb
Reed–Sternberg cells, 375–376
  EBV in, 382, 383, 384, 385
  T lymphocytes and, 386
Reticuloendotheliosis virus (REV), 229
Retinoblastoma tumor suppressor protein: *See* pRb

Retinoids
  EBV Z protein and, 335, 337
  reversal of HPV effects, 143
Retroviruses, interaction of herpesviruses with, 228–229
REV: *See* Reticuloendotheliosis virus
Rous sarcoma virus (RSV), MDV interaction with, 229
RSV: *See* Rous sarcoma virus

Simian virus 40 (SV40)
  general characteristics, 51–52
  lymphotropism, 4–5
  oncogenicity
    in humans, 60–61, 62, 81–86
    in primates, 54
    in rodents, 76–79, 84
    in polio vaccine, 60, 61, 62, 79–80
    transformation *in vitro* by, 82–83
  See also Large T antigen; Small t antigen
Small RNAs
  in EBV latent infection, 334, 350–351
  in HVS transformation of T cells, 245–246
  in lymphomas, 334, 357, 358, 360
Small t antigen
  of hamster polyomavirus, 11
  of human polyomaviruses, 52
  of murine polyomavirus, 11
  of SV40, 16–17, 38–39, 75, 76, 78–79
Steroids, as risk factor for cervical cancer, 101
SV40: *See* Simian virus 40

T antigens: *See* Large T antigen; Middle T antigen; Small t antigen
TATA box binding protein
  adenovirus E1A binding to, 199
  p300 association with, 203
  p53 binding to, 205
TGF-β: *See* Transforming growth factor-β

T lymphocytes
  adenovirus escape from, 207–209
  growth media for immortalization, 240
  HHV-6 in, 313
  immortalization by HSV, 239–240
TNF-α: *See* Tumor necrosis factor-α
Transformation, definition of, 253, 353
Transforming growth factor-β
  in chronic hepatitis, 174
  in HPV-infected cells, 142, 143
Tubular endosomes, of HCMV, 299–300, 306
Tumor necrosis factor-α
  adenovirus E1A and, 207
  in chronic hepatitis, 174
  EBV LMP-1 and, 383–384
  in HPV-infected cells, 143
  in liver cancer, 180
Tumor suppressor proteins
  role in oncogenesis, 15–16
  from Y chromosome, 55
  See also p53; pRb

Vaccines
  for HCMV, 300
  for Marek's disease, 227, 229
  for polio, SV40 contamination of, 60, 61, 62, 79–80
  See also Epstein–Barr virus: vaccines; Papillomaviruses, human: vaccines

Warts
  HPVs in, 91, 92, 125
  vaccines for, 166, 167
WHV: *See* Woodchuck hepatitis virus
Woodchuck hepatitis virus (WHV)
  liver cancer and, 172
  myc genes and, 182–184

X-linked lymphoproliferative syndrome, 356

Zinc finger, of adenovirus E1A protein, 199